THE DELIMITATION OF THE CONTINENTAL SHELF BETWEEN DENMARK, GERMANY AND THE NETHERLANDS

Alex G. Oude Elferink's detailed analysis of the negotiations between Denmark, Germany and the Netherlands concerning the delimitation of their continental shelf in the North Sea makes use of the full range of government archives in these three States. He looks at the role of international law in policy formulation and negotiations, and explores the legal context, political considerations and, in particular, oil interests which fed into these processes. The study explains why the parties decided to submit their disputes to the International Court of Justice and looks at the preparation of their pleadings and litigation strategy before the Court. The analysis shows how Denmark and the Netherlands were able to avoid the full impact of the implications of the Court's judgment by sidestepping legal arguments and insisting instead on political considerations.

ALEX G. OUDE ELFERINK is a senior lecturer at the School of Law, Utrecht University, the Netherlands, where he is also Deputy Director of the Netherlands Institute for the Law of the Sea and Adjunct Professor at the K. G. Jebsen Centre for the Law of the Sea, University of Tromsø, Norway. He has worked in the field of public international law for over twenty years, focusing in particular on maritime boundary delimitation.

THE DELIMITATION OF THE CONTINENTAL SHELF BETWEEN DENMARK, GERMANY AND THE NETHERLANDS

Arguing Law, Practicing Politics?

A.G. OUDE ELFERINK

CAMBRIDGE
UNIVERSITY PRESS

CAMBRIDGE
UNIVERSITY PRESS

University Printing House, Cambridge CB2 8BS, United Kingdom

Published in the United States of America by Cambridge University Press, New York

Cambridge University Press is part of the University of Cambridge.

It furthers the University's mission by disseminating knowledge in the pursuit of education, learning, and research at the highest international levels of excellence.

www.cambridge.org
Information on this title: www.cambridge.org/9781107041462

© Alex Oude Elferink 2013

First published 2013

Printed in the United Kingdom by Clays, St Ives plc

A catalog record for this publication is available from the British Library

ISBN 978-1-107-04146-2 Hardback

Don't mention the War
("The Germans," *Fawlty Towers*, Season 1, Episode 6; original air
date: 24 October 1975)

CONTENTS

FIGURES

xiii

ACKNOWLEDGMENTS

There were times when I thought that what is now this book would never be finalized. Before I could start writing I had to do research in archives in Denmark, Germany and the Netherlands. I decided to start in the Dutch archives because these were easily accessible to me and my first visit was at the end of 2002. Research in Denmark and Germany took more time to organize and in both cases I had to return twice because I had identified further materials. I only completed that research by the middle of 2007. In the meantime, my research was delayed by other factors. Consultancy work was playing an ever-more important role in obtaining funding for our research and I had become involved in a couple of major projects involving the Republic of Suriname, a former colony and autonomous country of the Kingdom of the Netherlands. One of these projects was concerned with the delimitation of the continental shelf and required research in the same archives of the Dutch Ministry of Foreign Affairs that I also wanted to consult in connection with this book project. To keep my research in the different capacities distinct, I decided to put the latter research on hold. This meant that I finally finished my research for this project in the Dutch archives in 2008. At that time I had already started to analyze the materials I had at my disposal and write a first draft, but progress was exceedingly slow. Most of the time I could only steal a couple of hours between other work that kept piling up. Only in 2011 and 2012 did I finally get substantial time to write up my findings. This also allowed me to get a good grip on the project as a whole and piece together the many interrelated episodes from the thousands of documents I had collected, something which is just not possible if you are writing for short periods of time in-between all kinds of different assignments. I leave it to the reader to judge the outcome of my work, but I would like to stress that for me it was a great satisfaction to be able to carry out this project and see its final results.

Although it goes against much of the current ideas about doing legal research – with its focus on broad research programs and quantified output on a yearly basis – I am convinced that real progress in understanding international law and its relevance for international society is only possible on the basis of the kind of time-consuming research that resulted in this book. Otherwise we will never have more than a superficial understanding of the workings of the law.

This book has come about with the assistance of many people. It all started with the suggestion of Liesbeth Lijnzaad of the Dutch Ministry of Foreign Affairs, when I told her about my idea for this type of project, to take the delimitation between Denmark, Germany and the Netherlands as a case study. My work in the archives would not have been possible without the able assistance and advice of their staff. It would not be possible to mention everybody by name, but I would like to mention a couple. Mr. Den Hollander of the Dutch Ministry of Foreign Affairs who hosted me at the Ministry's archives and was always helpful in answering my questions concerning further archive folders that might be of relevance and providing me with photocopies. Mr. Knud Piening of the Political Archive of the German Foreign Office for providing me with information that allowed me to identify the relevant files in the Foreign Office before going to Berlin and helping me during my research in Berlin. Mr. Kurt Braband of the German Federal Archive and Mrs. Margit Mogensen and Mr. Peter Bundzen of the Danish National Archives for doing the same in connection with my visits to, respectively, Koblenz and Copenhagen. The staff of the British National Archives in London and the staff of the Dutch National Archive in The Hague. I would also like to acknowledge the financial support I received from the Netherlands Organisation for Scientific Research (NWO) and the School of Law of Utrecht University for my research in Denmark and Germany. I would like to thank Professor Rainer Lagoni for reading the manuscript and his encouraging words, and Mr. Bart Post for providing research assistance during his internship at the Netherlands institute for the law of the sea of Utrecht University. My thanks also go to the people who helped to get this book to its audience: the staff of Cambridge University Press for providing me with guidance during the editing of the manuscript and David Swanson for his work on the figures accompanying the text. Many of these figures were included in original documents that were of rather poor quality and they would not have been accessible to the readership without David's excellent job in improving their legibility. Similarly, I want to thank David Cox for further work on the figures during the production process of the book. My special appreciation goes to Professor Fred Soons, chair in public international law at Utrecht University. Thanks for encouraging me to finalize this project and for creating the conditions that allowed doing so.

Alex Oude Elferink

INTERNATIONAL INSTRUMENTS AND NATIONAL LEGISLATION

Act of 27 July 1956 concerning the territorial sea, continental shelf, fishery protection and airspace (Venezuela) (English text in A. Szekely (ed.), *Latin America and the Development of the Law of the Sea* (Dobbs Ferry, NY: Oceana Publications 1979-...), Vol. **, Booklet 20, issued January 1980, pp. 16–18)   pp. 51–52

Act on the provisional regulation of the rights over the continental shelf of 24 July 1964 (Germany) (F. Durante and W. Rodino (eds), *Western Europe and the Development of the Law of the Sea* (Dobbs Ferry, NY: Oceana Publications, 1979-...), Vol. *, L.24.7.1964 (English and German text); English text also available at www.un.org/Depts/los/ LEGISLATIONANDTREATIES/PDFFILES/DEU_1974_Act.pdf)   p. 78, n. 58, 81

Agreement between the Government of the Kingdom of Denmark and the Government of the Kingdom of Norway relating to the delimitation of the continental shelf of 8 December 1965 (634 UNTS p. 71)   pp. 133–137, 141, 160, 259, 379, 389, 452

Agreement between the Government of the Kingdom of the Netherlands and the Government of the United Kingdom of Great Britain and Northern Ireland relating to the delimitation of the continental shelf under the North Sea between the two countries of 6 October 1965 (509 UNTS p. 124)   pp. 140, 142, 151–155, 258–259, 276, 285, 297

Agreement between the Government of the Kingdom of the Netherlands and the Government of the Kingdom of Denmark concerning the delimitation of the continental shelf under the North Sea between the two countries of 31 March 1966 (604 UNTS p. 209)   pp. 155–157, 276, 285, 297

Agreement between the Government of the United Kingdom of Great Britain and Northern Ireland and the Government of the Kingdom of Denmark relating to the delimitation of the continental shelf between the two countries of 3 March 1966 (592 UNTS p. 207)   pp. 138, 276, 285, 297

British Guiana (Alteration of Boundaries) Order in Council, 1954 of 19 October 1954 (Statutory Instruments 1954 no. 1372; British Guiana)   p. 50

ABBREVIATIONS

AA	Auswärtiges Amt
AJIL	American Journal of International Law
B	Bundesarchiv
BIPM	Bataafsche Internationale Petroleum Maatschappij
CMS	Counter-Memorial of Suriname (*Guyana v. Suriname*)
CR	Common Rejoinder of Denmark and the Netherlands (ICJ Pleadings, *North Sea continental shelf* cases, Vol. I)
DAEB	Danish Embassy in Bonn
DCM	Counter-Memorial of Denmark (ICJ Pleadings, *North Sea continental shelf* cases, Vol. I)
DEC	Dutch Embassy in Copenhagen
DEB	Dutch Embassy in Bonn
Dis. op.	Dissenting opinion
DNA	Statens Arkiver; Rigsarkivet
DUC	Dansk Undergrunds Consortium
Ems-Dollard Treaty	Treaty between the Kingdom of the Netherlands and the Federal Republic of Germany concerning arrangements for co-operation in the Ems Estuary of 8 April 1960
FM	Foreign Minister
FO	Foreign Office
GEH	German Embassy in The Hague
GM	Memorial of the Federal Republic of Germany (ICJ Pleadings, *North Sea continental shelf* cases, Vol. I)
GR	Reply of the Federal Republic of Germany (ICJ Pleadings, *North Sea continental shelf* cases, Vol. I)
ICJ	International Court of Justice
ICJ Pleadings	International Court of Justice, Pleadings, Oral Arguments, Documents
ICJ Reports	Reports of Judgments, Advisory Opinions and Orders; The International Court of Justice
ICLQ	International and Comparative Law Quarterly
ILC	International Law Commission
K. I	Proceedings of the First Chamber of Parliament
K. I (BH)	Parliamentary Papers, First Chamber of Parliament

K. II	Proceedings of the Second Chamber of Parliament
K. II (BH)	Parliamentary Papers, Second Chamber of Parliament
LNTS	League of Nations Treaty Series
ME	Ministry for the Economy
MFA	Ministry of Foreign Affairs
MG	Memorial of Guyana (*Guyana v. Suriname*)
MiFA	Minister of Foreign Affairs
MS	Memorial of Suriname (*Guyana v. Suriname*)
NAM	Nederlandse Aardolie Maatschappij
NATO	North Atlantic Treaty Organization
NCM	Counter-Memorial of the Netherlands (ICJ Pleadings, *North Sea continental shelf* cases, Vol. I)
OP	Oral Pleadings (ICJ Pleadings, *North Sea continental shelf* cases, Vol. II)
PCIJ, Series A/B	Permanent Court of International Justice; Series A/B, Judgments, Orders and Advisory Opinions
RG	Reply of Guyana (*Guyana v. Suriname*)
RS	Rejoinder of Suriname (*Guyana v. Suriname*)
Sep. op.	Separate opinion
Special Agreements	Special Agreement for the submission to the International Court of Justice of a difference between the Kingdom of Denmark and the Federal Republic of Germany concerning the delimitation, as between the Kingdom of Denmark and the Federal Republic of Germany, of the continental shelf in the North Sea of 2 February 1967; Special Agreement for the submission to the International Court of Justice of a difference between the Kingdom of the Netherlands and the Federal Republic of Germany concerning the delimitation, as between the Kingdom of the Netherlands and the Federal Republic of Germany, of the continental shelf in the North Sea of 2 February 1967
Stb.	Staatsblad van het Koninkrijk der Nederlanden
UN	United Nations
UNCLOS I Off. Rec.	United Nations Conference on the Law of the Sea; Official Records
UNLS	United Nations Legislative Series
UNRIAA	Reports of International Arbitral Awards
UNTS	United Nations Treaty Series
Yearbook ILC	Yearbook of the International Law Commission
ZaöRV	Zeitschrift für ausländisches öffentliches Recht und Völkerrecht

1

Introduction

1.1 Origins and objectives

My interest in the role of international law in negotiations was first raised when I was working on my dissertation on the delimitation of the maritime boundaries of the Russian Federation in the first half of the 1990s. Most of the negotiations on the Russian Federation's maritime boundaries had been conducted by the Soviet Union. Negotiations with a number of neighboring States[1] were particularly complex and it took a long time to reach agreement. For instance, Sweden and the Soviet Union started their negotiations in 1969 and reached an agreement in 1988 and the negotiations with Norway that started in 1970 were only concluded in 2010. My impression of these negotiations was that international law did have an impact on State behavior. As I concluded at the time:

> the law indicates within certain margins what acceptable claims are. States can evaluate their dispute in legal terms and are not forced to fall back completely on political bargaining. This is not to say that political bargaining does not take place, but it is set in the framework of legal rules.[2]

I reached these conclusions on the basis of talks with people who had been involved in the negotiations and articles in newspapers and journals and other written sources. However, I did not have access to the records of the negotiations and the internal deliberations of the parties. This left me with the curiosity of what access to these documents might have added to my analysis. Some years later, I started considering the

---

[1] In this book, the word state with an upper case identifies independent states. If reference is made to a state that is part of a federation, such as the states of the Federal Republic of Germany, lower case will be used.

[2] A. G. Oude Elferink, *The Law of Maritime Boundary Delimitation: A Case Study of the Russian Federation* (Dordrecht: Martinus Nijhoff Publishers, 1994), p. 371 (footnote omitted).

possibility of a more detailed case study of one of the delimitations I had analyzed previously. My initial inquiries pointed out that getting access to all of the relevant archives would be difficult, if not impossible, and I decided to let the matter drop. Apparently, the idea lingered in the back of my mind and I revived this project when it was suggested to me that the delimitation of the continental shelf between Denmark, Germany and the Netherlands in the North Sea might be a good alternative, as these delimitations had been concluded in the early 1970s and archives would probably be open for research.[3] After I had ascertained that I could get access to the archives of the ministries of foreign affairs of the three States, I decided to embark on the project that eventually was to result in this book. Apart from the fact that these archives were accessible, a number of other considerations indicate that this choice was fully justified. Germany on the one hand, and Denmark and the Netherlands on the other, had advanced diametrically opposed views on the applicable law and were not able to reach a settlement of all of their boundaries through negotiations. This led to an agreement to submit the two disputes, between Germany and the Netherlands and between Germany and Denmark, to the International Court of Justice (ICJ). The three States committed themselves to determine the remainder of their continental shelf boundaries on the basis of the judgment. The outcome of the negotiations suggests that the judgment did not have a profound impact on it. This raised the question of what might explain this limited impact.

At the start of the project, I formulated its main purpose to be the determination of the role of international law in the negotiations between Denmark, Germany and the Netherlands. Answering this question required a number of things. First of all, it would be necessary to get a complete picture of the internal deliberations of the parties and the negotiations. Apart from looking at their bilateral negotiations, I considered it relevant to determine how the three States had contributed to the development of the general regime of the continental shelf. That development mostly took place in the 1950s and in 1958 resulted in the Convention on the continental shelf. The rule on the delimitation of the shelf between neighboring States contained in article 6 of the Convention was central to the legal controversy between Germany and its two

---

[3] At that time, Germany was still divided in the Federal Republic of Germany and the German Democratic Republic. All references to Germany in the present book concern the Federal Republic.

neighbors. It thus could be expected that their position on the formulation of the rule of article 6 had been influenced by their interests in the North Sea. Second, my case study of the Russian Federation had shown various examples of how the interests of a State in a specific delimitation may affect its overall position on the law and its other pending delimitations. This required looking into the other delimitations that the three States were facing in the period up to the final settlement of their disputes in the North Sea. To be able to determine the role of the law in the negotiations it would also be necessary to analyze the the applicable law. This again pointed to the importance of looking at the genesis of article 6 of the Convention on the continental shelf. In view of the significance of the judgment of the ICJ for the second stage of the negotiations, that judgment would also require careful review.

In looking at the above issues, and other legal matters that came up during the negotiations, the analysis is based on a legal positivist approach to the law: the existence and content of rules of international law can be determined in accordance with the relevant rules contained in such instruments as the Vienna Convention on the law of treaties and the judgments of the ICJ. This approach allows determination of the content of the relevant rules and to what extent they have an impact on the behavior of States. The negotiating record also shows that the parties looked at the law in the terms of rules that mandate specific outcomes. A description of the internal deliberations and negotiations from this perspective allows determination as to how the parties viewed the content of the law and how this relates to my own assessment of the content of the law, and the consequences they attached to their view on the law in determining their approach to the negotiations. However, I decided that the analysis should not stop at this point. International legal scholarship and international relations theory offers a wide range of explanatory models concerning the relationship between international law and State behavior. Chapter 11 of this book looks at a number of these models to establish what they can tell us about the case study and whether the case study allows conclusions to be drawn about the value of these different models.

1.2 Outline of the book

Chapter 2 provides some necessary background to the case study. It describes the development of the continental shelf regime in international law in the 1940s and 1950s. That regime provided the framework for the negotiations between Denmark, Germany and the Netherlands on the delimitation of the continental shelf in the North Sea. Chapter 2 also briefly describes the North Sea and in particular looks at the implications of the characteristics of the North Sea for the delimitation of the continental shelf between neighboring States. Finally, the chapter analyzes the other delimitations that Denmark, Germany and the Netherlands were facing, in order to assess if their interests in these other delimitations as regards a rule of delimitation were similar to, or different from, their interests in the North Sea.

Chapter 3 looks at the development of the rule on the delimitation of the continental shelf that is contained in article 6 of the Convention on the continental shelf. The focus of the analysis is on the contribution of Denmark, Germany and the Netherlands to the debate, but is also intended to determine the content of article 6 and its implications for the delimitations in the North Sea. In looking at the practice of Denmark, Germany and the Netherlands their broader interests in the delimitation of the continental shelf are also considered. A final section of Chapter 3 summarizes the main findings in respect of the approach of Denmark, Germany and the Netherlands in relation to the law and its linkages to the delimitation in the North Sea. A similar, final section is included at the end of the other chapters.

After the conclusion of the Convention on the continental shelf, Denmark, Germany and the Netherlands did not immediately engage in negotiations on the delimitation of the continental shelf in the North Sea. They did start considering whether to become a party to the Convention on the continental shelf and how to deal with a continental shelf claim in the North Sea. This matter is considered in Chapter 4. There is some overlap in time of this matter with the start of the negotiations on the delimitation of the continental shelf in the North Sea. This not only concerned the delimitation between Denmark, Germany and the Netherlands, but also bilateral delimitations with the other coastal States of the North Sea. In order to provide the reader with a clear picture of these different issues, they are presented in separate chapters. Chapter 5 looks at the first phase of the negotiations between the Netherlands and Germany and between Denmark and Germany. In both cases, the parties were able to reach agreement on a partial boundary and agreed to look into the options to settle the remainder of the boundary at a later stage. The second part of Chapter 5 looks at the negotiations on the delimitation of a continental shelf boundary between Denmark and the Netherlands and of both States with their other North

Sea neighbors. The conclusion of these other agreements was an important aspect of the shelf policy of Denmark and the Netherlands and raises
the question of their legal and political significance in relation to
Germany.

During the second phase of the negotiations between Denmark,
Germany and the Netherlands, which is considered in Chapter 6, there
were still some attempts to arrive at a negotiated settlement, but it was
predominantly concerned with agreeing on the modalities for submitting the two bilateral disputes to compulsory third party dispute settlement. Germany had a preference for arbitration, but the other two States
wanted to submit their dispute with Germany to the ICJ. The latter
option was eventually accepted by Germany. The continued uncertainty
about the location of Germany's boundaries with its two neighbors also
required consideration of an interim agreement on activities in the
disputed area. The negotiation of this interim agreement illustrates the
interaction between the applicable law and the impact of broader interests. The activities that were carried out in accordance with the interim
agreement were to have a significant impact on the further negotiations
between the parties after the ICJ had issued its judgment, even though
the interim agreement was intended to safeguard the rights of the parties.

Before turning to the pleadings in the *North Sea continental shelf*
cases, Chapter 7 picks up the analysis of the other delimitations of
Denmark, Germany and the Netherlands. While Chapter 3 looks at
developments in the 1950s, Chapter 7 deals with the 1960s. In particular
in the case of the Netherlands, there was a close linkage between its
delimitation with Germany and the delimitations of Suriname and the
Netherlands Antilles, the other parts of the Kingdom of the Netherlands,
and their neighbors.

In Chapter 8 we return to the case study. This chapter looks at the
preparation by the three States of their pleadings before the ICJ and sets
out the main arguments of these pleadings in order to determine how
they presented the law to the Court and what other factors possibly
played into that presentation. As will be seen, one of the most significant
aspects of the pleadings was Germany's cautious approach to its own
positive case. After the judgment, Denmark and the Netherlands would
successfully exploit that German approach to limit the judgment's consequences. Chapter 8 considers whether Germany could have taken a
different approach.

The judgment of the Court is analyzed in Chapter 9. The chapter first
looks at how the Court dealt with the arguments of the parties. A second

part of the chapter looks at the kind of guidance the judgment provided to the parties. The parties had agreed beforehand that they would negotiate the remainder of their boundaries on the basis of the judgment. The specificity of the judgment, or lack thereof, would impact on the scope for diverging views in these subsequent negotiations. As will be seen, the Court's pronouncements on the law are quite general in nature and leave much room for argument. At the same time the judgment and the views of individual judges that are appended to the judgment do allow us to establish what the Court considered the general characteristics of an outcome should be. This matter is considered in the third part of Chapter 9. As will be apparent, Chapter 9 is intended to determine the framework the judgment set for the further negotiations of the parties. The chapter is not intended to put it in the perspective of the subsequent development of the law or to criticize the Court's approach to the law. Those points are not relevant for the case study and the judgment has already been the subject of many such assessments.[4]

The negotiations between Denmark, Germany and the Netherlands that followed the judgment are considered in Chapter 10. Each State made an assessment of the judgment after it had been handed down and determined its approach to the further negotiations. After a description of these preliminaries, Chapter 10 turns to the actual negotiations, analyzing each of the negotiating rounds in turn. As this analysis points out, Denmark and the Netherlands were both intent on avoiding an

---

[4] See e.g. N. M. Antunes, *Towards the Conceptualisation of Maritime Delimitation* (Leiden: Martinus Nijhoff Publishers, 2003), pp. 46–63; E. D. Brown, *Sea-bed Energy and Minerals: The International Legal Regime* (Dordrecht: Martinus Nijhoff Publishers, 1992), pp. 49–84; I. Foighel, "The North Sea Continental Shelf Case; Judgment by the International Court of Justice of 20 February 1969," 1969 (39) *Nordisk tidsskrift for international ret*, pp. 109–127; W. Friedman, "The North Sea Continental Shelf Cases – A critique," 1970 (64) *AJIL*, pp. 229–240; E. Grisel, "The Lateral Boundaries of the Continental Shelf and the Judgment of the International Court of Justice in the North Sea Continental Shelf Cases," 1970 (64) *AJIL*, pp. 562–593; E. Menzel, "Der Festlandsockel der Bundesrepublik Deutschland und das Urteil des Internationalen Gerichtshofs vom 20. Februar 1969," 1969 (14) *Jahrbuch für internationales Recht*, pp. 14–100; F. Münch, "Das Urteil des Internationalen Gerichtshofes vom 20. Februar 1969 über den deutschen Anteil am Festlandsockel in der Nordsee," 1969 (29) *ZaöRV*, pp. 455–475; A. Reynaud, *Les Différends du Plateau Continental de la Mer du Nord devant la Cour Internationale de Justice* (Paris: Librairie générale de droit et de jurisprudence 1975); H. M. Waldock, *The International Court and the Law of the Sea* (The Hague: T.M.C. Asser Institute, 1979), pp. 11–15; P. Weil, *The Law of Maritime Delimitation: Reflections* (Cambridge: Grotius Publications, 1989), *passim*.

outcome that would be based on the judgment of the Court, but they at times conflicted on how to go about this.

An overall assessment of the case study is provided in Chapter 11. After setting out its most salient points, Chapter 11.3 presents a number of theoretical perspectives on the relation between international law and State behavior. The outcomes of the case study are further assessed in the light of these perspectives in Chapter 11.4.

1.3   On documentary sources

This book would not have been possible without the documents held in the archives of various ministries in Denmark, Germany and the Netherlands. This section sets out how I approached the research in respect of materials contained in archives and other sources of information.

In view of the central role of the ministries of foreign affairs of the three countries in the negotiations, this provided a logical starting point for my research in the archives in all three cases. This had somewhat varying results. In the case of Denmark, my research in the archives of the Ministry of Foreign Affairs led me to conclude that research in the archives of other ministries was unlikely to add much of particular value.[5] I consider that this preliminary conclusion is confirmed by the present analysis. The available material from the archives of the Ministry of Foreign Affairs allowed me to address the central research question in relation to Denmark in detail. At the same time, I have little doubt that further research might have yielded additional information. However, there also was an important practical consideration to refrain from further research. At some point, additional research in further archives that may be less central to the research will involve going through large amounts of documents that sometimes will not yield any new information.

For Germany and the Netherlands, I concluded that I should not limit myself to the archives of respectively the Foreign Office and the Ministry of Foreign Affairs. In the case of Germany, this was mostly explained by two considerations. My research pointed out that in particular the Ministry for the Economy had been actively involved during various

---

[5] Access to a limited number of files of the MFA was not granted. This mostly concerned files that contained personal information on individuals that were part of the Danish delegations in negotiations with Germany and the Netherlands.

stages of the negotiations. Some key documents, such as a study by Professor Rudolf Bernhardt that seemed to have provided important input into Germany's approach to the pleadings before the ICJ, that were mentioned in documents in the archives of the Foreign Office could not be located in those archives. Second, there was a gap in the record in respect of the negotiations with Denmark and the Netherlands after the judgment of the Court.[6] Research in the archives of the Ministry for the Economy in the Bundesarchiv in Koblenz allowed me to locate Professor Bernhardt's study and other relevant documents. They also provided some additional information on the negotiations after 1969, but did not make up completely for the missing folders in the archives of the Foreign Office – quite understandably, as the latter had the primary responsibility for conducting the negotiations. Fortunately, as far as the actual negotiations are concerned, this gap in the records could to a large extent be filled in by documents from the archives of the Ministries of Foreign Affairs of Denmark and the Netherlands. While I was in Koblenz, I also used the opportunity to look at the relevant archives of other Federal Ministries. In general, these archives were not essential for the case study, but did yield some interesting information.

Apart from the Federal Government, four of the German states had an interest in the delimitation of the continental shelf of the North Sea. In general, the state of Lower Saxony acted on behalf of the four states and the archives of the Foreign Office and the Ministry for the Economy allowed to determine the impact Lower Saxony and the other states had on Germany's shelf policy.[7] I decided against research in the archives of Lower Saxony because the focus of my research did not concern the relationship between the Federation and the states or the role of the states, and the material that I already had access to allowed me to deal with these issues in sufficient detail for the purposes of the case study.

In the Netherlands, an important reason to check archives other than those of the Ministry of Foreign Affairs was that the archives of that Ministry on some points are far from complete. For instance, there is little information on the preparation of the Dutch pleadings in its case

---

[6] A likely explanation seems to be that files at some point had been renumbered and a couple of them in the process were misplaced. The last volume in folder B 80/966 of the Political Archive of the FO is number 18 and runs to December 1969. It is indicated that the file is continued in number 19. The next available folder, Zwischenarchiv-193914, starts with volume 25, covering the period July 1971 and beyond.

[7] I have chosen to refer to the Federal Ministries without including the term federal and to specifically indicate in the text if a ministry of one of the states is concerned.

with Germany. The archives of the Ministry for the Economy, the only other Ministry that had an involvement during this stage, contained little additional information. This lack of information could be filled in, to some extent, from the archives of the Ministry of Foreign Affairs of Denmark. Those archives hold correspondence and other information on the cooperation between the Danish and Dutch Ministries of Foreign Affairs.

Apart from the above archives, a number of other archives were also potentially relevant for the case study. I visited the National Archives in London to consult the archives of the British Foreign and Commonwealth Office when I was there in connection with other business. This did provide some interesting information on the British participation in the drafting of the Convention on the continental shelf and contacts with Denmark, Germany and the Netherlands, but it also showed that this type of information was not critical to understanding the interactions between the latter three States. For that reason I decided to refrain from visiting the archives of the Ministries of Foreign Affairs of the other North Sea States.

Oil interests played a significant role in shaping the shelf policies of Denmark, Germany and the Netherlands and there were regular contacts between the Ministries of Foreign Affairs and specific companies. I eventually got permission to do research in the archives of Shell International in The Hague, but a keyword search in the index by the staff did not yield any hits. At that point I decided against putting further effort into this matter. Available documents already gave me a pretty good idea of the kind of contacts there existed between the companies and the Ministries and how the Ministries dealt with the companies. The archives from the companies would most likely not yield much information on the latter point, which would have been of most interest from the perspective of the case study.

Apart from archival materials a number of other sources were potentially relevant for trying to reconstruct events. When I carried out my research on the Russian Federation, newspaper reports, journal articles and interviews with Foreign Ministry staff of various countries were helpful in piecing together what had transpired. In the present case I have made less use of these other sources. For starters, in most cases they have little to add to the detailed information that is contained in archival materials. The couple of interviews I was able to organize after the many years that had passed since the negotiations and a couple of chance meetings were most interesting because they provided me with a feel for the perceptions of the participants to the negotiations, and allowed me to get a better idea of the general atmosphere during the negotiations. The interviews and talks did not lead to significant new information. In my experience, one should in any case be careful not to attribute too much weight to statements concerning the facts many decades after the events actually took place.

Articles in newspapers and journals from the 1950s and 1960s are mostly of interest because in a number of instances they had some impact on the course of events and because they provide an impression of perceptions in public opinion. Otherwise, they did not add anything substantial to the material contained in the archives. During my research I did come across one rather detailed personal account by someone who had been directly involved. That account, by ambassador Fack, the head of the Dutch delegation to the negotiations after the judgment of the ICJ, provides a good illustration that personal impressions may not always be a reliable source of information. For instance, at the end of his narrative, Fack contends that Denmark and the Netherlands were focusing on the areas of interest from an oil and gas perspective, whereas Germany instead was focusing on a geologically uninteresting area that would give it access to the center of the North Sea. According to Fack "[t]he Germans were as pleased as Punch with the result."[8] The delegation reports on the negotiations show that Germany actually was interested in the areas with a promising oil and gas potential and that Denmark and the Netherlands did everything to exclude Germany from those areas. Germany considered the outcome of the negotiations barely acceptable.[9] As far as the collaboration between the Netherlands and Denmark is concerned, according to Fack everything was hunky-dory.[10] One will look in vain in Fack's account for any hint concerning Danish frustrations about the Dutch approach to the negotiations or the Dutch attempts to reach a separate deal with Germany.[11] Notwithstanding these observations, a personal account like this is of significant interest if it can be compared to a detailed negotiating record. Omissions and discrepancies at times may be just as insightful as concordance.

---

[8] R. Fack, *Gedane zaken; Diplomatieke verkenningen* (Amsterdam: Sijthoff, 1984), p. 82. Translation by the author. The original text reads "De Duitsers waren in hun nopjes met het resultaat."

[9] See further Chapter 10.5.   [10] See Fack, *Gedane zaken*, pp. 77–82.

[11] For the frictions between Denmark and the Netherlands see e.g. Chapter 10.5.2.

1.4   Some miscellaneous points

Having dealt with the main issues in respect of the case study, there remain a number of miscellaneous points that merit some further clarification in this introduction. A couple of these are concerned with the reference to documents from archives. Most of these documents are either in Danish, Dutch or German. Instead of using the original language to identify the type and title of these documents, I have provided English translations as I considered that this would be more useful to most readers. In referring to documents contained in archives, I have always included a code in square brackets. That code identifies the archive and folder in which the document is contained. A list of the relevant archives and folders is included in the bibliography at the end of this book.

Most documents in the archives can be attributed to a specific section or subsection of a Ministry and in many cases it is also possible to identify the author. For reasons of readability I have not always specified all of these details in the text and instead have chosen to refer to "the Ministry" or "the Embassy." In some instances I have opted to be more specific. Generally speaking, I have taken the latter approach if I considered that this information had additional value. A similar approach has at times been used in referring to statements during negotiations. For instance, instead of referring to "one of the members of the Dutch delegation," the text may refer to "the Netherlands."

As will become apparent, the first phase of the negotiations of Germany with Denmark and the Netherlands centered on the question whether the so-called equidistance method was appropriate to delimit their continental shelf boundaries. Denmark and the Netherlands considered that the law mandated the application of this method. Germany rejected this. As the term equidistance implies, the equidistance method results in a line that is at equal distance. In the case of maritime boundaries this concerns equal distance from the baselines from which the extent of maritime zones of the States concerned is determined. The normal baseline is the low-water line along the coast of the mainland or islands.[12] International law also provides for a number of other methods to determine baselines.[13] States may differ about the legality of each other's baselines. In that case, it will not be possible to agree on an

---

[12] See e.g. Convention on the territorial sea and the contiguous zone, article 3.
[13] See e.g. *ibid.*, articles 4 and 7 to 11.

unequivocal definition of the equidistance line before that difference will have been settled. However, if two States accept each other's baselines, there will be no difficulty in objectively defining the equidistance line. Another advantage of the equidistance line is that in many instances it will result in a broadly equal division of maritime zones between neighboring States. This idea is reflected in the term "median line" that is often used to refer to an equidistance line between States with coasts that are opposite each other. Technically, a median line and an equidistance line are exactly the same. Notwithstanding these advantages there are also certain drawbacks to the equidistance line. This method gives prominence to salient points on the baseline along the coast. For instance, in a delimitation in accordance with equidistance a small island of one State in front of the much longer mainland coast of another State might be attributed more extensive maritime zones than that mainland coast. Because of such cases, the equidistance method has never been accepted as a generally applicable rule of delimitation. Instead, it has always been accepted that delimitation rules in respect of maritime zones should allow for some flexibility.[14] How much flexibility this should entail has always been, and still remains, a controversial issue.

In the case of Germany, the combined result of the two equidistance lines between Germany and its direct neighbors, Denmark and the Netherlands, hemmed Germany in in the southeastern corner of the North Sea. Germany considered that this led to an unfair result and made the equidistance line inappropriate. Denmark and the Netherlands considered that there was nothing exceptional about their bilateral delimitation with Germany on the basis of the equidistance method and that it was not necessary to look beyond that bilateral framework. This set the stage for the two delimitation disputes that are the subject of the present case study.

---

[14] See e.g. the discussion in Chapter 3.

The setting

2.1 The development of the continental shelf regime

Until the 1940s, the law of the sea was characterized by the dichotomy between the territorial sea and the high seas.[1] The coastal State only had sovereignty over the territorial sea. There was no agreement on the breadth of the territorial sea. A large number of States claimed a territorial sea of 3 nautical miles but other States claimed a territorial sea of 4, 6 or 12 nautical miles.[2] There existed general agreement that the seas and oceans beyond the outer limit of the territorial sea were part of the high seas. The high seas were governed by the regime of freedom of the high seas. The exploitation of fisheries resources by foreign fleets had given some coastal States already reason to extend their jurisdiction to the 12-nautical-mile limit in the first half of the twentieth century and just after the Second World War, Chile, Ecuador and Peru claimed a 200-nautical-mile zone. This claim was not widely supported at the time.

The presence of non-living resources, in particular oil and gas, in the seabed of the high seas provided another driver for the extension of coastal State jurisdiction beyond the territorial sea in the 1940s and 1950s. In this case, claims in general did not apply to a zone of a specific breadth, but relied on the nature of the seabed. The continents do not stop at the coastline, but extend under the sea. The first part of this seaward extension is the continental shelf. In the 1940s and 1950s it was assumed that the continental shelf in general extended to a water depth of 200 meters. Beyond the continental shelf, the continental slope, a second element of the seaward extent of the continents, commences. The horizontal extent of the continental shelf varies widely. In some places,

---

[1] For a short summary of the historical development of the law of the sea see e.g. D. R. Rothwell and T. Stephens, *The International Law of the Sea* (Oxford: Hart Publishing, 2010), pp. 2–20.

[2] In addition, certain States claimed jurisdiction over a zone adjacent to the territorial sea for purposes such as customs control. One nautical mile equals 1,852 meters.

such as the western coast of Latin America, the continental shelf ends within a short distance from the coast. In other areas, like the North Sea, it extends for tens or even hundreds of nautical miles from the coast.

Reliance on the continental shelf concept offered one significant advantage for major maritime States. While bringing all readily accessible oil and gas resources under the jurisdiction of the coastal State, the continental shelf concept made it possible to detach the regime of the seabed and subsoil from the regime of the superjacent waters. This limited the risk that a regime of coastal State control over seabed resources would spill over into other areas and might lead to a limitation of other high seas freedoms, such as the freedom of navigation or fishing.

A 1942 Treaty between the United Kingdom and Venezuela on the delimitation of the seabed and subsoil of the Gulf of Paria beyond the territorial sea and a proclamation on the continental shelf of President Truman of the United States of 1945 are considered to have started the development of the continental shelf regime.[3] These instruments involved the two major maritime powers of the time and in both instances it was made clear that the claim over the mineral resources of the seabed did not affect the status of the overlaying waters.[4]

The Truman Proclamation was quickly followed by unilateral declarations of other States claiming continental shelf rights and in 1949, when the International Law Commission (ILC), a subsidiary body of the United Nations General Assembly charged with the promotion of the codification and progressive development of international law,[5] discussed its program of work, the regime of the continental shelf was one of the topics included in its work on the regime of the high seas.[6]

The ILC eventually adopted draft articles on the law of the sea in 1956. This draft included a number of articles on the regime of the continental

---

[3] Treaty between His Majesty in respect of the United Kingdom and the President of the United States of Venezuela relating to the Submarine Areas of the Gulf of Paria of 26 February 1942 (1942 Treaty); Presidential Proclamation no. 2667, concerning the Policy of the United States with respect to the Natural Resources of the Subsoil and Sea Bed of the Continental Shelf of 28 September 1945.

[4] See 1942 Treaty, article 6. In the case of the Truman Proclamation, an accompanying White House press release provided "the proclamation in no wise abridges the right of free and unimpeded navigation of waters of the character of high seas above the shelf nor does it extend the present limits of the Territorial waters of the United States." The Truman Proclamation itself indicated a number of justifications as to why reliance was placed on the continental shelf concept to exercise rights over mineral resources.

[5] See Statute of the International Law Commission, article 1(1).

[6] See Yearbook ILC, 1949, p. 43.

shelf.[7] The draft articles prepared by the ILC were further considered by a diplomatic conference, the United Nations Conference on the law of the sea, which was convened in Geneva in 1958. The conference led to the adoption of four separate conventions on specific aspects of the law of the sea. One of these conventions was the Convention on the continental shelf. The provisions on the regime applicable to the continental shelf have led to relatively little controversy, and they have subsequently been included almost *verbatim* in the United Nations Convention on the law of the sea, which currently provides the basic legal framework for the oceans. On the other hand, the provisions on the extent of the continental shelf and its delimitation between neighboring States proved to be much more problematical. The imprecise definition of the continental shelf in article 1 of the Convention on the continental shelf had already drawn criticism during its negotiation.[8] Developments in mining technology in the 1960s led to the realization that it would be necessary to determine the extent of the continental shelf precisely in relation to the area beyond the jurisdiction of the coastal State. As the pleadings of Denmark, Germany and the Netherlands in the *North Sea continental shelf* cases showed, the determination of the exact nature of the basis of entitlement to the continental shelf was also important for the rules applicable to the delimitation between neighboring States. Denmark and the Netherlands submitted that continental shelf entitlement was based on proximity to the coast and that as a corollary the equidistance method was the primary method of delimitation. Germany considered that the continental shelf of the North Sea was common to all of its coastal States and that a delimitation should result in an equitable division between these States. The ICJ in its judgment of February 1969 rejected these contentions and instead concluded that entitlement to the continental shelf followed from the fact that the shelf constituted the natural prolongation of the territory of the coastal State. According to the judgment, delimitation was intended to leave to each State as much as possible of its natural prolongation. The judgment's pronouncements

---

[7] See Report of the International Law Commission covering the work of its eighth session, 23 April–4 July 1956, Section II Articles concerning the law of the sea, articles 67–73 (Yearbook ILC, 1956, Vol. II, p. 264).

[8] For a detailed analysis of article 1 of the Convention on the continental shelf see D. N. Hutchinson, "The Seaward Limit to Continental Shelf Jurisdiction in Customary International Law," 1985 (56) *British Yearbook of International Law*, pp. 111–184 and B. H. Oxman, "The Preparation of Article 1 of the Convention on the Continental Shelf," 1972 (3) *Journal of Maritime Law and Commerce*, pp. 245–305, 445–472 and 683–723.

on entitlement to the continental shelf were to have a profound effect on the
further development of the law on this point.[9]

The delimitation of the continental shelf between neighboring States
had been a concern since the inception of the continental shelf regime.
The main purpose of the 1942 Treaty between Venezuela and the United
Kingdom had been the establishment of a boundary between the seabed
areas of parties in order to allow the orderly development of hydrocarbon resources. The Truman Proclamation had recognized that in a
case where the continental shelf of the United States overlapped with the
continental shelf of a neighboring State, the boundary had to be determined by the United States and the State concerned "in accordance with
equitable principles."

As will be discussed further in Chapter 3, one of the challenges facing
the ILC and the 1958 United Nations Conference on the law of the sea
was whether it would be possible to agree on more specific rules for the
delimitation of the continental shelf between neighboring States. The
major difficulty in this respect is that the delimitation of the continental
shelf is a zero-sum game. A specific formulation of a general rule that is
advantageous to one of two neighboring States will be disadvantageous
to the other State. One way to overcome the problem inherent in a
specific rule is to refrain from giving it detailed content. This has the
disadvantage of providing States who are involved in bilateral negotiations with limited guidance. Article 6 of the Convention on the continental shelf seeks to steer a middle ground between these two
approaches. It makes reference to a specific method of delimitation,
the equidistance method. At the same time, article 6 recognizes that
the application of equidistance may not be called for in
certain situations. If another boundary is justified by special circumstances, the equidistance method is not applicable. Article 6 fails to
specify what constitute special circumstances and does not indicate
how to assess the consequences of special circumstances. These questions were broached during the drafting of article 6 by the ILC and at the
Geneva Conference on the law of the sea of 1958, but this still left a large
measure of uncertainty.

Article 6 of the Convention on the continental shelf played a central
role in the negotiations between Denmark, Germany and the

---

[9] See e.g. A. G. Oude Elferink, "Article 76 of the LOSC on the Definition of the Continental
Shelf: Questions Concerning its Interpretation from a Legal Perspective," 2006 (21)
International Journal for Marine and Coastal Law, pp. 269–285, at pp. 272–273.

Netherlands on the delimitation of the continental shelf in the North Sea. Germany in the 1960s decided against becoming a party to the Convention. The Danish and Dutch interpretation of article 6 and the implications for their bilateral delimitations with Germany were the principal reason for this decision. Because Germany did not become a party to the Convention, its rules did not bind Germany as treaty law. Instead, the delimitation between Germany and its neighbors was governed by customary international law. However, there still existed the possibility, and this was strongly argued by Denmark and the Netherlands, that the Convention reflected customary law. If the Danish-Dutch thesis was correct it would have been necessary to decide on the question of how the rules contained in the Convention had to be interpreted and applied. On the other hand, if the Convention did not reflect customary law, it had to be decided what the content of customary law was and how it had to be applied to the specific case. These complex questions of law figured prominently in the negotiations between the three States before they submitted their dispute to the ICJ and during their pleadings before the Court.

The judgment of the Court initialed the start of a tortuous journey for the equidistance method and the rule contained in article 6 of the Convention on the continental shelf. The judgment rejected that article 6 of the Convention had any relevance for the determination of customary law. The Court instead found that under customary law no method was obligatory and emphasized that the main purpose of delimitation is to achieve an equitable result. During the negotiations at the third United Nations Conference on the law of the sea (1973–1982), proponents of the approach contained in article 6 of the Convention on the continental shelf were pitched against supporters of the rules propagated by the Court in 1969. After protracted negotiations, the outcome was a rule on the delimitation of the continental shelf that provided virtually no guidance to States. The law of maritime delimitation has been further developed by the ICJ and arbitral tribunals. Whereas jurisprudence in the early 1980s largely ignored equidistance, the method has subsequently made a remarkable comeback. The case law of the last decade or so indicates that the equidistance method in principle should always be applied as a starting point[10] and adjustments of this provisional line in recent cases have been limited. This view of delimitation comes closer

---

[10] See e.g. *Maritime Delimitation in the Black Sea (Romania v. Ukraine)*, Judgment of 3 February 2009, paras 115–116.

to the Danish and Dutch arguments in the 1960s than the Court's 1969 judgment.

2.2   The North Sea

The North Sea is located between the northwestern part of the European continent and the British Isles. To the north and northwest it borders on the Northeast Atlantic Ocean, to the south it is connected with the English Channel and to east through the Skagerrak, Kattegat and the Danish Straits to the Baltic Sea. The coastal States of the North Sea are Belgium, France, Denmark, Germany, the Netherlands, Norway and the United Kingdom. The North Sea is a relatively shallow sea, with a mean depth of less than 100 meters. Only in the Norwegian Trough, which is a submarine depression in front of the coast of Norway, are average water depths much more than 100 meters. At its deepest point the Norwegian Trough measures over 700 meters. The general location of the Norwegian Trough can be ascertained from the 200-meter depth contour included in Figure 2.1.

When the definition of the continental shelf was first considered in the 1940s and 1950s, one of the parameters to determine its extent was the water depth of 200 meters. This criterion of water depth implied that the entire North Sea apart from the Norwegian Trough would, in legal terms, form part of the continental shelf. But what about the Norwegian Trough? A negative answer to this question would have seriously limited Norway's part of the continental shelf. If the continental shelf of Norway would not have extended beyond the Norwegian Trough, the area seaward of the Trough would have been part of the continental shelf of Denmark and the United Kingdom and would have had to be divided between them.

In geophysical terms, the Norwegian Trough can be considered a channel in the continental shelf and not an area of oceanic origin.[11] That view had already been accepted in the late 1940s,[12] when Denmark considered this matter. [13] The fact that the Norwegian Trough was part of the geophysical shelf would make it more difficult to argue that it should be excluded from the regime of the continental shelf solely

---

[11] A. Judd and M. Hovland, *Seabed Fluid Flow* (Cambridge University Press, 2007), p. 10.

[12] See *Scientific considerations relating to the continental shelf* dated 20 September 1957 (Document A/CONF.13/2 and Add.l (UNCLOS I Off. Rec., Vol. I, pp. 39–46, at p. 41, para. 20)).

[13] See further Chapter 3.2.1.

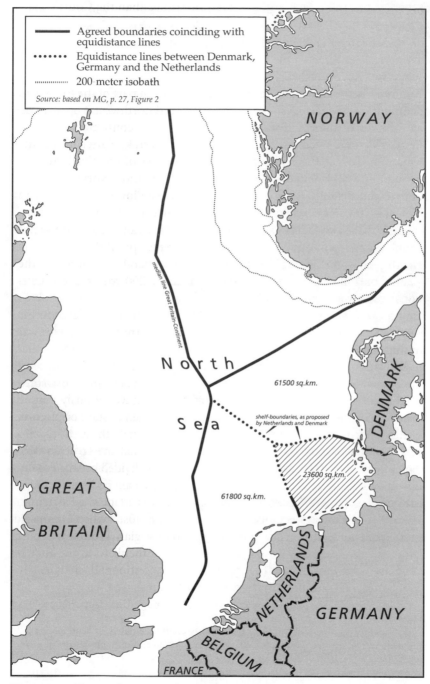

Figure 2.1 The North Sea: Equidistance lines and the location of the Norwegian Trough

because it did not meet the water depth criterion. As will be set out in Chapters 3.2.1 and 3.3, the drafting history of article 1 of the Convention on the continental shelf indicates that depressions like the Norwegian Trough are covered by the regime of the continental shelf.

As the entire North Sea was part of the continental shelf, it had to be delimited between the coastal States. A logical starting point to consider delimitation scenarios is looking at the division resulting from the application of the equidistance method. As is illustrated by Figure 2.1 the equidistance line between the European continent and the United Kingdom divides the North Sea roughly into two halves. The division between the continental States is quite uneven. The equidistance area of France and Belgium is limited in comparison to that of the other States. However, it can be easily argued that this outcome is not unreasonable because these two States have a relatively short coastline and are located in the narrow southern part of the North Sea. If only that part of the North Sea is considered, Belgium and France do not receive a much smaller part of it than the other States concerned. Similar arguments are not as readily available in the case of Germany. Its coast is comparable in length to that of Denmark, the Netherlands and Norway, but the German equidistance area is much smaller than the equidistance area of these other States. All four States border the central North Sea.

As Figure 2.1 indicates, Germany is hemmed in in a corner of the North Sea because of the convergence of the two equidistance lines with its direct neighbors. As is borne out by the negotiations on the delimitation of the continental shelf of the North Sea, this situation has a number of consequences. First, to the extent Germany would be dissatisfied with a delimitation on the basis of equidistance it would first of all have to deal with its direct neighbors. The equidistance line of Norway with Denmark does not directly affect the extent of Germany's continental shelf if the equidistance method is employed. Second, Denmark and the Netherlands had a common interest in sticking to equidistance. This suggests that both States might profit from collaboration and presenting Germany with a common front.

2.3   Other delimitation issues of Denmark, Germany and the Netherlands

2.3.1   Introduction

It is assumed that States, in formulating their preferences for specific interpretations of the applicable legal rules and making arguments in respect of the relevance of these rules, will take into account the totality of their interests, which might be affected by their preferences. If a State strongly argues for equidistance in one case, it may be difficult to argue against its application in another situation. There are a number of ways to deal with such dilemmas. A State may conclude that its overall interest is best served by sticking to one particular method of delimitation and accept that this method has certain drawbacks in some instances. This approach may help to reinforce the impression that the State is acting consistently and is not picking and choosing what best suits its interest without particular regard for the applicable law. A State may also seek to distinguish between the different delimitations it has to settle. There obviously is a limit to this approach. For instance, it will not look convincing if it is argued that the same special circumstance should be treated differently in two delimitations. But then again, when are two situations sufficiently similar to treat them in like fashion? The following subsections briefly look at the other delimitation issues Denmark, Germany and the Netherlands were facing. The actual handling of these other delimitation issues will be considered in subsequent chapters.

Before turning to these other delimitation issues, it is necessary to set out the standards which have been adopted to make this evaluation. In view of the central place of the equidistance in delimitation practice, this method is always taken as the point of departure in looking at a specific delimitation. Subsequently, it will be assessed if one of the States concerned had reason to reject the application of the equidistance method. In that respect, it should be realized that the law now is radically different from the law as it stood in the 1950s. Although at present it is still difficult to exactly predict the outcome of a delimitation in accordance with international law, the law is articulated in much more detail than in the 1950s. At that time there were only some general indications as to when a deviation from the equidistance line might be appropriate. Arguments, which at present would be immediately considered by lawyers, at that time may not have been that obvious. One example, which will be discussed below, concerns the Aru Islands of Indonesia and the territory of Netherlands New Guinea. These days a similar delimitation might give rise to the argument that there is a disproportion between the lengths of the relevant coasts facing the delimitation area. In the 1950s these concepts had not yet been developed and it might have been generally considered that equidistance in such a case was appropriate.

So what criteria will be used in the next sections to assess the appropriateness of the equidistance? First, it will be assessed if the coasts of

States involved in a delimitation are broadly similar and if the equidistance line divides the continental shelf broadly speaking in equal halves. If that is the case, it is assumed that the States concerned will find, in principle, the equidistance line acceptable. If either of the two preceding conditions is not met, it is further assessed what kind of arguments either State might have for rejecting the equidistance method.

2.3.2   Denmark

In the North Sea, Denmark was faced with three other delimitations if the equidistance method is taken as a starting point. It shared a short boundary with the United Kingdom. In view of the fact that this concerned a short segment, lying between the opposite coasts of Denmark and the United Kingdom, the delimitation of this boundary by the equidistance method would be a logical choice, which would accord both parties equal treatment. Similar considerations could, at first sight, be said to apply in the case between Denmark and the Netherlands, but the following observations on the equidistance line of Denmark with Norway in the North Sea are also relevant to the equidistance line with the Netherlands.

As far as the Danish equidistance line with Norway is concerned a distinction can be made between the Skagerrak and the North Sea. In the Skagerrak, the coasts of Denmark and Norway are clearly opposite and an equidistance line divides the area equally. In the North Sea the equidistance line between Denmark and Norway might, at first sight, also seem to be unproblematical. However, if this equidistance line is viewed in the broader context of Denmark's other delimitations in the North Sea, it would seem that an equidistance line is disadvantageous for Denmark. Whereas equidistance lines give Norway and the Netherlands a broad opening on the North Sea up to the equidistance line with the United Kingdom, the two equidistance lines between Denmark and these two neighbors converge and give Denmark a limited opening on the equidistance line with the United Kingdom. Denmark might have developed an argument that the combined coastal configuration of the three coastal States constituted a special circumstance, which required an adjustment of the equidistance line. Apart from the fact that it was uncertain whether this argument would have stood a chance at success, the presence of Germany, Denmark's southern neighbor, pointed to a major drawback of this approach. If this argument would have been made by Denmark, it could have been made with much greater force by Germany. Germany's equidistance area does not even come close to the equidistance line with the United Kingdom because of the convergence of the equidistance lines with its two direct neighbors.

Denmark in relation to Norway might also have considered raising the presence of the Norwegian Trough to reject the equidistance line as a boundary. In this respect, two options would have been open to Denmark. It could have taken the position that the continental shelf of Norway did not extend beyond the Trough. As will be discussed subsequently, Denmark did consider this option. Alternatively, Denmark could have argued that the Norwegian Trough, even if it did not limit the extent of Norway's continental shelf, constituted a feature, which had to be taken into account in determining a boundary between the two States. However, it would seem to be unconvincing to argue that a feature that is not relevant to determining the extent of entitlement is relevant in the delimitation between neighboring States.[14]

One equidistance line between Denmark and Sweden runs from the Skagerrak and the Kattegat through the Danish Straits into the Baltic Sea. A second equidistance line exists between the Danish island of Bornholm and the Swedish mainland. The most significant characteristic of these equidistance lines is that on the part of Denmark they are in large part determined by basepoints on islands, and on the part of Sweden in general by basepoints on its mainland coast. This concerns the small islands of Læsø, Anholt and Hesselø in the Kattegat and in the Baltic Sea the larger island of Bornholm. In the area of Bornholm the equidistance line is not only controlled by basepoints on Bornholm itself, but also by basepoints on the smaller island of Christiansø and its dependencies. Christiansø is some 20 kilometers to the north of Bornholm and only measures approximately 22 hectares (0.22 km²). Close to it are a number of smaller islands, the largest of which is Græsholm, which is half the size of Christiansø. On the Swedish side part of the equidistance line in relation to Christiansø is controlled by basepoints on the small islands of Utklippan, which are some 7 kilometers seaward of larger Swedish islands lying closer to the Swedish

---

[14] See e.g. *Continental Shelf (Tunisia/Libyan Arab Jamahiriya)* case, Judgment of 24 February 1982, [1982] ICJ Reports, p. 57, para. 66; *Continental Shelf (Libyan Arab Jamahiriya/Malta)* case, Judgment of 3 June 1985, [1985] ICJ Reports, p. 35, para. 39; see also *Anglo-French continental shelf* arbitration, Decision of 30 June 1977 (UNRIAA, Vol. XVIII, pp. 3–271), para. 107.

mainland. Utklippan is about half the size of Christiansø. The delimitation between Bornholm and the Swedish mainland coast can be divided in two sectors. In the western sector, the coast of Bornholm can be said to be similar to the opposite Swedish coast and the equidistance line divides the area concerned broadly equal. In the eastern sector, Christiansø and Utklippan could be considered to be special circumstances. In this sector, it could also be argued that Bornholm itself is a special circumstance as its coast facing the delimitation area is clearly shorter than the opposite Swedish mainland.

Denmark and Germany also share a maritime boundary in the Baltic Sea. Before the reunification of the Federal Republic and the German Democratic Republic in 1991, the most eastern part of this boundary was located between the German island of Fehmarn and the Danish island of Lolland. This boundary is mostly located between opposite coasts, which are relatively close to each other. In this situation, equidistance would seem to lead to a reasonable outcome. There would have been relatively little scope to adjust such a line. Until 1991, Denmark also shared two maritime boundaries with the German Democratic Republic. One of these boundaries was located between the Danish islands of Lolland, Falster and Møn and the German mainland and the island of Rügen. This case is similar to the one just described. The other boundary is located between the islands of Bornholm and Rügen. At first sight, an equidistance line between these opposite coasts would seem to be a reasonable outcome. Although Bornholm is detached from the Danish mainland and Rügen is closely connected to the German mainland, the lengths of the coasts of the islands facing the delimitation area are not that dissimilar. On balance, the German Democratic Republic could have argued for an adjustment of the equidistance line rather than Denmark. One argument in that respect could have been that its mainland coast southeast of Rügen also would have to be taken into account in looking at the overall coastal relationship. Denmark in this case had nothing to gain from rejecting the application of the equidistance method. This consideration applies with even more force in the case of the delimitation between Bornholm and the Polish mainland coast. There can be little doubt that the Polish coast on any count is much longer and that only Poland has arguments at its disposal to reject that a maritime boundary should be delimited in accordance with the equidistance method.

Denmark also had to consider the continental shelf of its two dependent territories, Greenland and the Faroe Islands. When Denmark initially looked into this matter in the late 1940s, it determined the extent of the continental shelf on the basis of the location of the 200-meter isobath. This implied that there was no need for the delimitation of the continental shelf of these territories with neighboring States.[15]

On balance, it can be concluded that it was only in the North Sea that Denmark possibly stood to gain from the rejection of the equidistance method. However, arguing against the equidistance method in this case almost certainly would have weakened Denmark's position against Germany in the North Sea and could also have been used against Denmark in respect of a number of its delimitations outside of the North Sea.

2.3.3 Germany

Germany's major concern without doubt was the effect equidistance would have on its share of the continental shelf in the North Sea. Prior to German reunification in 1991, Germany in the Baltic Sea only shared a maritime boundary with Denmark. As was submitted in the preceding section, in this case equidistance would seem to lead to a reasonable outcome.

There was one other concern for Germany in respect of the continental shelf boundaries in the Baltic Sea. The Federal Republic until 1969 took the position that it had the sole right to represent Germany within the boundaries as they existed in 1937. In the Baltic Sea, those boundaries extended to former East Prussia, the present day Russian territory of Kaliningrad. Germany wanted to prevent Denmark or Sweden from entering into agreements on the continental shelf with the German Democratic Republic, Poland or the Soviet Union, which might prejudice the German position. The German policy in this respect was only concerned with the negotiation of such agreements and did not consider substantive issues related to these continental shelf boundaries. Denmark and Sweden in general were willing to take the German interests into consideration in dealing with their other neighbors in the Baltic Sea.[16] At no time did this issue play a role in the delimitation of Germany's continental shelf in the North Sea.

---

[15] See Chapter 3.2.1.
[16] See e.g. the note II A 1–81.40 VS-NfD dated 9 May 1967 [AA/44], p. 4.

2.3.4   The Netherlands

In the North Sea, the application of the equidistance method would, apart from a boundary with Germany, lead to boundaries of the Netherlands with Belgium, Denmark and the United Kingdom. In all these cases, the Netherlands had little to gain from the application of a method other than equidistance. As far as the United Kingdom was concerned, the equidistance line is located between opposite coasts and divides the area broadly equally. In relation to its other three neighbors, these other States might rather have developed arguments against the application of equidistance.[17]

A possible problem in relation to application of the equidistance line constituted the starting point of the continental shelf boundaries with Belgium and Germany. In the case of Belgium, the Netherlands claimed historic rights over the territorial sea in the Wielingen, the southern main channel giving access to the Scheldt River and the Dutch (naval) port of Flushing and the Belgian port of Antwerp. This claim, which was rejected by Belgium, extended well beyond the equidistance line between Belgium and the Netherlands.[18] Germany and the Netherlands differ about the location of their land boundary in the Ems estuary. Germany maintains that the boundary runs along the western shore of the estuary whereas the Netherlands maintains that the boundary is located in the main channel.[19] These positions also have an impact on the delimitation of the territorial sea and the first part of a continental shelf boundary. These two issues only had minor importance for the delimitation of most of the continental shelf of the Netherlands in the North Sea.

Apart from its territory in Europe, the Kingdom of the Netherlands also comprised a number of overseas territories. In the Americas this concerned Suriname and the Netherlands Antilles. In Asia, the Netherlands remained in possession of the western part of the island of New Guinea after it had recognized the independence of Indonesia in 1949.

Like Denmark, the Netherlands, in assessing the continental shelf boundaries of its overseas territories, looked at the areas within the

---

[17] See further Chapters 2.2, 2.3.2 and 2.3.3.

[18] A summary of the Dutch position is contained in the letter from the Permanent Delegation of the Netherlands to the United Nations of 8 May 1953, which comments on the ILC's draft articles on the territorial sea (reproduced in Yearbook ILC, 1953, Vol. II, pp. 82–83, at p. 83).

[19] A summary of the Dutch position is contained *ibid.*

200-meter isobath.[20] For the Netherlands Antilles this implied that only a very limited area was taken into account in comparison with the present day situation of 200-nautical-mile zones. Of the islands of Aruba, Bonaire and Curaçao, which are close to the mainland of Venezuela, only the continental shelf within the 200-meter isobath of Aruba merges with the continental shelf within the 200-meter isobath of Venezuela. On the part of Venezuela, this equidistance line in part was controlled by the peninsula of Paraguaná. The western segment of an equidistance line would be controlled by the small barren islets of Los Monjes. Los Monjes are some 75 kilometers from the Venezuelan mainland coast and some 30 kilometers from the Colombian mainland coast. As far as can be ascertained, the Netherlands never considered arguing that Los Monjes should be treated as a special circumstance. This was the case even before it became apparent that Venezuela rejected the equidistance line as the boundary between Venezuela and Aruba.[21] After Venezuela had rejected the equidistance line, a Dutch reference to Los Monjes as a special circumstance likely would have weakened the Dutch position in respect of the equidistance line between Aruba and the Venezuelan mainland.

In the case of the leeward Netherlands Antilles (Saba, Saint Eustace and Saint Martin), the island of Saba is located on a separate shelf, if the 200-meter isobath is the point of reference. On the basis of this same criterion, the island of Saint Eustace shares a continental shelf with the island of Saint Kitts. The islands are just over 12 kilometers apart and an equidistance line, which would only measure some 6.7 nautical miles, seems a perfectly unobjectionable boundary. The Dutch, southern, part of the island of Saint Martin, which is also called Saint Martin, lies on a larger shelf area comprised within the 200-meter isobath. In this area

[20] See Chapter 3.2.3. The location of these 200-meter isobaths can be ascertained on nautical charts. Atlases in general also include information on the 200-meter isobath. Present-day knowledge of the location of the 200-meter isobath does not differ significantly from the information that was available in the middle of the last century. For the Dutch overseas territories see e.g. J. R. Feith, *De betekenis van de "Continental-Shelf" theorie voor de exploitatie van onderzeese gebieden* (Mededelingen van de Nederlandse Vereniging voor Internationaal Recht, no. 26 (1948)), pp. 23–25.

[21] The Netherlands may not have been aware of the location and impact of Los Monjes on the equidistance line in the 1950s. Reviews of the boundary between Aruba and Venezuela in the 1950s did not refer to Los Monjes and seemed to focus on the Venezuelan mainland coast (see e.g. Report of the Interdepartmental Commission of Experts concerning the establishment of the territorial waters and the continental shelf of the Kingdom dated 3 June 1954 [NNA/46], p. 6).

there are two equidistance lines between Saint Martin and neighboring French territories. The western equidistance line already intersects the 200-meter isobath within 3 nautical miles from the baseline and thus ends within the territorial sea.[22] The other equidistance line is also in large part located inside 3 nautical miles from the baselines. The relevant basepoints are mostly located on small islets and rocks, and there might have been room to argue that an equidistance line should be adjusted in favor of Saint Martin because of the location of some of these basepoints. Due to the geography of the area concerned this would only concern relatively small adjustments.

In the case of Suriname the 200-meter isobath is on average some 65 to 70 nautical miles from the coast. This area overlapped with the continental shelf of Suriname's two direct continental neighbors, the British territory of British Guiana and the French territory of French Guiana. In the case of French Guiana, the land boundary with Suriname is located in the boundary river, the Marowijne. The equidistance line between Suriname and French Guiana is controlled by basepoints which are in, or close to, the mouth of the Marowijne River. This implies that relatively minor changes in the baselines, due to processes of sedimentation and erosion at the river mouth may result in significant changes in the equidistance line further off shore. As far as can be ascertained this problem was not considered by the Netherlands when it assessed the boundary lines resulting from the application of the equidistance method. This may be due to the fact this problem was not acknowledged at the time and the equidistance line at the time approximately coincided with a line which took into account the general direction of the coasts of both territories. Such a line would have a bearing of about 25° east of true north. This would seem to constitute a reasonable boundary for both States and there would not seem to be arguments available to support a radically different boundary.

The equidistance line between Suriname and British Guiana is constituted by a number of straight line segments.[23] The equidistance line could be said to lead to a reasonable result for both Suriname and British Guiana. Arguments to reject the equidistance line would seem to be more readily available to Suriname. A continental shelf boundary would, in any case, have to take into account that the territorial sea boundary

---

[22] In the 1950s and 1960s, both France and the Netherlands claimed a 3-nautical-mile territorial sea.

[23] For a depiction of the equidistance line up to the 200-meter isobath on the basis of calculations by the United Kingdom from the 1950s see MG, Vol. 5, Plate 7.

had become established along a line of 10° E of true north. This boundary assured that the navigational channels giving access to the Corantijn River, like the river itself, were located in Suriname territory. Because of the adoption of the 10° line, the terminus of the territorial sea boundary, which provided the starting point of the continental shelf boundary, was closer to British Guiana than Suriname. The first part of the continental shelf boundary would thus, in any case, have to deviate from the equidistance line. Apart from this relatively minor issue, it could be argued that the protruding headland on the British Guianese side of the Corantijn River had the effect of pushing the continental shelf boundary in front of the coast of Suriname, and as such required an adjustment of the equidistance line.

The Dutch territory of Netherlands New Guinea shared maritime boundaries with the Territory of Papua and New Guinea, which was administered by Australia and became independent as Papua New Guinea in 1975, Australia and Indonesia. Most of the seabed area within the 200-meter isobath around the coasts of the territory was located to its south. An equidistance line between Netherlands New Guinea and the Territory of Papua and New Guinea and Australia generally could be said to be a reasonable solution. It might be argued that the Wessel Islands on the northern coast of Australia, which have the form of a long thin promontory, constitute a special circumstance. In the case of Indonesia, an equidistance line existed between the Aru Archipelago and the southern and southeastern coast of Netherlands New Guinea. Depending on the view of the relevant coasts, it could be argued that the equidistance line should have been shifted in the direction of the Aru Archipelago to arrive at a continental shelf boundary. However, it should be realized that in the 1950s maritime delimitation law was in its infancy and most of the complex arguments in relation to coastal lengths and adjustments of the equidistance line still had to be developed. The uncertain status of Netherlands New Guinea – Indonesia vehemently opposed the continued Dutch presence – was probably also not conducive to considering complex boundary scenarios.

Taking all the potential continental shelf boundaries the Netherlands was confronted with in the 1950s together, equidistance could be said to be an appropriate method. This was not only the case for the Kingdom in Europe but also held true for the Netherlands Antilles. Suriname was not necessarily best served by the equidistance method and this implied the possibility of a conflict of interest with the other parts of the Kingdom.

To complete the picture for the Kingdom of the Netherlands it is necessary to shortly consider its internal political organization. During the Second World War, the Dutch Government had indicated that the Dutch colonies after the war would be given a greater say in their internal affairs. The new structure of the Kingdom was formalized through the adoption of the Statute for the Kingdom of the Netherlands of 1954. The Statute provided that the three countries of the Kingdom, the Netherlands, the Netherlands Antilles and Suriname were autonomous in respect of their internal affairs and would take care of their common interests on a footing of equality.[24] Each of the countries had its own Government and Parliament.[25]

Under the Statute a number of matters were the concern of the Kingdom, instead of the separate countries. One of those was the conduct of foreign affairs.[26] The Statute laid down a procedure to deal with matters of concern to the Kingdom. Such matters had to be dealt with by the Council of Ministers of the Kingdom, and not the Councils of Ministers of the separate countries.[27] The three countries of the Netherlands did not have an equal representation in the Council of Ministers of the Kingdom. The Council was constituted by the Dutch Council of Ministers, to which were added a Plenipotentiary Minister of the Netherlands Antilles and a Plenipotentiary Minister of Suriname.[28] The Plenipotentiary Ministers were not entitled to participate in all matters of concern to the Kingdom, but only in those matters of concern to the Kingdom which affected the country concerned.[29] In the case of foreign affairs, the Statute provided that a country was affected if it was particularly involved in a matter or if there would be important consequences for it.[30] In addition, the Governments of Suriname and the Netherlands Antilles could indicate if they considered that there were other matters of foreign affairs which affected them and in their view had to be reviewed in the Kingdom Council of Ministers.[31] International agreements that affected the Netherlands Antilles and/or Suriname

---

[24] Statute for the Kingdom of the Netherlands, article 1.
[25] The Statute did not grant equal status to Netherlands New Guinea, which remained under the direct control of the Netherlands.
[26] Statute for the Kingdom of the Netherlands, article 3(1)(b).
[27] In the Dutch constitutional system a distinction is made between the Government and the Council of Ministers. The Government consists of the Ministers and the King (Queen). The Council of Ministers does not include the latter.
[28] Statute for the Kingdom of the Netherlands, article 7.   [29] Ibid., article 10(1).
[30] Ibid., article 11(4).   [31] Ibid., article 11(6).

would not only be submitted to the Dutch Parliament for approval, but also to the Parliaments of the Netherlands Antilles and/or Suriname.[32]

The Plenipotentiary Ministers of Suriname and the Netherlands Antilles did not have a veto power over decisions in the Council of Ministers of the Kingdom, but could object to a decision. In that case, a decision would be taken in a restricted committee including the Plenipotentiary Minister(s) and the Prime Minister. The Prime Minister had the decisive vote and the Council of Ministers had to follow the decision of the committee.[33]

The arrangements of the Statute in respect of the conduct of foreign affairs implied that Suriname and the Netherlands Antilles had a major say in the determination of the boundaries of their continental shelf. The Statute provided that as far as possible the organs of the countries had to be involved in such affairs.[34] An agreement concerning a continental shelf boundary required the approval of the Parliament of the country concerned. On the other hand, the impact on policy formulation, including the interpretation of the applicable rules, of the Netherlands Antilles and Suriname in respect of their own boundaries was more limited. As an issue of foreign policy, in the case of disagreements, this was a matter which had to be decided in the Council of Ministers of the Kingdom. As far as the negotiation of the continental shelf boundaries of the Netherlands was concerned the question might arise if this was a matter which affected the interests of the Netherlands Antilles and Suriname. The initiative to classify this matter lay with The Hague. If the Netherlands were to exclude the Netherlands Antilles and Suriname, they could employ the procedures of the Statute, but they might be hesitant to invoke these formal procedures, which in any case gave the Dutch the final say.

---

[32] *Ibid.*, article 24.
[33] *Ibid.*, article 12; Explanation to the Statute, articles 7–12 (reproduced as an Annex to K. II 3517 (1953–1954), no. 2).
[34] Statute for the Kingdom of the Netherlands, article 6(2).

The development of the delimitation rule of the Convention on the continental shelf

3.1 Introduction

Between 1949 and 1956, the ILC prepared draft articles on the law of the sea. This included articles on the definition and delimitation of the continental shelf. The work of the ILC on the continental shelf had been preceded by State practice, but the efforts of the ILC to systematize this practice into a set of rules in turn spurred States to further consider the shelf's legal regime. In particular, States considered preliminary draft articles prepared by the ILC and submitted observations. After the ILC had finalized its work on the law of the sea in 1956, the United Nations General Assembly decided to convene an international conference. The conference was charged to examine the law of the sea and to adopt "one or more international conventions or such other instruments as it may deem appropriate."[1] The United Nations Conference on the law of the sea was convened in Geneva from 24 February to 27 April 1958.

During the preparation for, and at, the Conference, States further discussed the articles on definition and the delimitation of the continental shelf. The substance of the draft articles of the ILC was adopted as part of the Convention on the continental shelf, one of the four conventions resulting from the 1958 Conference. Article 1 of the Convention defines the continental shelf as the seabed and subsoil adjacent to the coast. The outer limit was defined by reference to the 200-meter isobath. Beyond that limit the continental shelf also comprised the adjacent areas where the water depth admitted the exploitation of the natural resources. Article 6 of the Convention, in two virtually identical paragraphs applicable to States with respectively opposite and adjacent coasts, provides that the boundary of the continental shelf between neighboring States has to be established through agreement. In the absence of agreement,

---

[1] United Nations General Assembly resolution 1105 (XI) International conference of plenipotentiaries to examine the law of the sea adopted on 21 February 1957, para. 2.

and unless special circumstances justified another boundary, the boundary was defined as an equidistance line between the baselines of the States concerned.

The current chapter first focuses on the role of Denmark, Germany and the Netherlands in the debate on what was to become article 6 of the Convention on the continental shelf. Chapter 3.2 looks at the internal deliberation of the three States prior to the 1958 Conference. Chapter 3.3 looks at their contribution to the debates during the Conference. Chapter 3.4 compares the Danish, Dutch and German approach to this issue. In the early 1960s, when Denmark, Germany and the Netherlands further considered the definition of their continental shelf boundaries, it became apparent they had diverging views on the implications of article 6 of the Convention on the continental shelf. This first of all concerned the relationship between the reference to agreement and the equidistance principle. Does article 6 entitle a coastal State to unilaterally define a boundary in accordance with the equidistance principle in the absence of agreement with a neighboring State? Second, this question also points to a further question of interpretation. Is the equidistance principle the default rule and does the State invoking the presence of special circumstances have the burden of proof to establish that they justify a departure from the equidistance line? The above questions also point to a further issue. What is the regime applicable to the continental shelf in the absence of agreement on the boundary? What activities are States entitled to carry out in such a case and what response is allowed to activities of another State they deem to have gone beyond what is permitted? These questions are addressed in Chapter 3.5.

3.2   The initial reception of the continental shelf regime

3.2.1   Denmark

The Danish Ministry of Foreign Affairs established a committee to investigate the political, legal, economic and scientific questions in connection with the sovereign rights over the continental shelf in December 1948.[2] The committee was chaired by ambassador Georg Cohn and its membership included representatives of a number of ministries and the

---

[2] *UM's tidsskrift* no. 2 [DNA/25], p. 26; for another overview of the activities of the committee see Report of the Committee of 13 December 1948 to investigate the question concerning the continental shelf (Cohn-Committee), KSU/R.1 dated 24 April 1967 [DNA/132].

local government of Greenland.[3] At a meeting on 6 April 1949, the committee considered the possible extent of the continental shelf of Denmark, the Faroe Islands, and Greenland taking into account the 200-meter depth line. Applying that criterion, the continental shelf of the Faroe Islands and that of Greenland, to the extent it had been surveyed, did not overlap with the continental shelf of neighboring States. The committee concluded that this allowed extending the rights over the continental shelf for Greenland and the Faroe Islands through unilateral declarations.[4] In the case of Denmark, this was not considered to be possible. The 200-meter depth line did not have any real connection with the Danish land territory. The only area off the Danish coast in which there was a sudden change in depth was the Skagerrak – i.e. in the area of the Norwegian Trough.[5] The committee held that in the case of Denmark and its neighboring States overlapping areas could be divided through agreements.[6]

Possible methods of delimitation of the continental shelf between neighboring States in the North Sea and the Baltic Sea were further considered by two members of the committee.[7] The report of Professor Nørlund of the Geodetic Institute observed that the Norwegian Trough was not an oceanic area, but was generally considered to be a channel in the continental shelf. Nørlund submitted that Norway probably would have difficulty in accepting the criterion of water depth to define the extent of sovereign rights. Nonetheless, Nørlund proposed that in a delimitation based on seabed topography the delimitation line could be the median line between the 200-meter isobaths on both sides of the Norwegian Trough.[8] The report noted that for other parts of the North Sea and the Baltic Sea depth lines did not provide possible boundaries and concluded that it was difficult to find an equitable basis for those delimitations.[9] Commander Petersen of the Ministry of the Navy

---

[3] See *UM's tidsskrift* no. 2 [DNA/25], p. 26.

[4] Committee of 13 December 1948 to investigate the question concerning the continental shelf (Cohn-Committee), Report of the Committee's second meeting on 6 April 1949 3 PM [DNA/30], pp. 1–2.

[5] *Ibid.*, p. 2.   [6] *Ibid.*

[7] See letters of G. Cohn dated 28 September 1949 to respectively Commander O. Petersen of the Ministry of the Navy and Professor N. E. Nørlund of the Geodetic Institute [DNA/30].

[8] Report on the possibilities for a possible determination of a delimitation line of Danish waters; attachment to a letter dated 5 November 1949 from N. E. Nørlund to G. Cohn [DNA/30], pp. 1–2.

[9] *Ibid.*, pp. 3–5.

likewise concluded that it was impossible to delimit the continental shelf in the North Sea and the Baltic Sea using the criterion of water depth as these areas were generally of shallow or moderate depth.[10] Petersen did not make a separate reference to the Norwegian Trough. His report proposed the median line between opposite States and the perpendicular to the general direction of the coast in the case of adjacent States.[11]

The committee elaborated its views on the delimitation of the continental shelf in a report of 2 February 1952, commenting on the ILC's 1951 draft articles on the continental shelf.[12] The report criticized the fact that the draft articles only provided for delimitation of the continental shelf through agreement between neighboring States and recourse to arbitration or the ICJ in the absence of agreement.[13] A workable solution required the inclusion of guidelines or directives concerning the delimitation of the continental shelf in the proposed treaty. The committee observed with approval that the commentary of the ILC's special rapporteur already referred to the median line and the perpendicular to the general direction of the coast. In cases involving three or more States, the committee's report proposed a method that, although it was not identified as such, is the equidistance method.[14] This is also apparent from a map of the North Sea and a part of the Baltic Sea on which the methods proposed by the committee were indicated for illustrative purposes.[15] The committee's report was submitted with minor adjustment as the Danish Comments on the Draft Articles on the Continental Shelf circulated by the ILC.[16] The report did not refer explicitly to the significance of the Norwegian Trough, but accepted that the equidistance line could also be applied in this case.

---

[10] Considerations concerning criteria and possibilities for a division of the North Sea and the Baltic Sea between the respective coastal States dated 7 November 1949 [DNA/30].

[11] *Ibid.*, p. 1. In view of the relevant geography, it is likely that the perpendicular to the general direction of the coasts of Denmark and Germany in the North Sea was similar to an equidistance line.

[12] The draft articles are reproduced at Yearbook ILC, 1951, Vol. II, p. 141.

[13] Danish Experts' Report on the Draft articles on the Continental Shelf and Related Subjects prepared by the International Law Commission dated 2 February 1952 [DNA/30], pp. 2–3.

[14] *Ibid.*, pp. 3–5.

[15] A copy of this map is reproduced in DCM, Annex 9; a copy of the map is also included in [DNA/9].

[16] *Regime of the High Seas; Comments by Governments on the Draft Articles on the Continental Shelf and Related Subjects* (Doc A/CN.4/55 of 16 May 1952) at p. 15 and following.

For its part, Norway did address the issue of the Norwegian Trough. Commenting on article 1 of the ILC's 1951 draft articles, Norway observed that:

> If there are to be any rules governing the continental shelf, article 1 ought to be redrafted so that it is beyond doubt that the term "continental shelf" refers to the sea-bed and subsoil of the submarine areas lying off the coast, even if these submarine areas are separated from the coast by stretches of deep waters. The best thing would probably be not to use the term "continental shelf."[17]

On the other hand, the United Kingdom in its observations objected to any system which would allot to the coastal State shallow areas separated from the coast by waters of more than 100 fathoms (approximately 183 meters) in depth.[18]

Special rapporteur François considered the comments of Norway and the United Kingdom in his fourth report on the regime on the high seas. He took note of the position of the United Kingdom but, on the basis of the Norwegian observations, concluded that it would evidently be wholly contrary to equity if the existence of the elongated and narrow Norwegian Trough would exclude Norway from the division of the North Sea beyond the trench.[19] François proposed that it should be made clear that draft article 1 defining the continental shelf also applied to the submarine area adjacent to the coasts if they were separated from the coast by a narrow channel.[20] This proposal received support in the Commission.[21] Article 1 of the revised draft articles on the continental shelf adopted by the ILC only referred to the 200-meter isobath to define the extent of the continental shelf.[22] The commentary to the article observed that the rule was subject to equitable modifications if submerged areas of a depth of less than 200 meters, which were situated in considerable proximity to the coast, were separated by a narrow channel

---

[17] Comments of the Government of Norway transmitted by a letter dated 3 March 1952 from the permanent delegation of Norway to the United Nations, reproduced at Yearbook ILC, 1953, Vol. II, pp. 260–262, at p. 261.

[18] Comments of the Government of the United Kingdom transmitted by a letter dated 2 June 1952 from the permanent delegation of the United Kingdom to the United Nations, reproduced at *ibid.*, pp. 266–268, at p. 267.

[19] See Regime of the High Seas; Quatrième rapport (Le plateau continental et les sujets voisins); Yearbook ILC, 1953, Vol. II, pp. 1–51, at p. 11.

[20] See Yearbook ILC, 1953, Vol. II, pp. 1–51, at p. 11.

[21] See *ibid.*, Vol. I, p. 74, paras 6 and 8 (Lauterpacht); pp. 79–80, para. 8 (Pal); p. 80, para. 17 (Sandström).

[22] See *ibid.*, Vol. II, p. 212.

deeper than 200 meters from the part of shelf adjacent to the coast. Such areas had to be considered to be part of the continental shelf.[23] The Commission further held that it would be on the State relying on this exception to establish its claim to an equitable modification.[24] The discussion in the Commission leaves no doubt that it was considered that the Norwegian Trough would fall under this modification. The above understanding on narrow channels was included in the commentary on the ILC's 1956 draft articles on the law of the sea, which were considered at the Geneva Conference in 1958.[25]

The Danish continental shelf committee again considered the ILC's draft articles at the end of 1953. The committee concluded that the changes that had been introduced in the draft articles during the fifth session of the ILC were a significant improvement.[26] The article on the delimitation of the continental shelf apart from the need for agreement between the interested States now also referred to equidistance and special circumstances.[27] Most observations of the Danish government were reflected in the new draft and the committee considered that it would be desirable to establish general rules on the regime of the continental shelf. Although these general rules might have certain shortcomings, they were to be preferred to the currently existing uncertainty, which had resulted in major international difficulties and conflicts. The committee recommended the Government to support the proposal in the General Assembly.[28]

As far as can be established, the ILC's draft articles on the continental shelf were only revisited by the Ministry of Foreign Affairs in January 1958. A report commenting on the draft articles noted that article 72 on the delimitation of the continental shelf reflected the concerns concerning substantive delimitation rules, which had been expressed by the Danish continental shelf committee in its 1952 report. Further changes

---

[23] See *ibid.*, p. 214, para. 66.   [24] See *ibid.*

[25] See Report of the International Law Commission to the General Assembly, reproduced in Yearbook ILC, 1956, Vol. II, pp. 253–301, at p. 297.

[26] Note dated 13 November 1953 prepared by G. Cohn [DNA/85] (Note Cohn 13 November 1953 [DNA/85]), p. 1.

[27] The draft articles are reproduced at Yearbook ILC, 1953, Vol. II, pp. 212–213.

[28] Note Cohn 13 November 1953 [DNA/85], p. 1; *Comments on the draft articles on the continental shelf, fisheries and the contiguous zone adopted by the International Law Commission at its fifth session, transmitted by the Government of Denmark* (Doc. A/CN.4/86 of 13 May 1954) reproduced at Yearbook ILC, 1954, Vol. II, pp. 18–19.

to the draft article should not be pursued.[29] The instructions for the Danish delegation to the Geneva Conference did not make any specific reference to article 72.[30]

3.2.2   Germany

The German authorities were initially reluctant to exercise rights over the continental shelf beyond the territorial sea. At the end of 1951 the Foreign Office had been consulted by the Ministry for the Economy as to whether the coastal State under international law was entitled to drill for oil on the continental shelf and had concluded that this question was controversial.[31] In October 1953, the Foreign Office circulated the ILC's draft articles on the continental shelf and related subjects to other directly interested ministries,[32] but this apparently did not lead to an immediate follow-up. This may at least in part be explained by the limited interest in the work of the United Nations at the time.[33]

In 1955, the Ministry for the Economy again consulted the Foreign Office as to whether Germany could regulate oil activities beyond the territorial sea after an oil company had inquired about a concession in the Baltic Sea.[34] In that connection it was observed that more than 30 States had already claimed rights over the continental shelf. It could be argued that this development allowed the coastal State to regulate oil activities beyond its territorial sea. It was however left to the Foreign Office to decide if recent developments had made oil exploitation per-missible or whether it was rather advised to await the conclusion of an international agreement.[35] The Foreign Office indicated that there

---

[29] Report; The continental shelf dated 25 January 1958 [DNA/29], p. 12, and hand-written notes in the margin.

[30] Instruction for the Danish delegation to the United Nations Conference on the law of the sea that will convene in Geneva on 24 February 1958 dated 19 February 1958 [DNA/89], p. 34.

[31] See draft of a letter of the FO to the Minister for the Economy dated 24 November 1951 [AA/88], p.10; letter of the Minister for the Economy to the FO dated 6 June 1955 [AA/88], p. 2 (quoting from the letter dated 24 November 1951).

[32] Circular letter of the FO dated 16 October 1953 [B102/11955].

[33] See E. Czempiel, *Macht und Kompromiß; Die Beziehungen der Bundesrepublik Deutschland zu den Vereinten Nationen 1956–1970* (Düsseldorf: Bertelmans Universitätsverlag, 1971), p. 34; see also H. Dröge, F. Münch and E. Von Puttkamer, *The Federal Republic of Germany and the United Nations* (New York: Carnegie Endowment for International Peace, 1967), pp. 24–25.

[34] Letter of the Minister for the Economy to the FO dated 6 June 1955 [AA/88].

[35] *Ibid.*, p. 2.

existed reasonable doubt that the unilateral declarations of coastal States had led to the formation of customary international law: a number of States had indicated that particular care should be taken while developing new rules in this respect and that the matter could only be settled through the negotiation of a multilateral convention. The Foreign Office considered it appropriate to wait for the further treatment of the matter by the 11th session of the General Assembly of the United Nations in 1956.[36] The Foreign Office also raised a theme that was to play a significant role in the further deliberations of the German authorities on the continental shelf regime. That regime might lead to an encroachment on the freedom of the high seas "because the conceptual boundary between the territorial sea and the high seas might, through the continental shelf, be easily erased."[37] Fisheries interests, especially, would advance this argument to oppose the continental shelf regime.

The delimitation of the continental shelf in the North Sea was probably first considered by the German authorities in the second half of 1955. An article in the *Bremer Nachrichten* of 10 March 1955 seems to have triggered this interest.[38] The article included a figure that reportedly indicated the Dutch view on the delimitation of the continental shelf. The lines in the figure are broadly similar to equidistance lines and indicated that Germany would only receive a limited part of the North Sea near its coast. The matter was brought to the attention of the Foreign Office by the Minister of Transport, who submitted that a Dutch proposal along these lines in no case should be accepted.[39] The Minister

---

[36] Letter of the FO to the Minister for the Economy dated 20 June 1955 [AA/88], p. 3.

[37] *Ibid.* Translation by the author. The original text reads "da die begriffliche Grenze zwischen Küstenmeer und hohem Meer über den Festlandsockel leicht verwischt werden kann."

[38] The article "Holland verteilt die Nordsee" (Holland divides the North Sea) is annexed to a letter of the Minister of Transport to the FO dated 11 August 1955 [AA/3]. This issue was raised at about the same time by Professor Böhmert of the University of Kiel, who later acted as an advisor to the German Ministry for Food Supply, Agriculture and Forests on matters related to the development of the law of the sea. Böhmert not only pointed out that Germany had a very limited continental shelf, but also argued that acceptance of the regime of the continental shelf might affect other high seas freedoms, such as fisheries and scientific research, in which Germany had a vital interest (letter of Böhmert to Raiser dated 8 July 1955 [AA/2], pp. 2–3). In an article of 1956 Böhmert made a forceful plea for defending the freedom of the high seas. He among others submitted that the assertion that unilateral declaration on the continental shelf had led to customary law lacked all grounds (V. Böhmert, "Meeresfreiheit und Schelfproklamationen," 1956 (6) *Jahrbuch für internationales Recht*, pp. 7–99, at p. 98).

[39] Letter of the Minister of Transport to the FO dated 11 August 1955 [AA/3].

argued that the geology of the shelf, which indicated that tectonic lines generally extended in a northwestern direction, indicated that the delimitation proposal contained in the article was without basis.[40] The Foreign Office instructed the Embassy in The Hague to inquire with the Dutch Ministry of Foreign Affairs if the article in the *Bremer Nachrichten* was factually correct.[41] The matter was broached in a meeting with the Director of the European Department of the Ministry, who did not directly address the issue of delimitation. The Netherlands, like other States, was investigating the economic potential of the continental shelf, but those investigations had not yet led to concrete results.[42] Actually, the delimitation of the continental shelf in the North Sea had already been considered by an interdepartmental commission, which had concluded that the equidistance method led to a reasonable solution.[43]

At the end of February 1956 the Foreign Office informed other government agencies that it considered that the time had come to clarify the questions relating to the outer limits of the territorial sea and the continental shelf and that it would, to that end, arrange a meeting.[44] This meeting, which eventually took place at the end of November 1956, revealed that only the Ministry for the Economy was inclined to support the continental shelf regime. The resources of the shelf could contribute to covering the German energy demand. German companies had already expressed their interest.[45] The exploitation of the resources of the continental shelf under an international regime was not considered to be practicable.[46]

Most other ministries voiced doubts about the proposed regime.[47] The Ministry for Food Supply, Agriculture and Forests strongly argued against the new regime. Germany would only receive a very small part of the continental shelf in the North Sea. The risk existed that the rights of the coastal State would subsequently be extended to other activities.

---

[40] *Ibid.*, p. 2.
[41] Note with the original of a number of letters of the FO dated 1 September 1955 [AA/3]. The draft of the letter to the GEH is contained at p. 3.
[42] Letter of the GEH to the FO dated 28 September 1955 [AA/3].
[43] See further Chapter 3.2.3.
[44] Circular letter of the FO dated 21 February 1956 [B141/22020].
[45] Minutes of the interagency meeting of 30 November 1956 at the Foreign Office dated 14 December 1956 [B102/119551] (Minutes meeting 30 November 1956 [B102/119551]), pp. 8–10.
[46] *Ibid.*, p. 9.   [47] *Ibid.*, pp. 8–10.

This would be detrimental to German fishing interests.[48] The Foreign Office held on to the view that it was not expedient to formulate an official position. Although the subject was moving ahead, it would still take considerable time until generally recognized principles would develop. Since Germany was not represented in the ILC, it was preferable to see what direction further developments would take.[49] A further explanation for the reserved position of the Foreign Office at that time, or rather its legal department, which was handling this issue, was that it had thus far refrained from raising the matter with the Foreign Office's political department. In that context the position of Poland and the Soviet Union in the Baltic Sea would also have to be discussed.[50] The meeting decided that a further interagency meeting to establish a common position for the Federal Government could take place once the agencies represented at this first meeting had formulated their positions.[51]

During the November 1956 interagency meeting, the Foreign Office presented an assessment of the support for the continental shelf regime in the international community. It was argued that a considerable number of States expressed reservations about the ILC's draft articles on the continental shelf.[52] In this connection the Foreign Office relied on the comments, which States had submitted on the draft articles the ILC had adopted in 1951.[53] However, a comparison of the actual comments with the summary provided by the Foreign Office indicates a selective reading. For instance, Belgium and Denmark are listed as opposing any extension of sovereignty. That is certainly true, but neglects that both States expressed support for the regime of the continental shelf contained in the ILC draft articles.[54] The Netherlands was reported as

---

[48] *Ibid.*, p. 8; see also circular letter of the Minister for Food Supply, Agriculture and Forests dated 27 April 1956 [B141/22020], pp. 2–3.

[49] Minutes meeting 30 November 1956 [B102/119551].

[50] Note dated 3 December 1956 [B102/119551], p. 2. The reference to Poland and the Soviet Union likely was concerned with the German position that it had the sole right to represent Germany within the boundaries as they existed in 1937 (see further Chapter 2.3.3).

[51] Minutes meeting 30 November 1956 [B102/119551].   [52] *Ibid.*, pp. 3–5.

[53] The commentary in the Minutes indicates that this concerns the 1953 draft articles. However, the comments of States which were reviewed were submitted in 1953 and were concerned with the 1951 draft articles.

[54] See Comments of the Government of Belgium transmitted by a note dated 1 March 1953 from the permanent delegation of Belgium to the United Nations (reproduced at Yearbook ILC, 1953, Vol. II, p. 241). Communication of the Permanent delegation of Denmark to the United Nations (reproduced at *Regime of the High Seas; Comments by*

supporting the continental shelf regime with the reservation that exploitation should be carried out under international control. Actually, the Dutch Government said that it might in theory perhaps have been preferable to give jurisdiction to the international community as a whole, but that it felt that the practical difficulties of doing so would prove insuperable. The international body to which the Netherlands referred in its commentary was only to have an advisory role.[55] No further detailed assessment of the positions of States seems to have been made before the start of the Geneva Conference.[56] In that light, it can be concluded that the German government did not have a realistic view of the amount of support for the ILC's draft articles in defining the German position.[57]

In the months following the interagency meeting of November 1956, the ministries further considered their positions. The Minister for the Economy recognized that the regime of the continental shelf had not yet become customary international law, but he considered that Germany would not be able to prevent the further development of that regime. The orderly development of the continental shelf required regulation by the coastal State. A rejection of the continental shelf regime might furthermore give rise to other coastal State claims that would have more far-reaching implications for shipping and fishing. The Minister favored regulating the matter in a multilateral treaty, for which the ILC's draft provided a sufficient basis.[58] He also submitted that it might not be necessary to await the development of a global regime. The suggestion of the Minister for Food Supply, Agriculture and Forests to address the matter through regional agreements looked very promising for the areas in which Germany was interested. It could be considered if the idea of a monopoly of the coastal State over the continental shelf could be

Governments on the Draft Articles on the Continental Shelf and Related Subjects (Doc A/ CN.4/55 dated 16 May 1952) at p. 15 and following. Denmark commented even more favorably on the 1953 ILC draft articles (see Comments on the draft articles on the continental shelf, fisheries and the contiguous zone adopted by the International Law Commission at its fifth session, transmitted by the Government of Denmark (Doc. A/ CN.4/86 dated 13 May 1954) in Yearbook ILC, 1954, Vol. II, pp. 18–19).

[55] Comments of the Government of the Netherlands transmitted by a letter dated 24 March 1952 from the permanent delegation of the Netherlands to the United Nations (reproduced at Yearbook ILC, 1953, Vol. II, pp. 259–260).

[56] A review of the support for the continental shelf regime from September 1957 seems to be based on the assessment presented to the meeting of November 1956 (note dated 5 September 1957 [B102/119551]).

[57] But see text at footnotes 75 and following.

[58] Circular letter of the Minister for the Economy dated 6 June 1957 [AA/91], pp. 7–9.

abandoned in favor of a regime that would give all the coastal States of a region equal access to the area concerned to the exclusion of third States.[59] It is questionable whether other coastal States of the North Sea would have accepted such an approach, as it obviously implied that they would get a lesser share of the resources of the continental shelf. The suggestion that this constituted a viable approach may have been intended to convince other Ministries to give up their resistance against the continental shelf regime.

The Minister for Food Supply, Agriculture and Forests did not buy into the arguments of the Minister for the Economy and remained opposed to the continental shelf regime.[60] Proclamations of the coastal States in the North Sea should be avoided, as it would give Germany only a small part of the North Sea. Instead a regional agreement should be considered.[61] The Ministry for Transport and the Ministry for Defense also expressed reservations using similar arguments. Acceptance of the shelf regime posed a risk of creeping jurisdiction and Germany would only get a limited part of the continental shelf in the North Sea.[62] The Foreign Office, at least until the first half of 1957, refrained from adopting a specific position[63] and, after that time, remained cautious in its approach.[64]

States had been invited by the Secretary-General of the United Nations to submit comments to the ILC's final draft articles on the law of the sea by 31 July 1957. The German authorities at the time had not yet agreed on a common position on the regime of the continental shelf and during an interagency meeting in June 1957 it was agreed to postpone the consideration of a position on the continental shelf and the territorial sea.[65] Germany, for the time being, did not submit observations on the regime of the continental shelf to the Secretary-General.[66]

---

[59] Ibid., pp. 9–10.
[60] Letter of the Minister for Food Supply, Agriculture and Forests to the Minister for the Economy dated 28 June 1957 [AA/88].
[61] Ibid.
[62] See letter of the Minister for Defense to the FO dated 1 April 1957 [AA/88]; letter of the Minister for Transport to the FO dated 4 April 1957 [AA/88].
[63] See Minutes of the interagency meeting dated 11 June 1957 at the Foreign Office dated 13 June 1957 [AA/96].
[64] See e.g. Minutes of the interagency meeting of 20 February 1958 at the Foreign Office dated 21 February 1958 [AA/96], p. 3.
[65] Minutes of the interagency meeting of 11 June 1957 at the Foreign Office dated 13 June 1957 [AA/96].
[66] Germany submitted comments on the ILC's draft articles on the law of the sea through a note verbale from the Office of the Permanent observer of the Federal Republic to the

At the end of 1957 the question of the continental shelf regime was again taken up in interagency meetings. At a meeting of the Ministries for the Economy, Transport and Food Supply, Agriculture and Forests, the Ministry for the Economy expressed its support for the ILC's draft articles. A regional approach was no longer seen as an alternative. After agreement on the ILC's draft would have been reached, a regional agreement for the North Sea might be used to negotiate a larger German share in the continental shelf.[67] On the other hand, the Ministry for Food Supply, Agriculture and Forests considered that Germany should reject the draft articles and remained in favor of a regional approach.[68] The Ministry for Transport could support both options, but as a compromise preferred the latter option. An alternative might be that the coastal State would have exclusive rights up to the 12-nautical-mile limit. Beyond that distance, the continental shelf could be managed by an international institution.[69]

After two further interagency meetings had not led to agreement on a common position on the continental shelf regime,[70] the Foreign Office presented a draft position of the Federal government at an interagency meeting of 22 January 1958, just over a month before the start of the Geneva Conference on the law of the sea. The draft rejected the regime proposed by the ILC and instead envisaged that the freedom of the high seas would be maintained. The coastal State would only have a right to

United Nations dated 18 September 1957 (reproduced at UNCLOS I Off. Rec., Vol. I, p. 85). The note at the outset observed that: "In view of the short time at its disposal, the Federal Government has not been able to study more than a few of the questions to which the draft articles relate. Accordingly, the Federal Government reserves the right to make further comments at a later stage, particularly regarding the problems of the continental shelf and the territorial sea."

67 See letter of the Minister of Transport to the FO dated 5 December 1957, reporting on a meeting of 29 November 1957 [AA/88], p. 2. One factor contributing to this stronger support for the ILC's draft articles probably was the fact that the Federation of German Industries supported that regime. The ME had requested the views of the Federation in September 1957 (see letter of the ME to the Federation of German Industries dated 12 September 1957 [B102/119551]; letter of the German Fishing Industry Association to the Federation of German Industries dated 26 November 1957 [AA/88], p. 1). The German Fishing Industry Association disagreed with the arguments set out by the Federation of German Industries and requested the Federation to reconsider its position (letter of the German Fishing Industry Association to the Federation of German Industries dated 26 November 1957 [AA/88], p. 1).

68 Letter of the Minister of Transport to the FO dated 5 December 1957 [AA/88], p. 2.

69 Ibid.

70 See Minutes of the interagency meeting of 10 December 1957 at the Foreign Office dated 14 December 1957 [AA/96]; Minutes of the interagency meeting of 19 December 1957 at the Foreign Office dated 2 January 1958 [AA/96].

supervise installations to protect its interest in the areas of high seas in the vicinity of its coast.[71] The proposal led to a clash between the Ministry for the Economy and all other ministries present at the meeting.[72] The former submitted that the exploitation of the shelf could only take place under the framework elaborated by the ILC. The coastal State was in the best position to regulate activities on the shelf and it was moreover unlikely that other States would agree to a completely new approach.[73] The other ministries rejected the ILC's approach because they feared that coastal States would also seek to limit other high seas freedoms and Germany only had a small continental shelf.[74]

Whether the Foreign Office wholeheartedly supported its own proposal may be open to some doubt. On more than one occasion representatives of the Foreign Office suggested that the development of the continental shelf regime was irreversible. A flat-out rejection of that development was not likely to serve Germany's interests.[75] It was also submitted that if the German proposal was not to gain support at the Geneva Conference, it would be more effective to support than to reject the ILC draft. It was to be expected that the practice of unilateral declarations would continue in the latter case. There, moreover, was little prospect that States which had started the exploitation of their continental shelf would renounce those claimed rights.[76] However, the Foreign Office did little to advance these views and seems to have considered itself mostly an intermediary between the Ministries having a substantive interest in the matter.[77]

Subsequent to the meeting of 22 January 1958, the Ministry for the Economy dropped its objections against the draft prepared by the Foreign Office. It took a number of meetings to finalize the draft, and the last meeting took place just a couple of days before the start of the Geneva Conference on 24 February 1958.[78] Ambassador Pfeiffer, the head of the German delegation to the conference, indicated that the German position

---

[71] See the document First draft of a position of the Federal Government on the articles 67–73 of the ILC draft dated 21 January 1958 [AA/91].

[72] According to a note on the meeting the debate "ignited" (entzündete) when the regime for exploitation was considered (see note dated 23 January 1958 [B102/119551], p. 1).

[73] See *ibid.*, pp. 2–3.   [74] See *ibid.*, p. 3.

[75] See note dated 10 December 1957 [B102/119551], p. 3; note dated 11 February 1958 [AA/88], p. 2.

[76] Note dated 11 February 1958 [AA/88], p. 4.

[77] See also note dated 10 December 1957 [B102/119551], p. 3.

[78] Minutes of the interagency meeting of 20 February 1958 at the Foreign Office dated 21 February 1958 [AA/96].

would in any case not be made public before its start. The German delegation was to await developments at the Conference.[79]

3.2.3   The Netherlands

The Dutch authorities first considered the continental shelf regime in an interdepartmental meeting at the beginning of 1952.[80] The matter had already been brought to the attention of the secretary-general of the Ministry of Foreign Affairs – the highest ranking civil servant in the ministry – at the end of October 1948.[81] The law concerning the exploitation of the shelf was developing, and this offered the Netherlands immense possibilities. It was advised that the Dutch Government develop a position soon.[82] At that time the matter was not further pursued. François, who acted as counsel at the Ministry, had advised to let the matter rest until the first ILC report on the matter would become available.[83] François himself had been elected as rapporteur for the topic "regime of the high seas" by the ILC in 1949, and in that capacity was also looking at the regime of the continental shelf. At the 1952 interdepartmental meeting François explained that the ILC draft intended to accord the coastal State rights over the continental shelf, without attributing it any rights over the superjacent waters.[84] Most participants in the meeting expressed support for this approach, but some of them considered that the proposed regime, although having practical use, constituted an infringement on the freedom of the high seas.[85] The Dutch comments on the 1951 ILC draft articles reflected these views.[86] The Netherlands endorsed the principles underlying the ILC's draft articles on the continental

---

[79] *Ibid.*, p. 3.

[80] See Minutes of interagency meeting on the International Law Commission, held at the Foreign Office, 8 January 1952 [NNA/7].

[81] See note for the secretary-general dated 19 October 1948 [NNA/6].

[82] In this connection the note not only referred to the North Sea, but also to the waters of the Indonesian archipelago. Indonesia had proclaimed its independence from the Netherlands on 17 August in 1945, but the Netherlands only recognized Indonesia's independence on 27 December 1949.

[83] Note dated 17 April 1951 [NNA/7].

[84] Minutes of interdepartmental meeting on the International Law Commission, held at the Ministry of Foreign Affairs, 8 January 1952 [NNA/7], pp. 1–2.

[85] *Ibid.*, pp. 2–4.

[86] Comments transmitted by the Government of the Netherlands to the Secretary-General of the United Nations by a letter dated 24 March 1952 from the Permanent Delegation of the Netherlands to the United Nations (reproduced at Yearbook ILC, 1953, Vol. II, p. 259–260).

shelf. Like Denmark, the Netherlands argued that there was a need for specific rules for the delimitation of the continental shelf between neighboring States.[87] Article 7 of the ILC's draft only provided that boundaries had to be established by agreement between the interested States.

The consideration of the limits of the continental shelf became a matter of some urgency for the Netherlands in the fall of 1953, following the proclamation by Australia on its continental shelf of 11 September 1953 and the enactment of the Pearl Fisheries Act (no. 2) later that same month.[88] Part of the area covered by the Pearl Fisheries Act was adjacent to the territory of Netherlands New Guinea. In an internal memorandum to the government commissioner for Indonesian affairs,[89] François suggested a study on the appropriateness of a similar proclamation for Netherlands New Guinea, even though the boundary Australia had established was acceptable. Australia had not extended its shelf beyond the median line with Netherlands New Guinea.[90] Another reason to consider a proclamation on the continental shelf of Netherlands New Guinea was that Indonesia seemed to be considering such a step.[91] The note cautioned to only adopt a proclamation if that would be required by the circumstances. A continental shelf proclamation could lead to renewed frictions with Indonesia concerning the sovereignty over Netherlands New Guinea. François moreover advised to refrain from unilateral steps before the United Nations General Assembly would have pronounced itself on the regime proposed by the ILC.[92] François considered that a similar urgency to adopt a continental shelf proclamation for the Kingdom in Europe, Suriname and the Netherlands Antilles did not exist.[93]

The consideration of a proclamation on the continental shelf of Netherlands New Guinea was taken up by an interdepartmental commission. This commission, which was chaired by François, presented its findings in September 1954.[94] The ILC's draft was considered acceptable

---

[87] *Ibid.*, p. 37.

[88] Commonwealth of Australia Gazette 1953 no. 56 and 38. For further background information on the proclamation and the Pearl Fisheries Act see e.g. L. F. E. Goldie, "Australia's Continental Shelf: Legislation and Proclamations," 1954 (3) *ICLQ*, pp. 535–575; S. V. Scott, "The Inclusion of Sedentary Fisheries Within the Continental Shelf Regime," 1992 (41) *ICLQ*, pp. 788–807.

[89] The government commissioner for Indonesian affairs was among others charged with conducting the negotiations with Indonesia over the status of Netherlands New Guinea.

[90] Note dated 22 October 1953 [NNA/8], p. 1.   [91] *Ibid.*   [92] *Ibid.*, pp. 1–2.   [93] *Ibid.*

[94] See letter of J.P.A. François and C.W. van Santen to the MiFA dated 10 September 1954 [NNA/46].

for the Netherlands and the overseas parts of the Kingdom and offered the possibility to find a satisfactory solution for any problems that might arise.[95] The interdepartmental commission observed that the median line principle that had been proposed by the ILC provided a useful and clear concept that seemed to offer the most reasonable solution.[96] The report did not provide much detail in this connection and there is no indication that the meaning and possible implications of the special circumstances clause contained in the ILC's draft was considered. The commission considered that it was likely that the coastal States in the North Sea would adopt the median line principle in consultations concerning their boundaries.[97] A similar view was expressed in the case of Suriname and the Netherlands Antilles. Only in the case of Aruba did the median line result in some advantage in relation to Venezuela, but it was not expected that Venezuela would cause any problems.[98] The commission concluded that for the time being it was not necessary to react to Australia's continental shelf legislation, as it did not extend the shelf beyond the median line in relation to Netherlands New Guinea.[99] If Australia were to propose negotiations, these could take place if the point of departure would be that the median line in principle provided the fairest boundary.[100] Similar considerations applied in respect of Indonesia.[101] A proclamation on the continental shelf of Netherlands New Guinea was in any case not necessary since the ILC's draft articles implied that the coastal State had rights over the shelf *de jure*.[102]

The regime of the continental shelf and its delimitation with neighboring States were further considered in an interdepartmental consultation in February 1955. The consultation had a wider participation than the interdepartmental commission, which had concluded its work in the previous year and, unlike the commission, also considered policy questions.[103] The continental shelf regime was accepted by all participants, but representatives of the Ministries of agriculture, fisheries and food supply and of transport and public waterworks did express concerns that

---

[95] Report of the Interdepartmental Commission of Experts concerning the establishment of the territorial waters and the continental shelf of the Kingdom dated 3 June 1954 [NNA/46], p. 10.

[96] *Ibid.*, p. 3.   [97] *Ibid.*, p. 5.   [98] *Ibid.*, p. 6.   [99] *Ibid.*, pp. 3 and 10.

[100] *Ibid.*, pp. 3–4 and 10.   [101] *Ibid.*   [102] *Ibid.*, p. 4.

[103] The participants in the consultation are listed in Minutes of the interdepartmental consultation concerning the "continental shelf" and the territorial sea of the Kingdom held at the Ministry of Foreign Affairs on 9 February 1955 [NNA/46] (1955 Interdepartmental consultation [NNA/46]), p. 1.

the regime as currently proposed did not sufficiently protect the interests of shipping and navigation.[104] The equidistance method was considered acceptable for the delimitation of the continental shelf of the Kingdom in Europe and all overseas territories.[105] Any final decision on that matter, or any other matter discussed at the consultation, would however require agreement with the governments of the Netherlands Antilles and Suriname.[106] The delimitation of the continental shelf in the North Sea was considered in some detail, following a remark that equidistance lines would give Belgium and Germany very little in comparison to Denmark, the Netherlands and the United Kingdom. It was accepted that article 7 of the ILC's draft articles, to address such difficulties, rightly provided "that the course of the boundary in the first place depended on mutual agreement and on equity in the case of special circumstances."[107]

The possibility of a proclamation or other steps concerning the continental shelf of Netherlands New Guinea, which had occasioned the first detailed discussions of the continental regime in 1954, had receded into the background by 1955.[108] At the same time, this matter became urgent for Suriname at the end of 1954 because of the interest of oil companies in its continental shelf and that of neighboring British Guiana.[109] A proclamation concerning the continental shelf of British Guiana had been issued on 19 October 1954.[110] The Government of Suriname requested that a similar proclamation be issued for Suriname.[111] After consultations between the Ministry of Foreign Affairs and the Ministry of Overseas Territories, the Government of Suriname was informed that it was already competent to issue a concession for the continental shelf,

---

[104] 1955 Interdepartmental consultation [NNA/46], pp. 2, 5–7 and Annex 2.

[105] *Ibid.*, p. 5. [106] *Ibid.*, p. 11.

[107] *Ibid.*, pp. 3–4. Translation by the author. The original text reads "dat het grensverloop in de eerste plaats laat afhangen van onderlinge overeenkomsten en van de billijkheid in bijzondere omstandigheden."

[108] There remained little support for a proclamation because of the absence of any urgent need and because it might burden relations with Indonesia (see note no. 153 dated 22 October 1954 [NNA/8]; note no. 152 dated 5 July 1955 [NNA/16]; note of the legal advisor to the secretary-general dated 11 July 1955 [NNA/16]; note no. 186 dated 11 July 1955 [NNA/16]; note no. 198 dated 26 August 1955 [NNA/16]; note concerning the continental shelf dated 1 February 1956 [NNA/43], p. 2).

[109] See also 1955 Interdepartmental consultation [NNA/46], pp. 2 and 7. Suriname had been contacted by an American company, the Heep Oil Company, in 1954 concerning a concession, which also concerned the continental shelf (see the folder [NNA/42]).

[110] British Guiana (Alteration of Boundaries) Order in Council, 1954.

[111] Letter of the acting chairman of the Government Council to the Governor of Suriname dated 24 November 1954 [NNA/44].

as the shelf belonged to the coastal State *de jure*.[112] The Minister of Foreign Affairs moreover planned to contact the British Government shortly to ascertain whether it also accepted the median line.[113] In granting a concession the Government of Suriname could already employ the median line as the boundary.[114]

The Council of Ministers of Suriname took note of the fact that it was already entitled to issue licenses, but nonetheless insisted on an official proclamation as this had also been done for British Guiana. The absence of a proclamation for Suriname might lead interested companies to draw the wrong conclusions. Moreover, a proclamation would create more certainty for these companies.[115] Suriname's continued insistence on a proclamation led to further consultations between the three countries of the Kingdom and a draft was eventually prepared in the beginning of 1956.[116] The draft provided that the boundary of the continental shelf with British and French Guiana was to be established through consultations – the assumption being that the principle of the median line would be utilized.[117]

The draft declaration attracted objections on the part of the Netherlands Antilles.[118] It feared that a proclamation might draw the attention of Venezuela and lead to a claim in relation to the continental shelf between Venezuela and Aruba.[119] The Netherlands Antilles had voiced concerns in this respect earlier, and expressed the wish to conclude a delimitation agreement with Venezuela as soon as possible.[120]

---

[112] Letter of the Minister of Overseas Territories a.i. to the Governor of Suriname dated 14 March 1955 [NNA/47], p. 1; letter of the Governor of Suriname to the acting Prime Minister of Suriname dated 27 April 1955 [NNA/44]. The MFA set out its views in a letter of the MiFA to the Minister of Overseas Territories dated 19 February 1955 [NNA/47].

[113] Letter of the Minister of Overseas Territories a.i. to the Governor of Suriname dated 14 March 1955 [NNA/47], p. 1.

[114] *Ibid.*, pp. 1–2.

[115] Letter of the acting Prime Minister of Suriname to the Governor of Suriname dated 5 May 1955 [NNA/44]; letter of the Governor of Suriname to the Minister of Overseas Territories dated 14 May 1955 [NNA/47].

[116] The draft proclamation is attached to the note concerning the continental shelf dated 1 February 1956 prepared by the MFA [NNA/43].

[117] *Ibid.*, p. 2 of the draft proclamation.

[118] See note no. 349 dated 11 February 1956 [NNA/16]; note – Proclamation concerning the continental shelf of Suriname dated 13 February 1956 [NNA/47].

[119] Note no. 349 dated 11 February 1956 [NNA/16], p. 1; note – Proclamation concerning the continental shelf of Suriname dated 13 February 1956 [NNA/47].

[120] See Overview of questions relating to the Dutch attitude with respect to the continental shelf [NNA/16], p. 1. The overview is not dated, but was submitted to the

The Ministry of Foreign Affairs acknowledged that Venezuela might consider invoking the presence of special circumstances and it should not be excluded that such arguments, which had been accepted by the ICJ in relation to the territorial sea, would also be adopted for the continental shelf.[121] It was concluded that Venezuela's views concerning the ILC's draft should be ascertained. If Venezuela supported the draft unconditionally, the Netherlands Antilles could drop its objections to a proclamation.[122] After the Government of the Netherlands Antilles had agreed to this approach,[123] the Ministry of Foreign Affairs considered how to deal with the matter. The ILC's draft articles should, in any case, form the basis of an arrangement. Venezuela had expressed support for the draft articles at a conference of Latin American States concerning the law of the sea earlier in 1956. It was also considered that the 1942 Agreement between Venezuela and the United Kingdom, delimiting the continental shelf in the Gulf of Paria, indicated that there would be a reasonable chance that the delimitation of the continental shelf between Venezuela and the Netherlands Antilles could be effected on the basis of the ILC's draft. According to the note, this implied that the boundary would follow the median line.[124]

The hopes of agreement with Venezuela on a median line boundary soon evaporated. On 27 July 1956 Venezuela adopted new legislation on its maritime zones.[125] The legislation did accept the ILC's definition of the continental shelf,[126] but a number of other elements suggested that the delimitation of the continental shelf between Venezuela and Aruba might be problematic. A provision on the delimitation of the continental shelf with neighboring States was lacking. The act

secretary-general of the MFA through a note of the ministry's legal department dated 26 September 1955 [NNA/16].

[121] Overview of questions relating to the Dutch attitude with respect to the continental shelf [NNA/16], p. 3. Although the analogy with the judgment of the Court in the *Anglo-Norwegian fisheries* case, where it referred to the economic interests of the States concerned, might seem to be beside the point to a present-day observer, in the 1950s, when the applicable law was in its first stages of development, this certainly was not obvious.

[122] See note no. 349 dated 11 February 1956 [NNA/16]; note – Proclamation concerning the continental shelf of Suriname dated 13 February 1956 [NNA/47].

[123] Letter of the Prime Minister of the Netherlands Antilles to the Plenipotentiary Minister of the Netherlands Antilles dated 2 May 1956 [NNA/43].

[124] Note dated 5 July 1956 [NNA/16], pp. 1–2.

[125] Act of 27 July 1956 concerning the territorial sea, continental shelf, fishery protection and airspace.

[126] *Ibid.*, article 4.

established a 12-nautical-mile territorial sea and envisaged that in this case the delimitation was to be resolved through agreements with the interested States or other means recognized by international law.[127] A reference to the median line was conspicuously absent. The shortest distance between Venezuela and Aruba is less than 15 nautical miles. Because the Netherlands only claimed a 3-nautical-mile territorial sea, the territorial sea of Venezuela extended a considerable distance beyond the median line without overlapping with the territorial sea of the Netherlands Antilles. Faced with this new situation, the Netherlands pursued its attempt to engage Venezuela in discussions on the delimitation of the continental shelf on the basis of the ILC's draft. Because the territorial sea and the contiguous zone raised numerous questions, it seemed highly commendable to first reach an agreement on the continental shelf. The acceptance of a median line boundary for the continental shelf – if it would be forthcoming – might also facilitate an agreement on the boundary of Venezuela's territorial sea.[128] The Netherlands proposed that Venezuela, referring to article 72 of the ILC's draft, take the median line as the point of departure for the negotiations on a boundary for the continental shelf.[129]

The initial reaction of the Minister of Foreign Affairs of Venezuela to the Dutch proposal led to cautious optimism.[130] However, at the end of November it became clear that a delimitation in accordance with the median line would not be acceptable to Venezuela.[131] Venezuela indicated that there existed methods other than the equidistance line, which had to be adapted to the circumstances of each specific case.[132] In a reaction, the Netherlands reaffirmed that it remained of the view that the continental shelf should be delimited on the basis of the median line rule

---

[127] *Ibid.*, article 1.
[128] Note concerning Venezuela: territorial sea and continental shelf [NNA/19] dated 24 September 1956; see also note concerning the territorial sea dated 27 September 1956 [NNA/19], p. 3.
[129] Note no. 4089 of the Ambassador of the Kingdom of the Netherlands to the Ministry of Foreign Affairs of the Republic of Venezuela dated 25 September 1956 [NNA/39], p. 1.
[130] See received message in cipher (ref. no. 17783) of the Dutch Embassy in Caracas to the MFA dated 28 September 1956 [NNA/39]; note dated 29 September 1956 [NNA/39].
[131] Received message in cipher (ref. no. 17783) of the Dutch Embassy in Caracas to the MFA dated 23 November 1956 [NNA/43].
[132] Note no. 2550 of the Ministry of Foreign Affairs of the Republic of Venezuela to the Ambassador of the Kingdom of the Netherlands dated 31 December 1956 [NNA/19].

and also objected to the 12-nautical-mile territorial sea established by Venezuela.[133]

In the meantime, the consultations between the Netherlands, the Netherlands Antilles and Suriname concerning a proclamation of Suriname's continental shelf had continued. While the Netherlands Antilles remained opposed to a proclamation,[134] Suriname expressed a renewed interest in a proclamation in the second half of 1957 because it was expected that exploratory activities would start on the shelf of British Guiana in the border region with Suriname.[135] To address the concerns of Suriname and the Netherlands Antilles it was decided to prepare a new draft, which would follow as much as possible the ILC's draft and stress the declaratory nature of the proclamation.[136] At the same time, the Ministry of Foreign Affairs considered that a proclamation just before the envisaged law of the sea conference would be highly undesirable as the Netherlands had, on more than one occasion, stressed the need for an international regime to put an end to the practice of unilateral regulation.[137] In view of this objection Suriname was again advised to address the matter in its mining legislation or confirm Suriname's rights over the continental shelf in a letter addressed to the interested oil companies.[138] The Government of Suriname informed the Colmar oil company accordingly, which held a license to Suriname's continental shelf,[139] and also agreed to the suggestion of a proposal to the United Kingdom, to delimit

---

[133] Note no. 3433 of the Ambassador of the Kingdom of the Netherlands to the Ministry of Foreign Affairs of the Republic of Venezuela dated 30 October 1957 [NNA/19].

[134] See for instance the letter of Plenipotentiary Minister of the Netherlands Antilles to the Plenipotentiary Minister of Suriname dated 4 September 1957 [NNA/43]. All of the relevant correspondence, which covers the period up to March 1958, is contained in the archive folder [NNA/43].

[135] See letter of the Prime Minister of Suriname to the Governor of Suriname dated 17 August 1957 [NNA/43], p. 2.

[136] Short report on the interdepartmental consultation concerning the protection of the maritime boundaries of Netherlands New Guinea and concerning the continental shelf of Suriname (MFA, 18 October 1957) [NNA/18], p. 2; note concerning a proclamation on the continental shelf of Suriname dated 19 October 1957 [NNA/18], pp. 2–3. A draft of the proclamation is annexed to the report.

[137] Note concerning a proclamation on the continental shelf of Suriname dated 19 October 1957 [NNA/18], p. 3; note concerning the continental shelf of Suriname dated 21 October 1957 [NNA/18].

[138] Note concerning a concept proclamation on the continental shelf dated 29 October 1957 [NNA/18], p. 1; note concerning the continental shelf of Suriname dated 8 January 1958 [NNA/18].

[139] Letter of the Prime Minister and the acting Governor of Suriname to the N.V. Colmar Surinaamse Oliemaatschappij dated 14 April 1958 [NNA/18].

the continental shelf between Suriname and British Guiana on the basis of the equidistance principle through an exchange of notes.[140] The United Kingdom accepted applying the equidistance line, but suggested that the delimitation of the continental shelf could be included in a draft treaty also settling the land boundary between the two countries.[141] The United Kingdom presented such a draft treaty at the end of 1961.[142] A proposal to delimit the continental shelf between Suriname and French Guiana in accordance with the equidistance method was made to France at about the same time as the proposal to the United Kingdom.[143] France eventually accepted the proposal to negotiate an agreement in May 1963.[144]

The delimitation between the Netherlands Antilles and Venezuela also figured in the instruction for the Dutch delegation to the 1958 Geneva Conference on the law of the sea. The delegation in general was to base itself on the draft articles of the ILC of 1956.[145] It should be guaranteed that neighboring States could never extend their territorial sea beyond the median line, even if the territorial sea of one of the States concerned did not reach up to that distance.[146] This covered the case of Venezuela and Aruba. The instruction did not explicitly address the provision on the delimitation of the continental shelf. Commenting on the 1956 draft articles of the ILC, the Dutch Government had emphasized "the necessity of an internationally accepted rule for [delimitation], together with adequate safeguards for impartial adjudication in the case of disputes." It would not be enough "simply to express the hope that the States concerned will reach agreement on this matter."[147]

---

[140] Letter of the Prime Minister of Suriname to the Governor of Suriname dated 11 March 1958 [NNA/18]; letter of the Governor of Suriname to the Minister for Overseas Territories dated 17 March 1958 [NNA/18]; letter of the Plenipotentiary Minister of Suriname to the Prime Minister of Suriname dated 18 February 1958 [NNA/44].

[141] The Dutch proposal is contained in an aide-mémoire dated 6 August 1958 [NNA/18]; the British reply is contained in a letter from the FO to the Dutch ambassador in London dated 13 January 1959 [NNA/14].

[142] See further Chapter 7.3.2.

[143] See note of the Dutch Embassy in Paris to the MFA dated 20 August 1958 [NNA/18].

[144] Note of the MFA to the Dutch Embassy in Paris dated 11 May 1963 [NNA/18].

[145] Draft instruction dated 30 January 1958 [NNA/8A], B.1. The instruction was approved by the Kingdom Council of Ministers on 7 February 1958 (Minutes of the Kingdom Council of Ministers dated 7 February 1958 [NNA/1], item 5).

[146] Draft instruction dated 30 January 1958 [NNA/8A], B.3.

[147] Letter from the Permanent Mission of the Netherlands to the United Nations dated 17 October 1957 (UNCLOS I Off. Rec., Vol. I, pp. 105–110, at p. 110).

3.3   The 1958 Conference on the law of the sea

The United Nations Conference on the law of the sea was convened in Geneva between 24 February and 27 April 1958. The 1956 draft articles on the law of the sea of the ILC formed the basis of discussions at the conference. The conference adopted four conventions and an optional protocol on the settlement of disputes.[148] One of those was the Convention on the continental shelf. Its provisions on the definition of the continental shelf and its delimitation between neighboring States are closely modeled on the draft of the ILC. That result was only obtained after a debate that revealed the existence of diverging views on the role of equidistance and special circumstances. This debate did not lead to a narrowing of these divergences.

During the general debate on the draft articles on the continental shelf, Germany introduced its proposal for an alternative regime for the exploitation of the shelf's resources.[149] Germany considered that the ILC's draft detracted too much from the freedom of the high seas. According to current international law, the coastal State had no rights over the continental shelf. The German proposal envisaged that exploitation beyond the territorial sea would become a regulated high seas freedom. The coastal State should act on behalf of the international community and would only bear responsibility for securing internationally-agreed rules in respect of installations which would be closest to its coast.[150] The German proposal was immediately rejected by a large number of delegations.[151] After informal talks organized by Germany also revealed that the proposal stood no chance of adoption, the matter was dropped.[152] In its final report on the conference, the German delegation suggested that there had been quite some sympathy for its initiative, but that no delegation had wished to support it because

---

[148] Convention on fishing and conservation of the living resources of the high seas; Convention on the continental shelf; Convention on the high seas; Convention on the territorial sea and contiguous zone; Optional Protocol of signature concerning the compulsory settlement of disputes.

[149] UNCLOS I Off. Rec., Vol. VI, pp. 7–8, paras 1–5; the proposal is contained in *Federal Republic of Germany: Memorandum concerning draft articles 67 to 73* dated 4 March 1958 (Doc. A/CONF.13/C.4/L.1; reproduced in UNCLOS I Off. Rec., Vol. VI, pp. 125–126).

[150] UNCLOS I Off. Rec., Vol. VI, p. 8, para. 4.   [151] See *ibid.*, pp. 9–30.

[152] See note dated 23 March 1958 [B102/119551], p. 6; Ministerie van Buitenlandse Zaken, 56 *Verslag over de Conferentie van de Verenigde Naties over het Zeerecht* (The Hague: Staatsdrukkerij-en Uitgeverijbedrijf, 1958), p. 77; see also annotation; Call to Dr. Dreher on 18.3 [1958] [AA/7].

they had all arrived with strict instructions, which could no longer be changed.[153] Without saying this in so many words, the report thus admitted that the German initiative had been poorly prepared because there had been no attempt to secure support for it prior to the conference. Moreover, the assessment that there, in principle, might have been a considerable amount of support for the proposal seems overly optimistic.[154] As was observed in another German assessment of the conference, the fact that thirty-five States had already claimed a continental shelf made this development probably irreversible.[155]

After the failure to secure support for the German initiative for an alternative regime for the continental shelf, the German delegation decided to insist on a precise definition of the rights of the coastal State over the continental shelf.[156] There was no mention that it would attempt to influence the debate on the delimitation provision applicable to the continental shelf.

The debate at the conference on article 72 on the delimitation on the continental shelf of the ILC's draft is of interest for two reasons.[157] It provides some indication on the interpretation of article 6 of the Convention on the continental shelf and on the views of Denmark, Germany and the Netherlands. A number of States had submitted proposals to amend article 72 of the ILC draft. On the one hand, a number of States sought to limit the role of equidistance. For instance, a Venezuelan proposal referred to delimitation by agreement between the interested States or by other means recognized by international law.[158] Venezuela argued that the failure to make due provision for special circumstances, which were frequently imposed by geography, could not result in a fair solution.[159] At the other extreme was

---

[153] The International Conference on the Law of the Sea in Geneva; Final Report of the German Delegation dated 23 May 1958 [B141/22024], Annex 7, p. 5.

[154] See also H. Meyer-Lindenberg, "Das Genfer Übereinkommen über den Festlandsockel vom 29. April 1958," 1959 (20) *ZaöRV*, pp. 5–35, at pp. 17–18; F. Münch, "Die Internationale Seerechtkonferenz in Genf 1958," 1959–1960 (8) *Archiv des Völkerrechts*, pp. 180–208, at p. 206.

[155] See note dated 23 March 1958 [B102/119551], p. 6.   [156] See *ibid.*, pp. 6–7.

[157] This debate mainly took place in the Fourth Committee of the Conference (see UNCLOS I Off. Rec., Vol. VI, pp. 91–98). There was some further debate on the article as adopted by the Fourth Committee in the Plenary (see UNCLOS I Off. Rec., Vol. II, p. 15).

[158] *Venezuela: proposal* dated 26 March 1958 (Doc. A/CONF.13/C.4/L.42; reproduced in UNCLOS I Off. Rec., Vol. VI, p. 138).

[159] UNCLOS I Off. Rec., Vol. VI, p. 92, para. 19.

Yugoslavia, which proposed to delete the reference to special circum-stances.[160] The Yugoslav delegation observed that this reference was unacceptable on legal grounds and "[i]t was both vague and arbitrary, and likely to give rise to misunderstanding and disagreement."[161]

The Netherlands occupied a middle ground and supported the ILC's draft. The Netherlands had submitted a proposal in respect of article 72, but this only entailed certain minor changes.[162] The Netherlands criticized both the proposal of Venezuela because it omitted a reference to equidistance and special circumstances and a proposal by the United Kingdom because it departed from the spirit of the ILC's text.[163] The United Kingdom, commenting on its proposal, had explained that for its delegation "the adoption of the median line as a boundary was the fundamental principle and the most equitable solution, to be departed from only if special circumstances so required."[164] This explanation attracted the comment that the British proposal came down to the same thing as the Yugoslav proposal.[165] The Netherlands recognized that it might be difficult to decide whether or not special circumstances existed. For that reason it had proposed additional language to draft article 73 on dispute settlement procedures, hoping that this might meet the objection of some delegations to the reference to special circum-stances.[166] The Dutch proposal provided that, in the case of proceedings relating to article 72, the ICJ "shall have the power to decide *ex aequo et bono* whether a boundary line other than that defined in that article is justified by special circumstances."[167] This proposal, like draft article 73 itself, was not adopted by the Fourth Committee. Germany expressed support for both the original text of article 73 and the Dutch amend-ment. According to the German delegate, some of the articles adopted by

---

[160] *Yugoslavia: proposal* dated 19 March 1958 (Doc. A/CONF.13/C.4/L.16 and Add.1; reproduced in UNCLOS I Off. Rec., Vol. VI, p. 130).

[161] UNCLOS I Off. Rec., Vol. VI, p. 91, para. 4.

[162] *Netherlands: proposal* dated 19 March 1958 (Doc. A/CONF.13/C.4/L.23; reproduced in UNCLOS I Off. Rec., Vol. VI, p. 132).

[163] UNCLOS I Off. Rec., Vol. VI, p. 94, para. 14 and p. 96. For the British proposal see *United Kingdom of Great Britain and Northern Ireland: proposal* dated 21 March 1958 (Doc. A/CONF.13/C.4/L.28; reproduced in UNCLOS I Off. Rec., Vol. VI, p. 134). See also text at footnotes 187 and following.

[164] UNCLOS I Off. Rec., Vol. VI, p. 96, para. 5.

[165] *Ibid.*, p. 97, para. 12; see also p. 96, paras 3 and 4.   [166] *Ibid.*, p. 95, para. 15.

[167] *Netherlands: proposal* dated 3 April 1958 (Doc. A/CONF.13/C.4/L.62; reproduced in UNCLOS I Off. Rec., Vol. VI, p. 143). A decision *ex aequo et bono* is not based on the strict application of the law but aims to achieve a result which is fair to the parties concerned in the light of the circumstances of the specific case.

the committee lacked clarity and undoubtedly would lead to differences and disputes.[168]

The Fourth Committee eventually adopted the proposal of the Netherlands concerning article 72 with some minor amendments. The whole of the article was adopted by the Fourth Committee by thirty-six votes to none, with nineteen abstentions. In the Plenary, the article was adopted by sixty-three votes to none and two abstentions.[169] After the vote in the Fourth Committee, the German delegate indicated that his delegation would have preferred the Venezuelan amendment. The German delegation explained this by a somewhat enigmatic reference to the inexact nature of the outer limit of the continental shelf.[170] The final report of the German delegation on the conference confirms the obvious explanation for this support: the equidistance method would attribute Germany a limited area in the North Sea.[171] In the Fourth Committee, Germany further stated that it had accepted the views of the majority of the Committee "subject to an interpretation of the words 'special circumstances' as meaning that any exceptional delimitation of the territorial waters would affect the delimitation of the continental shelf."[172]

In the plenary of the conference Germany voted against the adoption of the Convention on the continental shelf. Germany had first attempted to set the number of ratifications required for the entry into force of the Convention at fifty, arguing that in adopting the Convention the international community was disposing of common property in favor of coastal States.[173] In light of the opposition to this high number, and support for a Canadian proposal to set the figure at twenty-two, Germany proposed forty ratifications. This proposal was rejected by forty-nine votes to six, with six abstentions.[174] The Convention as a whole was adopted by fifty-seven votes to three, with eight abstentions. In its explanation of vote Germany referred to a number of considerations, which were not directly related to article 6 of the Convention. Germany considered the exploitability criterion included in article 1 on the definition of the

---

[168] UNCLOS I Off. Rec., Vol. VI, p. 101, paras 28–29.
[169] See respectively UNCLOS I Off. Rec., Vol. VI, p. 98; UNCLOS I Off. Rec., Vol. II, p. 15.
[170] UNCLOS I Off. Rec., Vol. II, p. 15.
[171] Final report of the German delegation dated 23 May 1958 [B141/22024], pp. 23–24; see also note dated May 1958 [B102/119551], p. 23.
[172] UNCLOS I Off. Rec., Vol. VI, p. 98, para. 38.
[173] UNCLOS I Off. Rec., Vol. II, p. 55, para. 2.   [174] Ibid., pp. 55–56, paras 1–8.

continental shelf incorrect and could not accept the Convention without a provision on compulsory dispute settlement.[175]

The conference also further considered the issue of shallow areas separated from the coast by a narrow channel, such as the Norwegian Trough between Denmark and Norway. Canada argued for the inclusion of such "shelf islands" in the definition of the continental shelf.[176] The United Kingdom, which had previously rejected this approach,[177] now also supported it and explicitly referred to the case of Norway.[178] Norway thanked the United Kingdom delegation for its statement and indicated that it confirmed its delegation's view that the Norwegian Trough was part of the continental shelf.[179] Denmark did not intervene in this debate.[180]

3.4 A comparison of the Danish, Dutch and German approach

In broad terms, the approach of Denmark and the Netherlands to the developing regime of the continental shelf was similar. They both accepted the regime of the continental shelf was becoming a part of international law, considered that this regime was generally acceptable and that it would be best to address it in the framework of the United Nations as this would prevent unilateralism and excessive claims. Germany had a decidedly different take on the status of the continental shelf regime and, until the end of the 1958 Conference, considered that the exploitation of the seabed beyond the territorial sea should be arranged on a different basis.

Denmark started to consider the regime of the continental shelf at the end of the 1940s. As far as the delimitation of the shelf with neighboring States was concerned the Norwegian Trough probably constituted the most important issue Denmark had to assess. A couple of reasons explain why Denmark at an early stage accepted that the Norwegian Trough should not be taken into consideration in determining the extent of its continental shelf in relation to Norway. A first consideration of the

---

[175] *Ibid.*, p. 57, paras 31 and 35.   [176] UNCLOS I Off. Rec., Vol. VI, p. 30, para. 37.
[177] See Chapter 3, text at footnote 18.
[178] UNCLOS I Off. Rec., Vol. VI, p. 41, para. 10.   [179] *Ibid.*
[180] The Danish delegation at the conference did not have instructions in respect of this matter. Neither was the matter considered by the Danish delegation (see Report; The Norwegian Trough's significance for the division of the continental shelf in the North Sea and the Skagerrak between Denmark, Norway and Great Britain dated 22 February 1963 [DNA/2], p. 6).

matter by the Danish continental shelf committee pointed out that the Norwegian Trough was not a discontinuity in the shelf. Although one committee member suggested that the Trough might provide the basis for a bilateral delimitation, this point was not picked up by the committee. Second, Norway from the outset argued that the regime of the continental shelf should also be applicable to features like the Norwegian Trough. The Norwegian argument was immediately picked up by the ILC in connection with the draft article on the definition of the continental shelf, which considered that reasons of fairness required the inclusion of areas like the Norwegian Trough as part of the continental shelf. If Denmark had wanted to reject this approach it would have led to conflict with Norway. Denmark likely would have had difficulties with developing arguments in favor of such an approach. The Norwegian Trough was part of the physical continental shelf and, as the ILC had indicated, fairness required that such an area should not lead to cutting a State off from more seaward continental shelf areas. Finally, a more assertive Danish approach might also have risked weakening Denmark's position in relation to Germany, and it might have opened the door for a general debate on an equitable division of the North Sea.

At the 1958 Conference, the issue of channels in the continental shelf was again discussed and there was general support for the approach suggested by the ILC. This also included the United Kingdom, Norway's other neighbor potentially affected by the presence of the Norwegian Trough. The United Kingdom recognized that Norway's rights extended beyond the Norwegian Trough. The ILC and the 1958 Conference settled this issue. From a political perspective it would have been difficult for Denmark to later revisit it and a legal assessment would likely have led to the outcome that article 1 of the Convention on the continental shelf was also applicable to the Norwegian Trough.

Denmark was among the States encouraging the ILC to provide for a substantive rule of delimitation. After the ILC had included the reference to equidistance and special circumstances in the rule on the delimitation of the continental shelf, Denmark no longer actively contributed to debate on this matter. As Denmark's comments to the ILC and its initial assessment of its own situation suggest, Denmark was generally satisfied with the outcome of the application of the equidistance method, but apparently at the same time accepted that there was a need for flexibility.

Germany was confronted with the possibility of regulating activities on the continental shelf in the early 1950s. From the outset the German Foreign Office was hesitant to take a firm view. At first it considered that international law did not permit the coastal State to regulate mining activities beyond the territorial sea. At that time, this may have been a correct assessment of the applicable law. However, the analysis of internal German documents points out that Germany consistently took too optimistic a view of the amount of opposition to the continental shelf regime and the amount of support for a different regime.

When the Ministry for the Economy again raised the issue in 1955 and referred to the significant practice on the continental shelf, the Foreign Office maintained its original position, but now also pointed to encroachment on other high seas freedoms that the continental shelf might cause. In this period the delimitation provisions of the ILC's draft also became an issue. When the effects of equidistance were first considered there was a suggestion that the geology of the shelf pointed to a very different outcome, but this argument was never picked up and it was practically taken as a given that the shelf in the North Sea would be divided by the equidistance method under the rules proposed by the ILC. The opposition of most ministries to the regime of the continental shelf provides a large part of the explanation as to why alternatives were not considered. Opponents obviously had no interest in arguing that the operation of the special circumstances clause might give Germany a larger share of the North Sea. For instance, the delimitation provision was considered by Professor Böhmert of the University of Kiel, who acted as an advisor to the Ministry for Food Supply, Agriculture and Forests. Böhmert considered that the special circumstances clause would at best only lead to minor changes to the advantage of Germany and that the equitable nature of the equidistance line, which would give Germany only one twentieth of the North Sea, was irrefutable.[181] The logic of this reasoning seems, to say the least, questionable. The Ministry for the Economy was the only proponent of the continental shelf regime and only at a late stage embraced it wholeheartedly. The Ministry did not explicitly look at alternative delimitation methods but only suggested that a more favorable solution might be achieved with Germany's North Sea neighbors.

The German authorities took a long time to work out an agreement on the German position on the regime of the continental shelf. In large part, this can be attributed to the failure of the Foreign Office to work out a clear approach on how to respond to the development of the law. As a consequence, Germany did not have any impact on the debate on the

---

[181] Letter of Böhmert to Dierks of the Chamber of Commerce of Bremerhaven dated 4 January 1958 [B141/22023], p. 1; letter of Böhmert to Hillger dated 15 January 1958 [AA/88], p. 6.

regime of the continental shelf. In 1955 Germany did try to find out more about the views of the Netherlands on the delimitation of the North Sea. These contacts did not clarify that Germany was opposed to the equi-distance method. Again, the fact that Germany had not established its own position on the continental shelf regime provides an obvious explanation. All the same, this was something of a missed opportunity. At the time, Dutch civil servants had concluded that the equidistance principle did not lead to an equitable result for Germany and an early discussion of this matter could have prevented the Netherlands from committing itself to the equidistance principle to the extent it did after which it found it difficult to extract itself from this position.

Only weeks before the 1958 Conference, the German authorities finally agreed on an approach to the negotiations on the continental shelf. This left Germany no time to sound out other States before the start of the conference. In view of the broad support for the continental shelf regime, it is hardly surprising that Germany's alternative proposal fell on deaf ears. The German proposal had, in any case, little to offer to coastal States. It attributed responsibilities to the coastal State, but did not give it a share in the benefits.

After the German initiative failed, it supported a Venezuelan proposal to delete the reference to equidistance from the delimitation provision. However, Germany failed to clearly express that it supported this pro-posal because of the consequences of equidistance in the North Sea. Although it was clear that Germany had difficulties with the continental shelf regime until the end of the conference, Denmark and the Netherlands had not really been put on notice about the German objec-tions to the equidistance method. There is little doubt that it would have been more difficult for Denmark and the Netherlands to insist on the equidistance method in the 1960s if Germany had opposed it more clearly at the 1958 Conference. So, why did this not happen? Most importantly, the issue simply was not on the radar of the German delegation. There is a striking difference between the German involve-ment in the discussions on the provisions on the delimitation of the territorial sea and the continental shelf. Prior to the Conference, the legal advisor of the Foreign Office considered the content of the provision on the lateral delimitation of the territorial sea of the ILC's draft articles because of the dispute with the Netherlands over their land boundary.[182] At the conference, this point was brought up by the German delegation and led to the inclusion of the reference to "historic title" in article 12 of

---

[182] See note dated 19 February 1958 [AA/94].

the Convention on the territorial sea and the contiguous zone. The Foreign Office failed to make a similar review of the draft article on the delimitation of continental shelf, and the German delegation had no instructions in respect of the article.[183] This was in line with the Foreign Office's passivity as regards the formulation of an approach to the continental shelf regime. The resistance of most German ministries to that regime might have made it difficult to secure agreement on that point in any case.

Like Denmark, the Netherlands was generally supportive of the ILC's draft on the continental shelf and it considered that this draft in general reflected the applicable law. The Netherlands also favored the inclusion of a substantive delimitation provision. After the ILC had introduced the reference to equidistance and special circumstances, the Netherlands in 1954–1955 further reviewed the delimitation of the continental shelf. Equidistance in general was considered to be an appropriate method of delimitation for the Netherlands, but particularly in respect of the North Sea it was recognized that the equidistance method might need to be adjusted in relation of Belgium and Germany for reasons of equity.

Developments in the 1950s required the Netherlands to already consider in more detail the continental shelf boundaries of Netherlands Antilles in relation to Venezuela, and those of Netherlands New Guinea and Suriname. In general, the Ministry of Foreign Affairs adopted a cautious position and did not want to take any initiatives that might give the impression of unilateralism prior to the Geneva Conference. The equidistance method was considered to be an acceptable outcome for all Dutch delimitation issues. In the case of the Netherlands Antilles, Venezuela rejected this approach after bilateral contacts in 1956. In the North Sea there was no urgent need to further consider the delimitation with neighboring States and when Germany contacted the Netherlands in 1955 it was given a noncommittal reply.

The Dutch contribution to the debate at the 1958 Conference might give the impression of stronger support for special circumstances than might be expected on the basis of the interests of the Netherlands. The

---

[183] It could of course be questioned if a German initiative would have had stood a chance of success. The rejection of a Venezuelan proposal to delete the reference to equidistance was defeated by thirty-two votes to five, with nineteen abstentions. At the same time, a Yugoslav proposal to delete the reference to special circumstances was rejected by thirty-nine votes to nine, with eight abstentions (see UNCLOS I Off. Rec., Vol. VI, p. 97). A German initiative would at least have clarified the German position on this point.

Netherlands could be said to have a certain interest in the special circumstances clause to justify a deviation from the equidistance line in the case of the boundary with Belgium in the North Sea and between Suriname and British Guiana, because of the terminus of the (potential) territorial sea boundary that was in both cases located beyond the equidistance line. However, the Dutch proposal to accord the ICJ with the power to decide *ex aequo et bono* on the presence and impact of special circumstances seemed to open the door for significant departures from the equidistance line.[184] The proposal is likely to have been inspired by the wish to have a third party rule on those issues, instead of having to deal directly with neighboring States in the case of disagreement, such as Venezuela in the case of the Netherlands.[185] The Dutch position at the conference certainly did not give Germany reason to think that the Netherlands would later be unwilling to consider a departure from the equidistance line in determining their bilateral boundary.

3.5   Questions in relation to the implications of article 6 for the delimitation in the North Sea

Article 6 of the Convention on the continental shelf played an important role in the negotiations between Denmark, Germany and the Netherlands on the delimitation of the continental shelf in the North Sea. This, first of all, concerned the relationship between the reference to agreement and the equidistance principle. Does article 6 entitle a coastal State to unilaterally define a boundary in accordance with the equidistance principle in the absence of agreement with a neighboring State? Second, is the equidistance principle the default rule and does the State invoking the presence of special circumstances have the burden of proof to establish that they justify a boundary different from the equidistance line? Finally, what is the regime applicable to the continental shelf in the absence of agreement on the boundary? What activities are States entitled to carry out in such a case and how can they respond to the activities of another State that they deem to have gone beyond what is allowed?

---

[184] See also Dutch delegation to the UN Conference on the law of the sea, Twenty-sixth meeting of the delegation held in Geneva on 17 March 1958 [NNA/10], p. 2, where it is observed that article 73 of the ILC was drafted too narrowly, because the establishment of an international boundary in general was not effected on the basis of legal principles but on an equitable basis.

[185] See also UNCLOS I Off. Rec., Vol. III, p. 192, para. 22 and Chapter 4, text at footnote 121.

The first sentence of the paragraphs 1 and 2 of article 6 provides that the boundary of the continental shelf "shall be determined by agreement." The second sentence of these paragraphs then refers to respectively the median line and the equidistance principle for determining the boundary in the absence of agreement and unless another boundary is justified by special circumstances. It could be argued, as would be done by Denmark and the Netherlands, that article 6 implies that the boundary is the equidistance line, as long as there is no agreement and a neighboring State has not proven the existence of special circumstances that justify a different boundary. This interpretation would seem to be problematic and to be contrary to the general rule of treaty interpretation that a treaty

> shall be interpreted in good faith in accordance with the ordinary mean-
> ing to be given to the terms of the treaty in their context and in the light of
> its object and purpose.[186]

The above interpretation of article 6 would imply that a coastal State could impose its preference for the equidistance line on its neighboring States and could refuse to negotiate a boundary on a different basis. This would make the requirement that the boundary has to be determined by agreement an empty shell. The language of article 6 also does not indicate that the burden of proof in respect of the presence of special circumstances justifying a different boundary rests with the State invoking special circumstances. In view of the uncertainty about the meaning of the term special circumstances and their implications for the delim-itation of the continental shelf, it would be difficult to provide any indication as to when this burden would have been met.

The drafting history of article 6 of the Convention on the continental shelf indicates that the equidistance rule was not considered a default rule that would become automatically applicable in the absence of agree-ment unless one of the States concerned proved the existence of special circumstances that would justify a different boundary. First, the first sentence of paragraphs 1 and 2 of article 6 was agreed upon by the ILC at its 1956 session. The debate on this provision indicates that the new wording was adopted to make it explicit, as was observed by the chair-man of the Commission, that "in case of dispute, the boundary of the continental shelf should be settled by agreement."[187] At the 1958

---

[186] Vienna Convention on the law of treaties, article 31(1).

[187] Yearbook ILC, 1956, Vol. I, p. 152, para. 36; see also p. 152, paras 29–35 and 37. Prior to this amendment, paragraph 1 of the article read: "Where the same continental shelf is contiguous to the territories of two or more States whose coasts are opposite to each

Conference, the United Kingdom proposed to revert to the language the Commission had employed previously.[188] The Dutch representative considered that this proposal "departed from the spirit of the International Law Commission's text. Agreement between the States concerned must be the cornerstone of the article."[189] Article 6 of the Convention on this point is identical to the ILC's 1956 draft.

The debate in the ILC furthermore indicates that the Commission considered that a rule on the delimitation of the continental shelf should be accompanied by a provision obliging States to settle any dispute concerning delimitation by third party settlement.[190] From the discussion on this point it is clear that the Commission considered that it would be on the judge or arbitrator to assess the presence and implications of special circumstances.[191] In other words, it was not envisaged that there would be a burden of proof for one of the parties to a dispute in this respect.

At the 1958 Conference there was broad opposition to a provision on compulsory dispute settlement in respect of the articles dealing with the continental shelf regime. This discussion at the conference on dispute settlement points out that it was acknowledged that the determination of the presence of special circumstances was a matter that would have to be decided by a court or tribunal.[192] There is no indication that the omission of a provision on compulsory dispute settlement was intended to

---

other, the boundary of the continental shelf appertaining to such States is, in the absence of agreement between those States or unless another boundary line is justified by special circumstances, the median line every point of which is equidistant from the base lines from which the width of the territorial sea of each country is measured" (Yearbook ILC, 1953, Vol. II, p. 213). Paragraph 2 of the article used identical language in this respect (see *ibid.*).

[188] *United Kingdom of Great Britain and Northern Ireland: proposal* dated 21 March 1958 (Doc. A/CONF.13/C.4/L.28; reproduced in UNCLOS I Off. Rec., Vol. VI, p. 134), article 72.

[189] UNCLOS I Off. Rec., Vol. VI, p. 96, para. 6. See also *ibid.*, pp. 96–97, paras 3–4, 6 and 9.

[190] See the discussion in the ILC during its sessions of 1951, 1953 and 1956 (Yearbook ILC, 1951, Vol. I, pp. 288–294; Yearbook ILC, 1953, Vol. I, pp. 106–108 and 125–130; Yearbook ILC, 1956, Vol. I, pp. 151–153).

[191] See Yearbook ILC, 1953, Vol. I, p. 131, para. 14; p. 131, para. 17; p. 132, para. 28; p. 132, para. 29; p. 133, para. 33; p. 133, para. 35; p. 133, para. 39; Yearbook ILC, 1953, Vol. II, p. 216, para. 82; see also Yearbook ILC, 1956, Vol. II, p. 300. The Commission's 1956 draft articles contained the following article: "Any disputes that may arise between States concerning the interpretation or application of articles 67–72 shall be submitted to the International Court of Justice at the request of any of the parties, unless they agree on another method of peaceful settlement" (Yearbook ILC, 1956, Vol. II, p. 264, article 73).

[192] See UNCLOS I Off. Rec., Vol. III, p. 192, para. 22; UNCLOS I Off. Rec., Vol. VI, pp. 94–95, para. 14, p. 95, para. 16, p. 97, para. 10, p. 99, para. 11, p. 102, para. 2; see also *ibid.*, p. 10, para. 14, p. 91, para. 4.

change that fact and place the burden of proof on one of the parties to the dispute.

Article 6 does not specify the relationship between equidistance and special circumstances. The commentary of the ILC on the draft articles on the continental shelf and the territorial sea probably provides a fair summary of the uncertainties surrounding these provisions. In its commentary on the final draft of the article on the delimitation of the continental shelf the Commission observed:

> As in the case of the boundaries of the territorial sea, provision must be made for departures necessitated by any exceptional configuration of the coast, as well as the presence of islands or of navigable channels. This case may arise fairly often, so that the rule adopted is fairly elastic.[193]

The debates at the 1958 Conference do little to clarify the ILC's commentary on the interpretation and application of special circumstances and their relationship to equidistance. The record of the conference does show that there was general support to retain the scheme proposed by the ILC and not change the balance either in the direction of proponents or opponents of the equidistance rule. It has been suggested that a statement of the United Kingdom that the equidistance line would form the best starting point for delimitation reflects how the rule was envisaged to operate.[194] Apart from the fact that this still does not provide a clear answer on the role of special circumstances, it can be noted that this statement was made in explaining a British proposal concerning article 72.[195] As was noted earlier, that proposal was criticized because it disturbed the balance contained in draft article 72.[196] The adoption of article 6 by an overwhelming majority at the conference also suggests that there was no common understanding regarding its effects in the specific case.

The regime applicable to the continental shelf in the absence of agreement on its delimitation was considered during an early stage of the discussions in the ILC. This topic was brought up by Scelle during the 1951 session of the Commission. Scelle submitted that:

> the Commission should state that, if two governments could not reach agreement as to the partition of the continental shelf, neither State was

[193] Yearbook ILC, 1956, Vol. II, p. 300; see also *ibid.* p. 271.
[194] Antunes, *Conceptualisation*, p. 31 and note 97.
[195] UNCLOS I Off. Rec., Vol. VI, p. 93, para. 2.
[196] See text at footnote 188 and following.

> entitled to exploit it. They must either maintain the *status quo* or they would be under an obligation to refer the question to the International Court of Justice.[197]

The proposition that neither State would be entitled to exploit the continental shelf in the absence of agreement was explicitly supported by three other members of the Commission.[198] None of the members of the Commission expressed disagreement with Scelle's proposition. In order to address this specific problem the ILC proposed that in case of disagreement a dispute over the delimitation of the continental shelf could be submitted to third party settlement.[199] After the Commission had taken the decision on including a provision on compulsory dispute settlement, the regime applicable to the disputed areas was not further considered. A likely explanation is provided by an observation of Spiropoulos during the Commission's 1951 session. He did not consider it "necessary to envisage a return to the *status quo*. The [ICJ] could after all make provisional arrangements."[200] During the 1958 Conference, the ILC's proposal on compulsory dispute settlement did not find its way into the Convention on the continental shelf. This did not lead to a renewed debate on the regime of disputed areas. This is hardly surprising as the conference was pressed for time and this was an issue that had not been addressed in the draft articles of the ILC.

---

[197] Yearbook ILC, 1951, Vol. I, p. 288, para. 5; see also *ibid.*, p. 292, para. 65.
[198] See *ibid.*, p. 291, paras 45–46; see also *ibid.*, p. 292, para. 64.
[199] See *ibid.*, pp. 288–292, paras 4–72 especially at p. 291, para. 46 and p. 292, para. 62.
[200] *Ibid.*, p. 292, para. 64.

Digesting the outcome of the 1958 Conference

4.1   Introduction

The adoption of the Convention on the continental shelf at the 1958 Conference faced States with a clear choice: was this an acceptable outcome or not? And, if it did not provide an acceptable outcome, what alternatives did a State have to best protect its interests? The present chapter considers how Denmark, Germany and the Netherlands dealt with these questions. This chapter does not look at developments in relation to specific delimitations of the continental shelf, which are considered in subsequent chapters.

4.2   Denmark

Denmark signed the four Geneva Conventions and the Optional Protocol on dispute settlement on 29 April 1958, the day they were opened for signature. The consideration of the Danish ratification of the Conventions was initially postponed to await the outcome of the second Conference on the law of the sea, which took place in Geneva in March–April 1960.[1] The Ministry of Foreign Affairs did not, in any case, consider it expedient that Denmark would proceed with ratification without taking into account the position of the major western maritime powers and the other Nordic countries.[2] After the second Conference, objections of the Ministry of Defense against the ratification of the

---

[1] The second Conference on the law of the sea had been charged to further consider the breadth of the territorial sea and the limits of fisheries jurisdiction. The 1958 Conference had not reached agreement on the former point. The question of the limits of fisheries jurisdiction was linked to a decision on the breadth of the territorial sea. A proposal at the second Conference for a 6-nautical-mile territorial sea and an adjacent fishery zone also measuring 6 nautical miles by Canada and the United States failed to attract the required majority by one vote (fifty-four delegations voted for the proposal, twenty-eight against and five abstained).

[2] See letter of the MFA to the Ministry of Public Works dated 20 August 1959 [DNA/8].

Convention on the territorial sea and the contiguous zone at first halted further consideration of the ratification of all four Conventions.[3] At the end of 1961, the Ministry of Foreign Affairs requested the Ministry of Defense to reconsider its position. A number of States had ratified one or more of the conventions. The United States had now ratified all four conventions and the UK had ratified all conventions but the Convention on the continental shelf. Moreover, Norway was also considering this question.[4] After the Ministry of Defense had withdrawn its objections, the Ministry of Foreign Affairs further considered whether the time for ratification had come.[5] A further deferral was caused by a number of questions relating to fisheries off the Faroe Islands and the extent of its territorial sea.[6] The ratification of the Convention on continental shelf had in the meantime been raised by the Ministry of Public Works in January 1963.[7] In July 1962, the Danish Government had granted an exclusive concession to prospect and recover hydrocarbons for the territory of Denmark to the Danish ship-owner A.P. Møller and two companies of the A.P. Møller Group.[8] It was envisaged that the concession could be extended to the continental shelf.[9] The concessionaire had requested the Ministry of Public Works that ratification of the

[3] See letter of the Ministry of Defense to the MFA dated 22 December 1960 [DNA/8].
[4] See letter of the MFA to the Ministry of Public Works and the Ministry of Fisheries dated 13 December 1961 [DNA/8], pp. 1–3.
[5] See Report; 1958 Geneva Conventions on the legal order of the sea dated 28 June 1962 [DNA/8]; circular letter of the MFA dated 9 July 1962 [DNA/8].
[6] See circular letter of the MFA dated 9 July 1962 [DNA/8]; Report; 1958 Geneva Conventions on the legal order of the sea dated 28 June 1962 [DNA/8], p. 2. For an overview of this matter see F. Laursen, *Small Powers at Sea; Scandinavia and the New International Marine Order* (Dordrecht: Martinus Nijhoff Publishers, 1993), pp. 104–108, see also fn 13.
[7] See Report; Danish ratification of the Convention on the continental shelf dated 30 January 1963 [DNA/8] (Report dated 30 January 1963 [DNA/8]), p. 5.
[8] The concession was only applicable to the metropolitan area of Denmark and not to the Faroe Islands and Greenland. For further background information on the concession see M. Hahn-Pedersen, *A.P. Møller and the Danish Oil* (Copenhagen: Schultz Forlag, 1999), pp. 25–58. Møller subsequently transferred his share of the concession to a newly formed company of the A.P. Møller Group. The A.P. Møller Group collaborated with two international oil companies, Gulf and Shell, in the Dansk Undergrunds Consortium (DUC) in the actual exploration for and exploitation of hydrocarbons.
[9] See Protocol relating to exclusive concession of July 8th 1962, to prospect for and recover hydrocarbons etc. in Denmark's underground [DNA/145]. The Protocol provided: "If the area of sovereignty of the Government should be extended under new generally recognized standards of international law during the currency of the exclusive concession, the Minister [of Public Works] will be willing, at the request of the concessionaires, to consider a corresponding extension of the area covered by the exclusive concession."

Convention on the continental shelf would take place to ensure that Danish rights would be extended to the continental shelf. According to the concessionaire that area contained significant resources and companies from other North Sea coastal States had already shown an interest in the offshore.[10] The Ministry of Foreign Affairs concurred with the Ministry of Public Works that in view of these foreign activities it was desirable that Denmark ratify the Convention.[11]

Just a month after the decision to proceed with the ratification of the Convention on continental shelf had been taken a proposal was submitted to Parliament to obtain consent for ratification,[12] and Denmark ratified the Convention on the continental shelf on 12 June 1963.[13] Almost simultaneously, Denmark enacted legislation on its continental shelf, which defined the extent and lateral limits of the shelf by reference to the Convention.[14] The exclusive concession for the exploration and exploitation of hydrocarbons of the companies of the A.P. Møller Group was extended to the continental shelf of metropolitan Denmark in October 1963.[15]

Before the Convention on the continental shelf had been ratified, the Ministry of Foreign Affairs had also considered whether the Convention reflected customary international law. The legal advisor of the Ministry in 1962 had concluded that the Convention was not a codification, but had created new law. However, he was inclined to conclude that there existed a customary rule, which allowed a State to proclaim a continental shelf.[16] On the basis of this assessment it had been concluded that a

---

[10] See Report dated 30 January 1963 [DNA/8], p. 5.

[11] See *ibid*. The recommendation to this effect was accepted by the Foreign Minister on 14 February 1963 (see *ibid.*, p. 8, hand-written note).

[12] See Proposal for a decision of the Folketing on Denmark's ratification of the Convention on the continental shelf signed at the United Nations Conference on the law of the sea of 1958 dated 14 March 1963 (Tillæg A til Folketingstidende (99) Folketingsåret 1962–63, p. 1570).

[13] Negotiations with the United Kingdom concerning fisheries around the Faroe Islands continued until 1964. After the negotiations broke down in early 1964, Denmark revised the straight baselines along the Faroe Islands and established a 12-nautical-mile fishery zone (see Laursen, *Small Powers*, p. 108). The other three Geneva Conventions were only ratified by Denmark in September 1968.

[14] Royal Decree dated 7 June 1963 concerning the exercise of Danish sovereignty over the Continental Shelf. On the content of the Decree see below text at footnote 34.

[15] Decree of 5 October 1963 [DNA/145].

[16] Note; The question of a Danish ratification of the Convention on the continental shelf dated 4 October 1962 [DNA/8], p. 2 (quoting a report of the legal advisor dated 1 June 1962).

proclamation offered an alternative to ratification of the Convention.[17] Shortly after the Danish ratification of the Convention, its status was again considered following an inquiry of the United Kingdom Foreign Office into how Denmark intended to act in relation to neighboring States, which had not ratified the Convention. The legal advisor suggested that Danish legislation could be applied in relation to these States because the Convention's general principles represented customary international law.[18]

In connection with the pending ratification of the Convention on the continental shelf, its delimitation with neighboring States was also reconsidered. This assessment focused on the continental shelf of metropolitan Denmark. The continental shelf of the Faroe Islands, defined by reference to the 200-meter isobath, did not overlap with the continental shelf of any other State. An area of continental shelf between Greenland and Canada was less than 200 meters in depth but a more precise delimitation in this area was not required in the foreseeable future due to Arctic conditions.[19]

The assessment of Denmark's continental shelf boundaries was mainly concerned with two issues: the implications of article 6 of the Convention on the continental shelf in general and the role of the Norwegian Trough in the delimitation between Denmark and Norway. A first evaluation only briefly discussed the Norwegian Trough, referring to a report which had been drawn up for the Danish continental shelf committee in 1949. That report had concluded that the Norwegian Trough in general was considered to be a trench in the Norwegian continental shelf.[20] This conclusion was accepted by the 1963 report. The report also concluded that Denmark, in accordance with article 6 of the Convention on the continental shelf, could delimit its continental shelf in the North Sea and the Baltic Sea on the basis of the equidistance method. Denmark was not obliged to raise this matter with other States. If other States did not agree with the Danish view they should raise this matter themselves.[21]

---

[17] Ibid.
[18] Note; The continental shelf dated 23 August 1963, comment dated 26 August 1963 [DNA/12].
[19] See Report dated 30 January 1963 [DNA/8], p. 6; Report; The continental shelf dated 28 March 1963 [DNA/2], p. 2.
[20] Report; Delimitation of Denmark's continental shelf dated 21 January 1963 [DNA/2], p. 2; see also Chapter 3.2.1.
[21] Ibid., p. 7.

A more detailed consideration of the role of the Norwegian Trough in the delimitation of Denmark's continental shelf was made after the issue had been raised by the A.P. Møller Group.[22] This second assessment reviewed the drafting history of the Convention on the continental shelf and concluded that Denmark, Norway and the United Kingdom had all taken the position that the Norwegian Trough should not be taken into account in defining the extent of the continental shelf.[23] Consequently, the continental shelf had to be delimited by the median line.[24] The Danish Ministry of Foreign Affairs also submitted that a change in the Danish position inevitably would lead to a conflict with Norway. Norway could rightly consider such a change inequitable and not in conformity with the special standards applicable between the Nordic countries.[25] The characteristics of the Norwegian Trough and the analysis of the drafting history of article 1 of the Convention indicate that, apart from these political considerations, Denmark hardly had a convincing legal case to argue that the Trough limited the extent of Norway's continental shelf.[26]

The Ministry of Foreign Affairs informed the Ministry of Public Works of its assessment of the Norwegian Trough and that the only reasonable approach would be to accept the equidistance line.[27] The Ministry of Public Works fully agreed and would not support the concessionaire if it was going to push for a different boundary.[28] When Director Hoppe of A.P. Møller requested that the Ministry of Foreign Affairs consider presenting the equidistance line as a minimum demand in relation to Norway, he was told that the overall assessment of Danish interests required that Denmark would not go beyond the equidistance line. If the United Kingdom were to extend its claim to the Norwegian Trough, Denmark would rather support Norway, than support the United Kingdom in the hope of gaining something at the expense of

---

[22] See Report; The Norwegian Trough's significance for the division of the continental shelf in the North Sea and the Skagerrak between Denmark, Norway and Great Britain dated 22 February 1963 [DNA/2] (Report; The Norwegian Trough dated 22 February 1963 [DNA/2]), p. 1, hand-written note dated 25 February 1963; letter of O. Borch of the MFA to J. Bang Christensen of the Ministry of Public Works dated 26 February 1963 [DNA/9].

[23] See Report; The Norwegian Trough dated 22 February 1963 [DNA/2], pp. 2–6; see also Chapters 3.2.1 and 3.3.

[24] *Ibid.*, p. 6.   [25] *Ibid.*, pp. 6–7.   [26] See Chapters 2.2, 2.3.2, 3.2.1 and 3.3.

[27] Letter of O. Borch of the MFA to J. Bang Christensen of the Ministry of Public Works dated 26 February 1963 [DNA/9].

[28] Letter of J. Bang Christensen of the Ministry of Public Works to O. Borch of the MFA dated 4 March 1963 [DNA/9].

Norway.[29] Subsequently, the concessionaire came around to the view that it would be in Denmark's best interest to stand firm on the equidistance principle. The argument that the longer Denmark left the matter open, the more it would be exposed to the risks of a German claim probably had been convincing.[30]

Denmark also carried out a general assessment of the principles applicable to the delimitation of continental shelf boundaries. It was concluded that there was no need to reconsider the position Denmark had adopted in the 1950s, which recommended using the equidistance method.[31] What counted in this respect was the decision in principle. It was not considered necessary to arrive at a precise definition of the continental shelf, because that definition required negotiations and agreements with neighboring States. Moreover, that question for the time being did not have any practical relevance. The need for delimitation would only arise when exploitation started.[32] Denmark would not need to take the initiative in relation to other States, but in the absence of agreement could establish a boundary on the basis of article 6 of the Convention.[33] That is, Denmark in the absence of agreement would consider that the boundary with neighboring States was the equidistance line. The legislation that was adopted in connection with the ratification of the Convention on the continental shelf reflected this view:

> The boundary of the continental shelf in relation to foreign States whose coasts are opposite the coasts of the Kingdom of Denmark or are adjacent to Denmark shall be determined in accordance with article 6 of the Convention, that is to say in the absence of special agreement, the boundary is the median line, every point of which is equidistant from the nearest points of the baselines from which the breadth of the territorial sea of each State is measured.[34]

---

[29] Report; The continental shelf dated 22 March 1963, addition dated 28 March 1963 [DNA/2], pp. 3–4.

[30] See note dated 2 April 1964 [DNA/2], pp. 1–2.

[31] See e.g. Report dated 30 January 1963 [DNA/8], p. 3 and *ibid.*, continuation dated 11 February 1963, pp. 7–8; Report; The continental shelf dated 28 March 1963 [DNA/2], pp. 1–2.

[32] See e.g. Report dated 30 January 1963, continuation dated 11 February 1963 [DNA/8], pp. 7–8. The report was considered by the Foreign Minister who agreed to its recommendations (*ibid.* p. 8, hand-written note).

[33] See Report; Delimitation of Denmark's continental shelf dated 21 January 1963 [DNA/2], continuation dated 22 January 1963, p. 7; see also Report; The continental shelf dated 28 March 1963 [DNA/2], p. 2.

[34] Royal Decree of 7 June 1963 concerning the exercise of Danish sovereignty over the Continental Shelf, article 2(2).

4.3 Germany

The German authorities after the Geneva Conference continued to have doubts about the regime of the seabed beyond the territorial sea under customary international law. As a consequence, ratification of the Convention on the continental shelf was entertained as a possibility, or even held to be a necessity, to create certainty about Germany's rights in that area. Not surprisingly, the geographical situation of Germany in the North Sea, and its potential implications for a delimitation with neighboring States, influenced this debate, which was further complicated by the fact that the Federal authorities and the states disagreed over the division of competence between the Federation and the states in respect of the continental shelf. Four states, Lower Saxony, Schleswig-Holstein, Bremen and Hamburg, were involved in this matter. Lower Saxony and Schleswig-Holstein occupy most of Germany's North Sea coast. Bremen and Hamburg also had an interest in bottom fisheries and feared that the developing regime of the continental shelf might negatively impact on that interest.[35] Lower Saxony occupies the southern North Sea coast of Germany and is divided from Schleswig-Holstein, which lies on Germany's eastern North Sea coast, by the river Elbe.[36] Lower Saxony in general represented the other states in the discussions with the Federal authorities concerning the delimitation of continental shelf in the North Sea.

The possibility of signing the Geneva Conventions was considered at an interdepartmental meeting of 18 September 1958. The meeting agreed that it would not be appropriate to sign the Convention on the territorial sea and the contiguous zone and the Convention on fishing and conservation of the living resources of the high seas in light of the close connection of both these conventions with the pending question of the breadth of the territorial sea and a contiguous fishing zone.[37] It was agreed to sign the other two conventions and the Optional Protocol, although there existed certain reservations in respect of article 6 of the Convention on the continental shelf.[38] The Foreign Office subsequently

[35] See e.g. note dated 31 October 1958 [AA/57], p. 2.
[36] Schleswig-Holstein also has a coast in the Baltic Sea, and the dispute between the Federation and the states also concerned the Baltic Sea. The geography of that part of the Baltic implies that the extent of this continental shelf was limited. At present the entire area is part of Germany's 12-nautical-mile territorial sea.
[37] See note dated 20 September 1958 [AA/12], pp. 1–2.
[38] See *ibid.*, p. 3; see also circular letter of the FO dated 29 September 1959 [B141/22024], pp. 1–2.

indicated that it did not share those reservations. Article 6(2) of the Convention only provided for the application of the equidistance method in the alternative, when no agreement had been concluded between the neighboring States and when there did not exist special circumstances. In essence, article 6 did not regulate the delimitation of the continental shelf.[39] The Foreign Office also considered that the possibility to submit a dispute over the existence of special circumstances under the Convention to arbitration under the Optional Protocol offered an advantage.[40] Another argument in support of signing the Convention was that it would be easier to impact its interpretation.[41] Finally, the Convention could be used to resist further claims of coastal States over the water column superjacent to the shelf.[42] Germany signed the Convention on the continental shelf, the Convention on the high seas and the Optional Protocol on 30 October 1958. In the middle of 1959, the Ministry for the Economy argued that ratification of the Convention on the continental shelf was necessary to allow Germany to exercise sovereign rights over the shelf. Although the Foreign Office apparently considered that this question would only acquire practical significance in ten to twenty years, the Ministry for the Economy had reason to believe that activities were in the offing.[43]

At the beginning of 1962, all ministries involved agreed to a proposal of the Foreign Office that Germany should become party to all Geneva Conventions and the Optional Protocol, with the exception of the Convention on fishing and conservation of the living resources of the high seas.[44] It was expected that Germany would be among the parties to the Convention on the continental shelf upon its entry into force, which was envisaged by the end of 1964.[45] The Convention on the continental shelf actually entered into force on 10 June 1964. Germany was not yet a party at that time and also did not become a party subsequently. So, what happened?

A possible German ratification of the Convention was influenced by two issues. The implementation of the Convention required the adoption of legislation, but the Federation and the states differed over the

---

[39] Circular letter of the FO dated 29 September 1959 [B141/22024], p. 2.   [40] *Ibid.*
[41] See letter of the ME to the FO dated 8 October 1958 [AA/88].
[42] Note dated 31 October 1958 [AA/57], p. 2.
[43] Note dated 7 July 1959 [B102/119551].
[44] See Report on the interdepartmental meeting of 19 January 1962 dated 25 January 1962 [AA/57], p. 2.
[45] Note dated 17 July 1962 [AA/57], p. 2.

division of competences as regards the continental shelf.[46] Second, doubts about the implications of the Convention's article 6 for the delimitation of the continental shelf of Germany in the North Sea continued to exist. The resolution of these matters took a considerable time, and eventually the idea of ratifying the Convention was abandoned.

Oil companies started to show an interest in the German continental shelf in April 1962 and had contacted both the Ministry for the Economy and competent authorities of the four states with an interest in this matter about the possibilities of obtaining a concession.[47] The Federal authorities initially took the position that the area beyond Germany's territorial sea was part of the high seas. Germany would only be able to exercise sovereign rights over the continental shelf once it had ratified the Convention.[48] In talks with oil companies it was emphasized that Germany in any case would not issue a unilateral proclamation to claim rights over the shelf before the entry into force of the Convention,[49] and companies could not claim any rights in the absence of legislation.[50] At the same time, it was recognized that Germany should already consider what steps might be taken to safeguard its rights in the area.[51]

In July 1962, the Foreign Office indicated that it had absolutely no interest in starting the preparation of legislation. Discussions with the states concerning the continental shelf should avoid addressing the matter of legislation.[52] The Foreign Office considered that the Federation would be in a better position if this matter were to be taken up in the preparation of the ratification of the Convention. The latter issue also had implications for foreign policy,[53] which was an exclusive

---

[46] See e.g. letter of the FO to the Ministry of Transport dated 13 March 1963 [AA/57], p. 2; note dated 20 June 1963 [AA/3]. For further information on the dispute between the Federation and the states see e.g. J. Frowein, "Verfassungsrechtliche Probleme um den deutschen Festlandsockel," 1965 (25/1) ZaöRV, pp. 1–25; J. Kölble, "Bundesstaat und Festlandsockel," 1964 (17) Die öffentliche Verwaltung, pp. 217–225; E. Menzel, "Der deutsche Festlandsockel in der Nordsee und seine rechtliche Ordnung," 1965 (90/1) Archiv des öffentlichen Rechts, pp. 1–61; R. Willecke, "Der Festlandsockel – seine völker- und verfassungsrechtliche Problematik," 1966 (81/13) Deutsches Verwaltungsblatt, pp. 461–468.

[47] Note dated 20 March 1963 [B102/260125], p. 2.

[48] See note on the state of deliberations on the exploration and exploitation of mineral resources in the area of the continental self of the North Sea beyond the three-nautical-mile zone dated 20 March 1963 [B102/260125].

[49] Note dated 3 August 1962 [B102/260124], p. 3.   [50] Ibid., p. 3.

[51] Note dated 17 July 1962 [AA/57], p. 2.

[52] Note dated 10 July 1962 [B102/260124], p. 5.   [53] Ibid.

Federal competence. At the same time, the Ministry for the Economy had tried to reach an agreement with the states on a common approach in view of the growing interest in exploration.[54] When the states, after repeated invitations, finally agreed to meet in March 1963, it transpired that they had been actively building up their position.[55] The states informed the Ministry that they intended to give a consortium of German and foreign firms a conditional undertaking concerning the issuance of a concession. The Ministry rejected this proposal because it did not involve all foreign companies which had shown an interest. The activities of the consortium might also lead to difficulties because it intended to also work in an area that was claimed by the Netherlands.[56] The Ministry did consider it important that activities in the shelf start as soon as possible and that they would be regulated.[57]

On 25 March 1963, only a couple of weeks after the meeting between the Ministry for the Economy and the states, the Central Bureau of Mines at Clausthal-Zellerfeld issued a conditional concession to a number of companies, commonly referred to as the North Sea consortium, on behalf of the four states.[58] This prompted the Federal authorities to look further at questions of international and constitutional law and in particular at the internal competence to issue concessions.[59] The Foreign Office considered that this issue should be settled before the finalization of the preparations for the ratification of the Convention on the continental shelf. At a meeting of the ministries that were involved it was suggested that deferment of the ratification of the Convention might be considered.[60] However, the Foreign Office indicated that it remained of the view that the Convention and two of the other Geneva Conventions and the Optional Protocol should be submitted for approval jointly. The growing importance of the matters regulated in the Convention on the continental shelf made its ratification just as desirable. Deferral of the submission of that Convention for parliamentary

---

[54] Note dated 20 March 1963 [B102/260125], p. 2.

[55] See note dated 24 June 1963 [AA/3], p. 1.

[56] Note dated 20 March 1963 [B102/260125], pp. 2–3.   [57] Ibid., pp. 3–4.

[58] The conditional concession did not address the division of competence between the states and the Federation (note dated 16 April 1964 [AA/88], p. 1). For the question of the spatial scope of application of German legislation see below. For further background information on the consortium and the concession see Menzel, "Der Festlandsockel der Bundesrepublik," pp. 26–31.

[59] See circular letter of the ME dated 29 April 1963 [B141/22027]; note dated 24 June 1963 [AA/3].

[60] Ibid., p. 7.

approval should only be considered if the legal questions related to it would prove so difficult and time-consuming that it would no longer justify the deferral of the process of ratification of the other instruments.[61] Other ministries were requested to inform the Foreign Office which provisions of the Convention required implementation and to what extent this concerned competencies of the Federation.[62]

The situation came to a head at the end of 1963, when it became known that a subsidiary of Caltex, an American company, intended to start exploratory activities off the coast of Germany in the spring of 1964.[63] This move by Caltex, which did not participate in the North Sea consortium, forced Germany to reconsider whether or not it could regulate activities on the continental shelf before it would have become a party to the Convention on the continental shelf or the Convention would have entered into force. The North Sea consortium itself had also sought a legal opinion on these points. The opinion expressed the view that Germany had rights in the shelf under customary law. Moreover, it was argued that the concession had resulted in acquired rights of the consortium, which Germany was under a duty to protect.[64] In a meeting with a representative of the consortium, staff of the Foreign Office submitted that Germany could only acquire rights over the continental shelf through ratification of the Convention.[65] As was also observed by the representative of the consortium, that position was contradictory.[66] If the Convention did not represent customary law, Germany would only acquire rights in relation to other parties to the Convention. This point was conceded by the Foreign Office, but it for the moment did not want to consider its implications. By ratifying the Convention, it wanted to adapt to the new developments. The representative of the Foreign Office

---

[61] Circular letter of the FO dated 3 September 1963 [AA/57], p. 2.   [62] Ibid., p. 3.

[63] See e.g. note of the head of section III of the ME to the Minister dated 26 November 1963 [B102/260126], p. 2. For these developments see also I. von Münch, "Die Ausnutzung des Festlandsockels vor der deutschen Nordseeküste," 1963–1964 (11) Archiv des Völkerrechts, pp. 391–416, at pp. 397–400; D. von Schenk, "Die Festlandsockel-Proklamation der Bundesregierung vom 20. Januar 1964," in W.J. Schütz (ed.), Aus der Schule der Diplomatie; Beiträge zu Außenpolitik, Recht, Kultur, Menschenführung (Düsseldorf: Econ-Verlag, 1965), pp. 485–498, at pp. 486–488.

[64] E. Lauterpacht, Opinion "The German Continental Shelf" [B102/260126] dated 6 November 1963, pp. 1 and 12–13.

[65] Note dated 8 November 1963 [B102/260126], pp. 1–3. The note observes that this position had been approved by the Minister himself and that the ME held the same view (ibid., p. 3).

[66] Ibid., p. 2.

emphatically rejected the view that the consortium's concession had any legal consequences for Germany.[67] The Ministry for the Economy had also indicated to all firms that their activities in the North Sea would be at their own risk and that they could not claim any rights on that basis.[68]

The Foreign Office and the Ministry for the Economy considered that it had now become necessary to ratify the Convention as soon as possible.[69] At the same time, pending the ratification of the Convention, activities of foreign companies on the German shelf without the permission of the German authorities had to be prevented. The proposal of the Foreign Office to achieve that purpose through a unilateral proclamation received general support.[70] The proclamation was to be adopted on the understanding that Germany would ratify the Convention as soon as possible.[71]

Another issue to be resolved was the relationship between the Federation and the states. The Federal agencies had agreed that it should be avoided that the proclamation would lead to an escalation of the conflict between the Federation and the states.[72] To this end the states were informed beforehand of the proclamation and that it would be without prejudice to the question of the internal division of competence.[73] The proclamation provided that activities on the German shelf for its exploration and the exploitation of its resources without express permission of the "competent German authorities" were prohibited.[74] Following the proclamation, Caltex withdrew the drilling platform it had intended to employ in front of the German coast.[75]

---

[67] *Ibid.*, p. 4; see also note of the Head of Section III of the ME to the Minister dated 26 November 1963 [B102/260126], p. 2.

[68] See e.g. note of the Head of Section III of the ME to the Minister dated 26 November 1963 [B102/260126], p. 2.

[69] *Ibid.*, p. 3; circular letter of the FO dated 23 December 1963 [AA/84], p. 2.

[70] For the proposal see circular letter of the FO dated 23 December 1963 [AA/84], p. 2. For the outcome of the deliberations on the proposal see letter of the FM to the State secretary of the Office of the Federal Chancellor dated 8 January 1964 [AA/33].

[71] See letter of the FM to the State secretary of the Office of the Federal Chancellor dated 8 January 1964 [AA/33]; Proclamation of the Federal Government on the exploration and exploitation of the German continental shelf of 22 January 1964, preamble.

[72] See letter of the Foreign State secretary to the State secretary for the Economy dated 14 January 1964 [B102/260127], p. 2.

[73] Note dated 20 January 1964 [AA/33].

[74] See Proclamation of the Federal Government on the exploration and exploitation of the German continental shelf of 22 January 1964, para. 2; see also letter of the Foreign State secretary to the State secretary for the Economy dated 14 January 1964 [B102/260127], pp. 2–3.

[75] Von Schenk, "Die Festlandsockel-Proklamation," p. 487.

The proclamation did not reflect the view that only parties to the Convention on the continental shelf were allowed to exercise rights over their continental shelf.[76] Although the proclamation mentioned the signature of the Convention as an important step in the development of the law,[77] in hindsight the proclamation seems to have contributed to the loss of interest in ratification of the Convention. As will be seen, the evolving views on article 6 of the Convention also seem to provide a large part of the explanation for that development. The adoption of interim legislation on the continental shelf may also have facilitated that development.

The option to adopt interim legislation was discussed between representatives of the Foreign Office and the Ministry for the Economy shortly after the proclamation on the continental shelf. The choice was between interim legislation addressing the most urgent questions or proceeding directly with the introduction of legislation for the approval of the Convention. The former approach was agreed upon.[78] It was further agreed that the interim legislation should be without prejudice to the division of competence between the Federation and the states. It was considered that the prospect to reach agreement with the states on that point was favorable.[79] The interim legislation, which was adopted in June 1964, provided for involvement of the Central Bureau of Mines at Clausthal-Zellerfeld, which represented the states, and the Federal authorities in regulating activities on the continental shelf.[80] The spatial scope of application of the interim legislation was defined through a reference to the German Proclamation on the continental shelf.[81] This left the precise limits of the continental shelf open. A concession for the North Sea consortium, which was granted under the interim legislation, also did not further specify its area of application and only referred to the German continental shelf in the North Sea.[82] At the same time, the Foreign Office had made it clear that drilling to the west of

---

[76] The proposal for a proclamation by the FO had still indicated the opposite view and raised the question if Germany could rely on customary law, in view of the position it had thus far taken (circular letter of the FO dated 23 December 1963 [AA/84], p. 3).

[77] See Proclamation of the Federal Government on the exploration and exploitation of the German continental shelf of 22 January 1964, para. 1.

[78] See note dated 6 February 1964 [AA/32], p. 1.   [79] See *ibid.*, pp. 1–2.

[80] Act on the provisional regulation of the rights over the continental shelf of 24 July 1964, article 2. For a further discussion of this legislation see the literature mentioned in note 46.

[81] *Ibid.*, article 1.

[82] Concession for the exploration and exploitation of petroleum and natural gas of 30 October 1964 [B102/260133], section 1.a.

the equidistance line was out of the question, because it would disrupt the negotiations with the Netherlands and the Ministry for the Economy was requested to inform the North Sea consortium about this most emphatically.[83] There is no indication that the consortium ever planned to drill in the area beyond the equidistance line.[84] The interim legislation apparently provided a satisfactory framework. In the subsequent years, Federal and state authorities collaborated without major difficulties and consulted on questions concerning proposed exploratory activities of the consortium and the delimitation of the continental shelf with neighboring States.

Between 1958 and the middle of the 1960s the perception of the German authorities on the implications of the delimitation provision contained in article 6 of the Convention on the continental shelf changed dramatically. This development was probably in large part brought about by the views of the Netherlands and Denmark on the interpretation of article 6, which both stressed the centrality of equidistance and relegated special circumstances to a secondary position.

In the latter half of 1958, when Germany considered whether or not to sign the Geneva Conventions, it was argued that equidistance had a subsidiary role under article 6, which, it was submitted, left the matter of delimitation practically unresolved.[85] Early on, it had been suggested to defer the ratification of the Convention on the continental shelf until after a satisfactory arrangement would have been reached with neighboring States on the division of the shelf.[86] At a meeting in 1962, upon questions of the Ministry for the Economy, the Foreign Office, which was responsible for the delimitation of the shelf with neighboring States, responded that it would not take any initiatives in respect of its division. Talks with other coastal States were at present out of the question. Everything should be left to the Convention on the continental shelf and the steps envisaged by it.[87]

---

[83] See note dated 6 February 1964 [AA/32], p. 4.

[84] See also D. von Schenk, "Die vertragliche Abgrenzung des Festlandsockels unter der Nordsee zwischen der Bundesrepublik Deutschland, Dänemark und den Niederlanden nach dem Urteil des Internationalen Gerichtshofes vom 20. Februar 1969," 1971 (15) *Jahrbuch für internationales Recht*, pp. 370–391, at p. 377.

[85] See further text at footnote 37 and following.

[86] The International Conference on the Law of the Sea in Geneva; Final Report of the German Delegation dated 23 May 1958 [B141/22024], Annex 7, p. 18; see also letter of the Minister for the Economy and Transport of Lower Saxony to Minister for Economic Affairs dated 25 May 1961 [B102/119551], pp. 9–10.

[87] Note dated 10 July 1962 [B102/260124], p. 5.

During the renewed consideration of the possible ratification of the Convention in late 1963 the Foreign Office initially maintained its position that the delimitation with neighboring States could be addressed after the ratification of the Convention.[88] Other ministries argued that the ratification should be postponed until a satisfactory arrangement on the division of the shelf would have been reached with the Netherlands and Denmark.[89] One assessment pinpointed one potential problem with article 6. The advantage of negotiations outside the framework of the Convention was considered to be "that the automatism of the Convention, in a case in which negotiations would break down, would be absent and as a consequence the pressure under which the Federal Republic would have to conduct such negotiations would also be absent."[90] That is, under the Convention equidistance would be the default rule in the absence of agreement. Other perceived difficulties in respect of article 6 were that it was uncertain whether there actually were special circumstances and whether Germany would have the possibility to submit that question to a neutral third party, such as the ICJ.[91]

Germany was confronted with a clear illustration of the "automatism of the Convention" shortly afterwards. The German proclamation on the continental shelf did not define provisional limits of the shelf, but only provided that "[t]he detailed delimitation of the German continental shelf in relation to the continental shelf of foreign States shall be the subject of agreements with those States."[92] In a meeting at the German Foreign Office, the counselor of the Danish Embassy indicated that the proclamation's provision on delimitation was superfluous. The delimitation of the continental shelf was not a "diplomatic" problem, but only a "geodetic" problem, namely the determination of the

---

[88] See note dated 8 November 1963 [B102/260126], p. 5; see also circular letter of the FO dated 3 September 1963 [AA/57], p. 2.

[89] Note dated 19 December 1963 [B102/260126]; circular letter of the Minister for Finance dated 10 January 1964 [B141/22029], p. 2; see also letter of the Minister for the Economy and Transport of Lower Saxony dated 25 February 1964 [AA/32], p. 1.

[90] Translation by the author. The original text reads "daß die Automatik der Konvention bei Scheitern der Verhandlungen und damit auch der Druck, unter dem die Bundesrepublik solche Verhandlungen führen müßte, fehlen würde" (note dated 19 December 1963 [B102/260126], p. 2; see also circular letter of the Minister for Finance dated 10 January 1964 [B141/22029], p. 2).

[91] See note dated 19 December 1963 [B102/260126], p. 2; see also circular letter of the Minister for Finance dated 10 January 1964 [B141/22029], p. 2; note dated 27 February 1964 [B106/53685], p. 1.

[92] Proclamation of the Federal Government on the exploration and exploitation of the German continental shelf dated 22 January 1964, para. 1.

equidistance line. Denmark in any case was not going to take any initiative for negotiations in view of the favorable location of the equidistance line.[93] In response, the Foreign Office pointed out that the Convention had not yet entered into force, making clear that the application of its article 6 was not at issue, and that Germany thus far had not ratified the Convention. Most importantly, however, Germany's understanding of article 6 was that equidistance only had to be considered if there was no agreement and if there were no special circumstances justifying a different boundary.[94]

Internally, the Foreign Office did not make a case for the above interpretation of article 6 of the Convention.[95] Instead, the Foreign Office concluded that Germany should not ratify the Convention if its neighbors would dispute the existence of special circumstances and would indicate to be prepared to consider an agreement on an appropriate solution.[96] After the first half of 1964, the ratification of the Convention was never again taken up, because it was considered article 6 had the potential to damage Germany's position on the delimitation of the continental shelf in the North Sea.[97]

Germany's change in views on article 6 of the Convention took place when the discussions with neighboring States had already started. In the first phase of the negotiations with the Netherlands, which lasted until July 1964, Germany still couched its position in the terms of article 6. In the first round of talks with Denmark, which took place in October of 1964, Germany rejected the relevance of article 6 because it did not

---

[93] Note dated 30 January 1964 [AA/34], pp. 1–2. Danish legislation on the continental shelf adopted in 1963 reflected a similar view on equidistance. In the absence of agreements the boundary with neighboring states was the equidistance line (see Chapter 4.2).

[94] Ibid., p. 2.

[95] A note prepared in the ME at the end of January 1964 did express support for this interpretation and sticking to a swift ratification of the Convention. This support seems to have been inspired mainly by the fear that internal legislation on the continental shelf would not be adopted and uncertainty over the regime of the German shelf would continue (note dated 27 January 1964 [B102/260127], p. 3). The note among others argued that absence of ratification would offer Germany hardly any advantage in relation to its neighbors. Even if article 6 originally had not been a codification, it was now well under way to becoming customary law, and likely to be applied as such by the ICJ (ibid., pp. 3–4).

[96] Note (not dated; on basis of content dated between 17 and 21 April 1964) [AA/38], p. 4.

[97] See e.g. note dated 18 April 1967 [B102/318714], p. 2; Report of the interdepartmental meeting in the Foreign Office on 5 May 1967 (not dated) [AA/59], p. 1.

constitute customary international law.[98] The equidistance principle could not bind Germany as long as it would not ratify the Convention.[99] To be on the safe side, it was also argued that the configuration of the German coast constituted a special circumstance, excluding a schematic application of the equidistance principle.[100]

The rejection of the equidistance method implied that Germany needed to develop an alternative approach. At first, this issue was assessed from the perspective of article 6(2) of the Convention on the continental shelf, which is concerned with the situation of adjacent States. Germany was placed in a much more difficult situation than Denmark and the Netherlands in terms of justifying its position. Article 6(2) of the Convention only refers explicitly to one method of delimitation: equidistance. Denmark and the Netherlands considered that there were no special circumstances in the North Sea justifying a different boundary. This implied that they could determine the location of their boundaries in relation to their neighboring States without difficulty and could refer to a method that was explicitly mentioned in a treaty provision in support.

Article 6 is silent as to what is to happen if special circumstances are present. Only in respect of small islands does the drafting history of the Convention suggest a possible solution. The equidistance line could, in that case, be drawn without taking into account basepoints on such islands. Germany was faced with a different situation. The most important circumstance that might be considered a special circumstance justifying a departure from the equidistance method was the location of Germany at the back of a concave coast. There had been a reference to "*any* exceptional configuration of the coast" as a special circumstance in one of the reports of the ILC,[101] but no indication had been given as to what constituted an "exceptional configuration" and what impact its presence should have on the equidistance line.[102] Germany was thus

---

[98] See Report of the negotiations on the delimitation of the continental shelf in the North Sea between Germany and Denmark in Bonn on 15 and 16 October 1964 (not dated) [AA/34], p. 1.

[99] See note dated 29 November 1965 [AA/43], p. 2.   [100] *Ibid.*

[101] *Report of the International Law Commission to the General Assembly*, Yearbook ILC, 1953, Vol. II, p. 200, at p. 217, para. 82 (emphasis provided).

[102] Similarly, the Tunisian delegate at the 1958 Conference "considered that the delimitation of the continental shelf between States adjacent to or opposite each other should take account of the geographical configuration of the region, and that considerable flexibility would have to be used in applying" the delimitation provision (UNCLOS I Off. Rec., Vol. VI, p. 22, para. 35).

faced with the challenge of finding a method of delimitation that would assure it a larger share in shelf of the North Sea and a justification for that method.

A first attempt to define a German position stems from the middle of 1962.[103] The proposal had a couple of aspects in common with a number of subsequent proposals.[104] First, the proposed method of delimitation would give Germany access to the central part of the North Sea. Second, the proposal did not originate from the Foreign Office, which was in charge of the negotiations with Germany's neighbors.

Before looking at the methods of delimitation suggested by these proposals, some further consideration of the role of the Foreign Office is called for. At a meeting in July 1962, the Ministry had accepted a suggestion from the Ministry for the Economy that it should consider the possibilities of dividing the continental shelf of the North Sea.[105] However, the Foreign Office seems to have done little subsequently and at the end of 1963 it apparently had not yet developed specific

---

[103] See letter of Dr. H. Closs, Executive Director of the Federal Institute for Geological Surveys to Dr. Mollat of the ME dated 14 May 1962 [B102/260124]. Earlier documents in the archives only contain general references to the outcome a delimitation should achieve. For instance, the Final report of the German delegation on the outcomes of the 1958 Conference referred to an "angemessenes Schelfgebiet" (a reasonable/appropriate shelf area) (Final report of the German delegation dated 23 May 1958 [B141/22024], Annex 7, p. 18).

[104] Apart from the proposal by Closs mentioned in the preceding footnote, reference can be made to the following: H. Flathe, *Geodätische Bemerkungen zu deutschen Interessengebieten in der Nordsee* (31 January 1963) [AA/32] (Dr. Flathe was employed at the Federal Institute for Geological Surveys); letter of the Minister of Food Supply, Agriculture and Forests to the FO dated 4 November 1963 [AA/32]; H. Flathe, *Bemerkungen zur Abgrenzung eines deutschen Interessensgebietes in der Nordsee* (not dated; mentioned in note dated 27 January 1964 [B102/260127]) [B102/119553]; Aufteilung des Festlandsockels im Bereich der Nordsee unter benachbarte Staaten – geometrische Grundlangen hierfür dated 10 February 1964 [B102/260126] (according to a hand-written note drafted by DEA, one of the participants in the German North Sea consortium); *Überlegungen zur Abgrenzung der Interessengebiete in der Nordsee* dated 13 January 1963 [B102/260127] (*Überlegungen zur Abgrenzung* [B102/260127]) (prepared by Deilmann Bergbau, one of the participants in the German North Sea consortium); E. Menzel, *Gutachten zur Frage des kontinentalen Schelfs in der Nordsee; dem Deutschen Nordsee-Konsortium* (Kiel: March 1964).

[105] Note dated 10 July 1962 [B102/260124], p. 5. Later that year, the section of the ME dealing with the exploration and exploitation of the continental shelf concluded that it was appropriate that the FO would consider this matter further (see note dated 12 September 1962 [B102/260124], p. 1) and the FO was requested to communicate its views on the interpretation of article 6(2) of the Convention on the continental shelf (see letter of the Minister for the Economy to the FO dated 17 November 1962 [AA/32], p. 3).

arguments to arrive at a more advantageous delimitation for Germany.[106] At that time, the Ministry for the Economy had already started to look into the matter itself.[107] Only after a number of studies on the delimitation in the North Sea had already been carried out, and the negotiations with the Netherlands had started, the Foreign Office at last seems to have seriously considered the option of a thorough investigation of the matter.[108] One possible explanation for the Foreign Office's hesitance may be that it did not believe that a convincing case could be made that there were special circumstances justifying a significant departure from the equidistance line.[109] The Foreign Office may have been unwilling to embark upon a mission it considered unlikely to have success and might burden the relations with two of Germany's direct neighbors. The view of the Foreign Office that a German claim in relation to the Netherlands had to be watertight ("hieb- und stichfest"),[110] suggests that it was not willing to press a claim going significantly beyond the equidistance line. To anyone slightly familiar with the subject, it would have been obvious that it would be virtually impossible to develop a "watertight" proposal. As the Foreign Office itself had observed, article 6 of the Convention on the continental shelf hardly provided any guidance.[111] The Foreign Office also seemed indisposed to consider alternatives. For instance, a proposal of the Ministry for the Economy was summarily rejected as "not particularly constructive and realistic."[112] Interestingly, this proposal was in line with some of the ideas Germany later developed in its pleadings before the ICJ and with the judgment of the Court of 1969.[113]

---

[106] Note dated 8 November 1963 [B102/260126], p. 5.

[107] See note dated 24 June 1963 [AA/3], p. 7.

[108] See note, not dated (on basis of content dated between 17 and 21 April 1964) [AA/38], p. 4.

[109] For instance, in 1963 Meyer-Lindenberg, the deputy head of the legal department of the Foreign Office, in a talk with the Dutch Embassy reportedly expressed as his personal opinion that the delimitation with the Netherlands would have to be effected on the basis of the equidistance method (letter of the DEB to the MiFA dated 21 June 1963 [NNA/13]). Similarly, in a preparatory meeting for the first meeting with the Netherlands, Meyer-Lindenberg, who was to head the German delegation, declared that the configuration of the German coast did not constitute a special circumstance (note dated 27 February 1964 [B106/53685], p. 1).

[110] Note dated 26 May 1964 [AA/38], p. 3.   [111] See text at footnote 37 and following.

[112] Note dated 26 May 1964 [AA/38], p. 2. Translation by the author. The original text reads "wenig konstruktiv und realistisch."

[113] The proposal envisaged to calculate an equidistance line between the Dutch coast and a straight line between the German islands of Borkum and Sylt (see ibid.). For the German pleadings and the judgment of the Court see further Chapters 8.3.2, 8.5.2, 8.7 and 9.

There is little doubt that a large number of the proposals for delimitation methods, which were developed in the period between the middle of 1962 and the beginning of 1964, would not have stood a chance of success. An obvious example is a proposal for an alternative approach to calculate the equidistance method. This proposal is based on a mistaken interpretation of a suggestion by Boggs to simplify the strict equidistance line and has nothing to do with the equidistance method. Part of the boundaries of Germany are defined by reference to points on the coasts of Denmark and the Netherlands, while disregarding the German coast.[114]

The fact that a certain number of the delimitation proposals may not have been based on sound arguments should not detract from the value of the overall exercise. A significant number of delimitation methods and possible justifications were tabled. This provided the necessary groundwork for building the German case. As a matter of fact, important elements of the case presented to the ICJ were already present, although often in a rudimentary form. For example, an important aspect of the German case presented to the Court was that the equidistance method cut Germany off from the central part of the North Sea, although it was in a similar position as its neighbors.[115] This focus on the convergence of the continental shelves of the coastal States at the center of the North Sea had been suggested in a number of the early proposals.[116] The figure in the German pleadings used to illustrate this idea is almost identical to figures contained in a number of these proposals.[117] At the same time, the diversity of the proposed methods, their different outcomes and their sometimes tenuous link to the coastal geography of the North Sea, indicated that it might be difficult to chart a course between maximizing the German claim and a case that could credibly be presented in court if need be.[118]

---

[114] See Menzel, *Gutachten*, pp. 112–114 and figure 9 and S. Whittemore Boggs, "Delimitation of seaward areas under national jurisdiction," 1951 (45) *AJIL*, pp. 240–266, at pp. 262–263.

[115] See e.g. GM, p. 84 and p. 85, figure 21.

[116] See letter of the Minister of Food Supply, Agriculture and Forests to the FO dated 4 November 1963 [AA/32], p. 2 and annexed figure; *Überlegungen zur Abgrenzung* [B102/260127], pp. 5–6 and Annex 9a; Menzel, *Gutachten*, p. 109 and figure 7.

[117] See e.g. GM, p. 85, figure 21 and *Überlegungen zur Abgrenzung* [B102/260127], annex 9a; Menzel, *Gutachten*, figure 7. It can also be noted that a first detailed assessment by the German FO based itself on the earlier proposals (note dated 29 September 1964 [AA/34]).

[118] For example, three of the proposals contained in the report *Überlegungen zur Abgrenzung* [B102/260127], annexes 7, 8 and 9a; and Menzel, *Gutachten*, at figures 5–7 resulted in the following areas for Germany and its neighbors: Denmark respectively 40,210 km²; 66,300 km²; and 55,000 km²; Germany respectively 44,440 km²; 53,600 km²; and 42,500 km²; and the Netherlands 57,790 km²; 22,700 km²; and

4.4   The Netherlands

The Netherlands assessed the overall outcome of the Geneva Conference positively. The Minister of Foreign Affairs considered that, although there were many shortcomings, the result of the conference was of historical importance for the codification and development of the international law of the sea.[119] Shortly after the conference it was concluded that the general principles on the extent and regime of the continental shelf contained in the Convention on the continental shelf reflected general international law.[120]

When the signature of the Geneva Conventions was considered by an interdepartmental meeting, it was critical about the Convention on the continental shelf. The meeting paid particular attention to the delimitation between Venezuela and the Netherlands Antilles. Especially in view of the inclusion of indefinite language in the Convention, the lack of a provision on mandatory dispute settlement was troublesome.[121] Another drawback was that the Convention allowed reservations, including on the provision concerning the delimitation of the continental shelf between neighboring States. As a matter of fact, Venezuela had already indicated that it would make a reservation to article 6.[122] It was considered an advantage that Venezuela would be excluded from making a reservation to article 1 of the Convention, which recognized that islands had the same continental shelf entitlement as mainland coasts.[123] The meeting concluded that the Convention should be accepted as soon

---

52,000 km^2 (the totals differ because the area in the Skagerrak taken into account for Denmark is not the same in all three cases). Application of the equidistance method would lead to the following areas: 56,700 km^2 (Denmark); 24,600 km^2 (Germany); and 61,400 km^2 (Netherlands).

[119] See Ministerie van Buitenlandse Zaken, *Verslag*, p. 5.

[120] See letter of the MiFA to the Dutch ambassador in London dated 21 July 1958 [NNA/18], p. 1; letter of the MiFA to the Dutch ambassador in Paris dated 21 July 1958 [NNA/18].

[121] Meeting on certain questions concerning the law of the sea at the Ministry of Foreign Affairs on 8 August 1958 (draft; undated) [NNA/11], p. 10.

[122] *Ibid.* Upon signing the Convention on the continental shelf on 30 October 1958, Venezuela had declared: "with reference to article 6 that there are special circumstances to be taken into consideration in the following areas: the Gulf of Paria, in so far as the boundary is not determined by existing agreements, and in zones adjacent thereto; the area between the coast of Venezuela and the island of Aruba; and the Gulf of Venezuela" (http://treaties.un.org/pages/ViewDetails.aspx?src=TREATY&mtdsg_no=XXI-4&chapter=21&lang=en).

[123] Meeting on certain questions concerning the law of the sea at the Ministry of Foreign Affairs on 8 August 1958 (draft; undated) [NNA/11] p. 10. Venezuela on the other hand had taken the position that Aruba was located on the continental shelf of Venezuela.

as possible.[124] The Netherlands Antilles, which after the Conference had again indicated that it considered a median line boundary with Venezuela of the utmost importance,[125] was assured that it went without saying that the Kingdom would object against a Venezuelan reservation referring to the presence of special circumstances in the case of Aruba.[126] The Netherlands signed the four Geneva Conventions and the Optional Protocol concerning dispute settlement on 31 October 1958.[127]

A preliminary draft for the explanatory memorandum to the bill for parliamentary approval of the ratification of the 1958 Conventions was prepared at the end of 1961.[128] The process probably was not started earlier to await the outcome of the second United Nations Conference of the law of the sea at the end of 1960. That Conference had been charged to consider breadth of the territorial sea, the most important issue that had not agreed upon at the 1958 Conference. Moreover, the Netherlands was bringing its national legislation in line with the Geneva Conventions.[129] That process took considerable time. Mining legislation for the continental shelf, which was mainly intended to regulate oil and gas activities, was only adopted in 1965 and entered into force in 1967.[130] Finally, it was submitted that the process of ratifying the Convention had been put on hold deliberately, to find out what reservations other States might make upon becoming a party to the Convention.[131]

The bill for approval of ratification of the Geneva Conventions was submitted to Parliament in July 1964. In respect of article 6 of the Convention on the continental shelf, the government observed that it did not see any grounds to deviate from the principles contained in paragraphs 1 and 2 (i.e. the equidistance/median line). The government rejected the position of Venezuela that special circumstances justified a

---

[124] *Ibid.*

[125] Letter of the Prime Minister of the Netherlands Antilles to the Plenipotentiary Minister of the Netherlands Antilles dated 28 August 1959 [NNA/12].

[126] Letter of H. G. Schermers of the MFA to the Plenipotentiary Minister of the Netherlands Antilles dated 29 January 1960 [NNA/19].

[127] Signature was for all three countries of the Kingdom (the Netherlands, the Netherlands Antilles and Suriname).

[128] See note 98/61 dated 1 November 1961 [NMFA/4].

[129] See e.g. *ibid.*; note 146/63 dated 29 March 1963 [NMFA/4].

[130] This concerns the Continental shelf mining act of 23 September 1965, which entered into force on 1 March 1967 (Decree of 7 February 1967, concerning the date of entry into force of the Continental shelf mining act (Stb.1967, 73). See also Chapter 6, text at footnote 271 and following.

[131] K. II (1963–1964), p. 1647.

different boundary and remained of the view that between both Aruba and Venezuela and between Aruba and Colombia the median line had to constitute the boundary in the absence of an agreement.[132] The Ministry of Foreign Affairs had attempted to establish beforehand whether the area concerned was of interest from a resource perspective through an inquiry with the Bataafsche Internationale Petroleum Maatschappij (BIPM), a subsidiary of Shell.[133] A Dutch claim to as large as possible a part of the continental shelf for the Netherlands Antilles and Suriname could lead to a cooling in the relationship with the neighboring coastal State. The government would probably like to avoid this if the interest of the Netherlands would not be served by such a claim. The representative of the BIPM apparently was not forthcoming and did not want to express himself on possible claims of the Netherlands in the Caribbean region.[134]

During the process of approval by the Parliaments of the Netherlands, the Netherlands Antilles and Suriname, the delimitation provision of the Convention on the continental shelf was not questioned.[135] The Netherlands ratified the Geneva Conventions on 18 February 1966 for all three countries of the Kingdom. The Convention on the continental shelf had entered into force almost one and a half years previously. In depositing its instrument of ratification the Netherlands reserved all rights regarding the reservations in respect of article 6 made by the Government of Venezuela when ratifying the Convention.[136]

Unlike Germany, the Netherlands had no hesitation about regulating oil and gas activities on its continental shelf before it had ratified the Convention on the continental shelf or enacted mining legislation. The Netherlands Oil Company showed an interest in exploratory activities in

---

[132] Explanatory memorandum to the Bill of the Kingdom Act for approval of the Law of the Sea Conventions and the Optional Protocol of signature concerning the compulsory settlement of disputes done in Geneva on 29 April 1958 (K. II (BH) (1963–1964) 7723 (R425) no. 3), p. 7. The bill and explanatory memorandum were submitted to parliament on 30 July 1964 (see K. II (BH) (1963–1964) 7723 (R425) no. 1). It is not clear on what basis the reference to Colombia was included in the explanatory memorandum. An equidistance line between Colombia and Aruba would only be possible if the Venezuelan islets of Los Monjes would be ignored in establishing equidistance lines.

[133] Note 69/63 dated 18 February 1963 [NMFA/4], p. 2. On the part of the ministry the head of the treaty department and the deputy legal advisor took part in this discussion (ibid., p. 1).

[134] Ibid.   [135] See K. II (BH) (1963–1964) 7723 (R425) nos 5–7.

[136] Upon ratification of the Convention Venezuela had not provided a specific interpretation of article 6, as it had done in signing the Convention, but made an express reservation in respect of article 6. Venezuela had ratified the Convention on 15 August 1961.

the North Sea in 1959 and other companies followed in 1962.[137] The Minister of Economic Affairs allowed companies to carry out exploratory work on the continental shelf pending the adoption of mining legislation for the continental shelf.[138] The minister subsequently decided that pending the adoption of legislation no permission would be given for exploratory drilling beyond the territorial sea.[139]

4.5   The different views on the status of the Convention

Denmark and the Netherlands both found that the Convention on the continental shelf generally provided a satisfactory regime. Both States considered that it would be possible to regulate activities on the continental shelf prior to the ratification of the Convention, although Denmark probably had more hesitance about the customary law character of the Convention. Germany continued to have considerable doubt about the content of customary law. This also raised the question as to how Germany should act against nationals of States who were not a party to the Convention. In relation to those States ratification of the Convention would not give a basis for action and the matter would be governed by customary international law. Faced with the possibility that a foreign company would start activities on the continental shelf without Germany's permission, Germany in its proclamation of January 1964 came round to the view that general international law granted Germany exclusive sovereign rights over the continental shelf.

The German proclamation specifically referred to the signing of the Convention on the continental shelf as an important aspect of State practice and Germany had the intention to ratify the Convention subsequent to the proclamation. Germany's assessment of article 6 of the Convention was the main reason why this step was never taken. In 1958, the Foreign Office had taken the view that equidistance had a limited role

---

[137] The Netherlands Oil Company (Nederlandse Aardolie Maatschappij or NAM) was founded in 1947 as a joint venture between BPM, a subsidiary of Shell and the Standard Oil Company of New Jersey, which both had a 50 percent share. Since 1956 the NAM had been conducting gravimetric research in the North Sea and from 1959 also conducted seismic surveys (A. Koelmans, *Van pomp tot put in honderd jaar; Bijdrage tot de geschiedenis van de voorziening van Nederland met aardolieprodukten*, Doctoral dissertation (University of Amsterdam, 1970), pp. 118–120).

[138] Explanatory memorandum to the bill of the Continental shelf mining act (K. II (BH) (1963–1964) 7670, no. 3), p. 1; see also note 33/63 dated 5 April 1963 [NNA/13].

[139] Memorial of reply of the Minister of Economic Affairs (K. I (BH) (1963–1964) 7400 J, no. 65a), p. 4.

in this provision and that it should not be an obstacle to Germany's signature of the Convention. Later, the Foreign Office found that article 6 was problematic for Germany. This was mainly based on the interpretation that there was a certain "automaticity" to the equidistance line. It could be argued that equidistance applied in the absence of agreement between the interested States. Contacts with Denmark in 1963 confirmed that this was at least the view of Denmark, and the Netherlands in the middle of 1963 had also indicated to Germany that, in connection with the pending ratification of the Convention, it would apply the equidistance line in relation to Germany. In its first contacts with the Netherlands Germany still relied on article 6 of the Convention, referring to the presence of special circumstances, but subsequently took the view that article 6 was irrelevant because it did not reflect customary international law. After 1964, the ratification of the Convention was never again seriously considered because of the possibly negative consequences article 6 might have for Germany's position in the North Sea. Clearly, it was also considered that the ratification would not bring Germany other advantages as compared to customary law. Otherwise, it would have been perfectly possible to make a reservation in respect of article 6 while ratifying the Convention. This course of action had been followed by some other States which opposed the application of the equidistance line in certain cases.

Germany was slow in developing alternatives for the equidistance method. In 1962 the Foreign Office had accepted a suggestion from the Ministry for the Economy to consider this issue, but refrained from developing any ideas. The Foreign Office may not have been convinced that it was possible to construct a credible German case. The fact that at a later stage the basic arguments of the German case were developed in a relatively short period of time[140] is confirmation of the fact that the state of the law was likely not the main obstacle for the Foreign Office's failure to act.

Alternative schemes for delimitation were developed by the Ministry for the Economy and interested private parties. These were met with little enthusiasm by the Foreign Office. Although some of these proposals had little to do with geographic realities, others offered useful approaches and similar ideas figured in the pleadings of Germany before the ICJ. It is reasonable to assume that these initial proposals helped to move the debate on a German approach forward.

---

[140] See further Chapter 8.3.1.

Before ratifying the Convention on the continental shelf, Denmark revisited the significance of the Norwegian Trough for the delimitation between itself and Norway. A number of legal and political considerations contributed to the decision to accept the equidistance line in this case. One of the reasons was that it was considered that rejection of the equidistance line could weaken Denmark's position that its boundary with Germany had to be the equidistance line.

5

The first phase of the negotiations on the delimitation of continental shelf boundaries in the North Sea

5.1   Overture to bilateral negotiations of Germany with Denmark and the Netherlands

For different reasons, Denmark, Germany and the Netherlands did not take any initiatives to delimit their continental shelf boundaries in the North Sea for some time after the Convention on the continental shelf had been adopted in 1958. Initially, Germany took the position that the area remained high seas until the Convention would have entered into force. Consequently, there was nothing to delimit between coastal States. Denmark considered equidistance the appropriate method of delimitation of its continental shelf and concluded that in light of article 6 of the Convention it was not on Denmark to take the initiative to negotiate agreements with its neighbors. The Netherlands had a similar view.[1]

Germany received information that the Netherlands would insist on delimitation of their mutual boundary by the equidistance method by the end of 1962, but the Foreign Office concluded that no immediate action was required.[2] The Netherlands and Denmark only received signals that Germany was not intending to agree on the delimitation of the continental shelf by equidistance by April of the next year. Before that time, it had been clear that the Netherlands and Germany would have to deal with the impact of their dispute over the location of the land boundary in the Ems estuary in connection with the determination of the starting

---

[1] For the Netherlands see also e.g. Second report of the interdepartmental consultations concerning an act on more detailed rules concerning the exploration for and exploitation of the natural resources of the continental shelf of the North Sea, held on 8 May 1962 at the Ministry of Foreign Affairs (not dated) [NNA/15], p. 2.

[2] See note dated 15 November 1962 [B102/260124]; letter of the Minister for the Economy to the FO dated 17 November 1962 [AA/32], pp. 2–3; note dated 20 November 1962 [AA/32].

point of a continental shelf boundary. Germany and the Netherlands had sought to reach agreement on the former boundary in their negotiations to settle various issues related to the German occupation of the Netherlands during the Second World War. According to Germany, the boundary runs along the low-water line on the Dutch side of the Ems on the basis of an historic title, while the Netherlands considers that the boundary is the *thalweg* of the main channel of the estuary.[3] Since agreement on a boundary was not forthcoming, one of the treaties adopted in connection with the settlement of issues related to the Second World War established a joint regime for the area.[4] The discovery of hydrocarbons after the treaty had been concluded resulted in the conclusion of a supplementary agreement in 1962.[5] That agreement defines an area of application by geographical coordinates.[6] The lateral limits of this area generally accord with the views of the parties on the location of the land boundary.[7] The area extends to the outer limit of the territorial sea, which at the time of the conclusion of the Supplementary Agreement for both States was at three nautical miles. At the three-nautical-mile limit, the distance between the lateral limits of the area of application of the Supplementary Agreement is slightly over 20 kilometers (points L' on the western side and A on the eastern side). The Supplementary Agreement divides the area of application into two approximately equal halves. To the west of this line the exploitation of mineral resources is to take place by concessionaires of the Netherlands and to the east by right holders of Germany.[8] The dividing line ended on the three-nautical-mile limit at point C," which is approximately midway between points L' on the western side and A on the eastern side.

---

[3] For further information on this dispute see e.g. Menzel, "Der Festlandsockel der Bundesrepublik," pp. 62–63; G. J. Tanja, "A New Treaty Regime for the Ems-Dollard Region," 1987 (2) *International Journal of Estuarine and Coastal Law*, pp. 123–142, at pp. 124–125.

[4] Treaty between the Kingdom of the Netherlands and the Federal Republic of Germany concerning arrangements for co-operation in the Ems Estuary (Ems-Dollard Treaty) of 8 April 1960.

[5] Supplementary Agreement to the Treaty concerning arrangements for co-operation in the Ems Estuary (Ems-Dollard Treaty), signed between the Kingdom of the Netherlands and the Federal Republic of Germany on 8 April 1960 of 14 May 1962 (Supplementary Agreement).

[6] The seaward termini of these lateral limits are identified in Figure 5.2. A detailed depiction of the common area is included in the nautical chart annexed to the Supplementary Agreement.

[7] Note 30/64 dated 3 April 1964 [NNA/13], p. 2.   [8] Supplementary Agreement, article 6.

The existence of the common area of the Supplementary Agreement implied that, even if Germany and the Netherlands would agree on the equidistance method to delimit their continental shelf, they would need to work out an arrangement on the starting point of that boundary and the course of its first part. There existed different views on the part of the continental shelf boundary that might be affected by the dispute. In a discussion between the Dutch Ministries of Foreign Affairs and Economic Affairs it is mentioned that this could only concern the first 19 kilometers of the continental shelf boundary,[9] but a figure from May 1964 indicates that only after some 80 kilometers the equidistance line would not be affected by the dispute over the boundary in the Ems.[10] That point is the first point on the equidistance line that is determined by basepoints outside the disputed area.[11]

In March 1963, the Dutch Ministry of Economic Affairs learned that the German North Sea consortium was planning to drill an exploratory well, which might be located in the area that was affected by the dispute over the location of the land boundary. The ministry considered that this required that the German government would be informed of the Dutch position.[12] After this matter was discussed with the Ministry of Foreign Affairs, it was concluded that the Dutch and German views could only diverge for a limited area, affecting the first 19 kilometers of the continental shelf boundary. This view was obviously based on the assumption that Germany accepted the equidistance method. It was agreed that the Netherlands should indicate that the continental shelf boundary should be established on the basis of the Dutch view of the land boundary. If Germany would insist on a different approach, it might be agreed to extend the regime of the Supplementary Agreement to Ems-Dollard Treaty to this disputed area, which in any case was limited in size.[13]

---

[9] Annotation for the record; Conversation with DEU/ME on 2 April [1963] (not dated) [NMFA/4].

[10] Memorandum for the Council of Ministers dated 6 May 1964 [NNA/13], figure II.

[11] See ibid., p. 9; Report of discussions between delegations of the Netherlands and the Federal Republic concerning the delimitation of the continental shelf in the North Sea, held in Bonn on 4 June 1964 dated 9 June 1964 [NNA/13] (Report Dutch-German discussions 4 June 1964 [NNA/13]), p. 3.

[12] Note 41/63 of the Head of the Directorate of Mines to the Minister of Economic Affairs dated 15 March 1963 [NNA/41], p. 6.

[13] Annotation for the record; Conversation with DEU/ME on 2 April [1963] (not dated) [NMFA/4].

Shortly afterwards, the Netherlands became aware that its intention to use the equidistance method might not be shared by all North Sea States. The secretary-general of the Ministry of Foreign Affairs was alarmed by an article in the Financial Times of 2 April 1963.[14] The article reported that various oil companies were planning to start drilling in the North Sea, even though it was not yet known how the area would be divided. Reportedly, the companies intended to operate on the assumption that their rights would be respected by the coastal States. The article suggested that an agreement on the division of the North Sea could be based on a "total shoreline" basis.[15] Riphagen, the legal advisor of the ministry, informed the secretary-general that the Netherlands acted on the basis that the Convention on the continental shelf reflected the applicable legal regime. The Dutch continental shelf was defined by reference to the median line. The Netherlands did not allow the exploration or exploitation of this area without its permission. Riphagen did seem to have certain doubts about the propriety of applying the equidistance principle to the North Sea. Because of the special configuration of the coast, in a delimitation in accordance with the equidistance line Germany would get the worst of it and the Netherlands would do very well. The legal advisor was not aware of the implications of the "total shoreline" principle, but it stood to reason that it would benefit its inventors.[16]

At a meeting with representatives of the NAM in early May 1963, the Ministry of Foreign Affairs was informed that Germany might be planning to make an ambitious claim. Driessen of the NAM had been shown a map in Bonn with a boundary between Germany and the Netherlands that ran in a west northwestern direction up to the median line with the United Kingdom.[17] This claim concerned between a third and half of the Dutch equidistance area. The Ministry of Foreign Affairs had also been presented a chart identifying what might be a German claim to the continental shelf. This claim covered most of the Dutch equidistance area north of 54° N.[18]

---

[14] Letter from the DEB to Mr. F. Simons at the MFA dated 11 April 1963 [NNA/21], p. 1.
[15] "Is there Oil or Gas in the North Sea?," *The Financial Times*, 2 April 1963, p. 2.
[16] Note 33/63 of the legal advisor to the secretary-general dated 5 April 1963 [NNA/13].
[17] Note dated 13 May 1963 [NNA/13], p. 2.
[18] Chart "North Sea" (1963) contained in [NNA/13]. This chart had been prepared by the BIPM, a subsidiary of Shell.

Although initially there was some disagreement concerning the timing,[19] the Ministries of Foreign Affairs and Economic Affairs agreed that it was desirable to urgently inform Germany of the Dutch views concerning delimitation.[20] By informing Germany of the Dutch position, they wanted to prevent Germany issuing a concession for an area that was claimed by the Netherlands or allowing drilling in that area before it would have been formally notified of the Dutch position.[21] The Netherlands was particularly concerned about the area immediately to the north of the Dutch province of Groningen. In 1959 a major gas field had been discovered in the northeast of Groningen. It was expected that the offshore area to the north of Groningen, including the area of overlap to the north of the Ems estuary also had a promising resource potential.[22] The NAM had indicated that its activities pointed out that at least one geological structure in the offshore seemed to be of sufficient interest to carry out an exploratory drilling. This structure was also known to the German North Sea consortium and other oil companies.[23]

The Dutch Embassy in Bonn was instructed to present the Dutch position in the clearest possible manner.[24] A draft of the letter had further elaborated on this point, observing among others that it would be improper to suggest in any way that the Netherlands might be willing to consider a compromise solution and the Dutch position should be presented as the legally correct view.[25] An inquiry by the Embassy whether article 6(2) of the Convention on the continental shelf did not require as a first step that

[19] See letter of the Minister of Economic Affairs to the MiFA dated 1 May 1963 [NMFA/4], p. 1. The MFA did not want to raise this issue until the ratification of the treaties with Germany concerning the settlement of matters related to the Second World War would have been approved by both Chambers of Parliament. Another dispute related to the location of the land boundary – in this case its impact on the location of the continental shelf boundary – might further delay the process of approval of the treaties, which already had taken considerable time.

[20] See note 59/63 dated 9 May 1963 [NMFA/4]; note dated 13 May 1963 [NNA/13], p. 2.

[21] Letter of the Minister of Economic Affairs to the MiFA dated 1 May 1963 [NMFA/4], p. 1; letter of the MiFA to the Dutch ambassador in Bonn dated 6 June 1963 [NNA/13], p. 3.

[22] See e.g. note dated 13 May 1963 [NNA/13], p. 2; note no. 57 dated 14 May 1963 [NNA/13]; see also "North Sea – Continental Shelf; Boundary between Germany and the Netherlands" dated 7 March 1963 [NNA/13]; (this note had been presented to the Ministry of Economic Affairs by the NAM (letter of the Minister of Economic Affairs to the MiFA dated 1 May 1963 [NMFA/4] p. 1)).

[23] "North Sea – Continental Shelf; Boundary between Germany and the Netherlands" dated 7 March 1963 [NNA/13].

[24] Letter of the MiFA to the Dutch ambassador in Bonn dated 6 June 1963 [NNA/13], p. 3.

[25] Draft of a letter of the MiFA to the DEB dated 20 May 1963 [NNA/13], p. 6.

negotiations should be proposed to Germany was answered negatively.[26] The minister did not necessarily agree with the view that article 6(2) of the Convention obliged States as a first step to conclude an agreement.[27] The note verbale presented to the German Foreign Office indicated that the Netherlands considered that the eastern boundary of its continental shelf was the equidistance line starting from intersection of the *thalweg* of the Ems with the 3-nautical-mile limit of the territorial sea.[28] Germany was advised of this view in connection with the envisaged ratification of the Convention on the continental shelf by the Netherlands.[29] Any suggestion of negotiations on a boundary was absent from the note.

Denmark learned that Germany was possibly opposed to the equidistance method in the North Sea at about the same time as the Netherlands.[30] A sketch map which Hoppe of A.P. Møller, the concessionaire for the entire Danish shelf, had obtained through Shell, showed a German claim consisting of two lateral limits extending from the German coast in a northwestern direction to the median line between the United Kingdom and the continent.[31] It was concluded that this information did not justify any action on the part of Denmark but that it was on Germany to take the initiative.[32] The adoption of continental shelf legislation in early June 1963 did provide Denmark with an opportunity to present its views to Germany. Denmark notified Germany and its other neighbors that it had issued legislation in connection with the ratification of the Convention on the continental shelf.[33] The legislation

---

[26] See received message in cipher (ref. no. 4887) of the DEB to the MFA dated 14 June 1963 [NNA/13]; send message in cipher (ref. no. 2954) of the MFA to the DEB dated 20 June 1963 [NNA/13].

[27] *Ibid.*

[28] Note no. 7099 of the DEB to the German FO dated 21 June 1963 [NNA/24].

[29] *Ibid.* This reference was included because it was considered to provide a credible pretext to inform Germany of the Dutch position (letter of the MiFA to the Dutch ambassador in Bonn dated 6 June 1963 [NNA/13], p. 3).

[30] See Report; The Continental Shelf dated 15 May 1963 [DNA/9]; note dated 17 May 1963 [DNA/9].

[31] The figure is annexed to *ibid.* According to the note, the figure was drawn after Shell's view on a possible German claim. The representation of the claim shows a certain similarity to the figure contained in letter of Dr. H. Closs, Executive Director of the Federal Institute for Geological Surveys to Dr. Mollat of the ME dated 14 May 1962 [B102/260124] and figure 2 in H. Flathe, *Geodätische Bemerkungen zu deutschen Interessengebieten in der Nordsee* (31 January 1963) [AA/32]. The latter report had been presented at a meeting in Hanover at the beginning of 1963 (see report dated 20 May 1963 [AA/32]).

[32] Report; The Continental Shelf dated 15 May 1963 [DNA/9], p. 2.

[33] Note of the Embassy of the Kingdom of Denmark to the German FO (Journal nr. 119. A.6) dated 10 July 1963 [AA/3].

provided that the boundary of Denmark's continental shelf in the absence of agreement was the equidistance line. Like the Netherlands, Denmark did not make any suggestion that there was a need to negotiate a delimitation agreement.

Germany accorded priority to the delimitation of its continental shelf with the Netherlands. Whereas Denmark was informed that its position was being considered, the Netherlands was told that Germany wished to start negotiations on an agreement to delimit their continental shelf.[34] Germany rejected the Dutch position and observed that the German government considered that historical grounds, as well as other special circumstances in more than one respect justified a deviation from the equidistance line.[35] This formulation signaled that Germany not only differed from the Netherlands over the starting point of the continental shelf boundary. The German Foreign Office initially had suggested not raising these other circumstances. A draft of the note to be presented to the Netherlands only referred to the issue of the starting point of the continental shelf boundary related to the dispute over the Ems.[36] The text of the note as presented was based on a proposal from the Ministry for the Economy.[37] The hesitance of the Foreign Office had also been clear to the Dutch Embassy. When the Dutch note had been presented on 21 June 1963, Meyer-Lindenberg, the deputy head of the legal department of the Foreign Office, reportedly focused on the question of the starting point of the continental shelf boundary and, as his personal view indicated, that the boundary should be established on the basis of equidistance.[38]

The most likely explanation for Germany's preference to first deal with the Netherlands was that exploratory activities were focusing on the

---

[34] Note of the German FO to the Embassy of the Kingdom of Denmark (V 1–80/52/3) dated 16 August 1963 [AA/34]; Note of the German FO to the DEB (V 1–80/52/3) dated 26 August 1963 [NNA/13].

[35] Note of the German FO to the DEB (V 1–80/52/3) dated 26 August 1963 [NNA/13], p. 1.

[36] See draft note of the German FO to the DEB (V 1–80/52/3) of August 1963 (annexed to a note dated 30 July 1963) [AA/32].

[37] See letter of the Minister for the Economy to the FO dated 17 August 1963 [AA/32], p. 2. The ME had not proposed more precise language in respect of the German position because studies on possible methods of delimitation had not yet been completed (ibid.).

[38] Letter of R. Fack on behalf of the ambassador in Bonn to the MiFA dated 21 June 1963 [NNA/13]. An account of the meeting by the German FO does mention the latter aspect, but indicates that reference was only made to the difficulties involved in selecting the starting point of the continental shelf boundary (circular letter from the FO dated 10 July 1963 [AA/32]).

area in the vicinity of their possible boundary. Those impending activities had also motivated the Netherlands to inform Germany of the Dutch views and also led the Netherlands to suggest a meeting for a first exchange of views before the end of 1963.[39] The deliberations of the German authorities on how to deal with unauthorized activities on the German shelf led to some delay,[40] but a first meeting took place in March 1964.

Just before the first meeting between Germany and the Netherlands took place, the Dutch Deputy Minister of Foreign Affairs raised the question if it was in the interest of the Netherlands to seek the conclusion of an agreement in the short term. The absence of an agreement would impede activities in the shelf, whereas the Netherlands could exploit and export gas from its onshore gas field in Groningen.[41] In a reaction it was among others set out that article 6 of the Convention on the continental shelf contained an obligation to settle bilateral boundaries by mutual agreement. It was moreover argued that in the absence of agreement there would be uncertainty about a large area to the west of the equidistance line claimed by the Netherlands and Germany would have a free hand in that area.[42] These explanations proved to be sufficient for the Deputy Minister[43] and afterwards the matter was never revisited. It may be open to some doubt if Germany really would have had a free hand in the disputed area.[44] There is no indication that the Dutch Ministry of Foreign Affairs made a detailed assessment of the legal regime applicable to disputed areas. A couple of months later, when this matter was discussed in the Dutch Council of Ministers, German unilateral activities were no longer seen as a problem. In response, the Netherlands might engage in similar unilateral activities.[45]

---

[39] See note no. 113 dated 16 September 1963 [NNA/13], p. 3; note of the DEB to the German FO dated 28 November 1963 [B141/22028].

[40] See for instance, note of the DEB to the German FO dated 28 November 1963 [B141/22028]; received message in cipher (ref. no. 7945) of the DEB to the MFA dated 15 November 1963 [NNA/13]. On the issue of unauthorized exploratory activities see further Chapter 4.3.

[41] See note dated 20 February 1964 [NNA/13]; note dated 24 February 1964 [NNA/13].

[42] Note no. 26/64 dated 25 February 1964 [NNA/13], pp. 1–2.

[43] See the hand-written note of the Deputy Minister, *ibid.*, p. 1; also see the hand-written note of the MiFA on note dated 20 February 1964 [NNA/13], concurring with the view that an early settlement was called for.

[44] See also Chapter 6.6.1.

[45] Minutes of the Council of Ministers dated 22 May 1964 [NNA/2], p. 8, item 7.

5.2 The partial boundary between Germany and the Netherlands

5.2.1 The first stage of the negotiations

Prior to the first meeting with Germany on the delimitation of the continental shelf, the Netherlands already sought to reach agreement on a provisional arrangement pending the conclusion of a delimitation treaty. The Netherlands proposed that both States would not issue permits for drilling activities for the area directly to the west of the equidistance line proposed by the Netherlands in its note of 21 June 1963, and which might be the subject of negotiations between the parties.[46] The Netherlands had approached Germany because it had received signals that the German North Sea consortium was planning to start drilling activities west of the equidistance line, unless the Federal government clearly opposed this.[47] The Dutch Ministry of Economic Affairs supported the moratorium on drilling activities. The ministry had been approached by a foreign company with the offer to drill in potentially disputed areas to reinforce the Dutch position, but it had not taken up that offer. The ministry did not exclude that a similar offer had been made to the German authorities.[48]

Upon receiving the Dutch proposal for a provisional arrangement, the German Foreign Office assured that permits for the area in any case would not be issued until Germany would have given a written reaction.[49] The Dutch proposal put Germany in a difficult position. The acceptance of an arrangement that would be limited to the area directly to the west of the Dutch equidistance line would weaken the German position in negotiations. On the other hand, a counterproposal on a different area required a decision of the German authorities on the continental shelf boundary Germany would be claiming or the area of application of a provisional arrangement.

The German Foreign Office at first proposed to inform the Netherlands that for the time being the Federation and the states would not issue concessions for the area west of the Dutch equidistance

---

[46] Note no. 1523 of the DEB to the German FO dated 30 January 1964 [NNA/13].

[47] Send message in cipher (ref. no. 460) of the MFA to the DEB dated 30 January 1964 [NNA/24], p. 2.

[48] Letter of the Minister of Economic Affairs to the MiFA dated 10 February 1964 [NNA/13].

[49] See received message in cipher (ref. no. 1190) of the DEB to the MFA dated 30 January 1964 [NNA/13].

line.[50] On the other hand, the Ministry for the Economy held that the area concerned would have to be defined more clearly and should extend west of the comparatively small area affected by the dispute over the boundary in the Ems estuary.[51] The Foreign Office was against the definition of an area and considered that first of all the general approach to the delimitation should be clarified.[52] It suggested a possible basis for an agreement with the Netherlands, but obviously did not want to commit itself to a specific position.[53] The Foreign Office considered that the Netherlands should not yet be given any indication about the maximum claim Germany might pursue. This implied that the definition of specific points in a provisional arrangement was not possible as that would allow the Netherlands to reach conclusions concerning the German claim.[54] As a result, a provisional arrangement with the Netherlands was not pursued for the moment.[55] At the same time, the German authorities opposed any activities of the North Sea consortium in the area west of the Dutch equidistance line.[56] To the Netherlands, Germany indicated that it was committed to keep companies from starting activities in the area, but that in the long run it might be difficult to restrain them. The German government did not have the means at its disposal to prevent those activities and this might not only impact negatively on the delimitation talks, but also on their bilateral relationship.[57]

Talks between Germany and the Netherlands took place on 3 and 4 March 1964. Germany argued that the concavity of the German coast was a special circumstance in the sense of article 6(2) of the Convention on the continental shelf. Other circumstances to be taken into account were the broad equality of the coasts of Germany, Denmark and the

---

[50] Letter of the FO to the Minister for the Economy dated 4 February 1964 [AA/32], p. 2.
[51] See draft for a letter of the Minister for the Economy to the FO of March 1964 [B102/318712], p. 1.
[52] See hand-written note on p. 2 of ibid.
[53] See note dated 24 March 1964 [B102/260173], p. 1.   [54] See ibid., p. 2.
[55] See Report of the interdepartmental meeting dated 20 March 1964 at the Foreign Office [AA/34]; see also note dated 6 April 1964 [B102/119553].
[56] See note dated 6 February 1964 [AA/32], p. 3; see also letter of the FO to the Minister for the Economy dated 4 February 1964 [AA/32], p. 3.
[57] See Minutes of the German-Dutch negotiations on the delimitation of the continental shelf in the North Sea on 3 and 4 March 1964 in Bonn dated 16 March 1964 [AA/38] (Minutes German-Dutch negotiations of March 1964 [AA/38]), p. 4; Minutes of the Dutch-German discussions concerning the delimitation of the continental shelf, held in Bonn on 3 and 4 March 1964 [NNA/24] (Minutes Dutch-German negotiations of March 1964 [NNA/24]), p. 5.

Netherlands and the relative size of German investments in research in the North Sea. On this basis, Germany proposed an equal division of the shelf, giving all three States approximately 45,000 km².[58] Cutting Germany off from middle of the North Sea, which was the result of the Dutch proposal, was obviously inequitable. The Dutch delegation immediately rejected the German proposal and the suggestion that there might be a boundary between Germany and the United Kingdom. The Netherlands observed that there were no special circumstances and that the equidistance line between the United Kingdom and the continent was determined by basepoints on the Dutch and Danish coasts and not on the German coast.[59]

After further discussions on the principles of equity and proportionality the meeting was adjourned to allow the Dutch delegation to confer with the ministry in The Hague. The delegation thereafter indicated that the idea of an equal division of the continental shelf was completely new. The Netherlands had never looked into that matter and had expected that the discussion would have focused on different options for the starting point of the boundary and the equidistance line.[60] This Dutch stance on the German proposal made sense from a tactical perspective, but the Netherlands was already aware that quite ambitious claims had been advocated in Germany. The German note in reply to the Dutch note of June 1963 also had put the Netherlands on notice that more was at issue than the exact location of the equidistance line.[61]

The Dutch delegation obtained approval from the Ministry of Foreign Affairs to immediately reject the German proposal as unacceptable.[62] This was considered necessary to prevent that German authorities other than the Foreign Office, such as the Ministry for the Economy and the state of Lower Saxony, might get the impression that the Netherlands would be willing to entertain the German proposal or other significant

---

[58] Minutes German-Dutch negotiations of March 1964 [AA/38], pp. 1–2; Minutes Dutch-German negotiations of March 1964 [NNA/24], p. 2. A depiction of the German proposal in a Dutch source is included in Figure 5.1.

[59] Minutes German-Dutch negotiations of March 1964 [AA/38], pp. 1–2; Minutes Dutch-German negotiations of March 1964 [NNA/24], p. 2.

[60] Minutes Dutch-German negotiations of March 1964 [NNA/24], p. 2.

[61] See further text at footnote 46.

[62] See Minutes Dutch-German negotiations of March 1964 [NNA/24], p. 3; received message in cipher (ref. no. 1934) of the DEB to the MFA dated 3 March 1964 [NNA/13].

departures from the equidistance line.[63] Although the German proposal was unequivocally rejected, the Dutch delegation at same time seemed to open the door for a compromise. It indicated that an equitable division of the shelf could not be considered bilaterally, but might be discussed in a multilateral context involving all North Sea States.[64]

The Netherlands employed a number of arguments to reject the German proposal. The German coastal configuration was not a special circumstance. As a matter of fact, the Netherlands did not profit from that configuration, but it was rather the position of Denmark which resulted in the relatively limited shelf area of Germany.[65] The Dutch delegation in addition submitted a wide range of arguments: the area of continental shelf would not in itself determine how much resources each State would get; the Netherlands could not be expected to compensate a bigger neighbor endowed with large quantities of natural resources; a concession in respect of Germany would create a precedent in relation to Belgium; and the progress in European integration meant that all member States of the European Community would in a sense profit from the resources of each individual member.[66]

After the German delegation had presented a further justification for its proposal for equal division – equidistance was only a subsidiary principle; Germany as an important coastal State should not be treated differently from its neighbors; it was inequitable that Denmark and the Netherlands, without sharing a land boundary would have a common shelf boundary; the common use of the North Sea justified parity; and continental shelf boundaries might in the future also become relevant for other purposes – the German delegation retreated to a second proposal, which envisaged dividing the shelf in proportion to the length of the coasts of Denmark, Germany and the Netherlands.[67] This second proposal was immediately tabled because the proposal for a division based on parity was not considered tenable. Moreover, this second proposal

---

[63] Received message in cipher (ref. no. 1934) of the DEB to the MFA dated 3 March 1964 [NNA/13].

[64] Minutes German-Dutch negotiations of March 1964 [AA/38], p. 2. Another document indicates that the Dutch delegation may have made the proposal for a multilateral conference during the second day of the negotiations in connection with the rejection of the German proposal in respect of the bilateral delimitation (received message in cipher (ref. no. 1934) of the DEB to the MFA dated 3 March 1964 [NNA/13], pp. 1–2).

[65] Minutes Dutch-German negotiations of March 1964 [NNA/24], p. 3.

[66] Ibid., pp. 2–3; Minutes German-Dutch negotiations of March 1964 [AA/38], pp. 2–3.

[67] Minutes German-Dutch negotiations of March 1964 [AA/38], pp. 3–4; Minutes Dutch-German negotiations of March 1964 [NNA/24], p. 4.

might also be more acceptable to the Netherlands because it would be less prejudicial to it in relation to Belgium.[68]

The second German proposal envisaged that Denmark, Germany and the Netherlands would get continental shelf areas of respectively 40,000, 44,000 and 57,000 square kilometers in direct proportion to their coastal lengths of respectively 245, 273 and 385 kilometers. The German head of delegation urged serious consideration of this proposal, because a solution based on equidistance would be unacceptable. It would be difficult to explain to the general public that despite the good neighborly relations and many statements of European solidarity an equitable solution could not be reached because each State insisted on its supposed rights. The Dutch delegation for its part pointed to the political and psychological difficulties of "giving away" areas, which were perceived to be Dutch. The discussions on the treaties between Germany and the Netherlands to settle issues related to the German occupation in the Second World War had shown that Parliament and public opinion had little willingness to make any concessions to Germany.[69] At the end of the meeting, the German delegation observed that the matter also might be settled through arbitration. Germany in that case could do no worse than equidistance and probably would do better.[70] The parties agreed that they would meet again at the end of March to discuss the starting point of the boundary. The Dutch delegation accepted to consider in the meantime if the principle of proportionality might be included in the deliberations.[71] Both parties indicated they would seriously try to find a satisfactory solution.[72]

5.2.2   The search for a compromise

After the meeting of 3 and 4 March 1964, the Dutch Ministry of Foreign Affairs considered how to move the matter forward. The focus was on the room for compromise and not so much on legal aspects.[73] A discussion on article 6 after a further meeting with Germany, suggests that

---

[68] Note dated 9 March 1964 [AA/32], p. 2.
[69] Minutes German-Dutch negotiations of March 1964 [AA/38], pp. 3–4.
[70] Ibid., p. 4; see also Minutes Dutch-German negotiations of March 1964 [NNA/24], p. 4.
[71] Minutes Dutch-German negotiations of March 1964 [NNA/24], p. 5.   [72] Ibid.
[73] These issues are discussed in note 18/64 dated 6 March 1964 [NNA/13; NNA/21]; note 30/64 dated 10 March 1964 [NNA/24]; and note no. 14 dated 13 March 1964 [NNA/13]. For a discussion see also below.

Riphagen, the legal advisor of the Ministry, remained of the position that the concavity of the German coast did not constitute a special circumstance.[74]

For Riphagen, the main reason to consider a compromise solution, instead of sticking to the equidistance line, was that the absence of agreement would result in continued uncertainty, which would only play into Germany's hand.[75] Germany in any case would never receive less than the equidistance line before the ICJ. Even more importantly, it was to be expected that German companies would start drilling in the area that the Netherlands considered to be its continental shelf. At the moment, the Netherlands had not adopted a legal framework to deal with that eventuality, but even after mining legislation would have been adopted, there might be circumstances in which the Netherlands would be forced to take military action against German companies.[76] In the light of these risks, it was suggested to offer a compromise to Germany. The offer entailed adding a corridor to the German equidistance area, which would extend in a northwest western direction up to the median line with the United Kingdom. This area covered about a third of the area that Germany had claimed on the basis of equal treatment (see Figure 5.1).[77]

The compromise proposal should only be made if a number of conditions were met. First of all, apart from the Netherlands, Denmark would also have to make a sacrifice.[78] As a matter of fact, the figure depicting Riphagen's idea for a compromise offer indicates that Denmark's contribution was to be significantly bigger, but a later figure suggests that Denmark and the Netherlands would offer Germany comparable areas.[79] Other conditions attached to the compromise proposal were that the starting point of the boundary should not be too far to the west (i.e. not beyond the terminus of the line dividing the area of application of the Supplementary Agreement to the Ems-Dollard Treaty), that the boundary should first follow the equidistance line for a considerable distance and that it would have to be established

---

[74] See note 30/64 dated 3 April 1964 [NNA/13], Annex, p. 1.
[75] Note 18/64 dated 6 March 1964 [NNA/13; NNA/21], p. 2.   [76] *Ibid.*, pp. 2–3.
[77] That implied an area of some 15,000 km². The figure is only attached to the copy of *ibid.* included in [NNA/21].
[78] *Ibid.*, p. 2.
[79] See note dated 25 March 1964 [NNA/21], Annex B. The discussion of the proposal does not shed any light on the reasons for the differences in this respect.

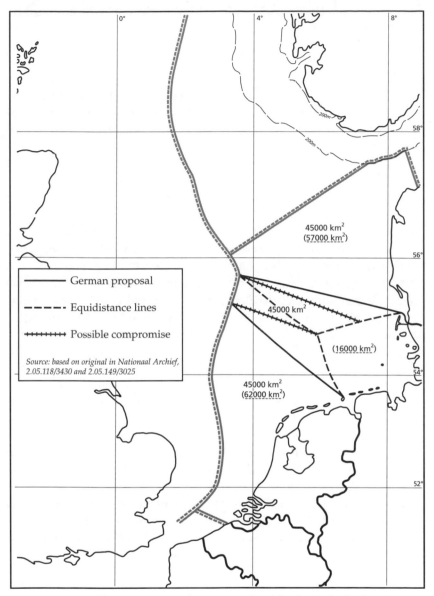

Figure 5.1 Possible compromise solution considered by the Netherlands and a German proposal of division based on parity

beforehand that the area to be offered up would not be rich in resources.[80] Finally, if Germany were to reject the offer, it could not remain on
the table. There was no "objective" justification for it and further discussions would only put the Netherlands at a tactical disadvantage.[81]
After some further items had been added to this shopping list – Belgium
should be prepared to accept that a compromise with Germany would
not form a precedent and the United Kingdom should be prepared to
offer some compensation – [82] the Minister of Foreign Affairs agreed to
the suggestion to submit the matter to the Council of Ministers in these
terms and upon approval contact Belgium, Denmark and the United
Kingdom.[83] However, matters were not to advance to that stage.

In reaction to the proposal of the legal advisor, the political section of the
ministry responsible for relations with Germany proposed a two-stage
approach. This was considered desirable in view of the complexities
involved in arriving at a compromise solution and the sensitivities of public
opinion and the German and Dutch parliaments. It was doubted that the
Dutch parliament would approve an agreement containing the compromise.[84] As a first step a bilateral agreement could address the starting point
of the boundary and delimit its first part in accordance with the equidistance method up to the point at which the boundary would veer to the
west. The remaining part of the boundary could at a later stage be settled in
a multilateral context. This solution would allow oil and gas activities in the
nearshore area and a multilateral approach would prevent the creation of
unwelcome precedents and could be expected to temper reactions in the
press and parliament in the Netherlands.[85]

The idea of an agreement on a partial delimitation and a subsequent
multilateral settlement was presented to the German Foreign Office in
advance of the next round of bilateral talks.[86] Although the Foreign
Office felt that there might be certain risks involved in this approach, it
was concluded that the Dutch government seemed interested in a
compromise and there still seemed to be a window of opportunity
because public opinion had not yet crystallized.[87] The Dutch Embassy
in Bonn informed The Hague in a similar vein. The talks indicated that

---

[80] Note 18/64 dated 6 March 1964 [NNA/13; NNA/21], p. 2.   [81] *Ibid.*
[82] Note 30/64 dated 10 March 1964 [NNA/24].
[83] Note 18/64 dated 6 March 1964 [NNA/13], p. 3, hand-written notes at the bottom.
[84] See note no. 14 dated 13 March 1964 [NNA/13].   [85] See *ibid.*, p. 2.
[86] See note dated 19 March 1964 [AA/38]; send message in cipher of the DEB to the MFA
    dated 19 March 1964 [NNA/21].
[87] Note dated 19 March 1964 [AA/38].

the Foreign Office hoped to reach a partial agreement, which left room for the view that a compromise settlement for the area further offshore would be possible at a later stage.[88] The Dutch offer actually was in line with the Foreign Office's thinking on the further negotiations. It was considered unlikely that the Dutch would accept the German proposal, but a partial agreement for the nearshore area deviating to some extent from equidistance was considered to be possible.[89]

At the second round of talks between Germany and the Netherlands on 23 March 1964, both parties repeated some of the arguments in support of their views on the boundary, but most of the discussion was concerned with the more prosaic issue of the possible starting point of the continental shelf boundary.[90] The Dutch submitted that this starting point should be point A defined in the Supplementary Agreement to the Ems-Dollard Treaty, which was an equidistance point. Point A is on the eastern, German, side of the common area of the Agreement and is on the line the Dutch considered to be the territorial sea boundary. The Dutch, moreover, contested that the Borkumer Riff,[91] a low-tide elevation to the northeast of the German island of Borkum could be used as a basepoint. Germany submitted that the starting point of the boundary was at least point L', the seaward terminus of the western limit of the area of the Supplementary Agreement to the Ems-Dollard Treaty.[92] At the end of the meeting, Meyer-Lindenberg, the head of the German delegation, proposed that as a working hypothesis the starting point of the boundary might be point C" of the Supplementary Agreement. Point C" was the seaward terminus of the line which divided the common area in two. Point C" could be connected by a straight line to an equidistant point. From the latter point, the boundary could follow the equidistance line for some time, after which it would turn westwards to link up with

---

[88] Send message in cipher of the DEB to the MFA dated 19 March 1964 [NNA/21].
[89] Note dated 9 March 1964 [AA/32], p. 2.
[90] For an account of the meeting see Report of the German-Dutch negotiations on the delimitation of the continental shelf in the North Sea on 23 March 1964 [AA/34]; Minutes of the continued discussions between the Dutch and German delegations concerning the delimitation of the continental shelf in the North Sea, held in The Hague on 23 March 1964 [NNA/23] (Minutes continued German-Dutch discussions 23 March 1964 [NNA/23]).
[91] The Borkumer Riff is also known by the name Hohe Riff.
[92] Minutes continued German-Dutch discussions 23 March 1964 [NNA/23], p. 3. The implications of different options to determine the equidistance line are indicated in Figure 5.2.

the equidistance line with the United Kingdom.[93] After the meeting, Meyer-Lindenberg indicated to his Dutch counterpart, Riphagen, that the turning point he had in mind would be located approximately at 55° N.[94] The reference to 55° N would seem to be a, one would be tempted to say Freudian, slip of the tongue. At 55° N, the equidistance line between Germany and the Netherlands already would have reached the equidistance line between Germany and Denmark, implying that a similar deal of Germany with both States would not leave any room for a compromise in the seaward area. Meyer-Lindenberg possibly intended to refer to 54° N, at which point the equidistance line is some 25 to 30 nautical miles from the coast. The terminus of the partial boundary Germany and the Netherlands agreed upon at the end of 1964 ends on that latitude.

Riphagen personally expressed some understanding that an attempt to negotiate an agreement using a different method than equidistance would be undertaken multilaterally. In that case, a bilateral agreement between Germany and the Netherlands would only deal with the land-ward part of their boundary affected by the issue of the starting point.[95] The suggestions of Meyer-Lindenberg in general coincided with the approach the Dutch Ministry of Foreign Affairs had considered inter-nally and that had been discussed with the German Foreign Office prior to the meeting. A next meeting was fixed for 20 April 1964, but even-tually took place at the beginning of June 1964.[96]

5.2.3   The Dutch assessment of the need for a compromise with Germany

In Germany, the possibility of agreement on a partial boundary was generally considered positively. One obvious condition was that such an agreement should not prejudice Germany's access to the center of the North Sea.[97] The Netherlands moreover should commit itself to sup-porting the German position in a subsequent multilateral setting. Agreement on a partial boundary had the advantage that oil companies

---

[93] *Ibid.*, pp. 3–4.   [94] *Ibid.*, p. 4.   [95] *Ibid.*

[96] This delay is explained by the fact the Dutch government was still in the process of establishing a position (see e.g. note dated 15 April 1964 [AA/38]).

[97] See e.g. note dated 6 April 1964 [AA/38], p. 2; note dated 8 May 1964 [B102/318712]; letter of the Minister for the Economy and Transport of Lower Saxony to the FO dated 22 May 1964 [AA/38], p. 2.

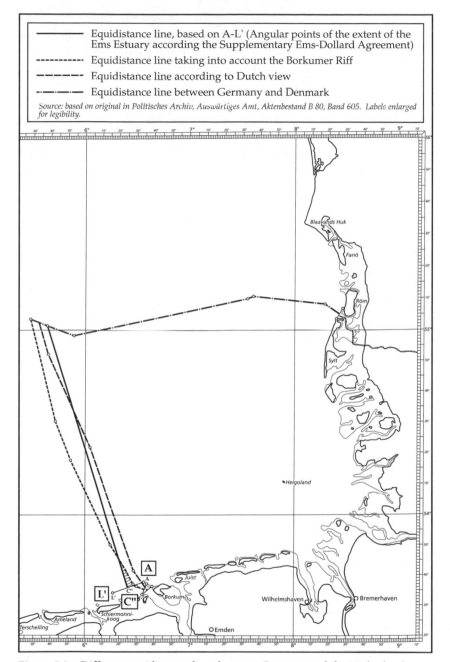

Figure 5.2 Different equidistance lines between Germany and the Netherlands

already could start activities in the interesting nearshore area.[98] Finally, it was agreed that Germany should insist on the Borkumer Riff as a basepoint. This implied the loss of some area nearer the coast, but would lead to a much more advantageous line further offshore.[99] In legal terms, the basepoint was considered to be problematic because it had not been used in earlier agreements and the Netherlands had rejected it for that reason.[100] However, the Borkumer Riff existed and had been included in German nautical charts, making it a valid basepoint under international law. The latter view was also implicitly adopted by the Netherlands in its preparation of the next round of negotiations with Germany. The Netherlands considered that one way to neutralize the dispute over the terminus of the boundary in the territorial sea would be to disregard the views of the parties in that respect altogether, and only take into account the basepoints of both States beyond the disputed area. That approach did require that the Borkumer Riff would also be taken into account.[101]

The German authorities also considered their options in case it might not be possible to reach an agreement with the Netherlands. It was suggested that in the case of that eventuality Germany should contact Denmark and the United Kingdom, should not hesitate to consider the option of arbitration, and should indicate that it might not be able to restrain the oil companies indefinitely – another question being if Germany should allow such activities.[102] As a matter of fact, a telephone call from the legal advisor of the Dutch Ministry of Foreign Affairs concerning the date of a next meeting was used to press home the latter point.[103] At the end of May 1964, the head of the legal department of the German Foreign Office again referred to the interest of the North Sea consortium to press for a rapid resumption of the bilateral talks, which

---

[98] Report of the interdepartmental meeting in the Foreign Office dated 16 April 1964 [AA/38], p. 1.

[99] See e.g. note dated 17 April 1964 [B102/119554], pp. 1–2; note dated 6 May 1964 [B102/318712], pp. 2–4; letter of the Minister for the Economy and Transport of Lower Saxony to the FO dated 22 May 1964 [AA/38], pp. 1–2. This line was also preferred by the North Sea consortium (note dated 6 May 1964 [B102/318712], pp. 2–3).

[100] Note dated 17 April 1964 [B102/119554], p. 2.

[101] Note 30/64 dated 3 April 1964 [NNA/13], Annex, p. 2.

[102] Note concerning the note dated 6 April 1964 to the State secretary (not dated, but not later than 14 April 1964) [AA/38], pp. 2–3; note (not dated; on basis of contents and notes in margin dated between 17 and 21 April 1964) [AA/38], p. 4.

[103] Note dated 15 April 1964 [AA/38].

had been postponed a couple of times by the Netherlands.[104] This may have had little impact. The Dutch Ministry of Foreign Affairs had been informed by a representative of a Shell affiliate that he had heard from a reliable source that the consortium, at least in 1964, would limit its activities to the area east of the equidistance line.[105]

German doubts about the willingness of the Netherlands to reach a mutually acceptable compromise were prompted by press reports on a statement by the Dutch Prime Minister in parliament.[106] Germany got the impression that the Dutch government was apparently seeking to create the impression that under international law the continental shelf had to be delimited by the equidistance principle and was thus seeking to exclude any opening for a compromise with Germany.[107] Moreover, it was feared that the Netherlands might seek to create a *fait accompli* by publishing its note of June 1963 concerning the Dutch view on the delimitation with Germany.[108] The actual pronouncements of the Prime Minister suggest a somewhat different picture.[109] After the Prime Minister gave what can be classified as an impartial rendering of article 6 of the Convention on the continental shelf, he pointed out that Germany maintained that there existed special circumstances, whereas the Dutch government did not see any circumstances justifying a significant departure from the equidistance line. The Prime Minister also pointed to the fact that the issue had multilateral aspects and involved other neighboring States. This clearly left open the possibility that a solution might be worked out in a multilateral context. A subsequent statement by Minister of Foreign Affairs in the First Chamber of Parliament at the end of May 1964 may have suggested that the Netherlands was not willing to depart from its position on the equidistance line.[110] The minister first pointed out that the Prime Minister

---

[104] Received message in cipher (ref. no. 4460) of the DEB to the MFA dated 26 May 1964 [NNA/13].

[105] Note no. 64/64 dated 20 May 1964 [NNA/21]. Discussions between the German authorities and the North Sea consortium in 1964 and 1965 confirm that the consortium was interested in drilling in the area to the north of the Ems as soon as possible, but this only concerned locations near, and generally to the east of, the equidistance line (see further text at footnote 156 and following).

[106] Note (not dated; on basis of contents and notes in margin dated between 17 and 21 April 1964) [AA/38].

[107] *Ibid.*; see also note dated 6 May 1964 [B102/318712], p. 1.

[108] Note (not dated; on basis of contents and notes in margin dated between 17 and 21 April 1964) [AA/38], p. 3.

[109] See K. II (1963–1964), p. 1647. [110] See K. I (1963–1964), p. 946.

had already informed parliament that the government did not see any special circumstances justifying a radical departure from the principle of equidistance contained in article 6 of the Convention on the continental shelf and he then indicated that the government, in searching for a solution of the matter, would be guided by the principles contained in the Convention. The government's interpretation implied that this should result in an equidistance boundary. The questions in parliament and press reports did, in any case, signal that it might be difficult for the government to dispel the impression that the Netherlands would be "giving away" a part of its continental shelf if it compromised with Germany. For instance, a note on the talks with Germany of a staff member of the Dutch Embassy in Bonn observed in passing that an article in a Dutch weekly entitled "Germans want a piece of our seabed – Should the Netherlands give in to save friendship?" in his view "did not exactly broaden the negotiating margin for the Dutch delegation."[111]

A detailed assessment of the options of the Netherlands to deal with the delimitation of the continental shelf in the North Sea – with a focus on Germany – was elaborated by the Ministry of Foreign Affairs in a memorandum for the Dutch Council of Ministers.[112] The memorandum started off with a discussion of article 6 of the Convention on the continental shelf, and identified agreement between the States concerned as the general rule. If the States concerned could not reach agreement, the Convention provided some guidelines, which however did not provide much assistance. The Convention did not provide a definition of special circumstances, did not indicate in which cases special circumstances would justify a boundary different from the equidistance line or on what basis such a boundary should be established.[113] The memorandum submitted that if a court or tribunal were to conclude that special circumstances existed it would have to establish a boundary *ex aequo et bono*.[114]

---

[111] Note dated 25 March 1964 [NNA/21], p. 4. Translation by the author. The original text reads: "Nu gas- en olierijkdom blijkt: Duitsers willen stuk van onze zeebodem. Moet Nederland toegeven om vriendschap te redden? . . . de onderhandelingsmarge voor de nederlandse [sic] delegatie bepaald niet verrruimt [sic]."

[112] Memorandum for the Council of Ministers dated 6 May 1964 [NNA/13]. A first draft of the memorandum, which in all major aspects is identical to the final memorandum, was prepared by the legal advisor of the MFA (see note dated 8 April 1964 [NNA/21]).

[113] Memorandum for the Council of Ministers dated 6 May 1964 [NNA/13], p. 2.

[114] *Ibid.*, p. 10; see also below on the memorandum's assessment of the risks involved in third party dispute settlement.

After this forceful argument on the indeterminate nature of article 6, the memorandum's assessment of the delimitation involving the Netherlands and Germany seems a *non sequitur*: "it was difficult to see why the location and concave configuration of the German North Sea coast actually should be considered a special circumstance."[115] The German wish of a common boundary with the United Kingdom was not considered to have any legal merit, but was thought to be most of all inspired by the fact that the area where the equidistance lines between Denmark, the Netherlands and the United Kingdom converged had promising resource potential.[116]

In view of the significant interest in oil and gas activities in the continental shelf in dispute between the Netherlands and Germany, it was considered necessary to settle this matter within the foreseeable future.[117] The memorandum next set out the options for the nearshore and offshore area. In the former case, this consisted of a solution along the lines discussed in the last round of talks between Germany and the Netherlands. If that solution was not feasible, a joint area might be considered.[118] For the offshore area two approaches were presented: a multilateral conference or submission of the dispute between the Netherlands and Germany to third party dispute settlement. It was advised to adopt the second option, as the first option carried more risks.[119] It was not to be denied that the equidistance method gave the Netherlands a significant part of the North Sea. This area, on the basis of available data, offered the best chances for resource exploitation. Other States might also be inclined to reject the equidistance method in a multilateral setting. Even if that was not the case, it would be likely that other States would hold that it was, first of all, on the Netherlands to offer something to Germany.[120] In view of the limited guidance provided by article 6 and its drafting history and State practice, third party settlement also implied certain risks, but those risks were more limited than those of a multilateral conference, at least if the question were submitted to the ICJ, instead of ad hoc arbitration.[121] The Court could be expected to be more sensitive about the implications of its

---

[115] *Ibid.*, p. 6. Translation by the author. The original text reads "valt niet goed in te zien, waarom de ligging en concave figuratie [sic] van de Duitse Noordzeekust nu eigenlijk een 'special circumstance' zou moeten zijn."

[116] *Ibid.*   [117] *Ibid.*, pp. 6–7.   [118] *Ibid.*, p. 9.   [119] *Ibid.*, pp. 7–8 and 10.

[120] *Ibid.*, pp. 7–8. For a further elaboration of those points see note no. 53 dated 9 April 1964 [NNA/24], pp. 2–3.

[121] Memorandum for the Council of Ministers dated 6 May 1964 [NNA/13], p. 8.

decision. If the sole fact of the concave nature of the German coast would be considered to be a special circumstance justifying a departure from the equidistance line, "delimitations in other parts of the world certainly also would not become easier!"[122] Although it would seem to be fair to assume that the outcome of ad hoc arbitration might be more uncertain, the suggestion that the ICJ would be inclined to reject the German position seems a misapprehension of the interests involved. The assumption that States generally were interested in a predictable rule giving priority to the equidistance principle is misplaced. As the negotiations at the 1958 Geneva Conference had shown, this certainly was not the case. It could be expected that this was a point which would also be highlighted by Germany before the ICJ. It is of course true that the equidistance principles in many instances would provide an acceptable outcome for both States concerned. However, in other cases this clearly was not the case. In that context, it seemed to be overly optimistic to expect that the Court, for the sole reason of not complicating matters, would adopt the view of the Netherlands. The Court, although it operates in a political environment, is required to apply the law, which may entail complexity rather than simplicity.

The memorandum also proposed an interim arrangement for as long as the second part of the boundary would not have been agreed upon. Both parties should, until that time, refrain from exploring or exploiting the area to the north and the west of the terminus of the first part of the boundary.[123] This is an interesting proposal on two counts. It shows that the Dutch Ministry of Foreign Affairs was prepared to entertain the possibility of a provisional arrangement, and it illustrates that the definition of the area of application of a provisional arrangement did not require the formulation of a specific claim in respect of the boundary as a condition for such an arrangement, as had been submitted by the German Foreign Office.[124] The willingness to compromise on this point contrasts with the Dutch government's public position in respect of article 6 of the Convention on the continental shelf. In connection with the approval of ratification of the Convention by parliament, the government had indicated that it did not see any grounds to deviate from

---

[122] *Ibid.* Translation by the author. The original text reads "zouden ook afbakeningen in andere delen van de wereld er niet eenvoudiger op worden!"

[123] *Ibid.*, p. 10.

[124] See text at footnote 54. The FO also took this position in later discussions between the German authorities (see further Chapter 6.6.2).

the principle contained in article 6 (i.e. the equidistance principle) and had taken the position that article 6 implied that in the absence of agreement the equidistance line served as a provisional boundary.[125]

The memorandum was discussed in the Council of Ministers at the end of May 1964. Apparently, the memorandum had been a little too successful in arguing the Dutch case. The council concluded that the Netherlands should hold on to the disputed area and, instead of endorsing the view that as a first step the Netherlands would seek a partial agreement with Germany, the council decided that the Netherlands should first try to reinforce its position through discussions with its other neighbors. This would provide the most effective guarantee against the German claim that special circumstances justified a departure from the equidistance method. Only afterwards might it be considered to go to the ICJ.[126]

The council's rejection of further discussions with Germany put the Ministry of Foreign Affairs in a tight spot. The ministry had repeatedly assured Germany that the bilateral talks would resume as soon as the Council of Ministers would have considered the matter.[127] To set things right, Luns, the Minister of Foreign Affairs, revisited the matter at the council's next meeting, indicating that he now had the possibility to look into all of its aspects and used the existence of the dispute over the boundary in the Ems as a justification.[128] Luns submitted that the determination of that boundary under the Ems-Dollard Treaty required bilateral consultations.[129] Luns at first indicated that the proposed talks would only concern the Ems-Dollard and not the continental shelf

---

[125] Explanatory memorandum to the bill of the Kingdom Act for approval of the Law of the Sea Conventions and the Optional Protocol of signature concerning the compulsory settlement of disputes done in Geneva on 29 April 1958 (K. II (BH) (1963–1964) 7723 (R425) no. 3), p. 7. The bill and explanatory memorandum were submitted to parliament on 30 July 1964 (see K. II (BH) (1963–1964) 7723 (R425) no. 1).

[126] See Minutes of the Council of Ministers dated 22 May 1964, p. 8, item 7 [NNA/2]. The minutes do not make it completely clear that the council for the moment rejected discussions with Germany. However, this is confirmed by a note of the MiFA in which he observed that the council did not agree to those discussions because they might result in concessions from the Netherlands. Luns had not been able to contradict this (note no. 54/64 dated 22 May 1964 [NNA/21]).

[127] See e.g. received message in cipher (ref. no. 4460) of the DEB to the MFA dated 26 May 1964 [NNA/13], p. 1.

[128] The discussion in the council is recorded in Minutes of the Council of Ministers dated 29 May/1 June 1964 [NNA/2], pp. 27–28, item 8a.

[129] Actually, the earlier talks between the Netherlands and Germany had indicated that there was no interest in revisiting the land or territorial sea boundary in the Ems.

beyond, but, upon further questioning, he indicated that article 6 of the Convention required that States should seek to agree on their boundaries through bilateral negotiations. Luns affirmed that the Netherlands would conduct these negotiations on the basis of equidistance. A couple of days after the Council of Ministers had met, the delegations of the Netherlands and Germany met again in Bonn.

5.2.4   The agreement on a partial boundary between Germany and the Netherlands

The negotiations in June 1964 confirmed that Germany and the Netherlands remained at loggerheads over the principles applicable to the delimitation of the continental shelf.[130] Apart from the daring assertion that the drafting history of article 6 of Convention on the continental shelf indicated that the concavity of the German coast was a typical case of special circumstances,[131] the German delegation sig-naled its resolve by stressing the unacceptability of equidistance for the German government, parliament and the general public. A solution along those lines would inflict very serious damage on the German economy and would no doubt burden the good relations between the two States.[132] The Netherlands limited itself to reasserting that it did not see any special circumstances justifying a departure from equidistance and expressed its willingness to submit the dispute to the ICJ. Germany indicated that it was not afraid to go to Court as it could not do worse than equidistance, but Meyer-Lindenberg, the German head of delega-tion, warned that this step might negatively impact the bilateral rela-tions.[133] Moreover, he considered that the reference to agreement between neighboring States in the Convention should be taken to

---

[130] An account of this round of negotiations is contained in Report of the German-Dutch negotiations in Bonn on 4 June 1964 dated 4 June 1964 [AA/38] (Report German-Dutch negotiations 4 June 1964 [AA/38]); note dated 5 June 1964 [B106/53685]; Report Dutch-German discussions 4 June 1964 [NNA/13].

[131] See Report German-Dutch negotiations 4 June 1964 [AA/38], p. 1; note dated 5 June 1964 [B106/53685], p. 2; Report Dutch-German discussions 4 June 1964 [NNA/13], p. 2.

[132] Note dated 5 June 1964 [B106/53685], p. 2; see also Report German-Dutch negotiations 4 June 1964 [AA/38], p. 1; Report Dutch-German discussions 4 June 1964 [NNA/13], p. 2.

[133] Report German-Dutch negotiations 4 June 1964 [AA/38], p. 2; note dated 5 June 1964 [B106/53685], p. 4; Report 1964 Dutch-German discussions 4 June 1964 [NNA/13], p. 3.

imply that a multilateral conference should take precedence over third party settlement.[134] The German delegation also urged consideration of the possibility of a compromise solution at the next meeting after the Dutch government would have been consulted. The Dutch delegation reacted by referring to Dutch public opinion and the fact that the government was obliged to maintain its legal position.[135]

After these preliminaries, the discussions turned to the definition of the first part of the boundary. Both sides agreed without any difficulty on the extension of this line to the parallel of 54° N. The partial boundary started from point C" of the Supplementary Agreement to the Ems-Dollard Treaty and the first line segment linked up with the equidistance line, which also took into account the Borkumer Riff, after some 10 nautical miles.[136] The boundary then continued along the equidistance line up to the parallel of 54° N. After the meeting, some German ministries expressed reservations about seeking to conclude a partial agreement with the Netherlands (or Denmark) and instead preferred to first contact the United Kingdom.[137] In February 1964, the United Kingdom had contacted all of its continental neighbors, including Germany. The identically worded notes had proposed preliminary discussions on a bilateral basis with a view to arriving at a boundary for the continental shelf on the basis of median line principle contained in the Convention on the continental shelf.[138] The Ministry for the Economy did not support the idea to contact the United Kingdom first because a compromise with the Netherlands seemed to be materializing.[139]

After a bilateral working group had filled in the technical details of the agreement on a partial boundary, the delegations completed a final draft and official minutes of the discussions. The latter were among others

---

[134] Report German-Dutch negotiations 4 June 1964 [AA/38], p. 2; note dated 5 June 1964 [B106/53685], p. 4; Report Dutch-German discussions 4 June 1964 [NNA/13], p. 3.

[135] Note dated 5 June 1964 [B106/53685], p. 5.

[136] Report German-Dutch negotiations 4 June 1964 [AA/38], p. 3; Report Dutch-German discussions 4 June 1964 [NNA/13], p. 4.

[137] See e.g. letter of the Minister of Transport to the FO dated 10 June 1964 [AA/34].

[138] For the proposal to Germany see note no. 9 of the British Embassy in Bonn to the Federal Foreign Office dated 14 February 1964 [AA/37]. It later transpired that the United Kingdom had included Germany by mistake in the list of neighbors to be contacted in this connection (see letter of D. D. Brown of the Foreign Office to E. Melville at the British Embassy in Bonn dated 31 August 1964 [FO 371/176338]). On the further contacts between Germany and the United Kingdom see text at footnote 273 and following.

[139] See note dated 30 July 1964 [B102/260131], p. 1.

intended to safeguard the legal positions of both parties.[140] The minutes indicated the absence of agreement on the boundary beyond the parallel of 54° N and that the Dutch delegation considered that the boundary beyond that point should also be based on the principle of equidistance.[141] Germany had agreed to this redaction of the Dutch position, which indicated that at least the last part of the partial boundary was based on equidistance, because the Netherlands had only been prepared to adopt a formulation which would allow pointing out to parliament that equidistance had been employed.[142] On the German side, it was asserted that the absence of a reference to the method of delimitation in the treaty was in itself a success for Germany.[143] It was however clear from the definition of points E1 to E3 of the partial boundary that these were equidistant from the relevant baselines and that only the first segment of the boundary between the outer limit of the area of the Supplementary Agreement to the Ems-Dollard Treaty and point E1 was not an equidistance line. The minutes also recorded that the German delegation intended to convene a conference of the coastal States of the North Sea to achieve an appropriate division of the continental shelf in the sense of article 6(1) and 6(2) of the Convention on the continental shelf. The Netherlands took note of this intention.[144] This fell well short of the original German intention to get Dutch support for the German position at a multilateral conference in exchange for agreement on a partial boundary.

The draft treaty was accepted by both governments without significant discussions and was signed on 1 December 1964.[145] The treaty was kept as simple as possible. Apart from the definition of the partial boundary (article 1), the preamble expressed the consideration that there was an

[140] For the outcome of the bilateral working group see Joint report dated 24 June 1964, with annexes [AA/38]; see also letter of the Dutch ambassador in Bonn to the MiFA dated 25 June 1964 [NNA/21]. For the meeting between the delegations see Report of discussions between delegations of the Netherlands and the Federal Republic concerning the delimitation of the Dutch and German part of the continental shelf in the North Sea, held in The Hague on 14 July 1964 dated 17 July 1964 [NNA/23]; report on application for official trip 1674/64 dated 23 July 1964 [B102/318712].

[141] Minutes of the consultations of a German delegation and a Netherlands delegation (Bonn, 4 August 1964) [NMFA/2].

[142] Note dated 20 August 1964 [B102/318712], p. 3.

[143] Note dated 2 February 1965 [AA/40], p. 2.

[144] Minutes of the consultations of a German delegation and a Netherlands delegation (Bonn, 4 August 1964) [NMFA/2].

[145] Treaty between The Federal Republic of Germany and the Kingdom of the Netherlands concerning the lateral delimitation of the continental shelf near the coast.

urgent need to determine the boundary of the continental shelf in the vicinity of the coast and pointed to the relationship with the Supplementary Agreement to the Ems-Dollard Treaty. Article 2 provides that the treaty does not affect the question of the boundary in the Ems estuary, but that a decision on that boundary also does not affect the present treaty. The treaty was subject to ratification.

Both sides considered that the treaty was the best deal they could have obtained bilaterally.[146] Indeed, both parties could maintain that the line that was agreed upon was to their advantage. The Netherlands had achieved an equidistance line for part of the boundary, whereas Germany obtained the most advantageous version of that line and could now work towards a different approach for the rest of the boundary. The German Foreign Office submitted that the special circumstances which in its view justified a departure from the equidistance line only came into play further offshore,[147] a view that is not unreasonable. Probably most importantly, oil companies wanted to start exploration activities on the continental shelf and the treaty allowed the orderly development of resources on both sides of the partial boundary.[148] At the same time, the partial boundary only provided a solution for a limited period. It was expected it delayed practical problems for a period of between one and two years.[149]

---

[146] See e.g. note no. 81 dated 16 July 1964 [NNA/24], p. 1, which states that the partial boundary "had to be considered as the most favorable possibility for compromise for the Netherlands" (translation by the author. The original text reads "moet worden beschouwd als de voor Nederland gunstigste compromismogelijkheid"); and Proposal to the Cabinet dated 10 August 1964 prepared by the FO [AA/39], which indicated that the treaty adopted the most advantageous version of the equidistance line. In Germany, the states had reservations about the agreement on a partial boundary, but did not actively oppose it (see e.g. letter of the Prime Minister of Lower Saxony to the FO dated 15 October 1964 [AA/39]). The FO rejected the criticisms on the agreement (see e.g. letter of the FM to the Prime Minister of Lower Saxony dated 25 January 1965). Lower Saxony and Schleswig-Holstein abstained in the vote on the approval of the agreement in the Federal Council (Bundesrat).

[147] Note dated 2 February 1965 [AA/40], p. 4; note dated 2 April 1965 [AA/42], p. 3.

[148] The treaty did not contain a provision on a regime for deposits straddling the boundary. The Dutch government in connection with the consideration of the treaty by parliament observed that after the conclusion of the treaty it had become clear that it was desirable to reach agreement on this matter and that it might be addressed in an additional agreement or in the framework of a final agreement on the area to the north of the partial boundary (K. I (BH) (1964–1965), 8093, no. 158a). The latter option would eventually be adopted.

[149] Note no. 108/64 dated 9 November 1964 [NNA/24], p. 1.

In both States, the treaty had to be submitted for approval to parliament. One of the major concerns of the German government was to avoid the impression that the treaty was considered to be a success for Germany, as this might burden further negotiations with the Netherlands and with Denmark.[150] Moreover, Dutch public opinion had to be prevented from turning against ratification.[151] The explanatory memorandum to the bill for approval of ratification was kept brief on purpose.[152] A more detailed explanation, which also set out the benefits for Germany, was provided orally to the relevant parliamentary bodies.[153] It was among others mentioned that the treaty allowed activities in the nearshore area, resulted in the most advantageous boundary line for Germany and did not prejudice Germany's position in respect of the rest of the boundary. The explanatory memorandum submitted to the Dutch parliament in general struck a neutral tone.[154] Although it was emphasized that the government did not see any special circumstances justifying a departure from the principle of equidistance and that the views of the Netherlands and Germany differed fundamentally, the difficulties inherent in the application of the rule contained in article 6 of the Convention on the continental shelf were recognized.[155]

The pending ratification of the delimitation treaty impacted on the German authorities' assessment of planned activities of the North Sea consortium. In November 1964 the consortium expressed an interest in a point (D1), which was located on the German side of the treaty boundary

---

[150] See draft of a circular letter of the FO dated 17 September 1964 [AA/39]; draft of a circular letter of the FO dated 18 January 1965 [AA/40]; note concerning German-Dutch treaty concerning the delimitation of the continental shelf in the vicinity of the coast dated 3 February 1965 [AA/40], p. 1.

[151] See e.g. note concerning German-Dutch treaty concerning the delimitation of the continental shelf in the vicinity of the coast dated 3 February 1965 [AA/40], p. 1; see also note dated 2 February 1965 [AA/40], p. 5.

[152] See draft of a circular letter of the FO dated 17 September 1964 [AA/39]; draft of a circular letter of the FO dated 18 January 1965 [AA/40]. The explanatory memorandum to the Bundestag dated 16 February 1965 is reproduced at Deutscher Bundestag – 4. Wahlperiode, Drucksache IV/3087 (AB-31109–3012/64).

[153] For speaking notes setting out the relevant points see note dated 2 February 1965 [AA/40] and note dated 2 April 1965 [AA/42].

[154] See Explanatory memorandum to the bill of the Kingdom Act for approval of the Agreement between the Kingdom of the Netherlands and the Federal Republic of Germany on the lateral delimitation of the continental shelf in the vicinity of the coast (K. II (BH) (1964–1965) 8093, no. 3).

[155] This concerned the same arguments as set out in the Memorandum for the Council of Ministers dated 6 May 1964 [NNA/13], p. 2; see text at footnote 113.

but west of the equidistance line that did not take into account the Borkumer Riff.[156] Whereas the Ministry for the Economy had objections, the Foreign Office responded favorably, but it did point out that a discovery might complicate the ratification of the treaty.[157] After further consultations between the ministries and the consortium it was agreed to grant permission if the consortium would accept the risks that might be involved.[158] The German authorities did oppose drilling at a location (P1) between the partial boundary and the most unfavorable version of the equidistance line, and another location (K1) that lay between these two versions of the equidistance line to the north of the terminus of the partial boundary[159] because it was considered they might have repercussions for the ratification of the treaty.[160] An additional objection against point K1 as compared to D1 was that it was located in the area that remained in dispute.[161] The consortium considered points K1 and P1 to be of special importance because they would allow establishing if recoverable quantities of natural gas were present. There was a particular urgency to start exploitation because natural gas from the Netherlands was already conquering the German market.[162] The Ministry for the Economy subsequently favored allowing an operation at point P1 prior to ratification of the treaty, but resigned itself to the view of the Foreign Office that a number of questions falling primarily under its competence were involved.[163] The objections

---

[156] Note dated 1 December 1964 [AA/35].

[157] See note dated 6 January 1965 [B102/119556], p. 1; letter of the FO to the Minister for the Economy dated 7 January 1965 [AA/34].

[158] Note dated 2 February 1965 [B102/318712].

[159] The consortium first expressed an interest in point K1 in April 1965 and in point P1 in August 1965. It is not always immediately clear from the documents concerned, which points are under consideration. At times reference is made to D1, P1 or K1, sometimes a point is defined by geographical coordinates or its location in relation to the equidistance lines. A comparison of the various files makes it possible to establish the relationship between these different descriptions.

[160] See e.g. note dated 11 May 1965 [AA/43]; note dated 24 August 1965 [B102/318713]. The consortium had already been informed about the views of the FO in respect of drilling between the partial boundary and the other version of the equidistance line (see note dated 23 December 1964 [B102/318712]). The Foreign Office had indicated that in such a case it would be appropriate to notify the Netherlands (see note dated 23 December 1964 [B102/318712]).

[161] Note dated 11 May 1965 [AA/43], p. 2.

[162] Note dated 3 June 1965 [AA/43], p. 1; note dated 24 August 1965 [B102/318713], p. 1.

[163] Note dated 24 August 1965 [B102/318713], pp. 1–2.

against point P1 were dropped once the treaty had been ratified,[164] but discussions over point K1 between the authorities and the consortium continued. The consortium did not take up the offer of the authorities to sound out the Netherlands concerning point K1.[165] In December 1965, the consortium sought approval for drilling at another point to the north of the terminus of the partial boundary but clearly within Germany's equidistance area. Just as the location at K1, this location might offer the possibility of a finding that would allow obtaining a share of the German gas market.[166] The German authorities opposed drilling at this location because they considered that at the time there again was good hope of a compromise solution. A successful exploratory drilling near the disputed area might complicate the further negotiations.[167] After January 1966 the interest of the consortium in finding gas quickly dwindled and it preferred to await the outcome of the negotiations on the delimitation of the continental shelf.[168]

The Netherlands was not confronted with similar problems, but it also took care to prevent that the approval of the treaty on a partial boundary might be jeopardized. Further steps to delimit the continental shelf boundary to the north of 54° N (the terminus of the partial boundary) before both parliaments would have dealt with the treaty were considered to be decidedly undesirable[169] – quite understandably, as the envisaged partners for a first stage of those talks were Denmark and the United Kingdom, not Germany.[170]

In both countries, parliamentary approval was obtained without any significant difficulties and the treaty was ratified on 17 September 1965 and entered into force the following day.

5.3   The agreement on a partial boundary between Denmark and Germany

Denmark and Germany had a first contact concerning the delimitation of their continental shelf at about the time negotiations between

---

[164] See letter of the Minister for the Economy to the Minister of Transport dated 20 September 1965 [AA/43].

[165] See e.g. note dated 16 September 1965 [B102/119557]; letter of the Minister for the Economy to the FO dated 22 October 1965 [AA/43]. For an explanation of the position of the consortium see note dated 3 June 1965 [AA/43], p. 2.

[166] See note dated 4 January 1966 [AA/46], pp. 1–2.   [167] See *ibid.*

[168] See e.g. note dated 15 April 1966 [B102/260030], p. 2.

[169] See note no. 108/64 dated 9 November 1964 [NNA/24], p. 2.

[170] See further Chapter 5.5.

Germany and the Netherlands started in March 1964. The Danish ambassador in Bonn informed the German Foreign Office that Denmark could not accept the arguments which had been advanced in the German periodical *Die Welt* to justify a division of the North Sea. The Foreign Office indicated that it had not inspired the publication, but at the same time observed that Germany could not accept a delimitation on the basis of equidistance. Rather, a reasonable solution had to be found, which would take into account the needs of Germany as an important North Sea coastal State.[171]

In August 1964, after Germany had reached agreement with the Netherlands on a partial boundary, Denmark was advised that Germany was willing to hold a first round of negotiations in October 1964.[172] Before the negotiations with Denmark took place, the German authorities further considered the legal arguments to reject delimitation on the basis of equidistance. The Foreign Office, for the first time, provided a more detailed view on the applicable law. It was submitted that, as yet, there were no rules of customary law for the delimitation of the continental shelf.[173] Consequently, delimitation should be carried out in accordance with equitable principles, which were also mentioned in the Truman Proclamation and proclamations on the continental shelf of Iran, Saudi Arabia and Bahrain. An equal share for all North Sea States would not be an equitable solution, as it would prejudice the United Kingdom and give too much to Belgium. On the other hand, an equal division between Denmark, Germany and the Netherlands would certainly be more equitable than a delimitation in accordance with equidistance. In that respect reference could be made to the almost equal coastal lengths of Denmark and Germany.[174] Too great a reliance on equidistance was not in accordance with the basis of entitlement to the continental shelf, which was not based on closest proximity to the coast. Equidistance gave too much weight to protruding parts of the coast. In particular at a greater distance from the coast this led to inequitable consequences. The concavity of the German coast moreover constituted a special circumstance in the sense of article 6 of the Convention on the continental shelf.[175] It was submitted that Denmark

---

[171] Note dated 20 March 1964 [AA/34], pp. 1–2.
[172] Note verbale of the Ministry of Foreign Affairs to the DAEB dated 6 August 1964 [DNA/133].
[173] Note dated 29 September 1964 [AA/34], p. 1; see also note dated 7 October 1964 [AA/34]; note dated 9 October 1964.
[174] Note dated 29 September 1964 [AA/34], p. 1. [175] *Ibid.*, pp. 3–4.

in particular should be sympathetic to this argument, as it had itself recognized that the geographical anomaly of the Norwegian Trough did not influence Denmark's delimitation with Norway.[176]

It was acknowledged that it was unlikely that Denmark would agree to a German proposal for equal division of the eastern part of the North Sea. In that case, it would be possible to seek agreement on the delimitation of the first part of the boundary in accordance with equidistance. For this part of the boundary, equidistance was similar to a perpendicular to the general direction of the coast and did not lead to an inequitable result.[177] At an interdepartmental meeting on 12 October 1964, Meyer-Lindenberg of the Foreign Office's legal department, who was to head the German delegation for the negotiations with Denmark, was pessimistic about the possibility that Denmark would accept the German arguments to deviate from equidistance.[178] The meeting accepted that Germany could try to reach agreement on a partial boundary and agreed that the equidistance line in this case was acceptable as it would lead to an equitable solution. The meeting showed a continued preference for a multilateral settlement of the delimitation beyond the partial boundaries with Denmark and the Netherlands.[179] In that sense, the negotiations with Denmark were only an interlude.[180] Whether the possibility of a multilateral settlement still was a realistic option – if it ever had been so – is doubtful.[181]

During their talks on 15 and 16 October 1964, Denmark and Germany had a lengthy exchange concerning their respective positions.[182] Germany submitted that article 6 of the Convention on the continental shelf did not constitute customary law. Moreover, under article 6 equidistance was subsidiary to the requirement of agreement and special circumstances. Germany also observed that the continental shelf was of vital importance to its coastal states and it had to represent their interests

---

[176] *Ibid.*, p. 4.   [177] *Ibid.*, pp. 2, 4–5.
[178] See Report of the interdepartmental meeting dated 12 October 1964 [AA/34], p. 1.
[179] See note dated 29 September 1964 [AA/34], pp. 2–3.
[180] See note dated 7 October 1964 [AA/34], p. 1; see also Report of the interdepartmental meeting dated 12 October 1964 [AA/34], p. 2.
[181] See further Chapter 5.5.
[182] See Report of the negotiations concerning the delimitation of the continental shelf in the North Sea between Germany and Denmark on 15 and 16 October 1964 in Bonn [AA/34] (German report negotiations October 1964 [AA/34]); Report on the negotiations in Bonn on 15th–16th October, 1964, on the demarcation of the continental shelf between Denmark and Germany dated 27 October 1964 [DNA/67] (Danish report negotiations October 1964 [DNA/67]).

in the best possible way. Denmark rejected the German contention that equidistance was not a part of customary law. The principle had been accepted by all North Sea coastal States at the Geneva Conference and the German delegation had not made a reservation. The idea of equal division of the continental shelf had no basis in law and there was no reference whatsoever to it in the Convention on the continental shelf. Denmark also submitted that its bargaining space was severely limited by the agreements it had already reached with the United Kingdom and Norway to delimit their common boundaries by equidistance.[183] The parties also had widely diverging views on what constituted special circumstances under article 6. Germany submitted that the concavity of the German coast was a typical example of special circumstances as had been recognized by "the no doubt unimpeachable witness François."[184] François, who had been closely involved in the drafting of article 6 as the ILC's special rapporteur in the 1950s, in an interview with a Dutch newspaper had, among others, indicated that the 1958 Conference had not provided a definition of special circumstances because it was realized that this would lead to difficulties in many cases and could bog down the discussions on matters of detail. He also submitted that the configuration of the German, Dutch and Danish coasts could be considered to be a special circumstance. The law remained largely indeterminate and in a third party settlement equitable considerations would be a major factor.[185] The Dutch Ministry of Foreign Affairs had not been exactly amused by the interview François had given.[186] After it had been checked that the views of François had been represented correctly, it was agreed to refrain from any reaction and in case of questions simply observe that "the journalist as usual had gotten things wrong."[187]

---

[183] German report negotiations October 1964 [AA/34], pp. 1–2; Danish report negotiations October 1964 [DNA/67], pp. 2–4 and 8.

[184] German report negotiations October 1964 [AA/34], p. 2 (translation by the author. The original text reads "dem gewiß unverdächtigen Zeugen François"); see also Danish report negotiations October 1964 [DNA/67], p. 6.

[185] "Nederlandse en Duitse belangen in botsing; Prof. François bepleit overleg van drie landen" (Dutch and German interests in conflict; Professor François argues for talks between three countries) Utrechtsch Nieuwsblad of 28 April 1964. The GEH had informed the German FO of the article (see letter of the GEH to the FO dated 8 May 1964 [AA/38]).

[186] See note no. 40 dated 5 May 1964 [NNA/13] and letter of the MiFA to the Embassies in Bonn, London, Copenhagen, Oslo and Brussels dated 19 May 1964 [NNA/13], p. 1.

[187] See note 40/64 of the legal advisor to the secretary-general dated 6 May 1964 [NNA/13]. Translation by the author. The original text reads "de betrokken journalist het wel weer verkeerd begrepen zal hebben."

Denmark argued that the drafting history of article 6 pointed out that special circumstances referred to very anomalous configurations and situations, such as islands, sand banks and particular delimitations in the territorial sea.[188] Denmark admitted that the equidistance principle was subsidiary to the duty to negotiate an agreement, but in the absence of agreement, equidistance was the only available principle and negotiations had to focus on how this principle had to be applied.[189] Germany also distinguished its situation from that of opposite coasts. Equidistance might lead to a reasonable result in the latter case, but in the other case it might lead to "quite unreasonable results."[190] Denmark did not provide a clear refutation of this point, apart from remarking that the geographical situation of adjacent and opposite coasts hardly allowed making such a sharp distinction.[191]

After arguing their respective positions, the delegations turned to the practical matter of the starting point of the continental shelf boundary. The selection of a starting point was complicated by the fact that the territorial sea boundary had been defined by reference to a navigational channel which was subject to change. Moreover, there existed different options to define the baselines in the vicinity of land boundary.[192]

In a private meeting of the heads of delegations, Meyer-Lindenberg informed Oldenburg, his Danish counterpart, that he saw no possibility of reaching an agreement on the entire continental shelf boundary in the North Sea, but that he considered a partial delimitation an option.[193] As was observed earlier, Germany accepted that this partial boundary could be based on equidistance. The German proposal for a partial boundary led to some discussion on the Danish side. Among others the Danish concessionaire was against further negotiations on a partial boundary as this might weaken the Danish position in respect of the remainder of the boundary.[194] The Ministry of Foreign Affairs favored negotiations on a partial boundary to manifest a willingness to reach a solution. Sørensen,

---

[188] German report negotiations October 1964 [AA/34], p. 2; Danish report negotiations October 1964 [DNA/67], p. 4.

[189] Danish report negotiations October 1964 [DNA/67], p. 7.   [190] *Ibid.*, p. 5.

[191] *Ibid.*, p. 9.

[192] See also note; Methods for the boundary delimitation according to equidistance and their consequences in the German-Danish boundary area (Sylt-Röm) of the North Sea dated 29 October 1964 [AA/34]; memorandum to department ZB7 dated 12 January 1965 [AA/41].

[193] Danish report negotiations October 1964 [DNA/67], p. 11.

[194] See Summary; Meeting held at the Ministry of Foreign Affairs, Copenhagen on 11 January, 1965, for discussions of the problems relating to the demarcation of

the ministry's legal advisor, in this connection submitted that article 6 of the Convention on the continental shelf indisputably established that the primary aim was delimitation through negotiations.[195] Sørensen considered that it might negatively reflect on Denmark in legal proceedings if it would transpire that it had rejected the conclusion of an agreement, even though Germany had been willing to accept the Danish position, as such a refusal was contrary to the wording and spirit of article 6 of the Convention.[196]

During their second round of negotiations in March 1965, Denmark and Germany refrained from further discussions on their legal positions, but immediately turned to the question of a partial boundary.[197] Notwithstanding the uncertainties surrounding the starting point of the boundary, the parties agreed on a specific point without difficulty. From this point, the boundary was defined as a straight line of some 30 nautical miles running to an equidistance point between a point on the German coast and Blaavands Huk on the Danish coast. The choice for the latter point is explained by the fact that further seaward, because of the presence of the headland of Blaavands Huk, the equidistance line turns in a more southerly direction. Germany considered an equidistance line beyond that point unacceptable and Denmark could not accept that the boundary after this point would not be based on equidistance.[198] A draft agreement, which was prepared by Germany and had been modeled on the agreement between Germany and the Netherlands, was accepted by the Danish side after some minor changes.[199] This draft treaty was approved by both governments, was signed on 9 June 1965 and entered into force on 27 May 1966.[200]

Denmark's continental shelf dated 11 February 1965 [DNA/134] (Summary meeting 11 January 1965 [DNA/134]), p. 3.

[195] *Ibid.*, p. 2.

[196] Note; Delimitation of continental shelf areas in the North Sea. The letter of counselor Thorsen of 21 November 1964 dated 30 November 1964 [DNA/3].

[197] Information on the negotiations is contained in Report on the negotiations in Copenhagen on the 17th–18th March, 1965, on the demarcation of the continental shelf between Denmark and Germany and Recommendation on the next step dated 31 March 1965 (translation) [DNA/67] (Danish report negotiations March 1965 [DNA/67]); Report of the German-Danish negotiations on 17 and 18 March 1965 in Copenhagen [AA/64].

[198] See note dated 29 September 1964 [AA/34], pp. 5–6; Summary meeting 11 January 1965 [DNA/134], pp. 4–5.

[199] Danish report negotiations March 1965 [DNA/67], p. 6.

[200] Treaty between the Federal Republic of Germany and the Kingdom of Denmark concerning the delimitation of the continental shelf near the coast.

During their negotiations in October 1964 and March 1965 Germany and Denmark also discussed the delimitation in the Baltic Sea. Both States agreed that in the Baltic Sea the equidistance principle resulted in an acceptable outcome. A complicating factor in this case was the status of the pre-war boundaries of Germany. The Federal Republic until 1969 took the position that it had the sole right to represent Germany within the boundaries as existed in 1937. In the Baltic Sea, those boundaries extended to former East Prussia, the present day Russian territory of Kaliningrad. Germany wanted to prevent that Denmark or other States would start bilateral delimitation talks with the German Democratic Republic and Poland and was interested in an early agreement with Denmark. In view of the complicated political situation, Germany found it difficult to conclude a formal agreement. As a consequence, the parties agreed that each side would unilaterally determine the equidistance line after consultations with the other party.[201]

5.4   Denmark's bilateral boundaries with Norway and the United Kingdom

Norway proposed Denmark to initiate discussions on the delimitation of their continental shelf at the end of 1963.[202] Norway observed that both States had accepted the principle of equidistance in their national legislation, but that it would be desirable to formally record this in a treaty.[203] The Norwegian proposal led to a reserved reaction on the part of Denmark, which had concluded previously that there was no urgent need to conclude bilateral agreements. It was suggested that Norway was possibly seeking to reinforce its position in relation to the United Kingdom. Although the United Kingdom at the 1958 Conference had accepted that the Norwegian Trough did define the extent of Norway's continental shelf, there were indications that Norway still might not be altogether sure about its position.[204] It was not in the interest of Denmark to take sides in a possible conflict between Norway and the United Kingdom.[205] The Ministry of Foreign Affairs instructed the Embassy in Oslo to play for time and preferred to leave the initiative to

---

[201] See note dated 23 October 1964 [AA/34], p. 3; Danish report negotiations March 1965 [DNA/67], p. 7; note dated 1 April 1965 [AA/41], p. 3.

[202] For the earlier Danish consideration of this delimitation see Chapters 3.2.1 and 4.2.

[203] See Report; Delimitation of the continental shelf between Denmark and Norway dated 23 November 1963 [DNA/2], p. 1.

[204] *Ibid.*, p. 6.   [205] *Ibid.*, p. 8.

the United Kingdom.[206] However, that strategy was short-lived. When the Minister of Foreign Affairs of Norway met with his Danish colleague, Hækkerup, on 29 November, he raised the matter, indicating that Norway (as had been assumed by the Danish Ministry of Foreign Affairs) would like to use a treaty with Denmark in future discussions with the United Kingdom. He reminded that Denmark had been able to use a Norwegian-Danish agreement on agricultural products in negotiations with among others the United Kingdom.[207] Hækkerup in principle wished to accommodate the Norwegian request. Sørensen, the legal advisor of the ministry considered that there was no objection to accommodating Norway and the ministry concluded that there were no objections to bilateral talks with Norway.[208]

Before a meeting between Denmark and Norway took place at the beginning of March 1964, the position of the United Kingdom had been clarified. The United Kingdom had proposed bilateral discussions to all of its continental neighbors concerning the delimitation of the continental shelf in accordance with the equidistance principle. As a consequence, the risk that Denmark, by agreeing on an equidistance boundary with Norway, might be drawn into a conflict between its two neighbors was allayed.[209] The meeting between Denmark and Norway confirmed that they agreed on using the equidistance method and that there were no other points of significant disagreement.[210] Negotiations on a delimitation treaty started a month after these first talks.[211] The technical definition of the equidistance line did not raise any difficulties. The meeting paid considerable attention to two other matters. One concerned the inclusion of a provision on deposits straddling the boundary. Norway favored a clause on the unity of deposits, but Denmark felt that it could not agree to this proposal without first consulting the

---

[206] See letter of the MFA to the Danish ambassador in Oslo dated 28 November 1963 [DNA/2].

[207] See letter of the Danish ambassador in Oslo to the MFA dated 2 December 1963 [DNA/2].

[208] See letter of the MFA to the Ministry of Public Works dated 12 December 1963 [DNA/2].

[209] See also Report on the talks concerning the question of the delimitation of the continental shelf between Denmark and Norway in the Ministry of Foreign Affairs on 5 March 1964 dated 20 March 1964 [DNA/2], p. 2.

[210] For an account of the meeting see *ibid.*

[211] For an account of these negotiations see Report; The question of the shelf dated 16 April 1964 [DNA/2]; Report on the Danish-Norwegian negotiations concerning the delimitation of the continental shelf of the two countries in the Ministry of Foreign Affairs on 16 April 1964 10 a.m. dated 23 April 1964 [DNA/2].

Danish concessionaire.[212] Second, Denmark proposed a confidential exchange of notes in connection with the treaty, which would provide for its revision in case the equidistance principle would not find general application in the North Sea.[213] This was accepted by Norway.[214] At the end of the meeting the parties agreed to aim to approve a provisional text by June and to ratify the treaty by the coming winter.[215]

Denmark and Norway did not even come close to meeting the time frame they had agreed upon in April 1964 for the ratification of their delimitation treaty. The provision on the unity of deposits after some discussion between the Danish authorities and the Danish concessionaire was accepted by the latter,[216] but the proposed exchange of notes on the eventuality of a renegotiation of the boundary treaty proved to be a major stumbling block. Although Norway initially agreed to the exchange of notes, it afterwards persistently worked towards the adoption of the treaty without the exchange of notes taking place. Norway in addition signaled to Denmark that it would seek to extract itself from participation in any future multilateral conference on the delimitation of the North Sea shelf. For instance, in a meeting with the Danish ambassador in Oslo, Evensen, the head of the legal department of the Norwegian Ministry of Foreign Affairs, pointed out that Norway in response to a German invitation would indicate that it had applied the equidistance principle with Denmark and the United Kingdom, implying that Germany would have to deal with those States, not Norway.[217]

Norway started to voice second thoughts about a confidential exchange of notes almost immediately after it had agreed to it. Evensen expressed concerns about the fact that the exchange of notes was not classified as secret. If Germany was to become aware of the exchange of

---

[212] See *ibid.*, pp. 5–7 and 9. One problem in this respect was that the clause proposed by Norway in certain cases would imply an obligation for the concessionaire to negotiate, which it did not have under Danish law (*ibid.*, p. 6).

[213] *Ibid.*, pp. 7 and 10.   [214] *Ibid.*, p. 10.

[215] See Report; The question of the shelf dated 16 April 1964 [DNA/2], p. 5.

[216] See note; "The continental shelf" dated 1 October 1964 [DNA/3], p. 3.

[217] See letter of the Danish Embassy in Oslo to the MFA dated 17 June 1964 [DNA/2], p. 3; (see also letter of the Dutch ambassador in Oslo to the MiFA dated 20 August 1964 [NNA/13], p. 2). Norway also suggested that it would be prepared to participate in a multilateral conference as an observer and in that capacity would strongly support Denmark, the Netherlands and the United Kingdom (letter of the Danish Embassy in Oslo to the MFA dated 9 September 1964 [DNA/3], p. 4).

notes it might pressure Denmark to give up the equidistance principle.[218] Evensen as strongly as possible argued that Denmark could not expect to be compensated by Norway, if Denmark would adopt a conciliatory approach towards Germany. Evensen preferred a gentlemen's agreement, as it would obviate the need for an exchange of notes.[219]

In a subsequent round of negotiations, the heads of delegations bilaterally drafted a new version of the notes, but Norway also indicated that it would only be prepared to depart from the equidistance principle if the United Kingdom would be prepared to do the same. The negotiations resulted in an initialed treaty text, but the text of the notes required further consideration.[220] While the text of the notes was further considered, the Norwegian Ministry of Foreign Affairs had already submitted the treaty to the government for approval without a reference to the planned exchange of notes. Denmark was informed that the government could not approve the treaty with the accompanying notes, as those might create difficulties in respect of Norwegian concessions, which would be located near the equidistance line.[221] In December 1964, the Norwegian Minister of Foreign Affairs brought the matter up, arguing that if Denmark would not conclude a similar deal with the United Kingdom it would be discriminating against Norway. The minister moreover considered that the exchange of notes was no longer required, as all North Sea States except Germany were working towards delimitation on the basis of equidistance.[222] A couple of days later the Danish ambassador was invited to meet with the minister and was informed that there existed further objections against the notes from other ministries.[223] Evensen maintained that he was still prepared to seek an acceptable solution in respect of the exchanges of notes, notwithstanding the

[218] Letter of the Danish Embassy in Oslo to the MFA dated 9 September 1964 [DNA/3], p. 3; see also letter of the Dutch ambassador in Oslo to the MiFA dated 20 August 1964 [NNA/13], pp. 2–3.

[219] Letter of the Danish Embassy in Oslo to the MFA dated 9 September 1964 [DNA/3], pp. 3–4.

[220] See Report on the negotiations concerning an agreement on the delimitation of the continental shelf between Denmark and Norway in Oslo between 6 and 8 October 1964 dated 10 October 1964 [DNA/3], pp. 4–6.

[221] See letter of the Danish Embassy in Oslo to the MFA dated 16 October 1964 [DNA/3].

[222] Telegram in cipher of the Danish Embassy in Oslo to the MFA dated 17 December 1964 [DNA/3].

[223] Telegram in cipher of the Danish Embassy in Oslo to the MFA dated 22 December 1964 [DNA/3]. This again concerned the issue of the uncertainty in relation to concessions located close to the equidistance line.

noticeable dislike of the other ministries and his own minister.[224] The Danish Ministry of Foreign Affairs at that point recognized that existing concessions might complicate the renegotiation of the boundary, but considered that this issue could be addressed in such negotiations.[225] The ministry apparently did not respond to the invitation of Evensen to further discuss the matter,[226] but reassessed its own options.[227] When the exchange of notes had been proposed there still was uncertainty about the delimitation of the North Sea. In the meantime a large measure of agreement had been reached between the neighboring States to delimit their boundaries by equidistance. An arrangement to satisfy the German aspirations involving all of the North Sea had become virtually impossible, particularly after the Netherlands had rejected multilateral negotiations. In the light of all these circumstances, it was considered to be politically and legally unrealistic to leave the boundary between Denmark and Norway undelimited to eventually seek compensation from Norway. On the other hand, having an agreement with Norway based on the equidistance method could contribute to strengthening Denmark's position in relation to Germany. After the idea of an exchange of notes was dropped, Denmark and Norway proceeded with the signature of the bilateral agreement on 8 December 1965, which entered into force on 22 June 1966.[228]

At the end of the 1960s it became clear that the area just to the north of the equidistance line between Denmark and Norway contained significant oil resources. This led to allegations that Minister of Foreign Affairs Hækkerup had given away this area "over a glass of whisky" in the meeting with his Norwegian counterpart on 29 November 1963.[229] It has also been suggested that the exchange of notes between Norway and Denmark was close to being accepted but did not materialize because Hækkerup and the

---

[224] See letter from the Danish Embassy in Oslo to the MFA dated 26 January 1965 [DNA/4A].

[225] See letter from the MFA to the Danish Embassy in Oslo dated 19 January 1965 [DNA/4A].

[226] See letter from the Danish Embassy in Oslo to the MFA dated 22 July 1965 [DNA/1].

[227] The note dated 9 August 1965 containing the initial assessment by the ministry could not be located in the files of the Danish National Archives that were consulted. It is mentioned in a letter from the MFA to the Ministry of Public Works dated 30 August 1965 [DNA/4B]. The following account is based on a note for the MiFA dated 18 October 1965 [DNA/1].

[228] Agreement between the Government of the Kingdom of Denmark and the Government of the Kingdom of Norway relating to the delimitation of the continental shelf.

[229] See N. Bøgh, *Hækkerup* (Aschehoug Dansk Forlag A/S, 2003), p. 7.

Foreign Affairs Committee of the Danish Parliament did not insist.[230] The current review of the negotiations between Denmark and Norway indicates that this view on the matter is not correct.[231] Before Hækkerup and his Norwegian counterpart met and the negotiations were started, the Ministry of Foreign Affairs had made an assessment of the Danish options and concluded that the Danish interests were best served by an equidistance line with Norway.[232] Second, the suggestion that the exchange of notes was close to being accepted is contradicted by the fact that Norway, after an initial acceptance of this idea, did everything to extract itself. Third, Denmark had an interest in an agreement with Norway because this would reinforce its position in relation to Germany and the rejection of the agreement would obviously have weakened Denmark's position. Finally, the criticism of Denmark's acceptance of the equidistance line in relation to Norway is based on the assumption that a better outcome would have been possible. In view of the drafting history of article 1 of the Convention on the continental shelf it was practically excluded that the Norwegian Trough would have been accepted as an alternative boundary under the applicable law.[233] The judgment of the ICJ in the *North Sea continental shelf* cases may have given the impression that things were otherwise. The Court indicated that the Norwegian Trough constituted a break in the continental shelf and observed that the shelf areas in the North Sea beyond the Trough "cannot in any physical sense be said to be adjacent to [the Norwegian coast], nor to be its natural prolongation."[234] However, this finding of the Court was not based on a thorough analysis and among others ignores the drafting history of article 1 of the Convention on the continental shelf. As the areas beyond the Norwegian Trough can be considered to be a part of the continental shelf of Norway, equidistance presents itself as a reasonable method of delimitation if the findings of the Court on the rules of the delimitation of the continental shelf in its judgment in the *North Sea continental shelf* cases were to be applied to it. Without going into too much detail, it can be noted that the natural prolongations of the southwestern coast of Norway and western coast of Denmark overlap up to the median line with the United Kingdom. The equidistance line divides this area more or less equally and cuts off part of the projections of both States in this area.[235]

---

[230] See Laursen, *Small Powers*, pp. 76–77.   [231] See also Bøgh, *Hækkerup*, pp. 7–12.

[232] See Chapter 4, text at footnote 20 and following.

[233] See further Chapters 3.2.1 and 3.3.   [234] [1969] ICJ Reports, p. 32, para. 45.

[235] See also Chapter 9 for a discussion of the judgment of the Court in the *North Sea continental shelf* cases.

The United Kingdom had proposed Denmark to start talks on the delimitation of their continental shelf on the basis of the median line principle in February 1964.[236] Denmark found this proposal acceptable, but waited to inform the United Kingdom until after it had addressed its delimitation with Norway.[237] Denmark considered that an agreement with the United Kingdom to employ the equidistance principle would provide support for Denmark against Germany.[238] The talks with the United Kingdom initially focused on the technical definition of the equidistance line, on which agreement was reached in April 1965. Subsequently, the parties worked out an arrangement for transboundary deposits.[239] Neither issue led to any particular difficulties. The resulting bilateral delimitation agreement was signed on 3 March 1966 and entered into force on 6 February 1967.[240]

5.5   The broader framework of the bilateral negotiations

Thus far the analysis in this chapter has, to a large extent, focused on the negotiations concerning the delimitation of bilateral boundaries between the individual North Sea coastal States. This analysis points to the potential interactions between their bilateral delimitations and the desirability or not of discussing the delimitation of the continental shelf in a broader framework. The present section takes a closer look at the consideration of these issues by the North Sea coastal States.

The Netherlands started to consider the broader context of its bilateral delimitation with Germany after the first bilateral talks at the end of March 1964. Immediately after the meeting with Germany, it was

---

[236] See aide mémoire of the British Embassy to the MFA of Denmark received on 24 February 1964 [DNA/2].
[237] See note verbale of the MFA of Denmark to the British Embassy in Copenhagen dated 24 March 1964 [DNA/2]; Report on the talks concerning the question of the delimitation of the continental shelf between Denmark and Norway in the MFA on 5 March 1964 dated 20 March 1964 [DNA/2], p. 7.
[238] See note; The continental shelf dated 10 April 1964 [DNA/2], p. 2. The author of the note expressed some concern that Denmark would not be able to secure support from the United Kingdom in time. This concern was probably caused by the fact that the United Kingdom had also proposed negotiations on the delimitation of the continental shelf to Germany (see Chapter 4, text at footnotes 137 and following).
[239] See, e.g. note; The continental shelf dated 19 July 1965 [DNA/6], p. 1.
[240] Agreement between the Government of the United Kingdom of Great Britain and Northern Ireland and the Government of the Kingdom of Denmark relating to the delimitation of the continental shelf between the two countries.

decided to contact Denmark,[241] which found itself in a similar position as the Netherlands.[242] The talks revealed a large measure of agreement. Denmark informed the Netherlands that it would only abandon the position that the continental shelf of the North Sea had to be delimited by equidistance "at gunpoint."[243] The talks led to the understanding that each party would stick to the equidistance line in their bilateral dealings with Germany.[244] Denmark hinted at the fact that it hoped that the Netherlands would not make any concessions to Germany that might prejudice the Danish position.[245] Denmark would, if necessary, be prepared to participate in a multilateral conference to discuss the German desiderata.[246] However, whereas the Dutch Ministry of Foreign Affairs at this point in time considered the idea of a multilateral meeting as an option to reach a compromise with Germany,[247] Denmark rather viewed it as a possibility to attract support from the other coastal States for the equidistance principle.[248]

The talks between Denmark and the Netherlands initialed their close cooperation concerning the delimitation of the continental shelf that was to last until final agreement had been reached with Germany. Already in the first phase of the negotiations with Germany, which were purely bilateral in nature, they informed each other of the progress and exchanged relevant documents.[249] However, this certainly did not imply that all specific information was shared,[250] and at times the interests of both States did not run parallel.

---

[241] See received message in cipher of the MFA to the DEC dated 25 March 1964 [NNA/20].

[242] One source suggests that the Netherlands took the initiative for these contacts after Shell had inquired with A.P. Møller whether the Danish concessionaire was interested in a direct contact between the Danish and Dutch governments (see note dated 2 April 1964 [DNA/2], p. 1). There are no indications that the Netherlands contacted Denmark because of a suggestion of Shell. The Netherlands had been considering its options in relation to Germany and in that context had concluded that a compromise solution would also require the involvement of Denmark (see note 18/64 dated 6 March 1964 [NNA/13; NNA/21]).

[243] See note no. 23 dated 8 April 1964 [NNA/13], p. 2.

[244] See ibid., pp. 1–2; Report of discussion at the request of the Netherlands concerning the Dutch views on the delimitation of the continental shelf at the Ministry of Foreign Affairs at 10.30 on 3 April 1964 dated 17 April 1964 [DNA/2] (Danish report discussion April 1964 [DNA/2]), p. 4.

[245] See note no. 23 dated 8 April 1964 [NNA/13], p. 2.

[246] See ibid.; Danish report discussion April 1964 [DNA/2], p. 4.

[247] See Chapter 5.2.2.   [248] Danish report discussion April 1964 [DNA/2], pp. 3–4.

[249] See e.g. note no. 43 dated 8 May 1964 [NNA/13].

[250] See e.g. letter of F. van Raalte of the MFA to the Dutch ambassador in Copenhagen dated 6 November 1964 [NNA/13].

The Netherlands initially took a more cautious approach in relation to the other coastal States of the North Sea than Denmark. In April 1964, the United Kingdom had inquired whether the Netherlands already had an answer to the British note of February 1964, which had suggested bilateral talks concerning the delimitation of the continental shelf.[251] The Dutch Embassy in Bonn had urged contact with the United Kingdom in order to clarify the British position at about the same time.[252] In February 1964, the United Kingdom had contacted all of its other continental neighbors, including Germany. The identically worded notes had proposed preliminary discussions on a bilateral basis with a view to arriving at a boundary for the continental shelf. It was proposed that the dividing line should be calculated on the basis of the median line principle contained in the Convention on the continental shelf.[253] Germany had immediately reacted positively to this proposal.[254] The Dutch Embassy in Bonn pointed out that staff of the German Foreign Office in enquiring about the date for the next round of negotiations with the Netherlands had repeatedly hinted that Germany might respond to the British proposal to start delimitation talks.[255]

Riphagen, the legal advisor of the Dutch Ministry of Foreign Affairs urged against opening discussions with the United Kingdom as he considered that the Netherlands in that case also would have to bring up the issue of a multilateral conference. All signs indicated that the United Kingdom would flat out reject such a conference. If Germany would get to know about such a rejection, it would even be less inclined to conclude a partial delimitation agreement with the Netherlands.[256] Instead, the United Kingdom could be informed about the Dutch views concerning the delimitation in the North Sea and the willingness of the Netherlands to enter into bilateral negotiations at a later date.[257] The United Kingdom was informed accordingly.[258]

---

[251] See note no. 53 dated 9 April 1964 [NNA/24], p. 1.
[252] See send message in cipher of the DEB to the MFA dated 7 April 1964 [NNA/21].
[253] For the proposal to Germany see note no. 9 of the British Embassy in Bonn to the Federal Foreign Office dated 14 February 1964 [AA/37].
[254] See note dated 17 February 1964 [AA/37]; note verbale of the Foreign Office to the British Embassy in Bonn dated 26 February 1964 [AA/37]. For further developments in respect of this British initiative see text at footnote 264.
[255] See send message in cipher of the DEB to the MFA dated 7 April 1964 [NNA/21].
[256] Note 31/64 dated 15 April 1964 [NNA/13], p. 2; see also *ibid.*, p. 3.   [257] *Ibid.*
[258] Letter of M. J. Wilmhurst of the British Embassy in The Hague to C. Samuel of the FO dated 22 May 1964 [FO 371/176337].

After its first meeting with Denmark in 1964, the Netherlands abandoned the idea that a multilateral conference would be an option to work out a compromise solution with Germany. Denmark had indicated that it did not favor that approach and the Netherlands, after its talks with Denmark, moreover became aware that the other North Sea States were well-advanced in concluding bilateral agreements based on the equidistance method.[259] In that light, the Netherlands came round to the view that it should not wait passively for Germany to take the initiative for a multilateral conference.[260] Shortly after the discussions with Germany on a partial boundary had been finalized, the Dutch Embassies in Brussels, Copenhagen, London and Oslo were instructed to convey to the local authorities that the Netherlands rejected the convening of a multilateral conference. It had to be prevented that the Netherlands might be forced into having to accept participation in such a conference.[261] Indeed, the Netherlands stood little to gain from a conference. At best the other States might argue that this matter only concerned Germany and its two direct neighbors. At worst, pronouncements of the other States might weaken the position of the Netherlands and Denmark. Denmark did not take any similar initiative. Denmark was already negotiating bilateral agreements with Norway and the United Kingdom based on the equidistance principle. Officials of both governments had informed Denmark that they had indicated to Germany that they did not consider that they had a continental shelf boundary with Germany.[262]

All other North Sea coastal States shared the Dutch view that a German initiative for a multilateral conference should be rejected.[263] The United Kingdom in addition informed the Netherlands that the

---

[259] See letter of the Dutch ambassador in Copenhagen to the MiFA dated 22 April 1964 [NNA/13].

[260] Note no. 60 dated 9 June 1964 [NNA/13].

[261] Original of letter of the MiFA to the Dutch ambassador in Brussels/London/Oslo/Bonn/Copenhagen dated 18 August 1964 [NNA/13].

[262] See Report dated 14 August 1964 [DNA/2]; letter of the Danish Embassy in London to the MFA dated 22 August 1964 [DNA/3].

[263] See original of letter of the MiFA to the Dutch ambassador in Brussels/London/Oslo/Bonn/Copenhagen dated 18 August 1964 [NNA/13], p. 2; letter of the Dutch ambassador in Oslo to the MiFA dated 20 August 1964 [NNA/13], pp. 2–3; letter of the Dutch chargé d'affaires in London to the MiFA dated 8 September 1964 [NNA/13]; letter of W. N. Hellier-Fry of the FO to M. J. Wilmhurst of the British Embassy in The Hague dated 21 September 1964 [FO 371/176338], p. 1; letter of the Dutch chargé d'affaires in Brussels to the MiFA dated 17 September 1964 [NNA/13], p. 1.

German impression that the British note of February 1964 had suggested that Germany and the United Kingdom shared a boundary was mistaken.[264] The British Embassy in Bonn had told the German government that talks on a boundary could only be started after the boundaries of Germany with the Netherlands and Denmark would have been agreed upon and those boundaries indicated the existence of a British-German boundary.[265] The suggestion that the United Kingdom might be involved in a multilateral conference once Germany had concluded its initial talks with Denmark and the Netherlands was rejected.[266]

In early September 1964, the United Kingdom sought to speed up the conclusion of an agreement with the Netherlands. The United Kingdom indicated that it was first of all interested in a delimitation with the Netherlands because major oil companies had been urging the issuance of concessions.[267] Just over two weeks later, the British Embassy in The Hague advised the Ministry of Foreign Affairs that the British government was interested to start bilateral talks on the delimitation of the continental shelf as soon as possible.[268]

Germany still had a more optimistic view of the possibility of the convening of a multilateral conference in 1964. During the discussions between Germany and the Netherlands in the middle of July 1964, Germany had gotten the impression that the Netherlands was remarkably conciliatory in respect of the idea of a multilateral conference.[269] In view of the Dutch thinking on the matter at this time, it seems more probable that the Netherlands rather did not consider it necessary to outrightly reject a multilateral conference and risk unsettling or delaying the agreement on the partial boundary.

Germany did not take an initiative for a multilateral conference immediately after it had successfully concluded its negotiations on a partial boundary with the Netherlands in August 1964. The Foreign Office preferred to first reach an agreement with Denmark along similar

---

[264] Letter of the Dutch chargé d'affaires in London to the MiFA dated 8 September 1964 [NNA/13], p.1; see also text at footnote 251 and following.

[265] Letter of the Dutch chargé d'affaires in London to the MiFA dated 8 September 1964 [NNA/13], pp. 1–2. The British government had communicated its view to Germany through a note of the British Embassy in Bonn dated 3 September 1964 [AA/37].

[266] Letter of D. N. Royce of the British Embassy in Bonn to D. D. Brown of the FO dated 4 September 1964 [FO 371/176338].

[267] Letter of the Dutch chargé d'affaires in London to the MiFA dated 8 September 1964 [NNA/13], pp. 1–2.

[268] Note no. 110 dated 25 September 1964 [NNA/17] p. 2.

[269] Report on application for official trip 1674/64 dated 23 July 1964 [B102/318712], p. 3.

lines as the compromise with the Netherlands beforehand.[270] It may be asked why this approach, which implied a further deferral of a multilateral conference, was preferred.[271] Contrary to the area to the north of the coasts of the Netherlands and Germany, there were no activities planned in the area to the west of the Danish-German coast. Most likely, it was considered that Denmark might reject a multilateral conference if prior bilateral consultations with Germany had not taken place. In March 1964 Germany had already promised Denmark that it would be contacted for bilateral discussions later on.[272]

Contacts with the United Kingdom in September 1964 made it clear to Germany that there was no interest, for the moment, in talks on a common continental shelf boundary.[273] The British note followed inquiries from Denmark and the Netherlands concerning the British position. The United Kingdom had also been informed that the head of the German delegation had been making play with the British offer to negotiate a boundary in the negotiations with the Netherlands and Denmark.[274] In considering a reply to the British note, the German Foreign Office cautioned against criticizing the British position because Germany did not have an interest in antagonizing the United Kingdom. It might be in the interest of Germany to argue that an equidistance line between the continent and the United Kingdom did constitute an acceptable outcome of a delimitation.[275] Under that approach the United Kingdom would be assured that it would not have to make any concessions, while Germany could use the general configuration of the North Sea to support its claim for a larger share of the continental shelf. This argument would eventually play a significant role in Germany's pleadings in the *North Sea continental shelf* cases.

---

[270] See e.g. Proposal to the Cabinet of 10 August 1964 prepared by the FO [AA/39], p. 2.
[271] The FO also had taken that position previously (see e.g. note dated 17 April 1964 [B102/119554], p. 3).
[272] Note dated 20 March 1964 [AA/34], p. 1.
[273] Note verbale of the British Embassy in Bonn to the FO dated 3 September 1964 [AA/37].
[274] See letter of D. D. Brown of the FO to E. Melville at the British Embassy in Bonn dated 31 July 1964 [FO 371/176338]; letter of the FO to the British Embassy in Copenhagen dated 11 August 1964 [FO 371/176337]. The reports on the meetings between Germany and the Netherlands indicate that Germany was not giving particular significance to the British offer (see Report Dutch-German discussions 4 June 1964 [NNA/13], pp. 3–4; see also letter of D. N. Royce at the British Embassy in Bonn to D. D. Brown of the FO dated 4 September 1964 [FO 371/176338]).
[275] See note dated 17 September 1964 [AA/37].

During the first round of bilateral talks between Denmark and Germany on 15 and 16 October 1964, Germany only briefly mentioned the possibility of a multilateral conference to discuss the continental shelf delimitations in the North Sea as a whole, giving the impression that this no longer was an important concern for Germany.[276] The Danish delegation indicated that a prerequisite for Denmark's participation in such a conference would be the participation of all North Sea States, including Norway and the United Kingdom.[277] This Danish position practically excluded the possibility of a multilateral conference.

After its consultation of the other North Sea States in the summer and fall of 1964, the Netherlands started planning a strategy to arrive at the delimitation of its entire continental shelf. The Ministry of Economic Affairs was interested in an early settlement in connection with the adoption of implementing legislation of the Mining Act continental shelf, which was before parliament.[278] The implementing legislation would allow starting the development of the oil and gas resources on the Dutch shelf and the ministry wanted to be able to establish a concession policy for the entire Dutch shelf.[279] This came as somewhat of a surprise for the Ministry of Foreign Affairs. It had assumed that the partial agreement with Germany, which had been concluded in August 1964, would have prevented any practical difficulties resulting from the absence of a boundary for the area further north for one or two years.[280] Seeing itself confronted with the request of the Ministry for Economic Affairs, the Ministry of Foreign Affairs concluded that it was in any case undesirable to resume the negotiations with Germany on the delimitation of the second part of the continental shelf boundary before the agreement on the partial delimitation would have been approved by Parliament.[281] Since it was not expected that this agreement would enter into force before the spring of 1965, it was proposed to suspend further negotiations with Germany until after the Netherlands would have concluded delimitation agreements with Belgium, Denmark and the United Kingdom.[282] An additional advantage of this negotiating scheme was that, as it was to be expected that the equidistance principle

---

[276] Report on the negotiations in Bonn on the 15th–16th October, 1964, on the Demarcation of the Continental Shelf between Denmark and Germany dated 27 October 1964 (translation) [DNA/67], p. 12.

[277] Ibid.

[278] See letter of the MiFA to the Prime Minister dated 28 January 1965 [NNA/24], p. 1; note no. 108/64 dated 9 November 1964 [NNA/24], p. 1.

[279] Ibid.   [280] Ibid.   [281] Ibid., p. 2.   [282] Ibid.

would provide the basis of these agreements, it would be possible to argue more forcefully in relation to Germany that the generally accepted delimitation principle of equidistance also had to be applied to their entire boundary.[283] In briefing the Council of Ministers on the proposed scheme, the Minister of Foreign Affairs did point out that it was to be expected that the further delimitation with Germany was likely to lead to great difficulties. As far as the wish to adopt implementing legislation for the Mining Act continental shelf by the middle of 1965 was concerned, it had to be realized that Germany could be expected to reject the Dutch position. This would lead to an impasse and uncertainty over a significant part of their common boundary.[284] The Council of Ministers agreed to the negotiating scheme proposed by the Minister of Foreign Affairs.[285] The Council did not consider how to deal with Germany if negotiations were not successful.

The Dutch Ministry of Foreign Affairs had discussed its proposed negotiating scheme with Denmark even before the Council of Ministers had agreed to it. The ministry wanted to ensure that Denmark, to the largest extent possible, would follow the same course of action as the Netherlands as a common approach would reinforce the position of both States in relation to Germany.[286] The Netherlands and Denmark met in January 1965 to discuss their shelf delimitation policy. Oldenburg, the deputy director-general of the legal and political department of the Danish Ministry of Foreign Affairs, was very much interested in the approach presented by the Netherlands to confront Germany with the common front of the other North Sea States and indicated that Denmark would probably be willing to adhere to the timetable proposed by the Netherlands. However, Denmark should first conduct further negotiations with Germany. The legal advisor of the ministry considered that the Convention on the continental shelf required "to go through the motions" of negotiations.[287]

On the initiative of the Netherlands another meeting with Denmark took place at the end of January of 1965 because the Dutch felt that the Danish authorities were somewhat uncertain and hesitant.[288] In this

[283] Ibid.; letter of the MiFA to the Prime Minister dated 28 January 1965 [NNA/24], p. 4.
[284] Ibid.   [285] Minutes of the Council of Ministers dated 12 February 1965 [NNA/3].
[286] Letter of the MiFA to the ambassador in Copenhagen dated 29 December 1964 [NNA/17].
[287] See letter of the ambassador in Copenhagen to the MiFA dated 12 January 1965 [NNA/40].
[288] Note no. 12 dated 5 February 1965 [NNA/26], p. 1.

meeting, the Dutch representative, Van Raalte, sought to bring across the resolve of the Netherlands by, among others, mentioning that the Netherlands would not hesitate to unilaterally define the remainder of its continental shelf boundary with Germany.[289] Whether Denmark actually had less resolve than the Netherlands is questionable. As Oldenburg, who headed the Danish delegation, observed, the positions of the Netherlands and Denmark largely coincided, but Denmark had not advanced to the same stage in its negotiations with Germany. Further negotiations were considered to be desirable because of general political considerations and the fact that they should prevent Germany being able to accuse Denmark, in proceedings before the ICJ, that it had not sought to reach an agreement through negotiations.[290] Denmark moreover was interested in a continental shelf boundary with Germany in the Baltic Sea.[291] Oldenburg pointed out that the opening of negotiations between Denmark and the Netherlands to delimit their continental shelf would be proof to Germany that its neighbors had abandoned the idea of further meaningful negotiations.[292] Or, as Oldenburg argued during an internal meeting, Germany would certainly see this as something akin to an ultimatum.[293] Denmark expected to be able to comply with the Dutch wish to start negotiations by May 1965.[294]

Denmark and Germany conducted a further round of negotiations in March 1965. As was set out in Chapter 5.3, during these negotiations they concluded an agreement on a partial boundary in the North Sea and a common approach to the delimitation in the Baltic Sea. In view of this large measure of agreement, Germany was informed that since Denmark considered that the continental shelf in the North Sea had to be divided by equidistance, Denmark shared a continental shelf boundary with the Netherlands. Denmark had not yet started negotiations with the Netherlands, but it was quite possible that they would commence in the near future.[295] Truckenbrodt, the head of the German delegation,

---

[289] *Ibid.*, p. 2.

[290] Résumé; Meeting in the Ministry of Foreign Affairs of 29 January 1965 concerning the delimitation of the continental shelf under the North Sea: The Netherlands continental shelf policy dated 12 February 1965 [DNA/4A] (Résumé meeting 29 January 1965 [DNA/4A]), p. 3.

[291] Note; The continental shelf; Line of demarcation between the Danish and German shelf areas (translation) [DNA/67] p. 10.

[292] Résumé meeting 29 January 1965 [DNA/4A], p. 3.

[293] Summary meeting 11 January 1965 [DNA/134], p. 5.

[294] Résumé meeting 29 January 1965 [DNA/4A], p. 3.

[295] Danish report negotiations March 1965 [DNA/67], pp. 7–8.

apparently was surprised by the idea of early Danish-Dutch negotiations.[296] Germany probably had not yet entertained this option and only seemed to be considering the possibility of further bilateral or multilateral negotiations with its neighbors or judicial settlement.[297] Arguing against early Danish-Dutch negotiations, Truckenbrodt set out that Germany had not ratified the Convention on the continental shelf and did not recognize the equidistance principle for the delimitation between the three States. Germany had expected further negotiations with Denmark and the Netherlands. If Denmark and the Netherlands would conclude an agreement it would be understandable that Germany had to react immediately. Germany still contemplated a multilateral conference as a next step and the matter could also be submitted to compulsory dispute settlement. However, it was most likely that Germany would invite Denmark and the Netherlands for trilateral negotiations at an early date.[298]

5.6   The bilateral boundaries of the Netherlands with Belgium and the United Kingdom and the boundary between Denmark and the Netherlands

5.6.1   Introduction

After an agreement on a partial boundary with Germany had been reached, the Netherlands turned its attention to the delimitation with its other neighbors. With Belgium, the first contacts were made in the latter half of 1964, when the Netherlands sounded out its neighbors in respect of their views on the prospect of a German initiative for a multilateral conference on delimitation of the North Sea. At that time the Netherlands had also contacted the United Kingdom. These contacts pointed out that both States rejected the idea of a multilateral conference and supported the application of the equidistance principle to delimit their boundaries with the Netherlands. Denmark and the Netherlands started to consult each other on the delimitation of the continental shelf in the North Sea in the first half of 1964. The negotiation of a

---

[296] *Ibid.*, p. 8; note dated 19 March 1965 [DNA/4A]; letter of the Dutch ambassador in Copenhagen to the MiFA dated 19 March 1965 [NNA/40], p. 2.

[297] See note dated 3 March 1965 [AA/43], p. 2.

[298] Danish report negotiations March 1965 [DNA/67], pp. 8–9; note dated 19 March 1965 [DNA/4A]; letter of the Dutch ambassador in Copenhagen to the MiFA dated 19 March 1965 [NNA/40], p. 2.

delimitation agreement started after Denmark, like the Netherlands before it, had reached agreement with Germany on a partial boundary in the North Sea and the entire boundary in the Baltic Sea.

5.6.2   The bilateral boundary between the Netherlands and Belgium

Although Belgium supported the application of the equidistance principle, there was one matter that had the potential to complicate a continental shelf delimitation. The Netherlands claimed a historic title to the Wielingen. This is a navigational channel giving access to the mouth of the western branch of the Scheldt River.[299] This claim was rejected by Belgium. A territorial sea boundary between Belgium and the Netherlands, taking into account the Dutch claim, would intersect the outer limit of the 3-nautical-mile territorial sea to the northwest of Blankenberge and would be located more than 10 nautical miles west of an equidistant starting point of a continental shelf boundary. A linkage of the two delimitations would pose the risk of jeopardizing any chance of the speedy conclusion of the negotiations on the delimitation of the continental shelf.

The internal consideration of the impact of the Wielingen-issue on the continental shelf delimitation with Belgium was, from the start, focused on finding a practical solution. The Ministry of Economic Affairs had been looking at this matter in connection with the definition of the scope of application of the mining legislation applicable in the territorial sea and the draft mining legislation for the continental shelf. The problems which had been encountered in defining the lateral boundary in relation to Belgium were also considered by the political section of the Ministry of Foreign Affairs. It was suggested to simply draw the continental shelf boundary ignoring the Dutch claim to the Wielingen. It was acknowledged that this certainly would not reinforce the position of the Netherlands in respect of the Wielingen, but the author of the memorandum "did not see another solution."[300] Riphagen, the legal advisor of the ministry, who had been asked for an opinion, explained that from a legal perspective it was not necessary to link the matter of the Wielingen and the continental shelf delimitation. The claim to the Wielingen was

---

[299] See also Chapter 2, text at footnote 18.
[300] Note no. 110 dated 25 September 1964 [NNA/17], pp. 1–2. Translation by the author. The original text reads: "zie echter geen andere oplossing."

only concerned with the delimitation of the territorial sea and left the baselines from which the breadth of the territorial sea was measured unaffected and as a consequence the matter of the Wielingen was also irrelevant for the delimitation of the continental shelf.[301] Riphagen continued that the matter of the Wielingen could become relevant if the Netherlands would like to argue that the non-equidistant boundary in the territorial sea constituted a special circumstance in the sense of article 6 of the Convention on the continental shelf justifying a departure from the equidistance line. That argument was, however, not strong, even if the claim over the Wielingen would be indisputable. The exploitation of the continental shelf was not a natural consequence of the historic use of the territorial sea. In addition, it had to be kept in mind that Belgium in accordance with the equidistance principle already had a very small part of the continental shelf.[302] Riphagen concluded that accepting an equidistance line for the continental shelf would not prejudice the Dutch claim to the Wielingen. A non-prejudice clause might be included in an agreement on the delimitation of the continental shelf. If a speedy conclusion of the continental shelf delimitation with Belgium was considered desirable it was preferable to seek to accomplish this without diverging from the equidistance principle.[303] That approach was also obviously in line with the general shelf delimitation policy of the Netherlands that was elaborated at the time.

Riphagen's approach to the Wielingen was incorporated in the letter outlining a proposed shelf delimitation strategy which the Minister of Foreign Affairs submitted to the Council of Ministers at the end of January 1965.[304] The letter indicated that a claim to the continental shelf beyond the equidistance line with reference to the historic claim to the Wielingen did not stand a chance and would only serve to damage the relationship with Belgium. The letter moreover emphasized the navigational interests of the Netherlands in the Wielingen, and the lack of a linkage of these interests to the delimitation of the continental shelf. The letter did emphasize that Belgium should recognize that the continental shelf delimitation was without prejudice to the matter of the Wielingen and observed that it could be attempted to convince Belgium to look for a solution of the territorial sea delimitation in an overall

---

[301] Note 71/64 dated 28 October 1964 [NNA/17], p. 1; see also note no. 125 dated 26 October 1964 [NNA/17].
[302] Note 71/64 dated 28 October 1964 [NNA/17], p. 1.   [303] Ibid., pp. 1–2.
[304] See Chapter 5.5.

settlement of pending issues concerning the land boundary between the two countries.[305] In the Council of Ministers, which agreed to the proposed shelf delimitation strategy, it was urged that the Netherlands should fully maintain its rights in respect of the Wielingen.[306]

The Belgian Ministry of Foreign Affairs and Foreign Trade accepted the Dutch proposal to negotiate an agreement on a continental shelf boundary in accordance with equidistance. The assurance that the dispute over the Wielingen should not complicate the matter was accepted with relief. The Netherlands reserved its rights in respect of the Wielingen, but it was also mentioned that this matter constituted an anomaly in the light of the, at present, excellent relations between the two countries. It was hoped that this issue could be resolved quietly in the framework of an overall regulation of boundary questions.[307] Subsequently, the continental shelf boundary was defined in coordinates by technical experts. In this connection agreement was also reached on the definition of the trijunction point between Belgium, the Netherlands and the United Kingdom.[308] The draft continental shelf agreement, which was submitted to Belgium in early November 1965,[309] would never be finalized. The Belgian Government considered that it could not formally agree to the agreement in the absence of an internal legal basis.[310] Belgium eventually adopted continental shelf legislation in 1969. At that time, finalization of an agreement was no longer a matter of urgency for the Netherlands. The ICJ had already rendered its judgment in the North Sea continental shelf cases earlier that year. In its pleadings, the Netherlands had been able to refer to the agreement in principle with Belgium to argue that Belgium also supported the

---

[305] See letter of the MiFA to the Prime Minister dated 28 January 1965 [NNA/24], pp. 3–4.

[306] Note dated 12 February 1965 [NNA/26].

[307] Letter of the Dutch ambassador in Brussels to the MiFA dated 13 April 1965 [NNA/40]; see also letter of the MiFA to the Dutch ambassador in Brussels dated 5 April 1965 [NNA/40].

[308] See note 83/65 dated 11 June 1965 [NNA/22]; letter of the Belgian Embassy in The Hague to the MFA dated 2 July 1965 [NNA/25]; letter of the Belgian Embassy in The Hague to the MFA dated 15 September 1965 [NNA/25].

[309] See letter of the secretary of the Dutch delegation to D.C. van der Hooft of the Ministry of Economic Affairs dated 3 November 1965 [NMEA/1]; letter of A. van der Essen of the MFA and Foreign Trade (Belgium) to W. Riphagen of the MFA dated 5 November 1965 [NMEA/1].

[310] See letter of the Belgian Embassy in The Hague to the MFA dated 15 September 1965 [NNA/25], p. 2.

equidistance principle.[311] The Netherlands and Belgium only finally settled their continental shelf boundary in 1996. Until that time the boundary which had been agreed upon at the technical level in 1965 had been used in practice. The bearing of the 1996 boundary deviates about 4 degrees from an equidistance line to the advantage of Belgium. Belgium had argued for this adjustment, referring among others to its geographical situation between the Netherlands and France, which it found to be similar to that of Germany in relation to the Netherlands and Denmark.[312] A second agreement of 1996 delimits the territorial sea applying the equidistance method and put to rest the Dutch claim over the Wielingen.

5.6.3 The bilateral boundary between the Netherlands and the United Kingdom

The United Kingdom had repeatedly indicated to the Netherlands that it was interested in early negotiations concerning their common boundary. Even before the Netherlands had responded to the latest British inquiry at the end of September 1964, the United Kingdom presented a proposal on a boundary in accordance with the equidistance principle. The proposal envisaged that the boundary to the south and north would end at trijunction points with, respectively, Belgium and Denmark.[313] There existed two minor differences between the British equidistance line and the equidistance line the Netherlands had determined. The United Kingdom had not taken into account the extension of the jetties of the port of IJmuiden, which were under construction, and the planned Europoort project, which would extend the baselines off the coast of Rotterdam seaward for a couple of kilometers. Technical experts of both countries, who met in January 1965, agreed that the extension of the jetties at IJmuiden was in such an advanced state that at the time of the conclusion of a treaty they would certainly exist and could be taken into account. The technical experts refrained from suggesting a solution in respect of the baseline in the area of Europoort. That project was in a preliminary phase and was not indicated at all on the relevant nautical

---

[311] See e.g. NCM, para. 17.

[312] See E. Franckx, "Belgium-Netherlands; Report Number no. 9–21" in J. I. Charney and L. M. Alexander (eds), *International Maritime Boundaries* Vol. IV (The Hague: Martinus Nijhoff Publishers, 2002), pp. 2921–2934, at pp. 2930–2931.

[313] Aide Mémoire presented by British Embassy in The Hague dated 12 November 1964 [NNA/17].

charts.[314] Article 6 of the Convention on the continental shelf read, in conjunction with article 3 of the Convention on the territorial sea and the contiguous zone, implies that the equidistance line mentioned in article 6 in principle should take into account the existing low-water line as indicated on nautical charts. There is no obligation for a State to accept envisaged shifts in the low-water line impacting on the baseline.

The Netherlands at first maintained the position that the future baseline in the area of the Europoort had to be taken into account. The United Kingdom opposed this approach because it would be contrary to article 6 of the Convention on the continental shelf. Subsequently it was added that acceptance of the Dutch request would imply that the reference to the equidistance principle in the preamble of the agreement would have to be dropped. As the Netherlands considered this reference of much importance, because it would allow pointing out to Germany that the equidistance principle was accepted by the other North Sea coastal States, it dropped its insistence on the use of the future baseline along Europoort.[315] The fact that the shift in the baseline would only have a limited impact on the equidistance line made this a small sacrifice.[316]

The fact that the Netherlands and Denmark were intent on using their agreements with the United Kingdom in their negotiations with Germany raised some concerns on the British side. It was feared that the United Kingdom would be open to criticism if Germany was left completely in the dark. Although there was no obligation to inform Germany, the excellent bilateral relations in respect of the law of the sea might justify some sort of approach. At the same time, it should be avoided that Germany was given a say in the matter. The United Kingdom also had an interest in concluding these agreements, both because it would allow opening the area for exploration activities and because the agreements reflected support for the equidistance principle.

---

[314] See Report of the meeting on the "median line" between the Netherlands and England held in London on 12 January 1965 [NNA/28], p. 1.

[315] Anglo-Netherlands continental shelf discussions; The Hague, 22–24 March 1965, report [FO/ 371/181320]; Proposed Anglo-Dutch Continental Shelf Boundary Agreement, note dated 28 May 1965 [FO 371/181320]; letter of the MiFA to the Minister of Economic Affairs dated 4 June 1965 [NNA/40]; letter of the Minister of Economic Affairs to the MiFA dated 26 June 1965 [NNA/40].

[316] Letter of the MiFA to the Minister of Economic Affairs dated 4 June 1965 [NNA/40]. Similarly, the British Ministry of Power observed that the area in dispute was "probably of little importance in itself and could no doubt be conceded" (letter of P. J. Parratt of the Ministry of Power to R. C. Samuel of the FO dated 6 April 1965 [FO 371/181320]).

In the end, it was agreed to informally notify Germany before the agreements would be signed.[317]

The German government formally reacted to the British agreements with Denmark and the Netherlands through two identically worded aide-mémoires.[318] The aide-mémoires indicated that the agreements could not prejudice the delimitation of the continental shelf of Germany with respectively Denmark and the Netherlands. The German Embassy was furthermore instructed to inform the Foreign Office that Germany did not object to extent of the British continental shelf as determined by the agreements and that it was accepted that the median line constituted the boundary between the United Kingdom and the continent.[319] Germany did object to the fact that the agreements implied that the continental shelf of Germany would not extend to this median line.[320] Further discussions between the British authorities and the German Embassy pointed out that the former were not prepared to state in writing that the agreements were without prejudice to the German claims.[321] In a written reply, the British Government only observed that it took note of the German aide-mémoires and that Dr. Kilian of the German Embassy had indicated that Germany did not have a claim to any part of the continental shelf of the United Kingdom as defined in the two agreements.[322] The German Embassy in London, in addition, was told that the United Kingdom had not had the intention to prejudice German claims.[323] Thus, no position was

---

[317] See the note Question of informing the Germans about our continental shelf agreements dated 3 June 1965 with subsequent comments [FO 371/181321].

[318] Aide-mémoire of the German Embassy in London, dated 12 July 1966 (GM, Annex 10); aide-mémoire of the German Embassy in London, dated 12 July 1966 (GM, Annex 13A). The German Embassy had already been informed about the pending conclusion of the agreements in July 1965. The staff member concerned had not expressed an opinion, but had left the impression that the United Kingdom was justified in its approach and indicated that he would pass on the information to the German FO (see note Continental Shelf: Germany dated 21 July 1965 [FO 371/181322]). The latter at the time apparently did not happen (see note dated 12 July 1966 [AA/37]).

[319] See letter of the FO to the German Embassy in London dated 18 July 1966 [AA/37], p. 4.

[320] See ibid., p. 2.

[321] See letter of the German Embassy in London to the FO dated 9 August 1966 [AA/37]; note dated 30 September 1966 [AA/37]; see also note dated 22 February 1967 [FCO 14/366], p. 1.

[322] See aide-mémoire of the British Embassy in Bonn dated 28 September 1966 [AA/37].

[323] See note dated 22 February 1967 [FCO 14/366], p. 1; letter of the German Embassy in London to the FO dated 9 August 1966 [AA/37]; note dated 30 September 1966 [AA/37].

taken on the actual effect of the agreements on the German position. After Germany had indicated its dissatisfaction about the fact that the British aide-mémoire did not confirm what the German Embassy had been told,[324] the United Kingdom considered a further response. In the circumstances, it was considered that the agreements did give some weight to the position of Denmark and the Netherlands and a positive response was considered desirable to avoid a further bone of contention in the Anglo-German relations.[325] At the same time, a response should not be damaging to Denmark and the Netherlands. To satisfy both these demands, a British aide-mémoire indicated that the government of the United Kingdom considered that it did not have any standing in respect of the delimitation of the continental shelf between Denmark, Germany and the Netherlands.[326] Denmark and the Netherlands were also informed about the content of the aide-mémoire.[327] Germany considered that the "no standing" formulation probably was the best that could be expected from the British.[328] If need be "this strict neutrality" could be used by Germany during its pleadings before the ICJ.[329] During the pleadings before the Court, Denmark and the Netherlands did refer to the agreements as expressing support for the equidistance principle, but they refrained from claiming that the United Kingdom had taken a position on their bilateral delimitations with Germany.[330] To that extent the German approach to the United Kingdom had been successful.

The Netherlands also used the negotiations with the United Kingdom to extract a formal recognition of the equidistance principle from Denmark. Talks between both Ministries of Foreign Affairs had already

---

[324] See note dated 25 January 1967 [AA/48].

[325] See note dated 22 February 1967 [FCO 14/366], pp. 1–3. It was observed that from a legal perspective invoking the agreements would not carry much weight, as Germany would only have to state that they constituted *res inter alios acta* and as such were not binding on Germany (p. 2, hand-written note dated 2 March 1967).

[326] See note dated 22 February 1967 [FCO 14/366], pp. 1–3; aide-mémoire of the British Embassy in Bonn dated 13 September 1967 [AA/48; FCO 14/366].

[327] See note dated 13 September 1967 [AA/48]; letter of the British Embassy in Bonn to the FO dated 14 September 1967 [FCO 14/366].

[328] See note dated 13 September 1967 [AA/48]; letter of H. Treviranus to G. Jaenicke dated 9 October 1967 [AA/61].

[329] See letter of H. Treviranus to G. Jaenicke dated 9 October 1967 [AA/61]. Translation by the author. The original text reads "diese strikte Neutralität."

[330] See further Chapter 8.3 and following. Von Schenk considered that although the agreements did not have any legal significance, they did provide some support for Denmark and the Netherlands or at least covered their back (Von Schenk, "Die vertragliche Abgrenzung", pp. 376–377).

pointed out that Denmark did not have a problem with commiting itself to a trijunction point with the Netherlands and the United Kingdom. Still, in view of creating as strong as possible a legal position in relation to Germany, the Netherlands attached some value to explicit Danish agreement to the inclusion of a trijunction point in the draft agreement with the United Kingdom as this would establish that there existed at least one common boundary point between Denmark and the Netherlands.[331] The Dutch proposal on the trijunction point was accepted by the Danish Ministry of Foreign Affairs without further discussions.[332] The draft agreement between the Netherlands and the United Kingdom was subsequently finalized, was signed on 6 October 1965 and entered into force on 23 December 1966.[333]

5.6.4   The bilateral boundary between Denmark and the Netherlands

In March 1965, shortly after Denmark had reached agreement with Germany on a partial boundary, Denmark and the Netherlands met again to discuss the coordination of their positions in relation to Germany.[334] Denmark was not willing to take an initiative for the negotiation of a bilateral boundary with the Netherlands, but was ready to negotiate if the Netherlands would take the initiative.[335] The Danish side also questioned whether Germany could argue that it should be involved in these negotiations because the Danish-Dutch boundary would also be concerned with the trijunction point with the boundaries of Germany with both States. Riphagen, the legal advisor of the Dutch Ministry of Foreign Affairs, submitted that Germany could be told that Denmark and the Netherlands were negotiating a boundary that was "open" in the direction of Germany and that a Danish-Dutch agreement in any case could, nor was intended to, prejudice the question of the

---

[331] See send message in cipher of the MiFA to the DEC dated 4 June 1965 [NNA/28].

[332] See note verbale of the DEC to the MFA presented on 8 June 1965 [DNA/4B]. Denmark presented a note in reply on 16 June 1965 [NNA/28].

[333] Agreement between the Government of the Kingdom of the Netherlands and the Government of the United Kingdom of Great Britain and Northern Ireland relating to the delimitation of the continental shelf under the North Sea between the two countries.

[334] An account of the meeting is contained in Report; The Continental Shelf; Danish-Dutch meeting in The Hague on 5 April 1965 dated 22 April 1965 [DNA/4A] (Danish Report meeting April 1965 [DNA/4A]).

[335] *Ibid.*, p. 5.

location of the trijunction point.[336] This reasoning is not completely convincing. Germany had indicated to Denmark and the Netherlands that it considered that its continental shelf should extend to the median line between the continent and the United Kingdom. This excluded the possibility of any boundary between the Netherlands and Denmark. Moreover, the negotiation of an "open" boundary, i.e. a boundary which would fall short of the equidistance point with Germany, might have entailed a certain risk. It could have suggested that Denmark and the Netherlands were having second thoughts about the strength of their legal position. The agreement which was eventually concluded established the boundary up to a point at equal distance between the three States.

After Danish and Dutch technical experts reached agreement on the definition of the equidistance line,[337] at the end of October 1965 the Netherlands proposed Denmark to conduct formal negotiations on an agreement.[338] The Netherlands expected that the conclusion of this agreement would have the desired result in Germany, after which the dispute on the delimitation of the continental shelf between the three States could be submitted to the ICJ.[339] Denmark was not particularly enthusiastic about formal negotiations, but it was concluded that in the absence of clear indications on how firmly Germany stood by its position, there was neither reason to abandon the existing Danish position on the delimitation of the continental shelf nor to reject the Dutch proposal. After the conclusion of an agreement it could be seen how Germany would react.[340] Denmark did oppose that the draft agreement would refer explicitly to the equidistant trijunction point with Germany, as this was considered an unnecessary provocation.[341]

Denmark and the Netherlands agreed on and initialed the text of a delimitation agreement at the end of November 1965.[342] The Netherlands accepted that the article referring explicitly to the equidistant trijunction point with Germany would not be included. The

---

[336] *Ibid.*   [337] Note 80/65 dated 24 September 1965 [NNA/26], p. 2.

[338] Note verbale of the DEC to the MFA dated 29 October 1965 [DNA/36].

[339] Note 80/65 dated 24 September 1965 [NNA/26], p. 2.

[340] Note; Danish/Dutch continental shelf negotiations dated 5 November 1965 [DNA/36], p. 2.

[341] Note verbale of the Ministry of Foreign Affairs to the Embassy of the Netherlands in Copenhagen dated 12 November 1965 [NNA/26]; telegram of the Dutch ambassador in Copenhagen to the MiFA dated 15 November 1965 [NNA/26].

[342] For reports on the meeting see note; The continental shelf dated 29 November 1965 [DNA/4B]; note no. 110/65 dated 24 November 1965 [NNA/26].

delimitation line defined in the agreement did, however, run between that point and the trijunction point with the United Kingdom. The delimitation agreement was signed on 31 March 1966.[343] With this agreement between Denmark and the Netherlands, all continental shelf boundaries in the North Sea had been agreed except for a part of Germany's boundaries with Denmark and the Netherlands. The Danish-Dutch agreement cut off Germany from the middle of the North Sea, but it was clear to the three States that the agreement had not settled their dispute but only added a further prop to set the stage for its resolution. Shortly before the agreement was to be signed, it had been brought up in the margins of a meeting between the three States. Truckenbrodt, the German head of delegation, indicated that the agreement constituted *res inter alios acta*, which would not prejudice the position of Germany. Denmark and the Netherlands expressed their agreement,[344] and indicated that the agreement would be adjusted if a subsequent court ruling would require that.[345] Truckenbrodt also expressed the hope that Denmark and the Netherlands would not give too much publicity to the signature of their bilateral agreement.[346] After the signature of the agreement, Germany formally notified Denmark and the Netherlands that Germany considered it to be without prejudice to its delimitation with both States.[347]

5.7 Concluding remarks

Denmark and the Netherlands differed fundamentally with Germany about how their mutual continental shelf boundaries should be delimited. This made it difficult to reach a compromise solution to start with. The course of the negotiations subsequently excluded all possibility of a negotiated settlement. The more Denmark and the

---

[343] Agreement between the Government of the Kingdom of the Netherlands and the Government of the Kingdom of Denmark concerning the delimitation of the continental shelf under the North Sea between the two countries.

[344] Report of the trilateral Danish-Dutch-German negotiations in The Hague on 28 February 1966 concerning the delimitation between the three countries' part of the continental shelf under the North Sea dated 21 March 1966 [DNA/4C] (Report trilateral meeting February 1966 [DNA/4C]), p. 11.

[345] Send message in cipher (ref. no. 1424) of the MFA to the DEB dated 2 March 1966 [NNA/40], p. 2.

[346] *Ibid.*

[347] Aide-Mémoire of the FO dated 25 May 1966 [NNA/27] (GM, Annex 15 and 15A).

Netherlands committed themselves to the equidistance principle the more a compromise solution with Germany became excluded.

When Denmark and the Netherlands learned that Germany did not accept the equidistance principle for the delimitation of the continental shelf in the North Sea, initially they both did not want to take the initiative for negotiations. Instead, when they were presented an opportunity, they informed Germany that they applied to equidistance principle to define their boundary with Germany in accordance with article 6 of the Convention on the continental shelf. With some imagination, article 6 could be said to leave room for the interpretation that, in the absence of special circumstances, the provisional boundary would be constituted by an equidistance line.[348] Germany obviously could not remain silent any longer after Denmark and the Netherlands had presented these views. Germany preferred to first deal with the Netherlands. The area to the north of the land boundary with the Netherlands was considered to offer good prospects to find hydrocarbons and the German concessionaire wanted to explore this area.

The first round of talks between Germany and the Netherlands in March 1964 did address the legal arguments of both States to support their positions on the boundary. In comparison to the first round of negotiations with Denmark, the legal argument went into less detail. Before the negotiations with Denmark, the German Foreign Office had had a closer look at the legal aspects of the matter. In both cases, the discussion of legal arguments basically remained limited to each party setting out its position and there was no attempt to arrive at a common understanding of elements of the case. Some of the elements which were presented at these first meetings would be central to the legal reasoning of the parties until the pleadings before the ICJ in 1967 and 1968.

During the first round of negotiations with the Netherlands, Germany also presented two specific delimitation proposals. A first proposal, which entailed the equal division of the continental shelf, was immediately dropped after the Netherlands had rejected it. This was followed by a proposal to delimit the continental shelf by reference to the coastal length of each State, which would give the Netherlands a larger share of the continental shelf than either Germany or Denmark. This proposal was likewise rejected by the Netherlands. This first round of negotiations already clearly indicated that the dispute over Germany's share over the continental shelf would mainly concern Denmark and the Netherlands. The Netherlands at first

---

[348] See Chapter 3.5 for an interpretation of article 6 on this point.

considered that a multilateral approach would offer a forum to reach a compromise solution with Germany, but this idea was subsequently dropped and the Netherlands and Denmark instead focused on concluding agreements with the other North Sea States on equidistance boundaries. These other States in any case had little interest in a multilateral conference and if it had taken place, they most likely would have taken the position that the German claim only concerned Denmark and the Netherlands. Germany for a longer time considered that a multilateral conference was an option but eventually also gave up this idea in view of the unreceptive attitude of Denmark and the Netherlands.

After the first meeting between Germany and the Netherlands in March 1964, the Netherlands considered options for a compromise solution. At first, a compromise solution for the entire boundary was entertained. An important reason to look for a compromise was that there might otherwise be uncertainty about the area in dispute, and that the applicable law did not provide the Netherlands with effective mechanisms to exclude Germany from this area, safe by military action. However, a compromise from the outset was accompanied by a number of conditions, which made it probably rather unrealistic to accomplish to start with. The idea of a compromise quickly receded into the background and instead the Netherlands worked towards reinforcing its position in relation to Germany. A position paper of the Ministry of Foreign Affairs, which was presented to the Council of Ministers in May 1964, set out the options in this respect. Although the paper stressed the indeterminate nature of article 6 of the Convention on the continental shelf, it, one would say surprisingly, at the same time assured that there were no special circumstances in the case of Germany and the Netherlands. The paper also argued that it was unlikely that the ICJ would reach a different conclusion. In the light of these latter submissions it is understandable that the Council of Ministers decided that the Netherlands should seek to hold on to its equidistance area. This policy was subsequently implemented by the Ministry of Foreign Affairs.

The Netherlands first continued its negotiations with Germany. A number of factors contributed to an agreement on a partial boundary. Both States wanted an agreement in the nearshore area to allow exploratory activities. At the same time equidistance in this area did provide a solution that was acceptable to both parties. During the negotiations, Germany sought to put some pressure on the Netherlands by suggesting that the North Sea consortium might start activities beyond the equidistance line. In actual fact, the German authorities did everything to persuade the concessionaire to remain inside the German equidistance area. The agreement on a partial boundary implied that for some time there was no need for a provisional regime for the disputed area.

Germany, in the first stage of the negotiations, still had the impression that the Netherlands would cooperate with Germany in the framework of a multilateral conference to achieve a mutually acceptable outcome. At that time the Netherlands had already abandoned this option and, after it had reached agreement with Germany on a partial boundary, the Netherlands started working towards the finalization of its other North Sea boundaries in accordance with the equidistance method. This whole episode shows a remarkable discrepancy between the legal and political side of these delimitations. The idea behind this policy was that it would put pressure on Germany. Germany's attempts to extract a commitment from the United Kingdom that it was not taking sides indicate that it was not impervious to these attempts. At the same time, from a legal perspective, these agreements did not have any effect for Germany and as much was recognized by the Netherlands and Denmark.

Denmark pursued the same policy in respect of its continental shelf boundaries as the Netherlands. As far as Germany was concerned, similar legal arguments were exchanged and a similar partial boundary was agreed upon. The other Danish bilateral boundaries were delimited in accordance with the equidistance method. In the case of Norway, Denmark for a considerable time sought to provide for a possible modification of this boundary if the equidistance method would not be applied between Denmark and Germany. Although Norway originally agreed to an arrangement to this effect, it backtracked on its commitment after it had secured agreement on a delimitation in accordance with the equidistance method. Denmark eventually accepted that the boundary in any case would be final for a number of reasons. Consistency between its different positions figured prominently in this respect.

Finding a way out of the deadlock – the submission
of the disputes to the International Court of Justice

6.1 Introduction

After Germany and Denmark had reached agreement on a partial boundary in the North Sea in March 1965, it had to be decided how the
remainder of Germany's boundaries with Denmark and the Netherlands
should be settled. In view of the diametrically opposed views on the
applicable law and the fact that neither party was prepared to consider its
opponent's position, two options remained open. One would be to reach
a mutually acceptable compromise solution without taking into consideration the applicable law. This was hardly an attractive option for the
Danish and Dutch governments as public opinion could be expected to
view this as giving Germany a part of the Danish and Dutch continental
shelf. The only other option would be to submit the dispute to arbitration
or the ICJ. A decision of an independent third party that the equidistance
principle was not applicable and that the law mandated a different
boundary would make it possible to sell such an outcome to public
opinion.

6.2 The slow road to agreement on the purpose of further talks

During the Danish-German talks of March 1965, Germany had indicated that it would soon take an initiative to settle the remainder of
Germany's continental shelf boundaries in the North Sea. Germany in
particular pointed to the possibility of trilateral talks, but left the possibility of a conference of all North Sea coastal States open. If these further
negotiations failed, Germany reserved the right to submit the matter to
arbitration.[1] Germany at that time had already come to the conclusion
that the possibilities of convening a multilateral conference were bleak

---

[1] Report dated 31 March 1965 (excerpt) [AA/64], pp. 7–8; see also Danish report negotiations March 1965 [DNA/67], pp. 8–9.

and probably would not lead to any tangible results.[2] A trilateral approach made more sense because the German claim for a larger share in the continental shelf mostly turned on the combined effect of the two bilateral delimitations of Germany with Denmark and the Netherlands.[3]

Germany tried to convene a trilateral meeting in May or June of 1965.[4] The Netherlands refused to accept this invitation, taking the position that a trilateral meeting was out of the question because the delimitation of the German-Dutch boundary was a purely bilateral question.[5] Denmark had been more inclined to accept the German invitation as it felt that article 6 of the Convention on the continental shelf implied that delimitation primarily had to be effected through negotiations.[6] The Netherlands considered that it had complied with its obligations in this respect because the German-Dutch negotiations had been concerned with the entire bilateral boundary.[7] Denmark was not willing to accept an invitation for trilateral talks if the Netherlands did not accept such an invitation.[8] After the refusal of the Netherlands and Denmark to accept its invitation, Germany concluded that a further attempt to reach a compromise solution through negotiations did not make any sense and that it should instead aim to reach agreement with the other two States on the modalities of submitting the dispute to a judicial body.[9]

The possibility of submitting the dispute over delimitation of the continental shelf to judicial settlement had already been suggested by Germany during the first negotiations with the Netherlands in March 1964.[10] At the beginning of 1965, when the other options for settling the dispute seemed to have been nearly exhausted, the legal department of the German Foreign Office started to consider this possibility in more

---

[2] Note dated 2 February 1965 [AA/43], pp. 9–10 and 12; see also express letter of the Minister for the Economy to the FO dated 15 March 1965 [B102/260173], p. 2.

[3] See also note dated 2 February 1965 [AA/43] p. 10.

[4] See note (draft) dated 15 April 1965 [AA/41], p. 9; telegram in cipher of the DAEB to the MFA dated 31 May 1965 [DNA/4A].

[5] See send message in cipher of the MFA to the DEB dated 28 May 1965 [NNA/26], p. 1.

[6] Danish Report meeting April 1965 [DNA/4A], p. 3; see also letter of the MiFA to the Prime Minister dated 8 June 1965 [NNA/40], p. 2.

[7] Danish Report meeting April 1965 [DNA/4A], p. 2.

[8] Note to the MiFA dated 28 May 1965 [DNA/4A], p. 2.

[9] Note dated 10 June 1965 [B102/260135], p. 1; express letter of the Minister for the Economy to the FO dated 14 June 1965 [AA/43], p. 2; telegram in cipher of the DAEB to the MFA dated 31 May 1965 [DNA/4A]; note dated 7 September 1965 [B102/119557].

[10] See Minutes German-Dutch negotiations of March 1964 [AA/38], p. 4.

detail.[11] At that point, the political department of the Foreign Office raised objections to the suggestion of the legal department to submit the disputes with Denmark and the Netherlands to an arbitral tribunal. The political department expressed concerns that the political side of the matter had not been taken into account sufficiently.[12] Because of the still very sensitive relations with Denmark – especially at the moment of the 25th commemoration day of the German occupation and the end of the Second World War twenty years previously – it was to be feared that the Danish government, parliament and public opinion would fiercely react to a German initiative to start legal proceedings. For reasons of internal and foreign policy, the German Chancellor wanted the upcoming visit of the Danish Prime Minister to Bonn to go smoothly. The political department considered that an escalation of the discussions on the continental shelf would have a significant impact on the atmosphere of the visit. It should be examined very carefully whether the probably problematic chances of success of legal proceedings really were that big that the important political risks had to be estimated to be less significant.[13] Although the political department in this analysis focused on Denmark, similar arguments could no doubt be said to apply for the Netherlands.

The opinion of the political department led to a detailed refutation by the legal department.[14] A number of conclusions could, and had to, be drawn after the negotiations with Denmark and the Netherlands on partial boundaries.[15] First, there was absolutely no prospect of reaching a satisfactory compromise in further bilateral or multilateral negotiations because Germany did not have any political or economic means to

[11] A detailed consideration of the options for instituting legal proceedings is contained in the note dated 31 March 1965 [AA/40], which is discussed subsequently.
[12] Note with reference IA4–82.70/94.04/65 dated 12 April 1965 [AA/41].   [13] Ibid.
[14] The archives of the FO contain three documents prepared in the legal department in this connection. A draft of a note to be submitted to the Foreign state secretary dated 15 April 1965 [AA/41]; a note from the legal department to the political department dated 22 April 1965 [AA/41], which however was not submitted; and a note dated 30 April 1965 [AA/41] for the head of the legal department in connection with the discussion of the matter with the head of the political department.
[15] Ibid., p. 1. This note had been prepared by Truckenbrodt, the head of the German delegation during the negotiations with Denmark in Copenhagen in March 1965, who apparently was taken aback by the position of the political department. Truckenbrodt referred to the unusual form of the reaction, which in his view evidently was based on insufficient knowledge of the course of the negotiations and contained the implicit criticism that Truckenbrodt had not taken the foreign policy aspect of the matter sufficiently into account (ibid.).

obtain concessions from Denmark and the Netherlands. At the same time, the delimitation of the continental shelf was too important for Germany to let the matter drop. Apart from the economic importance of the shelf the wider implications of the delimitation of the shelf (for instance in the field of defense) had to be considered. Third, other Federal ministries and the directly interested states had a significant interest in the matter and could be expected to object if the Foreign Office were to let the German claim lapse by refraining from any further action. The Federal government should, in particular, count on difficulties with the states.[16] In these circumstances the only remaining possibility to obtain a more favorable outcome would be to submit the disputes to a court to obtain a clarification on the content of the law.[17] There was, in any case, no risk of an escalation of the discussions. The talks with Denmark and the Netherlands thus far had been conducted in a businesslike manner and they both accepted it as perfectly normal that Germany defended its position the way it did. The submission of the dispute to legal proceedings should not lead to any difficulties, provided, however, that public opinion in the three States would be properly informed about the purpose of proceedings. The political department had failed to appreciate the political significance of proceedings. Proceedings would allow defusing a politically sensitive situation, which could not be resolved through political or other means, by using a neutral third party. In that respect, Denmark during the bilateral talks in March 1965 had pointed out that the decision of the Permanent Court in the *Eastern Greenland* case had been accepted without great difficulty by public opinion in both Norway and Denmark because the governments had skillfully managed public relations.[18] Denmark had signaled its willingness to take a similar

---

[16] *Ibid.*, pp. 1–2. The note dated 15 April 1965 [AA/41] at p. 4 in this connection refers to the diametrically opposed views of the Federal government and the states on the ownership of the continental shelf (on this point see further Chapter 4.3). The note also observes that inaction on the continental shelf question might even lead to a controversy between the parties which formed the governing coalition in Bonn (*ibid.*, pp. 4–5).

[17] Note dated 30 April 1965 [AA/41], p. 2.

[18] *Ibid.*, pp. 3–4. The *Eastern Greenland* case was concerned with the sovereignty over the eastern part of Greenland, which was in dispute between Denmark and Norway. The Danish colonization of Greenland stretched back several centuries. Norway had started to show an interest in the eastern part of Greenland in the nineteenth century and had issued a declaration of occupation in 1931, which was the direct cause of the case. In its judgment of 5 April 1933 the Permanent Court of International Justice, which was the predecessor of the ICJ, held that Denmark had sovereignty over all of Greenland and rejected the Norwegian declaration of occupation of 1931 as unlawful and invalid (Legal Status of Eastern Greenland, judgment of 5 April 1933 (PCIJ, Series A/B 53), p. 75).

approach in the present case.[19] After this assessment of the legal department, there were no further written exchanges on the desirability of
recourse to legal proceedings,[20] but the political department maintained
its opposition until at least September 1965.[21] It seems likely that the
political department dropped its opposition after an interdepartmental
meeting showed broad support for submission of the dispute to
arbitration.[22]

Denmark and the Netherlands agreed that recourse to legal proceedings offered a solution for the settlement of their dispute with Germany.
During the Danish-German talks of March 1965, the Danish delegation
had indicated that Danish public opinion in general could be expected to
accept an unfavorable judgment. On the other hand, Danish public
opinion would not show any understanding if the Danish Government
were to abandon the, in Danish eyes, advantageous legal position and
would make "concessions" to Germany without an objective clarification
of the disputed legal matters.[23] The Netherlands had expressed similar
views to Germany during their bilateral negotiations in June 1964.[24]

The submission of the delimitation dispute to judicial settlement
required consideration of a number of questions. First of all, it would
have to be decided to whom the dispute would be submitted. One
possibility would be to submit the dispute to the ICJ and the other to
an ad hoc panel of arbitrators. One argument to choose for the Court
would be that it might be more inclined to base its decision on strictly
legal considerations. This was the perception of the Dutch Ministry of
Foreign Affairs. In addition, a controversial decision might easier gain
general acceptance if it was adopted by the Court, which is the principal
judicial organ of the United Nations. Second, the parties would have to
decide how a case would be instituted. One of the States concerned could
bring a case unilaterally, while the other possibility would be to bring it
jointly. In the former case the impression might exist that one of the
parties was bringing the other party to court against its will, while joint

---

[19] Note from the legal department to the political department dated 22 April 1965
[AA/41], p. 4.
[20] The archives of the German FO do not contain any documents on further consultations
between the two departments.
[21] Note dated 16 September 1965 [B102/119557], p. 3.
[22] For a report on that meeting see e.g. note dated 7 September 1965 [AA/41].
[23] Note from the legal department to the political department dated 22 April 1965
[AA/41], p. 4.
[24] See Chapter 5.2.4; see also note dated 15 April 1965 [AA/41], p. 4.

submission would suggest cooperation. Finally, the specific question(s) to be submitted to the ICJ or an arbitral tribunal would have to be decided. The content of these questions could have an impact on the answers that would be formulated and on the negotiations following a judgment. For instance, one option would be to only ask the ICJ or a tribunal to determine whether the equidistance method was obligatory in the specific circumstances of the North Sea. However, in case of a negative reply, the parties would have no indication in respect of the applicable rules.

The Netherlands had already concluded early on that if the negotiations with Germany would not result in agreement on a boundary, the dispute should be submitted to the ICJ.[25] In June 1965, the Dutch Minister of Foreign Affairs again set out the options for dealing with Germany to the Council of Ministers. In view of the impasse in the negotiations the only option that remained was going to the Court. The minister remained of the view that the Court would be more likely to accept the Dutch position than an arbitral tribunal.[26] Still, it was not possible to make any prediction about the outcome of the proceedings before the Court.[27] Notwithstanding this uncertainty, the Council of Ministers did not look at alternatives but agreed to the proposal to discuss the submission of the dispute with Germany to the ICJ.[28] Interestingly, a similar uncertainty about the outcome of proceedings was expressed in Denmark. For instance, Sørensen, the legal advisor of the Ministry of Foreign Affairs, considered that there was a 50 percent chance that a judgment would go against Denmark and the Netherlands,[29] and he later added that the outcome was completely unpredictable and would, among others, depend on the continental shelf issues the countries of origin of the judges were facing.[30] On the

---

[25] See Memorandum for the Council of Ministers dated 6 May 1964 [NNA/13], p. 8. This suggestion did not lead to any observations in the Council of Ministers (Minutes of the Council of Ministers dated 22 May 1964 [NNA/2]). For a discussion of the memorandum see further Chapter 5, text at footnote 112 and following.

[26] Letter of the MiFA to the Prime Minister dated 8 June 1965 [NNA/40], p. 3. For the earlier discussion of this point in the Dutch Council of Ministers and an assessment of the likelihood that the Court would be more inclined to adopt the views of the Netherlands see Chapter 5, text at footnote 121 and following.

[27] Letter of the MiFA to the Prime Minister dated 8 June 1965 [NNA/40], p. 3.

[28] Minutes of the Council of Ministers dated 2 July 1965 [NNA/4], item 5.

[29] Note dated 28 June 1965 [DNA/4B], p. 1.

[30] Letter of J. Paludan of the MFA to the Danish ambassador in The Hague dated 19 August 1965 [DNA/4B].

other hand, Riphagen, the legal advisor to the Dutch Ministry of Foreign Affairs, was more optimistic. The Court would have difficulty in distancing itself from the equidistance principle as this would mean a departure from the letter and spirit of article 6 of the Convention on the continental shelf and customary law. He also did not consider that the position of the countries of nationality of the judges would have a major impact on the outcome of the case. The Court was more likely to look at the impact of its decision on a global scale.[31] The optimism of Riphagen suggests that the cautious views of the Dutch Minister of Foreign Affairs may have been intended to avoid raising expectations. Whether that optimism was justified was another matter.[32]

When Denmark and the Netherlands discussed the possibilities of dealing with Germany at the end of March 1965, Riphagen explained the Dutch preference for the ICJ to Denmark.[33] Denmark at that time had not made a choice between the Court and arbitration, but did have a preference for a strictly legal approach to dealing with the question of whether the equidistance principle was applicable and did not want a decision *ex aequo et bono*.[34]

The legal department of the German Foreign Office considered the possibilities for submitting the dispute to the ICJ or arbitration in March 1965.[35] The 1957 European Convention for the Peaceful Settlement of Disputes, to which Germany, Denmark and the Netherlands were parties, provided one basis for jurisdiction. Article 1 of that Convention provides that parties shall submit all international legal disputes, including those concerning any question of international law, to the ICJ. As the analysis of the legal department pointed out, the question of whether special circumstances existed in the North Sea was a question of international law.

Article 1 moreover allowed invoking the ICJ unilaterally if the parties could not agree on a special agreement to submit the dispute jointly.[36] However, it was submitted that the ICJ was not an appropriate forum to deal with this matter as it concerned a question of regional importance

---

[31] These views were reported at a meeting between M. Bendix of the Danish Embassy in The Hague and Mr. Fraser of the Dutch MFA (see letter of M. Bendix to J. Paludan of the Danish MFA dated 26 August 1965 [DNA/4B], pp. 3–4).

[32] See further Chapter 5 text at footnotes 121 and following.

[33] Danish Report meeting April 1965 [DNA/4A], p. 5.

[34] *Ibid.*; see also Danish report negotiations March 1965 [DNA/67], p. 13.

[35] Note dated 31 March 1965 [AA/40].   [36] *Ibid.*, pp. 2–3.

and the ICJ was a global institution. Denmark and the Netherlands were likely to agree that it was not in their interest to invoke the ICJ.[37]

It is difficult to agree with the view that the delimitation was only of regional interest. The interpretation of customary international law, or article 6 of the Convention on the continental shelf, is a matter of global interest, as is also witnessed by the impact of the *North Sea continental shelf* cases on the development of general international law. The note of the legal department also hints at another objection of Germany against the ICJ. It pointed out that the fifteen judges of the Court represented "*all* legal systems and *all* political blocs."[38] Later objections against submitting the dispute to the ICJ also emphasized that this could go against Germany because of its difficult international position, whereas this was not the case for Denmark and the Netherlands.[39] In the middle of the 1960s Germany did not maintain diplomatic relations with the Soviet bloc and the Soviet press was not averse to depicting Germany as a bulwark of revanchism and neo-Nazism. In that light the fear that judges from the Soviet bloc would, in any case, vote against Germany and might otherwise express views to Germany's detriment was not without basis. This problem would not exist in the case of an arbitral tribunal as it would be unlikely that Denmark or the Netherlands would appoint arbitrators from the Soviet bloc. In hindsight, it seems unlikely that the outcome of an arbitration would have been as advantageous to Germany as the judgment of the ICJ, which provided a groundbreaking interpretation of the law.[40]

Germany did seem to have an alternative for the ICJ. Two bilateral treaties of Germany with the Netherlands and Denmark of 1926 allowed the submission of any dispute concerning international law to arbitration.[41] Proceedings in this case also could be instituted unilaterally.[42] However, political and legal considerations argued against unilateral submission and joint submission was to be preferred. The submission of the two bilateral delimitations to two separate tribunals also entailed a

---

[37] *Ibid.*, p. 3.
[38] *Ibid.* (emphasis in the original). Translation by the author. The original text reads "alle Rechtssysteme und alle politischen Machtsblöcke."
[39] See e.g. note dated 16 February 1966 [AA/42], p. 1; note dated 10 December 1965 [DNA/ 4B], p. 2.
[40] I thank Rainer Lagoni for suggesting this point to me.
[41] Treaty of Arbitration and Conciliation between Germany and Denmark of 2 June 1926 (1926 Treaty); Convention of Arbitration and Conciliation between Germany and the Netherlands of 20 May 1926 (1926 Convention; jointly referred to as 1926 Treaties).
[42] Note dated 31 March 1965 [AA/40], p. 4.

certain risk of diverging findings.[43] The preference for arbitration did not immediately lead to a decision on this point and the option of submitting the dispute to the ICJ was also kept open.[44] This analysis did miss one important aspect of the 1926 Treaties. They contained a Final Protocol that, instead of unilateral submission to arbitration, envisaged unilateral submission to the PCIJ if Germany would have become a party to its Statute or would become a member of the League of Nations.[45] In view of the legal relationships between Germany and its two neighbors, this implied that in the 1960s unilateral submission of disputes in accordance with the 1926 Treaties would have to be to the ICJ.[46]

The Dutch Ministry of Foreign Affairs had also realized that there existed a risk that Germany might submit the dispute to arbitration in accordance with the German-Dutch Convention of 1926. Ratification of the Convention on the continental shelf by Germany and the Netherlands would not necessarily change the situation. If the Netherlands would submit the dispute to the Court under the Optional Protocol to the Geneva Conventions and Germany would invoke the 1926 Convention it was not certain that the Court would find that it had jurisdiction. A similar case of concurrent jurisdiction had never presented itself to the Court before, which made it difficult to predict the outcome. In that light it was preferable that Germany and the Netherlands would submit the dispute jointly to the ICJ.[47] Like the German analysis, the Dutch analysis had missed certain aspects of the 1926 Convention.[48]

After the Dutch refusal to participate in trilateral talks, the German Ministry for the Economy wanted to deal with the delimitation with urgency.[49] The Ministry was interested in a quick settlement because there was a risk that natural gas from the Netherlands would get access to the German energy market, excluding German producers from this

---

[43] *Ibid.*, pp. 6–7.
[44] See e.g. note dated 15 April 1965 [AA/41], p. 2; Circular letter of the FO dated 4 August 1965 [AA/40], pp. 3–4.
[45] Final Protocol to the Treaty of Arbitration and Conciliation between Germany and Denmark, para. 4; Final Protocol to the Convention of Arbitration and Conciliation between Germany and the Netherlands, para. 4.
[46] See further Chapter 6.3.
[47] Letter of the MiFA to the Minister of Economic Affairs dated 15 March 1965 [NNA/26].
[48] See further Chapter 6.3.
[49] Note dated 10 June 1965 [B102/260135], p. 2; letter of the Minister for the Economy to the FO dated 14 June 1965 [AA/43], p. 2.

market. The Netherlands had discovered a major natural gas field at Slochteren in the Dutch province of Groningen in 1959, which would allow it to export large quantities of natural gas. Export markets were being developed from the first half of the 1960s. The German industry considered that the disputed areas of the continental shelf offered the best chance of finding natural gas, and of competing with the Dutch.[50]

The Ministry for the Economy considered that it was not worthwhile to spend further time on negotiations.[51] Instead, it was suggested to formally inform the governments of Denmark and the Netherlands about the German position.[52] This step would serve to pave the way for the institution of arbitration.[53] This would guarantee that Denmark and the Netherlands would be fully aware of the German position before they started their talks on the delimitation of their bilateral boundary.[54] This boundary, which obviously would be based on the equidistance principle, would be located in its entirety in the disputed area. Although the possibility of this approach had been considered by the Foreign Office,[55] it instead preferred to first hold further talks with Denmark and the Netherlands. Only after those negotiations would have failed, should Germany take the initiative to have the matter settled by arbitration.[56] This proposal was discussed at an interdepartmental meeting on 7 September 1965.[57] The Ministry for the Economy at this time was squarely in favor of resorting to arbitration, notwithstanding the objections of the Foreign Office.[58] The meeting once more concluded that Germany in negotiations would not have anything to offer to Denmark and the Netherlands in exchange for getting access to the center of the North Sea.[59] The meeting agreed to further negotiations as a first step, but also agreed that the first contacts should serve to establish whether negotiations stood any chance of success to start with. Some pressure might be brought on Denmark and the Netherlands by indicating that

[50] Note dated 13 August 1965 [B102/318713], p. 1.
[51] Note dated 10 June 1965 [B102/260135], p. 1.
[52] See *ibid.*, p. 2; letter of the Minister for the Economy to the FO dated 14 June 1965 [AA/43], p. 2.
[53] See note dated 10 June 1965 [B102/260135], p. 2.
[54] Letter of the Minister for the Economy to the FO dated 14 June 1965 [AA/43], p. 2.
[55] Note dated 10 June 1965 [B102/260135], p. 2.
[56] Circular letter of the FO dated 4 August 1965 [AA/40], pp. 2–3.
[57] See note dated 7 September 1965 [AA/41].
[58] See note dated 7 September 1965 [B102/119557].
[59] See note dated 7 September 1965 [AA/41], p. 1; note dated 16 September 1965 [B102/119557], p. 1.

Germany would resort to arbitration if the negotiations would fail.[60] This constituted a compromise between the Ministry for the Economy, which favored immediate recourse to arbitration, and the Foreign Office, which was still internally divided about whether to pursue that course of action.[61]

A further initiative of the Foreign Office towards Denmark and the Netherlands took considerable time to materialize and the Ministry for the Economy expressed its concern about this delay a couple of times.[62] Still, the suggestion that the Foreign Office had been dragging its feet since April 1965[63] does not seem to be completely justified. It had approached Denmark and the Netherlands in May 1965 for a trilateral meeting, but this had been rejected by the Netherlands. After that, the Foreign Office had indeed been slow in dealing with the matter, notwithstanding the wish of the Ministry for the Economy for a speedy resolution. As was mentioned above, internal divisions in the Foreign Office likely were an important factor in this respect.[64]

A proposal on how to deal with the delimitation disputes was eventually cleared with Schröder, the Foreign Minister, in early December 1965.[65] It was explained to the minister that the equidistance principle could not be said to constitute a rule of international law. The interpretation that the concavity of the German coast constituted a special circumstance in any case prevented a mechanical application of the equidistance principle.[66] Germany did not have anything to offer to Denmark and the Netherlands in exchange for getting access to the center of the North Sea.[67] It was also excluded to let the matter rest. Apart from the economic interest of the continental shelf, the ministries concerned and the coastal states were particularly interested in achieving a favorable outcome and the states would not accept inaction of the Federal government.[68] Only two options remained: to propose once more to resume the negotiations; or to submit the matter to compulsory dispute settlement. Negotiations were hardly likely to lead to success, but could be used to signal that the second option would be pursued. There was no risk that arbitration would harm the relations with Denmark and

---

[60] See note dated 16 September 1965 [B102/119557], pp. 1–2.   [61] See *ibid.*, p. 2.
[62] See note dated 27 October 1965 [B102/260173]; note dated 18 November 1965 [B102/260173]; note; "To director-general Kling" dated 30 November 1965 [B102/260173].
[63] Note dated 30 November 1965 [B102/260173], p. 2.
[64] See also note; "To director-general Kling" dated 30 November 1965 [B102/260173].
[65] The proposal is contained in a note dated 29 November 1965 [AA/43].
[66] *Ibid.*, p. 2.   [67] *Ibid.*, p. 3.   [68] *Ibid.*

the Netherlands. The disputes were concerned with the clarification of a legal question, in which case arbitration was particularly apposite.[69] The minister agreed to present an aide-mémoire to Denmark and the Netherlands, and that he would soon afterwards discuss the matter with his Danish and Dutch colleagues.[70]

The aide-mémoires that were presented to Denmark and the Netherlands on 8 December 1965 proposed further bilateral negotiations on the delimitation of the continental shelf, which were to be without prejudice to the legal positions of both sides.[71] Germany was also prepared to submit the legal question in dispute to arbitration. In that case, Germany proposed to consider the matter trilaterally. The two 1926 Treaties of Germany with Denmark and the Netherlands envisaged this possibility when a dispute between the parties had implications going beyond the specific case being submitted to arbitration. Denmark and the Netherlands did not object to the German suggestion to join the two disputes as such.[72] As a matter of fact, they both preferred that approach.[73] It seems that it was considered that the joining of the cases would allow Denmark and the Netherlands to emphasize their common support for the equidistance method. The fact that Germany had been slow in taking further action in 1965 might allow Denmark and the Netherlands to conclude a delimitation agreement before a German initiative would materialize.[74]

The German suggestion of further bilateral talks on the delimitation of the continental shelf did not stand any chance of being accepted as the Netherlands and Denmark were not prepared to renounce their reliance on the equidistance principle.[75] They also objected to the submission of

---

[69] Ibid., pp. 4–5.
[70] Ibid., pp. 5–6; note; "To director-general Kling" dated 1 December 1965 [B102/260173], p. 1.
[71] The text of the aide-mémoire presented to the DEB is quoted in received message in cipher (ref. no. 10434) of the DEB to the MFA dated 8 December 1965 [NNA/26]; the text of the aide-mémoire presented to the DAEB is quoted in telegram in cipher of the DAEB to the MFA dated 8 December 1965 [DNA/4B].
[72] The joining of the cases did pose a risk that a court or tribunal would not look separately at the two delimitations, but consider their combined effect. This was something that Germany favored but Denmark and the Netherlands opposed (see further Chapter 8).
[73] See e.g. note dated 5 August 1965 [DNA/4B], p. 3; send message in cipher (ref. no. 4495) of the MFA to the Dutch Embassy in Oslo dated 27 August 1965 [NNA/26]; send message in cipher (ref. no. 4516) of the MFA to the Dutch Embassy in Oslo dated 30 August 1965 [NNA/26].
[74] Note dated 5 August 1965 [DNA/4B], pp. 2–3.
[75] Note dated 10 December 1965 [DNA/4B], p. 2; note 155/65 dated 10 December 1965 [NNA/26], p. 2.

the disputes to arbitration. The Netherlands and Denmark, separately and in consultations preceding the meeting between the three foreign ministers, considered how to counter the German proposal to go to arbitration.[76] They agreed that they would only accept submission of the disputes to the Court and that they would consult each other before making further arrangements with Germany.[77] The arguments for submission to the Court were not exactly convincing from a legal perspective – certainly not for Germany. The Netherlands considered referring to the fact that article 6 of the Convention on the continental shelf reflected general international law. In view of the role of the Court under the dispute settlement mechanisms of the Geneva Conventions it constituted the obvious forum to deal with this dispute.[78] Denmark in addition intended to refer to the fact that in case the Netherlands rejected arbitration, it would be impractical for Denmark to accept arbitration because the two disputes were concerned with the same legal questions.[79] The actual reason to prefer the ICJ was that Denmark and the Netherlands feared that an arbitral tribunal would not adhere as strictly to the law as the Court.[80]

At the North Atlantic Treaty Organization (NATO) summit in Paris on 15 December 1965, minister Schröder, during his meeting with Luns and Hækkerup, his Dutch and Danish colleagues, maintained the German view on article 6 of the Convention on the continental shelf and suggested that it might be possible to reach agreement on commercial cooperation. Alternatively, it would be possible to submit the legal dispute to arbitration. The ministers agreed that new talks would first be held bilaterally and subsequently might be held trilaterally.[81]

Schröder concluded that the meeting with Luns and Hækkerup opened the way for another attempt to reach a compromise solution. In that light it was no longer sufficient for Germany to claim an appropriate part of the North Sea, as had been done in the past, but it would be necessary to come up with specific claims and boundaries.[82] The

---

[76] See note 155/65 dated 10 December 1965 [NNA/26]; note 81/65 dated 14 December 1965 [NNA/26]; note dated 10 December 1965 [DNA/4B]; telegram in cipher of the Danish Embassy in The Hague to the MFA dated 15 December 1965 [DNA/4C].
[77] Note 155/65 dated 10 December 1965 [NNA/26], p. 3.   [78] Ibid., pp. 2–3.
[79] Note dated 10 December 1965 [DNA/4B], p. 3.
[80] Ibid., p. 2; see also above text at footnotes 26 and 33.
[81] Note (St.S.-2192/65) dated 18 December 1965 [AA/43]; note 86/65 dated 20 December 1965 [NNA/25]; note dated 1 March 1966 [DNA/4C].
[82] Circular letter of the FO dated 31 December 1965 [AA/43], pp. 2–3.

Ministry for the Economy considered a number of options for a com-promise. The starting bid of Germany could be an equal division of the continental shelf between the three States, i.e. a continental shelf area of about 45,420 km^2 for each State.[83] This solution, and two solutions which were slightly less advantageous for Germany, could hardly be expected to be acceptable to Denmark and the Netherlands.[84] A final fallback position could consist of drawing two lines to a central point in the North Sea from the end points of the partial boundaries with the other two States. This would give Germany access to the middle of the North Sea and the area had a good prospectivity for natural gas.[85] That solution would give Germany a continental shelf area of about 36,700 km^2.[86] If this fallback position could not be achieved through negotiations, resort was to be had to arbitration.[87] The Foreign Office should not yet be informed about this last option. Rather, it should be stressed in a meeting between the ministries that the preliminary pro-posals of Germany should be met by counterproposals of Denmark and the Netherlands.[88] The Ministry for the Economy apparently remained unconvinced that the Foreign Office would put up a fight for the German shelf.

An interdepartmental meeting agreed to present Denmark and the Netherlands with the maximum position proposed by the Ministry for the Economy and wait for counteroffers. A compromise in any case should give Germany access to the middle of the North Sea.[89] It was agreed that the minimum which would be acceptable for Germany should not yet be established, but the Ministry for the Economy sub-mitted that a solution giving less area than its fallback position, without specifying it as such, would be unacceptable and resort in that case should rather be had to arbitration.[90]

---

[83] Note; "To the head of section III through the head of subsection III D" dated 12 January 1966 [B102/260136], p. 2.

[84] Ibid., pp. 2–3.   [85] Ibid., p. 3.

[86] This figure is based on information in the figure included in ICJ Pleadings, North Sea continental shelf cases, Vol. II, p. 182.

[87] Note; "To the head of section III through the head of subsection III D" dated 12 January 1966 [B102/260136], p. 3.

[88] Ibid.

[89] Minutes of the interdepartmental meeting of 14 January 1966 in the Foreign Office dated 18 January 1966 [AA/46]; note; "To the head of section III through the head of subsection III D" dated 14 January 1966 [B102/260030].

[90] Minutes of the interdepartmental meeting of 14 January 1966 in the Foreign Office dated 18 January 1966 [AA/46], p. 2.

In the meantime, the thinking about the outcome of the ministerial meeting during the NATO summit on 15 December 1965 had taken a decidedly different turn in the Netherlands. Shortly after the meeting, Riphagen, the legal advisor of the Ministry of Foreign Affairs, advised minister Luns that the new talks he had agreed upon only made sense if the Netherlands would be willing to give up its stance on equidistance or would be prepared to accept arbitration. Both these options were contrary to decisions which had been taken by the Council of Ministers on the basis of proposals of Luns. Riphagen advised contacting the Danish Ministry of Foreign Affairs to seek to agree to a proposal for trilateral talks. The talks should be used to again explain to the Dutch position on equidistance and arbitration to Germany and to offer submitting the dispute jointly to the ICJ. It was considered that this proposal only stood a chance of being accepted if it was explained to the German government at the highest level that the Danish and Dutch governments would resist a German attempt to have recourse to the 1926 Treaties by any means.[91] This proposal was supported by the secretary-general of the ministry, who moreover noted that the risk existed that Germany might unilaterally resort to arbitration, making it even more difficult to invoke the ICJ.[92] Luns accepted Riphagen's proposal and presented it to the Council of Ministers, which agreed to it.[93]

6.3   The legal arguments for submission to arbitration or the Court

At this point, it is necessary to say a little more about the implications of the 1926 Treaties and the 1957 European Convention for the Peaceful Settlement of Disputes and their interaction. The 1926 Treaties provided that a legal dispute between the parties could be referred unilaterally by either party to arbitration. However, a final protocol attached to the treaties implied that if Germany was to become a party to the statute of the PCIJ or a member of the League of Nations either party could refer the dispute unilaterally to the PCIJ. This procedure in that case replaced the possibility of unilateral submission to arbitration.[94]

---

[91] Note 86/65 dated 20 December 1965 [NNA/25; NNA/26].

[92] Hand-written note on *ibid.*, p. 2.

[93] Excerpt from the minutes of the Council of Ministers dated 23 December 1965 [NNA/26], item 5.e.

[94] Final Protocol to the Treaty of Arbitration and Conciliation between Germany and Denmark, para. 4; Final Protocol to the Convention of Arbitration and Conciliation between Germany and the Netherlands, para. 4. See also the Explanatory memorandum

Germany in fact became a member of the League of Nations in September 1926. The final protocols furthermore provided that the aforementioned provision also applied in the case of the entry into force between Germany and a third State of a general arbitration treaty containing a corresponding provision.[95] The PCIJ was the predecessor of the ICJ. Article 37 of the Statute of the ICJ regulates its relation to the Permanent Court and entails that a reference in a treaty to the PCIJ has to be read as a reference to the ICJ.[96] Denmark and the Netherlands as members of the United Nations were a party to the Statute of the ICJ. Germany was not a member of the United Nations and was not a party to the Statute, but had accepted the jurisdiction of the Court on 29 April 1961 in connection with its ratification of the 1957 European Convention. The possibility to accept the jurisdiction of the Court for States which were not a party to the Statute had been created through United Nations Security Council Resolution 9 (1946) of 15 October 1946. The resolution provides that acceptance of the jurisdiction of the Court is in accordance with the terms and subject to the conditions of the Stature of the Court. This includes article 37 of the Statute on the relationship between the Court and its predecessor. This implied that in accordance with the final protocol of the 1926 Treaties unilateral submission under these treaties to arbitration was no longer an option.

After the German aide-mémoire of 8 December 1965, which proposed to submit the dispute to arbitration, Riphagen, the legal advisor of the Dutch Ministry of Foreign Affairs, had a closer look at the German-Dutch 1926 Convention. He acknowledged that article 7 allowed unilateral submission of a dispute. Interestingly, Riphagen did not refer to the final protocol of the treaty. Riphagen submitted that in the light of article 7, the Dutch government should maintain that the 1926 Convention was not applicable to the delimitation dispute between the Netherlands and Germany. However, he submitted that according to the letter of article 7 of the 1926 Convention, if a dispute was submitted unilaterally by one party, it would then be up to the arbitral tribunal to decide on its

---

to the bill for approval of the treaty submitted to the Dutch parliament (K. II (BH) (1925–1926), 371, no. 3, p. 10).

[95] Final Protocol to the Treaty of Arbitration and Conciliation between Germany and Denmark, para. 4; Final Protocol to the Convention of Arbitration and Conciliation between Germany and the Netherlands, para. 4.

[96] Article 37 provides: "Whenever a treaty or convention in force provides for reference of a matter to a tribunal to have been instituted by the League of Nations, or to the Permanent Court of International Justice, the matter shall, as between the parties to the present Statute, be referred to the International Court of Justice."

jurisdiction to deal with the dispute.[97] To avoid that Germany would take this step, it should be made clear to the German government that the Netherlands would oppose it with all possible means. However, it would not suffice to reject arbitration. If Germany were to be contacted, the Netherlands should also indicate that it was willing to submit the dispute to the ICJ and should even propose this course of action. In other words, the Netherlands had to take the initiative and should not wait for Germany to act.[98] However, during their discussions in Paris in December 1965, the three foreign ministers had actually agreed that Germany would take the initiative for further steps. It cannot be excluded that the attitude of Schröder, the German Foreign Minister, may have conveyed the impression that he was not particularly interested in the matter and may have led the Netherlands to conclude that it could take a different approach without too much political risk. According to Hækkerup, the Danish Minister of Foreign Affairs, Schröder "began by saying that he did not know much about the matter, and he did not go into details."[99]

Sørensen, the legal advisor of the Danish Ministry of Foreign Affairs, carried out a detailed analysis of the implications of the Danish-German 1926 Treaty and the 1957 European Convention for the Peaceful Settlement of Disputes and their interaction in January 1966.[100] Sørensen concluded that there, in any case, existed an obligation for Denmark to accept either arbitration or submission of the dispute to the Court. Denmark could not object to a German choice for arbitration as article 28 of the 1957 Convention implied a precedence for that procedure.[101] Article 28 provided that "[t]he provisions of this Convention shall not apply to disputes which the parties have agreed or may agree to submit to another procedure of peaceful settlement." The effects of this provision would depend on one's interpretation of the final protocol to the 1926 Treaties. Sørensen questioned whether the final protocol to the 1926 Treaty could be used to argue that precedence had to be given to the Court.[102] Basically, Sørensen argued that Germany, although it had accepted the competence of the Court in accordance

---

[97] Note 83/65 dated 15 December 1965 [NNA/26], p. 1.   [98] *Ibid.*, pp. 1–2.
[99] Hand-written observation of Hækkerup dated 12 March 1966 on p. 2 of note Danish-German negotiations on the continental shelf in The Hague on 28 February 1966 dated 1 March 1966 [DNA/134]; an English translation is included in the same file.
[100] Settlement of legal disputes between Denmark and Germany dated 10 January 1966 (Danish text in [NNA/40]; English translation in [DNA/4C]).
[101] *Ibid.*, Danish text, p. 12; English translation, p. 9.   [102] *Ibid.*

with the 1957 Convention, had not become a party to the Statute and this
made article 37 of the Statute on the relation between the Permanent
Court and the ICJ inapplicable to the 1926 Treaty with Denmark.[103] This
argument did not consider that United Nations Security Council
Resolution 9 (1946) dealing with the acceptance of the jurisdiction of
the Court by States that are not a member to the Court's Statute, provides
that such acceptance is in accordance with the terms and subject to the
conditions of the Statute of the Court.

Sørensen's analysis was shared with the Dutch. Riphagen reached a
different conclusion from Sørensen.[104] He also diverged somewhat from
his earlier assessment of the German-Dutch 1926 Convention. The central
tenet of Riphagen's analysis concerned the interpretation of article 28 of
the European Convention, which as was indicated above, deals with its
relationship to other agreements. According to Riphagen, article 28 did
not give an absolute priority to arbitration over submission to the Court.
After all, the Convention itself rather was based on the priority of the
Court as was among others clear from its articles 1 and 4. In that context,
the exception laid down in article 28 should be interpreted to be only
applicable in a case in which another agreement provided for the *exclusive*
competence of another procedure. The 1926 Convention could not be
considered as such.[105] If Germany were to invoke article 28 of the
European Convention, Denmark and the Netherlands could thus reply
that this exception was not applicable in the present case. The interpretation of article 28 was entrusted to the Court and Riphagen had little doubt
that the Court would accept his interpretation of article 28, at least in
respect of prior treaties.[106] Whether Riphagen's assessment of the relation
between the 1926 Convention and the 1957 Convention is completely
convincing is open to question. Article 28(1) does not give reason to
assume that it only excluded reference to the ICJ in case a treaty provided
for exclusive jurisdiction for another procedure. The second sentence of
article 28(1) provides:

> Nevertheless, in respect of disputes falling within the scope of Article 1,
> the High Contracting Parties shall refrain from invoking as between

---

[103] *Ibid.*, English text, pp. 2 and 9.
[104] Letter of W. Riphagen to M. Sørensen dated 31 January 1966 [NNA/40]; the analysis of
Riphagen in large part was based on a memorandum prepared by Maas Geesteranus of
the legal department of the MFA (note 6/66 dated 21 January 1966 [NNA/40]).
[105] Letter of W. Riphagen to M. Sørensen dated 31 January 1966 [NNA/40], p. 2.
[106] *Ibid.*, pp. 3–4.

themselves agreements which do not provide for a procedure entailing binding decisions.

The 1926 Treaties did provide for the possibility of binding decisions. However, Riphagen's analysis did indicate that the Netherlands had *prima facie* credible arguments to oppose a reference of the dispute to arbitration. The existence of such opposition would have given the impression that Germany was imposing its views on the Netherlands and would have been at odds with the German wish to convey the impression that the disputes were being settled in a spirit of mutual cooperation without straining the relations with its direct neighbors.

The deputy head of the political and legal department of the Danish Ministry of Foreign Affairs, who also looked into the matter, did not attach decisive weight to Sørensen's analysis, but concluded that there existed extremely good arguments of a political and legal nature to reject arbitration and argue for reference to the ICJ. He considered that both the 1926 Treaty read in conjunction with its final protocol and the 1957 Convention gave precedence to the Court.[107]

6.4 Agreement to go to the Court

Following the decision of the Dutch Council of Ministers that the Netherlands should take the initiative for further consultations in order to avoid submission of the delimitation dispute to arbitration, the Ministry of Foreign Affairs contacted its Danish counterpart.[108] Paludan, the deputy head of the political and legal department concluded that the Netherlands apparently was in a hurry because it wanted to start exploration activities in the disputed area. On the other hand, Denmark was not in a similar hurry and Paludan found that it was preferable to leave the initiative to Germany. The passage of time might reinforce the interests in Bonn that did not want to press for a confrontation with Denmark and the Netherlands. In that light it might be left to the Netherlands to battle with Germany about what constituted the right forum, but the Netherlands might consider this to be a lack of solidarity. All in all, if the Dutch would insist, Denmark should not oppose a Dutch initiative.[109]

---

[107] Letter of J. Paludan to ambassador H. Hjorth-Nielsen dated 17 January 1966 [DNA/4C].

[108] See message in cipher (ref. no. 44) of the MFA to the DEC dated 6 January 196[6] (the message is (erroneously) dated 6 January 1965) [NNA/27].

[109] Note dated 10 February 1966 [DNA/4C]; see also received message in cipher (ref. 2210) dated 11 February 1966 [NNA/21A].

When Denmark and the Netherlands met to discuss the matter in February 1966, they easily agreed that there existed good legal arguments to refuse arbitration and support submission to the ICJ.[110] They disagreed on tactics. Denmark considered that the ministers in Paris had agreed to first hold bilateral talks and Denmark was waiting for a German invitation. Denmark moreover was not in a hurry and did not feel inclined to take any initiatives. If Germany would invite Denmark for bilateral talks, Denmark would accept the invitation. Denmark did agree to a Dutch proposal that the Netherlands would invite Germany for trilateral talks.

Germany had previously been in contact with the Netherlands concerning bilateral talks, but the Netherlands had played for time because it still needed to coordinate its position with Denmark.[111] The Dutch Embassy in Bonn had first indicated, in informal contacts, that the crisis in the European Economic Community had prevented the ministry from looking into the matter.[112] When the German Foreign Office proposed to hold the bilateral talks, a diplomat at the Dutch Embassy referred to the marriage between the Dutch crown princess Beatrix and the German diplomat Claus von Amsberg, which was to be celebrated at the end of March 1966. Bilateral talks could throw a shadow over the German-Dutch relations.[113] To the surprise of the Foreign Office, the Netherlands shortly afterwards issued an invitation for trilateral talks in The Hague on 28 February 1966. Notwithstanding the short notice – the invitation

---

[110] Information on this meeting is contained in note dated 21 February 1966 [DNA/4C] and send message in cipher (ref. no. 1233) of the MFA to the DEB dated 23 February 1966 [NNA/40].

[111] Received message in cipher of the MFA to the DEB dated 26 January 1966 [NNA/22].

[112] Ibid. This crisis in respect of the predecessor of the European Union concerned the decision-making procedures in the Council of Ministers of the Community. France opposed the gradual replacement of unanimity voting by majority voting as was being envisaged by the Treaty of Rome. From 30 June 1965, and for seven months, France refused to participate in the meetings of the Council of Ministers.

[113] Telex dated 17 February 1966 [DNA/36]; see also note dated 8 March 1966 [AA/46], pp. 1–2. The relationship of Beatrix with Claus von Amsberg had led to commotion in the Netherlands in the first half of 1965 when it was reported that Von Amsberg had been a member of the Hitler Youth. On request of the government this matter was investigated by the Netherlands Institute for War Documentation. After the Institute concluded that this membership was not an issue – membership had been obligatory – the engagement between Beatrix and Von Amsberg was announced. This episode had highlighted the continuing Dutch sensitivities about the German occupation during the Second World War.

was extended on 24 February – the Germans accepted to avoid the impression that they did not consider the matter of urgency.[114]

Germany viewed the meeting of 28 February as a preparation for further negotiations on a compromise solution, as had been agreed upon by the three ministers in Paris. At least that had been the understanding of minister Schröder.[115] This was not directly contradicted by Denmark and the Netherlands, but they conveyed that they did not believe that there was any use in such negotiations.[116] A departure from the equidistance line in a negotiated settlement was unacceptable.[117] Denmark and the Netherlands wanted to turn as soon as possible to a discussion of the appropriate forum to submit the dispute to. In their view only the ICJ was an option.[118] The German delegation indicated that it did not have a mandate to discuss this issue and reserved the right to further discuss a compromise solution at a next meeting, at which it would present specific options.[119] Nonetheless, there was a preliminary discussion on the options for dispute settlement. Germany indicated its preference for arbitration under the two 1926 Treaties, while Denmark and the Netherlands pointed out that the 1957 European Convention accorded precedence to the ICJ. They also argued that public opinion would have less difficulty in accepting a decision by the ICJ. Denmark and the Netherlands would be prepared to work closely together with Germany to inform public opinion if the case were to be brought to the ICJ and they also submitted that a procedure before the ICJ need not take more time than arbitration.[120] The Netherlands moreover observed that it would oppose the jurisdiction of an arbitral tribunal.[121] At the end of the meeting, Truckenbrodt, the German head of delegation, indicated to be prepared to consider the proposal to submit the dispute to the ICJ.[122]

---

[114] Note dated 8 March 1966 [AA/46], p. 2.

[115] Report trilateral meeting February 1966 [DNA/4C], p. 1; note dated 8 March 1966 [AA/46], p. 2; send message in cipher (ref. no. 1424) of the MFA to the DEB dated 2 March 1966 [NNA/40], p. 1.

[116] Report trilateral meeting February 1966 [DNA/4C], pp. 2–4; note dated 8 March 1966 [AA/46], p. 2.

[117] Note dated 8 March 1966 [AA/46].   [118] Ibid.

[119] Ibid., p. 3; Report trilateral meeting February 1966 [DNA/4C], p. 5; send message in cipher (ref. no. 1424) of the MFA to the DEB dated 2 March 1966 [NNA/40], p. 1.

[120] Send message in cipher (ref. no. 1424) of the MFA to the DEB dated 2 March 1966 [NNA/40], pp. 1–2; Report trilateral meeting February 1966 [DNA/4C], pp. 5–10.

[121] Send message in cipher (ref. no. 1424) of the MFA to the DEB dated 2 March 1966 [NNA/40], p. 2.

[122] Ibid.; Report trilateral meeting February 1966 [DNA/4C], p. 10.

Sympher, of the staff of Truckenbrodt's section in the Foreign Office, prior to the meeting had already observed that Germany would not be able to prevent that Denmark and the Netherlands would invoke the ICJ under the 1957 European Convention. Although article 28 of the Convention in his view gave precedence to the 1926 Treaties, the Court would be competent to decide on the interpretation of that article. Whether to object to the jurisdiction of the Court before the Court was also a political question.[123] The final protocols to the 1926 Treaties moreover accomplished a linkage between those treaties and the competence of the Court.[124] Sympher concluded his analysis by pointing out that Germany would have to act quickly if it wanted to prevent that its arguments for arbitration would not have to be framed in the terms of an explicit rejection of the ICJ.[125] This was what actually happened at the trilateral meeting at the end of February 1966. The Netherlands and Denmark argued for the Court and Germany had to explain why it did not like that option. After the meeting, Truckenbrodt concluded that Germany as a party to the 1957 Convention would not be able to reject the competence of the Court because the dispute was concerned with a question of general international law. As a consequence, Germany could not oppose the wish of the other States to submit the dispute to the ICJ.[126] Truckenbrodt judged that the Danes and Dutch had put much effort into convincing Germany that a case before the Court would not have any negative implications for their bilateral political relations and that any hint of politics would be avoided.[127]

Denmark, Germany and the Netherlands agreed on another meeting in the first half of May of 1966.[128] The German Foreign Office concluded that this would be the last possibility to make progress on a negotiated settlement.[129] The Foreign Office considered that this would only be possible, if at all, if the ministries and the coastal states would be able to agree on a proposal on joint development.[130] The Foreign Office did not come up with any suggestions in this respect itself. A number of

---

[123] Note dated 16 February 1966 [AA/42], p. 2.   [124] Ibid.   [125] Ibid., p. 3.
[126] Note dated 8 March 1966 [AA/46], p. 3.
[127] Ibid.; see also Report on the trip to The Hague for the German-Dutch-Danish negotiations on 28.2.66 concerning the delimitation of the continental shelf of the North Sea [B102/260173], p. 3.
[128] Note dated 8 March 1966 [AA/46], p. 3.
[129] Circular letter of the FO dated 14 March 1966 [B102/260030].
[130] Note dated 8 March 1966 [AA/46], p. 3; see also circular letter of the FO dated 14 March 1966 [B102/260030].

proposals for joint development were developed by the Ministry for the Economy and Transport of the state of Lower Saxony.[131] These proposals were concerned with an area beyond the equidistance lines between Germany and the other two States, which measured some 20,000 km^2 and which was to extend to the median line with the United Kingdom. This area was proportional to the length of the German coast as compared to the other two coasts. It was proposed to divide this area between Germany, Denmark and the Netherlands in a proportion of 2:1:1 or in equal shares. If one of the States concerned would not carry out activities in its area, the other States would be entitled to carry out activities. These arrangements could be either permanent or provisional.

A meeting between the ministries, the coastal states and representatives of the German North Sea consortium led to the conclusion that the proposals for joint development did not serve the interests of the consortium, would be difficult to implement, and probably would be unacceptable to Denmark and the Netherlands.[132] The consortium originally had been interested in a quick discovery of natural gas in the North Sea to allow it to make use of the possibilities to sell gas on the German market it had already secured under provisional contracts. The Consortium considered that it was unlikely that it would still make such a discovery and was now interested in obtaining the largest possible concession area, even if this would take time. The meeting agreed that Germany, at the meeting with Denmark and the Netherlands in May, would ask whether they saw any room for a compromise solution. In case of a negative reply, Germany should immediately turn to the question of third party dispute settlement.[133]

As a matter of fact, the German Foreign Office had already run ahead of the outcome of the internal meeting. The Danish Embassy in Bonn had been informed in early April that the Foreign Office expected that the next trilateral meeting would not lead to a compromise and it was

---

[131] Draft of a letter of the Minister for the Economy and Transport of Lower Saxony to the FO contained in an annex to a letter of the Minister for the Economy and Transport of Lower Saxony to representatives of the other three coastal states dated 22 March 1966 [B102/260030].

[132] Note on the meeting in the ME on 18 April 1966 dated 29 April 1966 [B102/260030 and AA/45]; see also letter of Preussag to Dr. Obernolte of the ME dated 3 May 1966 [B102/260030].

[133] Note on the meeting in the ME on 18 April 1966 dated 29 April 1966 [B102/260030 and AA/45].

suggested that the meeting could already consider the modalities for the submission of the dispute to the ICJ.[134]

The Foreign Office was also ahead of the internal deliberations on the latter point. The meeting between the German authorities of 18 April 1966 had not made a definitive choice between the ICJ and arbitration.[135] The Foreign Office, shortly after that meeting, prepared a draft of a special agreement to submit the dispute to the ICJ[136] and informed the Danish Embassy of its content before the trilateral meeting scheduled for May 1966.[137]

At the trilateral meeting in Bonn on 13 May 1966, Denmark and the Netherlands indicated that it was impossible for them to accept a compromise solution which would be contrary to their legal position that their boundary with Germany had to be determined in accordance with the equidistance principle.[138] The Danish delegation moreover indicated that the Danish concessionaire was not interested in joint development and the Dutch delegation rejected such an approach as unnecessary and impractical.[139] Interestingly, an internal note of the Danish Ministry of Foreign Affairs had looked at different scenarios for a compromise, but had concluded that in essence the Danish authorities lacked experience and knowledge in respect of oil exploration and exploitation and its financing and consequently were not in a position to judge whether joint exploitation was more advantageous to Denmark than a compromise on the boundary.[140] This may have contributed further to the unwillingness of Denmark to compromise, as it otherwise risked opening itself to the criticism that it had struck a bad deal. The internal note also suggested that the interests of the Danish State and the Danish concessionaire did not necessarily run parallel. One could even imagine that a compromise

---

[134] Note; The boundaries between the Danish, Dutch and German shelf areas in the North Sea dated 13 April 1966 [DNA/134]; see also send message in cipher (Ag. no. 5502–1402) of the DEB to the MFA dated 25 April 1966 [NNA/22].

[135] Note on the meeting in the ME on 18 April 1966 dated 29 April 1966 [B102/260030 and AA/45], p. 3.

[136] Note dated 3 May 1966 [AA/45].

[137] Telegram from the DAEB (translation) dated 10 May 1966 [DNA/134].

[138] Reports on the meeting are contained in Report on the Danish-Dutch-German negotiations in Bonn on 13 May 1966 concerning the boundaries between respectively the Danish and German and Dutch and German continental shelf areas in the North Sea dated 16 June 1966 [DNA/56] (Report on Danish-Dutch-German negotiations May 1966 [DNA/56]); circular letter of the FO dated 17 May 1966 [AA/45]; send message in cipher (ref. no. 3642) of the MFA to the DEB dated 18 May 1966 [NNA/27].

[139] Report on Danish-Dutch-German negotiations May 1966 [DNA/56], pp. 2–3.

[140] Note dated 20 December 1966 [DNA/39], pp. 2–3.

solution might turn out better for Denmark than winning the case in The Hague.[141] For the Danish concessionaire joint development would mean that it would lose a part of its exclusive concession area.

6.5 The content of the Special Agreements submitting the disputes to the Court

After the three delegations concluded that a compromise was out of the question, the trilateral meeting turned to a discussion of the German draft for an agreement to submit the disputes to the ICJ and an interim arrangement in respect of the disputed area. The German draft agreement envisaged that the Court would be requested to decide whether there existed a rule of international law that required that the remainder of the continental shelf boundaries between Germany and Denmark and Germany and the Netherlands had to be determined in accordance with the equidistance principle.[142] If the Court would decide this question in the affirmative, the parties were to determine their boundaries on that basis. In the case of a negative answer, the parties were to reach an agreement through negotiations.[143] The inclusion of this latter provision had been suggested to avoid the Court having to consider the effects of its decision for the specific case and would also avoid Denmark and the Netherlands arguing that a negative answer to the applicability of the equidistance principles would not advance the resolution of the dispute.[144] Just prior to the meeting there had also been suggestions to reformulate article 1(1) and to ask the Court either to look into the possible impact of special circumstances or circumstances which made the equidistance line inequitable.[145]

The formulation of the question to be submitted to the Court that was suggested at the trilateral meeting would make it possible for Germany to focus on the inequitable nature of the equidistance principle without having to elaborate its views on the content of the applicable law. The development of a convincing alternative to the equidistance principle had been a constant concern for the Foreign Office. For instance, in April 1965 the Foreign Office had concluded that the question of the German share of the continental shelf in the central part of the North Sea and its

[141] Ibid.   [142] Text of Draft Agreement dated 5 May 1966 [AA/45], article 1(1).
[143] Ibid., article 1(2).   [144] Note dated 3 May 1966 [AA/45].
[145] See note dated 11 May 1966 [AA/45] and telex of the ME to the FO dated 12 May 1966 [AA/45].

delimitation in relation to Denmark and the Netherlands was hardly justiciable.[146] The Foreign Office had moreover not been particularly diligent in working out an affirmative case.[147]

Denmark and the Netherlands rejected the German formulation of the questions to be submitted to the Court. The Netherlands argued that Germany, in its declaration on the continental shelf of January 1964 and during the negotiations with the Netherlands, had given the impression to base itself on the Convention on the continental shelf. This made it appropriate to define the question for the Court by referring to the Convention and ask it if there were special circumstances which made the equidistance principle inapplicable.[148] Germany responded that it had not ratified the Convention and consequently was not bound by it. The declaration had only been intended to stake out Germany's claim to a part of the continental shelf in the North Sea and had to be seen in that context. In the negotiations with the Netherlands, the hypothetical case had been considered that both parties would have ratified the Convention.[149] At that point Denmark intervened, stating that it was in the interest of the parties to have a judgment which would assist them as much as possible in their subsequent negotiations. The German proposal could lead to a judgment which would not allow the parties to agree on their mutual boundaries.[150] These arguments suggest that Denmark and the Netherlands considered that the absence of specific guidance would be to the advantage of Germany.[151] Denmark proposed to ask the Court which legal principles ought to be applied in the

---

[146] Note from the legal department to the political department dated 22 April 1965 [AA/41], p. 3; see also letter of Benkendorff of the Ministry for the Economy and Transport of Lower Saxony to Truckenbrodt of the FO dated 9 May 1966 [AA/46], p. 1; R. Bernhardt, *Rechtsgutachten über Fragen der Abgrenzung des Festlandsockels unter der Nordsee* (Legal Report on question concerning the delimitation of the continental shelf under the North Sea) (prepared for the German FO) dated 15 October 1966) [B102/260031] (Bernhardt, *Rechtsgutachten* [B102/260031]), p. 9.

[147] See further Chapters 4.3 and 5. Interestingly, the original request for the advice from Bernhardt (Bernhardt, *Rechtsgutachten* [B102/260031]) had only asked to focus on the question whether the equidistance principle was applicable to the delimitation of the continental shelf between Germany and Denmark and the Netherlands and had refrained from requesting advice on the rules which would be applicable in case of a negative answer to this question (letter of Thierfelder, head of the legal department of the FO to Bernhardt dated 7 April 1966 [B102/260030]).

[148] Report on Danish-Dutch-German negotiations May 1966 [DNA/56], p. 3.

[149] *Ibid.*, pp. 3–4.   [150] *Ibid.*, p. 4.

[151] See also send message in cipher (ref. no. 3642) of the MFA to the DEB dated 18 May 1966 [NNA/27], p. 2.

determination of the boundaries between the parties. A further elaboration of this question was not considered to be necessary, but could be left to the proceedings.[152] Germany agreed that the question to the Court should be as specific as possible. However, the problem of the applicability of the equidistance principle in the North Sea had to be placed center stage. A rejection of its applicability would give rise to a wholly new situation and a judgment in any case would contain indications which could guide the parties in subsequent negotiations.[153] The Netherlands expressed its support for the Danish formulation.[154] Denmark and the Netherlands in all likelihood had already agreed on this approach to the negotiation of article 1 beforehand. Both delegations had met on the eve of the trilateral talks and this had allowed them to "work in tandem" during the meeting.[155] The German delegation agreed to further study the text for article 1(1) proposed by Denmark which, after amendments by the Netherlands, came to read: "what principles and rules of international law are applicable to the delimitation, as between the parties, of the areas of continental shelf in the north sea [sic] which appertain to each of them beyond [their] partial boundaries."[156] The second paragraph of article 1 was amended on the basis of the suggestions of Riphagen and Sørensen. Instead of being limited to a reference to negotiations to determine the further boundaries, it was added that the parties agreed to delimit their boundaries on the basis of the judgment of the ICJ.[157] This change was a logical consequence of the new formulation of article 1(1), which no longer only asked the Court to rule on the equidistance principle, but to pronounce itself on the applicable law.

Draft article 1 was the subject of lengthy discussions between the German authorities to determine whether it was sufficiently comprehensive. In the end it was decided to leave the article as it was and eventually argue in the alternative during the proceedings.[158] There are limitations to introducing new issues during the proceedings before the Court. A

---

[152] Report on Danish-Dutch-German negotiations May 1966 [DNA/56], p. 4.
[153] *Ibid.*   [154] *Ibid.*
[155] Send message in cipher (ref. no. 3642) of the MFA to the DEB dated 18 May 1966 [NNA/27], p. 1. Translation by the author. The original text reads: "goed samenspel kon worden gegeven."
[156] *Ibid.*, p. 2.
[157] See received message in cipher (ref. no. 6186) of the DEB to the MFA dated 8 June 1966 [NNA/27]; Draft agreement (annexed to a circular letter of the FO dated 17 May 1966) [AA/45], article 1.
[158] Note dated 20 July 1966 [B102/260031], p. 3.

party is not entitled to unilaterally introduce a new claim that would transform the subject of the dispute originally submitted to the Court if it entertained that claim.[159] At the same time, the Court has taken a liberal approach in assessing a linkage between a new claim and the original dispute. An additional claim is allowed if it is implicit in the application or arose directly out of the question which is the subject matter of the application instituting proceedings.[160] The broad formulation of draft article 1 should allow the Court to deal with all questions related to the law applicable to the delimitation of the continental shelf.

An additional argument for the German authorities to accept the proposed question for the Court was that it offered the best guarantee against an intervention of the German Democratic Republic in proceedings before the Court.[161] It had been suggested that the German Democratic Republic or Poland might seek to intervene for political reasons, such as the reconfirmation of the Oder-Neisse boundary between Poland and the German Democratic Republic or to stress the fact that the German Democratic Republic was a State.[162]

The Statute of the ICJ allows a State to submit a request to intervene in a case before the Court if it considers that it has an interest of a legal nature which may be affected by the decision in the case. The Court is to decide on such a request.[163] In 1966 there was no judicial experience with article 62.[164] That the interest of a State in the construction of a rule of customary international law is not sufficient to meet the requirement of an interest of legal nature is indicated by article 63 of the Statute, which allows the intervention of a State whenever the construction of a convention to which States other than those concerned in the case are parties.[165] In that light, the question as formulated, which in addition

---

[159] Phosphate lands in Nauru (*Nauru v. Australia*) (Preliminary Objections) [1992] ICJ Reports 240, para. 70.

[160] *Ibid.*, para. 67.

[161] Note dated 20 July 1966 [B102/260031], p. 3; see also letter of the ME to the FO dated 5 July 1966 [B102/260031], p. 3; note dated 18 July 1966 [AA/45], p. 3.

[162] See letter of the Ministry of Justice to the FO dated 21 June 1966 [AA/45], p. 3; letter of the ME to the FO dated 5 July 1966 [B102/260031], p. 3; note dated 18 July 1966 [AA/45].

[163] Statute of the International Court of Justice, article 62.

[164] See also S. Rosenne, *The International Court of Justice* (Leyden: A.W. Sythoff, 1961), p. 333.

[165] See also A. Zimmerman, "International Courts and Tribunals, Intervention in Proceedings," *Max Planck Encyclopedia of Public International Law* (online edition) (www.mpepil.com), para. 9.

referred explicitly to the North Sea, indeed offered the German Democratic Republic little chance for a successful intervention.

One might wonder why Denmark, Germany and the Netherlands did not decide to ask the Court to establish the continental shelf boundaries between them. This option as a matter of fact was advocated by Elihu Lauterpacht, who was advising the DUC. Lauterpacht considered that the question as formulated might result in an answer which would limit itself to paraphrasing article 6 of the Convention on the continental shelf.[166] The general wording of the question also meant that Denmark lost "the opportunity of framing the issue in terms that would oblige [Germany] in effect to be the plaintiff."[167] Asking for specific boundaries would oblige the Court to decide the dispute in such a way that it would be clearly and finally resolved and would eliminate the need for the subsequent delimitation by agreement between the parties.[168]

The questions for the Court, suggested by Lauterpacht, point to one difficulty in asking for specific boundaries.[169] Lauterpacht's questions implied that Germany would not share a boundary with the United Kingdom, something which was a fundamental concern for Germany. On the other hand, it requires little imagination to understand that Denmark and the Netherlands could not accept a formulation which might suggest that they did not share a common boundary. In addition, Denmark and the Netherlands did not have an interest in suggesting a question which might imply that the specific boundaries would deviate from the equidistance lines. On the other hand, Germany had difficulty in arguing its positive case. Asking for specific boundaries would make this task even more difficult and there was no reason for Germany to commit itself on this point.[170]

In a meeting on the outcome of the trilateral talks in the Continental shelf committee, which had been set up by the Danish Ministry of Foreign Affairs to advise it on the conduct of the case with Germany,[171] Paludan explained that Denmark would have preferred to ask the Court to draw the boundary. This had not been done because it was not something which the Court could be expected to undertake, and such a concrete formulation would have been unacceptable for

---

[166] E. Lauterpacht, North Sea Continental Shelf cases, memorandum (dated 16 July 1966) [DNA/56], pp. 8–9.

[167] *Ibid.*, p. 9.   [168] *Ibid.*, pp. 8–11.   [169] See *ibid.*, p. 10.

[170] See also the next paragraph.

[171] On this committee see also Chapter 8, text at footnote 92.

Germany.[172] Following the meeting, the Danish Embassy was requested to seek a written commitment from the German Foreign Office that the parties would conduct their case in such a way as to allow that as little doubt as possible would remain about the content of the applicable law. Truckenbrodt of the Foreign Office's legal department indicated that in his view it would also serve the German interest to have as clear as possible a judgment from the Court. At the same time, Germany had to obtain the best result from the Court and the approach Germany would take had not yet been established.[173] Members of the Danish Continental shelf committee were not satisfied with this outcome and insisted on an official German statement. Paludan and Sørensen, the ministry's legal advisor, considered that it was impossible to ask for such a statement. Moreover, there was little disagreement in reality and Paludan was not concerned that Germany would sabotage the case.[174]

Discussions at a conference in Cambridge in April 1967, in which representatives of Denmark, Germany and the Netherlands participated, again made Denmark wary that Germany might take the position that there were no rules and principles applicable to the delimitation of the continental shelf. At the conference, Treviranus of the German Foreign Office had submitted that it would be impossible for a State which invoked special circumstances to indicate another specific boundary. If equidistance did not apply, one had to fall back on equity in reaching a specific settlement.[175] Commenting on the conference, Paludan concluded that it was difficult to see how article 1 of the Special Agreement should have been drafted to prevent Germany from making this kind of argument.[176] The German Memorial, which was submitted in August 1967, would make it clear that Germany had chosen to be quite specific

---

[172] Report on the meeting of the advisory committee for the case with Germany on the continental shelf (Continental shelf committee) held on Monday 29 August 1966 (KSU/MR.1) [DNA/138], p. 7.

[173] Letter of Paludan of the Danish MFA to the DAEB dated 31 August 1966 [DNA/56]; letter of the DAEB to the Danish MFA dated 5 September 1966 [DNA/56].

[174] Report on the meeting of the advisory committee for the case with Germany on the continental shelf (Continental shelf committee) held on Monday 19 December 1966 (KSU/MR.2) [DNA/138] (Report Danish CSC December 1966 [DNA/138]), pp. 2–3.

[175] Report; Cambridge conference, April 1967 (KSU/R.2) [DNA/119] dated 20 April 1967, para. 30.

[176] Report on the meeting of the advisory committee for the case with Germany on the continental shelf (Continental shelf committee) held on Monday 8 May 1967 (KSU/MR.4) [DNA/39] (Report Danish CSC May 1967 [DNA/39]), p. 6.

about the kind of delimitation it considered reasonable in the light of the applicable law.

Denmark, Germany and the Netherlands also had to decide on a number of other issues in connection with the submission of their disputes to the ICJ. One of these was whether the two disputes should be submitted by one trilateral agreement or by two bilateral agreements. This did not concern a fundamental difference of views on the propriety of keeping the two disputes separate or joining them. At the trilateral meeting of 13 May 1966, the parties agreed that the submission to the Court had to be done jointly.[177] The German proposal that Denmark and the Netherlands should act as one party was not considered feasible because of practical difficulties of a procedural nature.[178] It was agreed that the Netherlands and Germany would informally contact the Registry of the Court concerning this matter.[179] The Registrar of the Court indicated that the practice of the Court did not provide any precedent on the question as to whether the joint submission of the disputes from a procedural perspective might be better achieved by one trilateral or two bilateral agreements. The Registrar was prepared to discuss the matter informally with the President of the Court.[180] The Registry subsequently informed the parties that it was not prepared to advise on the question of whether the Court would join the cases. The President did not want to take a position because he himself and a number of judges would resign from the Court before this matter would have to be decided.[181] At that time, the three delegations had already decided to conclude two bilateral special agreements and a trilateral protocol dealing with questions related to the request to the Court to join the two cases.[182] Apart from the request to join the cases, the protocol registered the agreement of the parties that Denmark and the Netherlands were to be considered parties in the same interest in

---

[177] Send message in cipher (ref. no. 3642) of the MFA to the DEB dated 18 May 1966 [NNA/27], p. 1; see also note dated 11 July 1966 [AA/45], p. 1.

[178] Send message in cipher (ref. no. 3642) of the MFA to the DEB dated 18 May 1966 [NNA/27], p. 1.

[179] Ibid.

[180] Note dated 11 July 1966 [AA/45], pp. 1–2. Apart from the Registrar of the Court, Aquarone, the meeting was attended by Riphagen, the legal advisor of the Dutch MFA, Fleischhauer of the German FO and Kisum of the Danish Embassy in The Hague (ibid., p. 1).

[181] Note no. 48/66 dated 3 August 1966 [NNA/27].

[182] P.M. dated 1 August 1966 [NNA/27]. The text of the Special Agreements and the trilateral protocol is included in the judgment of the Court ([1969] ICJ Reports, pp. 6–7).

the meaning of article 31(5) of the Statute of the Court. This implied that, for the purposes of proceedings before the Court, Denmark and the Netherlands would be considered as one party, implying among others that they could only jointly nominate one judge ad hoc.

The Court did not act immediately on the request of the parties to join the cases. They were listed separately in the General List of cases of the Court and a number of decisions in respect of the written pleadings were directed separately to the parties in each of the cases. Two orders of the President of the Court of 1 March 1968 had set 30 August 1968 as the date for submission of the Rejoinder, the second round of written pleadings for Denmark and the Netherlands in their respective case with Germany. Shortly afterwards, an Order of the full Court of 26 April 1968 held that Denmark and the Netherlands were parties in the same interest and joined the proceedings in the two cases. The order modified the two orders of 1 March 1968 and enjoined Denmark and the Netherlands to submit a Common Rejoinder.[183] The order of the full Court indicates that further consideration of the first written pleadings of Denmark and the Netherlands, which had been submitted on 20 February 1968, probably led to the order to submit a Common Rejoinder. After noting the trilateral protocol between the three States, the order observed:

> the Counter-Memorials ... confirm that the two Governments consider themselves to be parties in the same interest since they have set out their submissions in almost identical terms.[184]

Following the judgment of the Court in February 1969, it was suggested that it had been a mistake of Denmark and the Netherlands to join the cases.[185] So why did Denmark and the Netherlands agree to this? From the perspective of Germany, the joining of the cases was a logical decision. Equidistance cornered Germany in the southeastern part of the North Sea because of the combined coastal configuration of the three States. On the other hand, Denmark and the Netherlands had an interest

---

[183] The Order on the Common Rejoinder led to some discussions between Denmark and the Netherlands, which disagreed on how to deal with certain aspects of the case (see further Chapter 8.6.1).

[184] North Sea Continental Shelf (Denmark/Federal Republic of Germany; Federal Republic of Germany/Netherlands) Order of 26 April 1968 [1968] ICJ Reports, p. 9, at p. 10.

[185] See E. Lauterpacht North Sea Continental Shelf; Judgment of the International Court of Justice; Memorandum dated 24 February 1969 [DNA/39], para. 8; Report dated 25 February 1969 [DNA/81], p. 1.

in looking at their delimitations with Germany in a strictly bilateral context. Viewed in isolation the two bilateral boundaries could be argued to be reasonable. Each of these boundaries was more or less perpendicular to the general direction of the coast on both sides of the respective continental shelf boundary, suggesting that the coast of each party was given similar treatment.

The choice to act jointly was certainly a deliberate one. A Dutch position paper, which was presented to Denmark on 6 January 1966, noted that if further talks with Germany were inconclusive, "Denmark and the Netherlands could consider submission to the Court, preferably together."[186] This approach was a logical outcome of the cooperation between Denmark and the Netherlands as it had developed over time. Both States had started to consult with each other in 1964, shortly after Germany had indicated that it rejected the equidistance principle. Denmark and the Netherlands had both been engaged in a policy of negotiating boundaries based on equidistance, and they had closely coordinated their positions in relation to Germany. As the Dutch Minister of Foreign Affairs pointed out shortly after the conclusion of the Special Agreement, the history of the dispute "has led to a community of interests between both countries in regard of the German claims."[187] Going it alone might also have posed certain risks for Denmark and the Netherlands. In that case they might have developed diverging and possibly contradictory arguments, which could have been expected to weaken their position. Finally, Germany in any case could have simultaneously brought two cases against Denmark and the Netherlands. That approach would also have offered Germany the possibility to stress the linkages between the two cases. As was set out previously, Denmark and the Netherlands were aware that Germany might resort to arbitration and their aversion for this option had contributed to their acceptance to submit the two cases to the Court in collaboration with Germany.[188]

Denmark and the Netherlands, although agreeing on joining the proceedings in the cases, at the same time were intent on keeping their cases separate. The request was to join the proceedings, not the cases as such, and the two agreements to submit the case to the Court provided

---

[186] Untitled paper [DNA/4C], p. 1.
[187] Letter of the MiFA to the Prime Minister dated 19 August 1966 [NNA/40], p. 2. Translation by the author. The original text reads "dat tussen beide landen een belangengemeenschap is ontstaan tegenover de Duitse aanspraken." See also *ibid.*, p. 3.
[188] See further Chapter 6.2.

for separate pleadings of Denmark and the Netherlands. During the discussions with Germany about approaching the Court, they rejected a German suggestion that the cases as such should be joined,[189] and it was also mentioned that the Court should not be precluded from delivering two different decisions.[190] In the end the Court delivered one judgment. The reasons for, and implications of, that approach will be further investigated subsequently.[191]

Another point the parties had to decide upon in connection with the submission of the disputes to the Court was the order of written pleadings. Denmark in particular preferred that Germany would plead first, while Germany had a preference for simultaneous pleadings.[192] As a matter of fact, simultaneous written pleadings were envisaged by the Rules of Court if a case was submitted jointly by the parties,[193] and Germany invoked that provision to support its position.[194] For a number of reasons the Danish Ministry of Foreign Affairs did not want to accommodate Germany. Other ministries and the Danish concessionaire had opposed the conclusion of a special agreement and instead had wanted Germany to have brought a case against Denmark, as this would have placed Denmark at an advantage. The government had accepted the view of the Ministry of Foreign Affairs that the dispute should not harm the bilateral relations and agreed to a joint submission. Now that the willingness to agree to joint submission risked leading to simultaneous pleadings, the Ministry of Foreign Affairs considered that its position in relation to the other ministries and the concessionaire had become almost untenable.[195] In addition, the ministry considered that contradictory pleadings would be the only sensible approach. The Danish and Dutch views on the law were well-known, whereas the German position remained unclear. Denmark and the Netherlands would only be able to develop their argument once they had a better picture of the German position.[196] The German authorities in the meantime had agreed that they would accept contradictory pleadings if Denmark and the

---

[189] Report on Danish-Dutch-German negotiations May 1966 [DNA/56], p. 5.
[190] Note dated 30 June 1966 [DNA/56], p. 1.   [191] See further Chapter 9.2.
[192] See e.g. letter of the Danish Embassy in The Hague to the MFA dated 28 June 1966 [DNA/134; Danish original and English translation], p. 1; Report on Danish-Dutch-German negotiations May 1966 [DNA/56], pp. 5–6.
[193] Rules of Court as adopted on 6 May 1946, article 41(1).
[194] Report on Danish-Dutch-German negotiations May 1966 [DNA/56], p. 6.
[195] Letter of the MFA to the DAEB dated 21 June 1966 [DNA/56], pp. 1–2.   [196] Ibid., p. 2.

Netherlands would present this as a *sine qua non*.[197] After consultations between the Danish Ministry of Foreign Affairs and the German Foreign Office, in which the internal difficulties the ministry was facing had been set out, Germany gave up its objections against alternate pleadings.[198] Denmark and the Netherlands agreed to the German request that the two Special Agreements would explicitly provide that the fact that Germany was to plead first was without prejudice to any question of burden of proof which might arise during the proceedings.[199]

As was observed above, the delegations of Denmark, Germany and the Netherlands agreed to recommend their respective governments conclude two identically worded Special Agreements to submit the two cases to the ICJ, and a trilateral protocol to request the Court to join the proceedings in the two cases during a meeting in Copenhagen on 1 August 1966.[200] Draft texts of the agreements and the protocol were agreed upon during the meeting.[201] The two agreements and the protocol were signed on 2 February 1967 and entered into force on that same date.

The time between the agreement in principle and its formalization was not caused by differences over its substance.[202] After the meeting of August 1966 doubts had arisen about the propriety to include the so-called Berlin-clause in the Special Agreements. This clause, which Germany sought to have included in treaties it negotiated, provided for the equal applicability of the treaty concerned to the state of Berlin. This policy had been adopted to reconfirm the status of the western part of Berlin as a part of the Federal Republic. The status of Berlin had been a

---

[197] Note dated 20 July 1966 [B102/260031], p. 2.

[198] Note dated 30 June 1966 [DNA/56], p. 1; note no. 130/66 dated 27 July 1966 [NNA/27].

[199] Report on my official trip to Copenhagen in the period from 31 July to 2 August 1966 to participate in the German-Danish-Dutch negotiations to submit the question of the delimitation of the continental shelf to the International Court of Justice dated 2 August 1966 [B102/260031] (Report official trip dated 2 August 1966 [B102/260031]), p. 1; Report of the Danish-Dutch-German negotiations in Copenhagen on 1 August 1966 on the submission of the question concerning the boundary between respectively the Danish and German and Dutch and German areas of continental shelf beyond the coastal area to the International Court of Justice in The Hague [DNA/21 (Danish original)], p. 3, [DNA/67 (English translation)] (Report negotiations August 1966), p. 4. This provision is contained in article 2(3) of the two special agreements.

[200] See P.M. dated 1 August 1966 [NNA/27].

[201] Report negotiations August 1966 [DNA/21 (Danish original)], p. 5; [DNA/67 (English translation)], p. 6.

[202] See e.g. note no. 124 dated 21 October 1966 [NNA/27].

bone of contention since the end of the Second World War, which in 1961 culminated in the construction of the Berlin wall. Denmark objected that the inclusion of the Berlin-clause in the Special Agreements or the trilateral protocol might lead certain judges to object to that part of the agreements or even to object to the validity of the Special Agreements.[203] The German Foreign Office shared this concern, but at the same time for internal political reasons could not accept the inclusion of the Berlin-clause in an exchange of letters or notes.[204] In the end, it was agreed that the Berlin-clause would be incorporated in a separate protocol that would make reference to the other three documents. This separate protocol would not be submitted to the Court.[205]

6.6   Agreement on an interim arrangement

6.6.1   Introduction

Exploratory work on the continental shelf in the North Sea in the first half of the 1960s had been limited to areas which were relatively close to the coast. The agreements on partial boundaries of Germany with Denmark and the Netherlands had prevented these activities taking place in disputed areas and they had obviated the need for a provisional arrangement for the disputed area. When the three States negotiated the submission of their disputes to the ICJ, they revisited the possibility of an interim arrangement pending the delimitation of their continental shelf. Exploratory activities were rapidly expanding seaward and it was to be expected that it would be at least a couple of years before the Court would hand down its judgment.

Broadly speaking, interim arrangements are intended to allow certain activities in a disputed area, while preventing these activities leading to conflict between the parties to the dispute or prejudicing their legal positions in respect of the dispute. An interim arrangement in general will have to determine the area to which it is applicable and the kind of activities that will be allowed.

---

[203] Received message in cipher of the MFA to the DEB dated 25 August 1966 [NNA/22].
[204] See received message in cipher (ref. no. 8495) of the DEB to the MFA dated 31 August 1966 [NMFA/3]; received message in cipher (ref. no. 9737) of the DEB to the MFA dated 4 October 1966 [NNA/27], p. 1; note no. 100 dated 5 September 1966 [NNA/22], p. 1.
[205] See note no. 124 dated 21 October 1966 [NNA/27].

The law was not exactly clear on the kind of activities a State might carry out unilaterally in an area of disputed continental shelf. The Convention on the continental shelf did not address this issue. However, as was set out in Chapter 3.5, this matter was discussed during the early stages of the debate on the continental shelf in the ILC. This debate indicated that it was generally accepted that in the absence of agreement on the delimitation between neighboring States, the *status quo* should be maintained and no State would be entitled to exploit the continental shelf. Disputes over the delimitation of the continental shelf should be settled by recourse to third party settlement, in which case interim measures of protection could address activities in the area concerned. Subsequently, the ILC adopted a scheme envisaging compulsory dispute settlement and the regime applicable in the interim was not further considered. The availability of compulsory dispute settlement implied that the interim period pending the final delimitation would be brief and States could request interim measures of protection in the course of judicial proceedings. When compulsory dispute settlement was dropped at the 1958 Conference, the issue of the regime applicable in the absence of agreement was not revisited.

If a State were to engage in unilateral activities on the continental shelf in a disputed area it will be on the other State concerned to decide what reaction it considers appropriate. For instance, the German proclamation on the continental shelf of 1964 provided in this respect that:

> The Federal Government regards as inadmissible any activities under-taken within the area of the German continental shelf for the purpose of exploring and exploiting its natural resources without the express approval of the competent German authorities. It will, if necessary, take appropriate measures against such activities.[206]

This provision would have allowed Germany to take a wide range of measures, including enforcement actions at sea. The latter course of action would, however, in any case have had little appeal to the German authorities. In view of the continued sensitivities in Denmark and the Netherlands about the German occupation during the Second World War, action involving force could be expected to have serious repercussions on the bilateral relations. A diplomatic protest could also be sufficient to protect the legal interest of Germany, but in this case

---

[206] Proclamation of the Federal Government on the exploration and exploitation of the German continental shelf of 22 January 1964, para. 2.

activities nonetheless might create facts on the ground, which could impact on the course of future negotiations if non-legal considerations were to play a role.

Once the parties would have submitted the disputes to the ICJ, they would moreover have to take into account the principle enunciated by the Permanent Court to the effect that:

> the parties to a case must abstain from any measure capable of exercising a prejudicial effect in regard to the execution of the decision to be given and, in general, not allow any step of any kind to be taken which might aggravate or extend the dispute.[207]

If one of the parties to a dispute would consider that another party has not acted in accordance with this principle, it can request the Court to indicate interim measures of protection.[208] In the case of Germany, recourse to this procedure would have to be viewed in the broader context of its relations with Denmark and the Netherlands. Germany wanted to avoid the impression that the disputes burdened the bilateral relations. German recourse to the Court to obtain interim measures would have suggested that there were frictions between the parties, notwithstanding the joint submission of the disputes to the Court.

6.6.2   The negotiations between the parties

The initiative to conclude an interim arrangement clearly fell on Germany. Denmark and the Netherlands did not have an interest to propose a regime which would apply to any part of their equidistance area. The Netherlands had entertained the idea that if Germany would again raise the possibility of an interim arrangement, the Netherlands could propose that specific Dutch blocks would not be opened for licensing pending the decision of the Court, on the condition that blocks for which a license would have been issued and after the judgment would be located on the German side of the boundary would be taken over by Germany.[209] This approach would have allowed companies with a Dutch permit to operate up to the equidistance line in the blocks they would have been awarded. This proposal was not tabled at the subsequent

---

[207] PCIJ, Series A/B 79, p. 199.
[208] Statute of the International Court of Justice, article 41; Rules of Court (adopted on 6 May 1946), article 61; the current Rules of Court deal with interim protection in articles 73–78.
[209] Note dated 21 June 1966 [NNA/27].

discussions with Germany. One drawback was that in this specific form it, in any case, could not have been applied between Denmark and Germany as Denmark had issued a concession for all of its continental shelf up to the equidistance line with Germany to the DUC.

The main problem for Germany in coming up with a proposal for an interim arrangement was that it had never advanced specific claims to boundaries with Denmark and the Netherlands. This raised the question of how the area of application of an interim arrangement should be defined. It was, in any case, highly unlikely that Denmark or the Netherlands would agree to an interim arrangement applicable to the entire continental shelf imposing significant restrictions on activities.

During the preparations for discussions with Denmark and the Netherlands in May 1966, the German authorities and the German North Sea consortium agreed that it was necessary to seek to obtain an interim regime for the duration of the proceedings before the ICJ.[210] The consortium could agree to the suggestion of the German authorities that an interim arrangement might consist either of a moratorium on activities or a regime of joint development, but it preferred the first option, because joint exploitation had proven to be complex.[211] It was acknowledged that this implied the need for a clearly defined disputed area.[212] The German authorities were not able to agree on a definition before the trilateral meeting of 13 May 1966. At that meeting Germany proposed that, as an interim arrangement, Denmark and the Netherlands would refrain from drilling on the German continental shelf in order to prevent that these activities would prejudice the decision of the Court.[213] The Dutch delegation argued that an interim arrangement was not possible as long as it remained unknown what the German claim entailed and that there was no practical need for an interim arrangement. Germany could always request interim measures from the Court if it considered that its rights might be prejudiced. Denmark added that it did

---

[210] Note; "To the head of section III through the head of subsection III D" [B102/260030], p. 2.

[211] Note on the meeting in the Ministry for the Economy on 18 April 1966 dated 29 April 1966 [B102/260030 and AA/45], pp. 3–4.

[212] Note dated 15 April 1966 [B102/260030], pp. 2–3.

[213] Send message in cipher (ref. no. 3642) of the MFA to the DEB dated 18 May 1966 [NNA/27], p. 2. Report on the Danish-Dutch-German negotiations in Bonn on 13 May 1966 concerning the boundaries between respectively the Danish and German and Dutch and German continental shelf areas in the North Sea dated 16 June 1966 [DNA/56], pp. 6–7; letter of the ME to the FO dated 4 July 1966 [AA/46; B102/260031] (the letter in the former archive is actually erroneously dated 4 June 1966), p. 2.

not have a legal basis to stop the Danish concessionaire from drilling in its concession area.[214] In the end the parties agreed that they would further consider this matter if Germany would come up with more detailed proposals.[215]

A specific proposal for an interim arrangement was formulated by the Ministry for the Economy and Transport of Lower Saxony with input from the North Sea consortium.[216] An interim arrangement was considered to be absolutely necessary to avoid that Germany might be confronted with a *fait accompli*. It would be extremely difficult to undo such facts and they later might work against Germany before the ICJ. The proposal envisaged that Germany should adopt the parity principle for an interim arrangement, which would give all three States an area of 45,420 km^2. The parity principle was adopted on the assumption that Germany would also base its claim on it during the proceedings before the Court. A proposal on the location of the German area had been prepared by the North Sea consortium.[217] It envisaged that from the partial boundaries with Denmark and the Netherlands two straight lines would be drawn, which terminated at respectively about 55° 20' N and 56° N on the equidistance line between the continent and the United Kingdom.[218] It was proposed that Germany should declare that it would not undertake any activities in the area beyond the partial boundaries with Denmark and the Netherlands. In exchange for the German commitment, Denmark and the Netherlands should show themselves prepared to refrain from exploration and exploitation in the German area based on the parity principle. The Netherlands probably would not have any problem accepting this arrangement as it had not yet opened up this area for activities. Securing acceptance from Denmark might be more problematic, as the DUC was engaged in exploratory drilling near the area concerned.[219] If Denmark and the Netherlands would not be

---

[214] Send message in cipher (ref. no. 3642) of the MFA to the DEB dated 18 May 1966 [NNA/27], p. 2; Report on Danish-Dutch-German negotiations May 1966 [DNA/56], pp. 6–7.

[215] Send message in cipher (ref. no. 3642) of the MFA to the DEB dated 18 May 1966 [NNA/27], p. 2.

[216] Letter of the Ministry for the Economy and Transport of Lower Saxony to the Minister of Economic Affairs dated 24 May 1966 [AA/46].

[217] See letter of Preussag to W. Obernolte of the ME dated 5 May 1966 [B102/260030], p. 1.

[218] See *ibid.*, Annex 1.

[219] The German ME apparently feared that the Danish concessionaire might drill in the disputed area (see note dated 30 June 1966 [DNA/24 and 56]).

prepared to accept this proposal, Germany at least had to declare that all Danish and Dutch activities in the area, which the Court would attribute to Germany, would be considered to be unlawful acts subject to compensation. A final option would be for the three States to request an interim arrangement for the disputed area from the ICJ.[220]

The Ministry for the Economy concurred with the proposal of the Ministry for the Economy and Transport of Lower Saxony for an interim arrangement and considered that a map with the claim of Germany should be shown at the next meeting with Denmark and the Netherlands.[221] If it would not be possible to agree on a formal arrangement this should be accomplished by unilateral declarations. If Denmark and the Netherlands could not agree to refraining from activities in the area concerned it should be indicated that Germany could no longer ensure that similar activities would not take place.[222]

The proposals for an interim arrangement were discussed at an interdepartmental meeting. The Foreign Office at first voiced some concerns, but in the end it was agreed that Germany would follow the approach suggested by the Ministry for the Economy and in that context should also present a map to the other two States to illustrate the German position.[223]

When Germany raised the issue of an interim arrangement at what was intended to be the last meeting to discuss the submission of the disputes to the ICJ on 1 August 1966, Denmark and the Netherlands again pointed out that no drilling activities were being planned in the disputed area and both delegations seemed to be prepared to accept the German proposal as far as a moratorium on drilling was concerned.[224] Difficulties apparently arose over the definition of the disputed area. What transpired exactly is not clear. According to a German source, Denmark and the Netherlands were only prepared to issue a statement on an interim arrangement if Germany would refrain from further explaining its views on the disputed area by showing the map with the division of the continental shelf in accordance with the parity

[220] Letter of the Ministry for the Economy and Transport of Lower Saxony to the Minister for the Economy dated 24 May 1966 [AA/46].

[221] Letter of the ME to the FO dated 4 July 1966 [AA/46; B102/260031] (the letter in the former archive is actually erroneously dated 4 June 1966), pp. 3–4.

[222] *Ibid.*, pp. 2–3.   [223] Note dated 20 July 1966 [B102/260031], p. 4.

[224] Report official trip dated 2 August 1966 [B102/260031], p. 3; see also note dated 10 August 1966 [AA/45], p. 2.

principle.[225] Dutch sources indicate that Germany was not prepared to indicate the extent of its claim.[226] Both accounts may be correct. The German delegation apparently had difficulty in agreeing on a position and held a long internal deliberation.[227] The head of the German delegation, Truckenbrodt, after the initial Danish and Dutch refusal to have a look at the German claim, was not prepared to indicate that Germany would no longer ensure that activities would not take place in the disputed area,[228] as had been previously agreed at an interdepartmental meeting. The German delegation moreover decided that it would only show the map if Denmark and the Netherlands would request that it be shown.[229] This approach was apparently not supported wholeheartedly by the entire delegation. One of its members ostentatiously unfolded the map containing the German claim.[230] Obviously, Denmark and the Netherlands did not request to be shown the map. The German vacillation to push its proposal had made it easier for Denmark and the Netherlands to reject it.[231] While the German delegation was recapitulating its position, the Netherlands indicated that the proposed solution could lead to difficulties because the delegations did not have a common view on the disputed area and agreement on this point was not forthcoming. The Netherlands therefore proposed that the parties would inform each other of all planned drilling on the continental shelf in the North Sea. After being notified, each party could then take the steps it considered appropriate. The Danish delegation immediately agreed to this proposal. The German delegation indicated that it reserved its position on the proposal but found it worth considering.[232] The Netherlands found that the commitment to inform Germany beforehand of proposed activities was of little significance. Information on

---

[225] Report official trip dated 2 August 1966 [B102/260031], p. 3. The fact that Denmark and the Netherlands were not prepared to identify a part of the continental shelf claimed by them as being in dispute is also reported in a letter of the German FO on the meeting (circular letter of the FO dated 5 August 1966) [AA/45], p. 3).

[226] Send message in cipher (ref. no. 5546) of the MFA to the DEB dated 8 August 1966 [NNA/27]; letter of the MiFA to the Prime Minister dated 19 August 1966 [NNA/40], p. 5.

[227] Send message in cipher (ref. no. 5537) of the MFA to the DEB dated 8 August 1966 [NNA/27], p. 1.

[228] Report official trip dated 2 August 1966 [B102/260031], pp. 3–4.   [229] Ibid., p. 4.

[230] Send message in cipher (ref. no. 5537) of the MFA to the DEB dated 8 August 1966 [NNA/27], p. 1; see also note dated 10 August 1966 [AA/45].

[231] Letter of the MiFA to the Prime Minister dated 19 August 1966 [NNA/40], p. 5.

[232] Report official trip dated 2 August 1966 [B102/260031], p. 4.

drilling activities was already publicly available before such activities started.[233]

It is not clear why the German head of delegation did not want to force the issue of the definition of the disputed area with Denmark and the Netherlands. If Germany would have indicated its claim, a clearly defined disputed area would have come into being. A number of explanations are available. The Foreign Office in any case was not an enthusiastic supporter of the proposed approach. The initial negative response of Denmark and the Netherlands may have been a welcome pretext to diverge from the originally agreed course of action. Second, the Foreign Office seemed to consider that the impossibility to reach agreement on the definition of a disputed area made the German proposal unfeasible from the start.[234] Although such agreement would not have been required to create a disputed area, German protests of activities which took place in an area which Denmark or the Netherlands did not consider to be disputed might possibly have led to political and public opinion fall-out. Third, Germany did not have a clear view of the claim it would be presenting to the ICJ and its legal underpinnings. The need for a positive linkage between the definition of the disputed area and the German pleadings before the ICJ had been argued previously.[235] Fourth, from a legal perspective agreement on an interim arrangement was not necessary to prevent prejudice of the legal position of Germany. To prevent such prejudice Germany could also deal with future activities on an ad hoc basis and did not have to commit itself to a position for the moment. Finally, Denmark and the Netherlands had assured Germany that they did not envisage any drilling activities on the continental shelf for the moment, which suggested that an interim arrangement was not urgently needed in any case.

At least Denmark may not have been completely candid on this latter point. On 5 August 1966, the Danish Embassy in Bonn notified that the DUC planned to start drilling at the position 5° 03' E 55° 24' N.[236] This point is just over 20 kilometers from the equidistance line between the Netherlands and Denmark and anybody even moderately familiar with

---

[233] Send message in cipher (ref. no. 5537) of the MFA to the DEB dated 8 August 1966 [NNA/27], p. 2.

[234] See letter of the FO to the German Embassies in Copenhagen and The Hague dated 17 August 1966 [AA/45], p. 1.

[235] See text after footnote 216.

[236] See note dated 4 August 1966 [AA/45]; letter of the FO to the German Embassies in Copenhagen and The Hague dated 17 August 1966 [AA/45].

the dispute and the German position could have understood that this point had to be in an area claimed by Germany.[237] The point is about 35 kilometers south of the northern limit of the German claim which had been developed prior to the meeting of 1 August 1966.

The Danish note indicated that it relied on the arrangement on prior notification that had been proposed at the meeting of 1 August 1966, and to which Germany at that time had not yet agreed. The Danish note was discussed at an interdepartmental meeting of 15 August 1966, during which agreement was reached on a German reply.[238] As far as can be ascertained it was not considered to still reject the proposed interim arrangement. The Foreign Office appreciated that Denmark evidently had chosen to play down the political significance of the matter by informing Germany through a note and Germany should follow this example.[239] The North Sea consortium indicated that it could agree to the proposal for an interim arrangement as long as notification would not lead to delays in its planned activities. It nonetheless regretted that the German proposal for an arrangement had not been accepted and feared that the activities of Denmark and the Netherlands might prejudice the position of Germany before the Court, even if Germany would protest them.[240]

The German reply to the Danish notification indicated that Germany was of the view that the proposed drilling could not prejudice the pending delimitation between Denmark and itself. Germany also reserved its rights in relation to the DUC.[241] The latter statement apparently led the DUC to reconsider whether to go ahead with drilling in an area which might be in dispute with Germany. The Danish authorities had made it clear to the concessionaire that it would be acting on its own risk if it would be operating in the disputed area.[242] In view of this situation, the concessionaire suspended its drilling program in the potentially disputed area.[243]

[237] For the location of the site and its relation to the equidistance line see Figure 6.1.
[238] Circular letter of the FO dated 16 August 1966 [AA/45].
[239] Letter of the FO to the German Embassies in Copenhagen and The Hague dated 17 August 1966 [AA/45], p. 3.
[240] Note dated 16 August 1966 [B102/260031].
[241] Note verbale dated 16 August 1966 [AA/45].
[242] This position of the Danish authorities is mentioned in a letter of E. Lauterpacht to I. Hoppe of A.P. Møller dated 31 January 1967 [DNA/56], pp. 2–3, which quotes from a note by Sørensen, the legal advisor of the Danish MFA dated 29 November 1966.
[243] See the opinion of E. Lauterpacht prepared for the DUC dated 18 May 1967 [DNA/123], pp. 1 and 17, which refers to the decision to drill at A 1. This is the location which

The fact that the Danish authorities were not overly supportive of the planned drilling of the DUC is also suggested by the Danish reaction to the German note of August 1966. The Danish Embassy in Bonn only presented a Danish reply in January 1967 and the German Foreign Office was left with the impression that it only had been presented because the DUC had insisted on it.[244] The Danish note itself was far from clear. It observed that the Danish position on the boundary had already been known to Germany for a long time and that Danish activities on its side of this boundary could not give rise to rights or claims and this possibility was consequently rejected.[245] The ambiguity of the note was observed by Elihu Lauterpacht, who had been retained by the DUC for legal advice. The note could either be read "as stating that Danish activities in the disputed area would not be relied upon by Denmark as justifying Denmark's claim. Alternatively, it could be read as saying that Danish activity could not give rise to any liability towards Germany."[246] The German reaction to the Danish note indicated that yet another view on the content of the note was possible, namely, that it suggested that Germany had only reserved its rights in respect of the notified drilling and not in respect of the Danish position in general. Germany, in a further note, made clear that its original note reserved Germany's rights in general.[247]

The DUC only went ahead with its drilling activities in the potentially disputed area after it had obtained the advice of Lauterpacht. Lauterpacht addressed the matter in two opinions for the DUC of January and May 1967.[248] In comparison, the second opinion was less optimistic about the interpretation of the applicable law and about the significance of the conduct of the parties for building up Denmark's case. On the latter point Lauterpacht had, in the meantime, received further information. In his first opinion Lauterpacht had concluded that a continental shelf boundary must have some objectively determinable

had been notified to Germany in August 1966 (see note dated 8 August 1969 [B102/260034]); see also the Danish note dated 17 July 1967 [AA/61], which informed Germany that the drilling at the site which had been notified in August 1966 would be resumed.

[244] On the latter point see note dated 10 April 1967 [AA/51], p. 2.

[245] Note verbale dated 11 January 1967 [AA/51].

[246] Letter of E. Lauterpacht to I. Hoppe of A.P. Møller dated 31 January 1967 [DNA/56], p. 1.

[247] Note verbale dated 17 April 1967 [AA/51].

[248] Opinion dated 17 January 1967 [DNA/123] and Opinion dated 18 May 1967 [DNA/123].

existence and that it *prima facie* had to be the median line. There was good support that the median line normally is an equitable line and the general law seemed to give little support to shift the median line.[249] In his second opinion Lauterpacht however concluded that, taking into account the drafting history of article 6, which in his view reflected customary international law, it would be optimistic to believe that special circumstances did not exist in the present case. If islands were included in this category it would be difficult to find that the situation of angles in the coast would be excluded.[250] He further noted that Denmark had first publicly presented its views on the delimitation of the continental shelf in the North Sea in a note to the United Nations. However, that note and the text accompanying a map, which had also been submitted to the United Nations in 1952 but had not been made public, suggested that this did not concern an unequivocal Danish claim. Apart from this material, the Danish claim to an equidistance continental shelf boundary had little priority in time over Germany's rejection of the equidistance method.[251] Lauterpacht concluded that if a decision to drill at A1, the site which had been communicated to Germany in August 1966, was dependent on certainty of success on the purely legal elements of the case, the decision would have to be in the negative.[252] However, there were also additional factors. If special circumstances were found to exist, the Court would have to consider what deviation from equidistance would be justified. A difficult task no doubt, as this would require the Court "to act by drawing an objectively verifiable line based on an essentially subjective exercise of its discretion."[253] In the end, although the Court could not decide on the basis of equity in terms of general law,[254] Lauterpacht was convinced that equity was bound to play a role and Denmark "ought to be ready to put her case in 'equity' [and] the basic equity lies in giving a small country like Denmark with incomparably less natural resources than Germany, at least that share in the continental shelf which is provided by the line of equidistance."[255] This latter consideration played an important role in setting out the arguments for and against drilling at the location A 1. The main argument against going ahead was that the DUC risked simply losing its investment if the area was held to be part of the continental shelf of Germany and "the risk of losing the case should not be underestimated."[256] Lauterpacht suggested that the Court might adopt

---

[249] Opinion dated 17 January 1967 [DNA/123], pp. 3 and 9.
[250] Opinion dated 18 May 1967 [DNA/123], pp. 3–4.   [251] *Ibid.*, p. 13.
[252] *Ibid.*, p. 14.   [253] *Ibid.*, p. 6.   [254] *Ibid.*   [255] *Ibid.*, pp. 15–16.   [256] *Ibid.*, p. 16.

a line which would be a continuation of the general direction of the first part of the continental shelf boundary between Denmark and Germany.[257] Such a line would place point A 1 on the German side of the boundary.

Lauterpacht also provided a number of considerations in favor of going ahead with the drilling. First of all, Denmark would not necessarily lose the case. Second, the question as put to the Court by the parties might prevent the Court from presenting a clear answer, which could be to the advantage of Denmark.[258] Presumably, in that case Denmark could play the card of Denmark's "basic equity" to attribute the area involved to Denmark. As a matter of fact, this was pretty much what happened after the judgment of the Court. Third, if the Court were to be specific it should not be excluded that it would recognize the rights of concessionaire:

> [I]f the Court were to know that oil had actually been found in commercial quantities at A 1 or elsewhere in the disputed area . . ., simply as a matter of human psychology, . . . it would be much less inclined to take the area away from Denmark; or if it did it would be much more inclined to make provision to protect the acquired rights of the consortium.[259]

Lauterpacht also pointed to the mildness of the German attitude in relation to activities in and around A 1 coupled with the terms of the interim regime of 1 August 1966 agreed between the parties. That arrangement could be taken to suggest that exploration could continue and that this would not be at the peril of the companies concerned. A final argument was the tactical importance of maintaining a consistent line of conduct. Denmark should show no sign of any doubt publicly.[260] As Lauterpacht pointed out, from a legal perspective conduct after the critical date of the dispute would still be relevant if it reflects a course of conduct consistently pursued before this date.[261] Finally, if necessary Denmark could make it clear to the Court that it would be absurd to have a moratorium on an area of continental shelf pending delimitation especially when one of the parties failed to specify the extent of its claims.[262] All in all, Lauterpacht's arguments seemed to weigh in favor of resuming drilling at point A 1.

There is no indication that the resumption of drilling by the Danish concessionaire was discussed with the Danish Ministry of Foreign Affairs.

---

[257] *Ibid.*, p. 14.   [258] *Ibid.*, p. 18.   [259] *Ibid.*, p. 19.   [260] *Ibid.*, pp. 20–21.
[261] Opinion dated 17 January 1967 [DNA/123], p. 9.
[262] Opinion dated 18 May 1967 [DNA/123], p. 21.

In July 1967 the ministry notified Germany that the concessionaire would resume drilling at location A 1.[263] Between September 1967 and October 1969 Denmark notified Germany of ten further sites at which the Danish concessionaire planned exploratory drillings.[264] The location of these drillings is identified on Figure 6.1. Six were located in the southern part of the continental shelf in the potentially disputed area.[265] Germany reacted to all of the Danish notifications by referring to the reservation of rights it had made in its note of 16 August 1966. In this way, Germany remained able to avoid the definition of a disputed area. This same approach was applied in relation to the Netherlands.[266] A number of the DUC's exploratory wells in the potentially disputed area were successful.[267] At least one of these finds was reported to the Foreign Office by the German Embassy in Copenhagen, which indicated that the well was located in the western part of the presumed disputed area between Denmark and Germany.[268] The Foreign Office did not take any action on the report of the Embassy but continued to routinely react to notifications of planned drillings by Denmark.

Exploratory activities on the Dutch continental shelf started later than those on the Danish shelf. The Dutch Ministry of Economic Affairs originally had intended to start issuing exploratory licenses by the middle of 1965.[269] However, the necessary regulatory framework for hydrocarbon activities was only finalized by the middle of 1967.[270] The

---

[263] Note verbale dated 17 July 1967 [AA/61].

[264] For a list of the sites notified by Denmark and the Netherlands see the Annex to this chapter.

[265] When Germany defined the area of overlapping claims after the Court had handed down its judgment in February 1969 all six locations were inside this area; see note dated 8 August 1969 [B102/260034]. The list of drilling locations of the DUC in this note generally corresponds with the information contained in the available diplomatic correspondence. The first entry on the list does not result from the diplomatic correspondence and the Danish note to which the list makes reference is not concerned with a specific location (see Danish note verbale dated 11 January 1967 [AA/51]).

[266] See for instance the note verbale dated 21 July 1967 [AA/61; NNA/22], which reserved Germany's rights in relation to Dutch legislation and concessions which would be granted in areas which would appertain to Germany after a boundary would have been established.

[267] According to the list in the note dated 8 August 1969 [B102/260034] this concerned five of the wells drilled in the disputed area as it had been defined by Germany after the judgment of the Court.

[268] See letter of the German Embassy in Copenhagen to the FO dated 9 November 1967 [AA/47].

[269] See Minutes of the Council of Ministers dated 12 February 1965 [NNA/26].

[270] The delay is explained by the controversial nature of certain aspects of the legislation. This in particular concerned the participation of the State in exploitation and the level of taxation of oil companies (see further J. de Jong and A. Koper, "Staat, bodemschatten

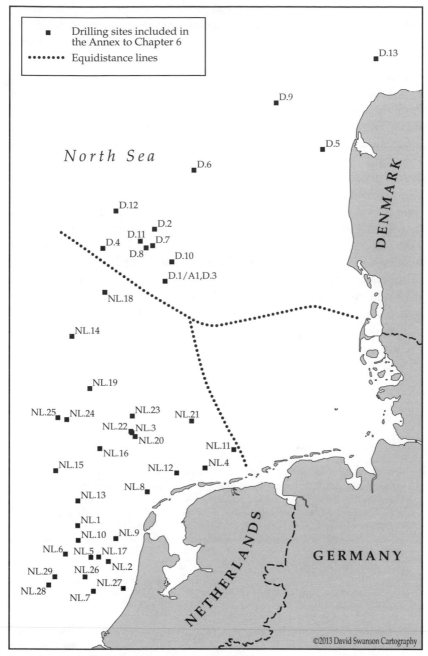

Figure 6.1   Drilling sites on Dutch and Danish equidistance area notified to Germany in accordance with the arrangement of 1 August 1966

Netherlands from the outset had applied the equidistance method to define the spatial scope of application of the proposed mining legislation.[271] The continued absence of agreement with Germany on the final part of the boundary did lead to a reconsideration of how to deal with the northern limit of the concession area. The Netherlands remained intent on opening up blocks to interested companies up to the equidistance lines with Denmark and Germany.[272] The implementing legislation to the Mining Act continental shelf defined the northern limit of the concession area by reference to the equidistance line and specific blocks, including a number of blocks located in the potentially disputed area were opened to interested companies.[273] The dispute with Germany played an important role in selecting specific blocks. A number of blocks were placed along the boundary with Denmark to avoid the impression that the Netherlands accepted that this area was in dispute with Germany and the boundary between blocks that would be opened for bidding and those that would not be should not show any relation with a possible German claim.[274] At the same time, the number of blocks near the boundary with Denmark to be opened for bidding should be kept low to preclude that too many blocks would be affected if the Court would indicate interim measures upon a request by Germany.[275]

Most exploratory drilling under Dutch licenses took place in the area to the south of the area of potential overlap with Germany.[276] The

en energiepolitiek: een analyse van de strijd om de Mijnwet continentaal plat," 1978 (2 (1)) *Tijdschrift voor politieke ekonomie*, pp. 7–51).

[271] For instance, when the draft Mining Act was under consideration in parliament the Minister of Economic Affairs after questions had indicated that there existed a dispute with Germany about the applicability of the equidistance method. At the same time, the minister had submitted a map and other information concerning the extent of the Dutch continental shelf using the equidistance method (see Explanatory memorandum of reply submitted on 19 February 1965 (K. II (BH) (1964–1965), 7670) no. 6), sections 4 and 9 and Annex (K. II (BH) (1964–1965), 7670, no. 7)).

[272] See e.g. note dated 21 June 1966 [NNA/27]; letter of the GEH to the FO dated 6 April 1967 [AA/51].

[273] More than half of the blocks, which were located to the north of the boundary which was eventually agreed upon with Germany, had been licensed by the Dutch authorities by 1970 (see map attached to the circular letter of the FO dated 10 June 1970 [B102/ 260037]).

[274] Note 59/66 dated 17 June 1966 [NNA/33], p. 2.   [275] *Ibid.*

[276] A note dated 11 August 1969 [B102/318714] lists twenty-nine exploratory wells drilled by Dutch concessionaires. Two of these wells are listed as being inside the area which Germany had defined as in dispute after the 1969 judgment of the ICJ and two as being on the limit of this area.

location of these drillings is identified in Figure 6.1. None of the wells in the area which Germany had defined as in dispute after the 1969 judgment of the ICJ had shown the presence of oil or gas.[277]

Exploratory drilling by the German concessionaire had almost come to a standstill by the time the agreement to submit the dispute to the Court had been reached.[278] The German authorities considered that there for the moment was no urgent economic need to speed up the exploration for natural gas in the North Sea, especially if it would mean that German companies would not be involved.[279] It is reasonable to conclude that Germany did not have an interest in creating a track record in the disputed area. Activities in the disputed area might also have had a negative impact on the German case and its relations with Denmark and the Netherlands. Especially if Germany allowed activities in the area of overlap with Denmark it would look like it was trying to take advantage of the work of the Danish concessionaire, which had shown that this area had a promising potential.

6.7   Was the framework for going to the ICJ satisfactory and were there any alternatives?

In looking at the question in the header of this section a number of issues can be distinguished. Apart from going to the ICJ, two other options for settling the disputes between the three States were considered: further negotiations to reach a compromise solution and arbitration. Second, going to the ICJ required formulating the questions to be submitted to the Court. Finally, the procedure before the Court would take a considerable time. This required considering a regime for the disputed area until the time the parties would have reached agreement on their boundaries on the basis of the judgment of the Court.

After the agreement on a partial boundary with Denmark, Germany still considered that trilateral talks might be used to reach a compromise solution on the remainder of its boundaries with Denmark and the Netherlands. After the refusal to engage in these talks on the part of in particular the Netherlands, Germany considered the possibility of compulsory dispute settlement. Germany from the outset preferred

---

[277] See *ibid.*

[278] See circular letter of the Minister for the Economy and Transport of Lower Saxony dated 29 November 1967 [B102/260138].

[279] See *ibid.*

submission to arbitration instead of the ICJ. Political considerations played an important role in this respect. It was considered that the composition of the Court might work to the disadvantage of Germany. In addition, the political department of the Foreign Office held that litigation might harm Germany's bilateral relations with Denmark and the Netherlands. Denmark and the Netherlands had a preference for submitting the dispute to the ICJ as it was expected that the Court was more likely to arrive at a favorable ruling.

Due to the internal deliberations in Germany on the way to proceed it took considerable time before the matter was further pursued with its neighbors. Talks between the three foreign ministers in December 1965 led to an agreement that Germany would take a further initiative to reach a settlement. Germany was left with the impression that the talk between the ministers had created another possibility to reach a negotiated settlement, but Denmark and the Netherlands remained uninterested in this option. As Germany did not want to push for a negotiated settlement over this opposition either, further talks shifted to the appropriate forum for settling the dispute. The existence of two bilateral treaties from 1926 providing for compulsory dispute settlement and the 1957 European Convention for the Peaceful Settlement of Disputes required the parties to consider whether these gave precedence to arbitration or the ICJ. The present analysis suggests that both sides may have failed to fully appreciate the implications of the coexistence of both instruments. The analysis of the three parties indicated that a discussion on this matter might have been time-consuming. In trying to reach a solution, the parties to a large extent steered clear of these legal arguments. A number of reasons explain why Germany was eventually prepared to accept the submission of the disputes to the Court and to give up its preference for arbitration. First of all, if Germany had attempted to submit the disputes to arbitration it was to be expected that Denmark and the Netherlands would object on jurisdictional grounds. This might have affected the German wish that the settlement of the disputes should not damage the bilateral relations. At the same time, Denmark and the Netherlands not only argued against arbitration but sought to abate the German concerns about the ICJ and actively pushed for this solution. To a certain extent this approach seems to be explained by the fear of both States that Germany might opt for a unilateral approach. Their cooperative stance made it easier for Germany to accept the ICJ.

The parties had relatively little difficulty in agreeing on the questions to be submitted to the Court. The questions were relatively open-ended – at least if the Court would decide that the equidistance principle was not applicable. Denmark and the Netherlands seemed to have considered that if the equidistance principle would not have been applicable, the absence of any specific guidance would be to the advantage of Germany, but they did not want to push Germany on this point. Interestingly, this contrasts with the situation after the judgment. Denmark and the Netherlands exploited the ambiguities in the judgment to avoid that Germany would use its positive aspects to its advantage. The apparent discrepancy can most of all be explained by one factor, namely the political dimension. The Dutch and Danish fear that Germany would exploit the absence of specific guidance might have come true if Germany had put its political weight behind its legal arguments. As the negotiations following the judgment would show, this was something Germany decidedly was not prepared to do.[280]

A final matter that had to be decided in connection with the submission of the disputes to the ICJ was an interim regime while the disputes were pending before the Court. The German authorities initially had been intent on presenting their view on the area in dispute. This would have allowed defining an interim regime for this specific area. In view of the Danish and Dutch opposition to this approach, the German Foreign Office did not insist, although Germany could have defined an area unilaterally. Instead, it agreed on a regime which in principle allowed unilateral activities to go ahead. This arrangement allowed Germany to reserve its rights, as was also acknowledged by Danish sources. However, Germany's acceptance of drilling by the DUC would still weaken its position in the negotiations after the judgment of the ICJ. The possibility that, politically speaking, Germany might find it difficult to claim an area in which the DUC had made a find was taken into account on the Danish side. The detailed assessment of this issue on the part of, in particular, the DUC contrasts with the absence of an in-depth discussion on the part of the German authorities.

The unwillingness of the German Foreign Office to present the German view on the definition of the disputed area is most likely explained by the absence of a firm legal basis for a specific German position on the delimitation of the shelf. The Foreign Office had expressed the view that there needed to be a solid legal basis for a

---

[280] See further Chapter 10.

German claim. However, this is not the whole story. The Foreign Office for a long time failed to make a detailed assessment of the applicable law, both as regards the delimitation of the shelf and interim arrangements. The Foreign Office had commissioned a detailed analysis from Professor Rudolf Bernhardt of the University of Frankfurt, but his study only became available in October 1966.[281] Bernhardt's study also briefly touched upon interim arrangements. He concluded that activities having a significant impact in a disputed area could not be carried out unilaterally and also referred to the debate in the ILC to the effect that neither party would be allowed to exploit a disputed area in the absence of agreement.[282] Bernhardt did not explicitly indicate how a disputed area had to be defined, but it is clear from the tenor of his analysis that each party to a dispute is entitled to unilaterally define its claim as regards the boundary. That is exactly what Denmark and the Netherlands had done. The disputed area would be located between the claimed boundaries of the parties.

Whether the German Foreign Office would have taken a different approach on the conclusion of an interim agreement if it had had Bernhardt's analysis prior to the meeting of August 1966 is impossible to tell. The fact that Bernhardt's analysis indicated that the law left Germany a large measure of discretion in formulating a specific claim suggests that the Foreign Office in this case also would not have been willing to face the risk of political fall-out because public opinion in Denmark and the Netherlands would hold that Germany was encroaching on the equidistance area without a clear legal rule to back this up.

Annex – List of drillings in equidistance areas of Denmark and the Netherlands after interim arrangement of 1 August 1966

Denmark[283]

D.1   55° 24' N 5° 03' E (notified in note dated 4 August 1966 [AA/45]; German reaction in note dated 16 August 1966 [AA/45]); also identified as A 1

---

[281] On this study see also Chapter 8.3.1.

[282] Bernhardt, *Rechtsgutachten* [B102/260031], p. 73.

[283] An overview of Danish drillings is contained in a note dated 8 August 1969 [B102/260034]. That note has been compared to the available Danish and German notes verbales on these sites.

D.2   55° 53' N 4° 53' E (reported in note dated 8 August 1969 [B102/ 260034])[284]

D.3   55° 24' N 5° 03' E – resumption of drilling (notified in note dated 17 July 1967 [AA/61]; reaction in German note dated 6 October 1967 [AA/47])

D.4   55° 42' 31" N 4° 02' 06" E (notified in note dated 30 September 1967 [AA/47])

D.5   56° 36' 43" N 7° 40' 02" E (notified in note dated 10 January 1968 [AA/48])[285]

D.6   56° 25' 27" N 5° 31' 57" E (notified in note dated 1 March 1968 [AA/48]; reaction in German note dated 7 March 1968 [AA/48])

D.7   55° 43' 53" N 4° 51' 05" E (notified in note dated 25 April 1968 [AA/48]; reaction Germany in note dated 7 June 1968)

D.8   55° 42' 32" N 4° 44' 40" E (notified in note dated 14 August 1968 [AA/48]; reaction Germany in note dated 26 August 1968 [AA/ 48])

D.9   57° 02' 02" N 6° 53' 46" E (notified in note dated 20 September 1968 [AA/48]; reaction Germany in note dated 25 September 1968)

D.10   55° 35' N 5° 10' E (referred to as approximate; notified in note dated 19 October 1968 [AA/48])

D.11   55° 46' 26" N 4° 38' 48" E (notified in note dated 9 December 1968 [AA/48])

D.12   56° 03' 10" N 4° 15' 00" E (notified in note dated 28 January 1969 [AA/48])

D.13   57° 26' N 8° 33' E (notified in note dated 31 October 1969 [AA/ 48])

Netherlands[286]

NL.1   53° 02' 33" N 3° 37' 00" E (notified in aide-mémoire dated 18 March 1968 [AA/48]; reaction Germany in note dated 2 May 1968)

---

[284] The note dated 8 August 1969 refers to notification through a note dated 11 January 1967. A note of that date in the files [AA/51] refers to a different matter.

[285] This site is not listed in the overview of Danish drillings in the note dated 8 August 1967. That list gives a non-notified site that is very close to this point.

[286] An overview of Dutch drillings is contained in a note dated 11 August 1969 [B102/ 318714]. That note has been compared to the available Dutch and German notes verbales on these sites.

NL.2   52° 41' 13" N 4° 7' 26" E (notified in aide-mémoire dated 10 June 1968 [AA/48])

NL.3   53° 57' 14" N 4° 30' 47" E (notified in aide-mémoire dated 20 June 1968 [AA/48])

NL.4   53° 36' 32" N 5° 43' 15" E (notified in aide-mémoire dated 5 July 1968 [AA/48])

NL.5   52° 43' 40" N 3° 50' 10" E (notified in *ibid.*)

NL.6   52° 45' 38" N 3° 24' 48" E (notified in *ibid.*)

NL.7   52° 23' 15" N 3° 52' 37" E (notified in aide-mémoire dated 5 August 1968 [AA/48])

NL.8   53° 22' 40" N 4° 46' 10" E (notified in *ibid.*)

NL.9   52° 54' 48" N 4° 14' 58" E (notified in aide-mémoire dated 9 September 1968 [AA/48])

NL.10   52° 53' 50" N 3° 37' 30" E (notified in *ibid.*)

NL.11   53° 47' 16" N 6° 11' 34" E (notified in *ibid.*)

NL.12   53° 33' 31" N 5° 14' 56" E (notified in aide-mémoire dated 21 October 1968 [AA/48])

NL.13   53° 17' 15" N 3° 37' 10" E (notified in *ibid.*)

NL.14   54° 53' N 3° 31' E (not notified but there was a German reaction on 7 November 1968 (note dated 11 August 1969 [B102/318714]))[287]

NL.15   53° 35' N 3° 15' E (not notified but there was a German reaction on 7 November 1968 (*ibid.*))

NL.16   53° 48' N 3° 59' E (notified 16 December 1968; German reaction 20 December 1968 (note dated 11 August 1969 [B102/318714]))

NL.17   52° 44' N 3° 58' E (notified 27 January 1969; German reaction 3 February 1969 (note dated 11 August 1969 [B102/318714]))

NL.18   55° 18' N 4° 04' E (notified 5 February 1969; German reaction 14 February 1969 (note dated 11 August 1969 [B102/318714]))

NL.19   54° 23' N 3° 49' E (not notified but there was a German reaction on 26 February 1969 (*ibid.*))

NL.20   53° 55' N 4° 34' E (not notified but there was a German reaction on 26 February 1969 (*ibid.*))

NL.21   54° 04' N 5° 30' E (not notified but there was a German reaction on 5 March 1969 (*ibid.*))

---

[287] According to the note dated 11 August 1969 this site was on the edge of the disputed area, although it actually is to the south of the partial boundary between Germany and the Netherlands agreed upon in 1964.

NL.22   53° 58' N 4° 30' E (not notified but there was a German reaction on 5 March 1969 (*ibid.*))

NL.23   54° 07' N 4° 31' E (not notified but there was a German reaction on 5 March 1969 (*ibid.*))

NL.24   54° 05' N 3° 26' E (notified on 12 May 1969; German reaction on 20 May 1969 (*ibid.*))

NL.25   54° 06' N 3° 17' E (notified on 20 June 1969; German reaction on 26 June 1969 (*ibid.*))

NL.26   52° 32' N 3° 44' E (notified on 27 June 1969; German reaction on 10 July 1969 (*ibid.*))

NL.27   52° 25' N 4° 22' E (notified on 8 July 1969; German reaction on 14 July 1969 (*ibid.*))

NL.28   52° 27' N 3° 08' E (notified on 7 August 1969; German reaction on 14 August 1969 (*ibid.*))

NL.29   52° 32' N 3° 14' E (notified on 1 September 1969; German reaction on 5 September 1969 (*ibid.*))

Interactions between the delimitation in the North Sea and other boundary issues of Denmark and the Netherlands in the 1960s

7.1 Introduction

As was set out in Chapter 2.3, Denmark, Germany and the Netherlands, apart from their continental shelf boundaries in the North Sea, also had to deal with the delimitation of the continental shelf in a number of other areas. These other delimitations have already been considered in connection with the negotiations of the Convention on the continental shelf and the Convention's ratification. The present chapter focuses on the interaction between these other delimitations with the delimitation in the North Sea between Denmark, Germany and the Netherlands in the 1960s. The present chapter in this connection only looks at Denmark and the Netherlands. As was set out in Chapters 2.3.3 and 5.3, the other delimitation issues of Germany did not significantly interact with the disputes in the North Sea.

7.2 Denmark – small islands, small problems

As was set out in Chapter 3.2.1, when Denmark determined its policy on the delimitation of the continental shelf in the early 1950s it concluded that its interests were best served by the general application of equidistance. In the first half of the 1960s Denmark was mainly involved in negotiations with its neighbors in the North Sea. In the case of Germany, agreement was also reached on the delimitation of the continental shelf in the Baltic Sea. This agreement also led to the first contacts with Sweden. In accordance with the agreement with Germany, Denmark had unilaterally determined the equidistance line.[1] This line consisted of two parts and three of its termini were also

---

[1] See Chapter 5.3.

equidistant with Sweden. When Denmark informed Sweden about the location of these points, it became clear that Sweden did not agree with Denmark that their common boundary had to be delimited by the equi-distance method. Sweden considered that the Danish island of Bornholm constituted a special circumstance and that consideration had to be given to the length of the coasts and the size of the continental shelf of each State. It would be inequitable if a small island would get a larger area of continental shelf at the expense of a large mainland.[2] This position was also likely to have consequences for the delimitation between Denmark and Sweden in the Kattegat.[3] In this case, the Danish islands of Anholt and Læsø, which are much smaller than Bornholm, affected the equi-distance line. The Swedish position had implications for Denmark's position in relation to Germany in the North Sea. If Denmark would give in to Sweden, this might be used by Germany to claim a deviation from the equidistance line.

Sørensen, the legal advisor of the Danish Ministry of Foreign Affairs, considered that Denmark should, in any case, stick to the position that equidistance was appropriate between Bornholm and the Swedish main-land coast. As far as Anholt and Læsø were concerned the situation was more problematic. This matter had, under all circumstances, to be postponed until the judgment of the ICJ on the continental shelf boun-dary with Germany would be known.[4] Sørensen considered that any concession on this point to Sweden prior to the judgment would preju-dice Denmark's case. In essence, Sweden's and Germany's positions came down to the same thing, namely that the equidistance principle did not give two neighboring States an equal share of the continental shelf.[5] Denmark might consider offering Sweden to negotiate a settle-ment after the judgment. It could also be decided, but this was a political question, that Denmark would in addition offer that in this case it would be prepared to depart from the equidistance line in favor of Sweden even if the Danish view would be upheld in the case with Germany.[6]

The matter was further discussed with Sweden by Paludan, the head of the political and legal department of the Danish Ministry of Foreign Affairs. Paludan, as his personal opinion, expressed that he found it

---

[2] Note; Shelf boundary in the Baltic Sea dated 7 December 1966 [DNA/42], p. 1.
[3] *Ibid.*   [4] *Ibid.*, p. 2, hand-written note.
[5] Note; The Danish-Swedish boundary relation in the Kattegat dated 19 June 1967 [DNA/42], p. 5.
[6] *Ibid.*

difficult to see that Sweden had a case as far as Bornholm was concerned. Bornholm was big enough to have its own shelf. On the other hand, it would be possible to talk about Christiansø.[7] Christiansø is a small island some 20 kilometers to the north of Bornholm, which measures some 22 hectares. Close to it are a number of smaller islands, the largest of which is Græsholm, which is half the size of Christiansø. Paludan considered that Sweden perhaps had a better case in the Kattegat. It would not be fair to disregard the small Danish islands completely, but it might be possible to consider the option of joint development. The Swedish side showed a considerable interest in the latter suggestion and the content of a possible regime was discussed in some detail. Paludan also indicated Denmark's interest in postponing negotiations until the case with Germany would have been finalized. There was understanding for this position, but this should not disadvantage the Swedish interests.[8] Further discussions on the delimitation of the continental shelf between Denmark and Sweden did not take place before the Court's judgment in the *North Sea continental shelf* cases.

In its pleadings in the *North Sea continental shelf* cases, Denmark argued that small islands constituted one of the most important examples of special circumstances.[9] This implied an acceptance of the Swedish view on the weight of small islands. Denmark considered it essential that the Court was informed on this point. This is contrasted by the Dutch position. The Netherlands did not want to associate itself with the Danish argument on islands because it held that this might negatively impact on the position of the Netherlands Antilles in relation to Venezuela. There is no clear-cut answer as to why Denmark and the Netherlands had different views on how to argue islands in their pleadings. The size of the islands is not an explanation. Aruba, which was the main focus of the delimitation between the Netherlands Antilles and Venezuela, and Bonaire and Curaçao are similar to Bornholm as far as size and coastal lengths are concerned. The same applies as regards the division of the shelf by the equidistance method. This suggests that the difference between Denmark and the Netherlands is not necessarily explained by legal considerations. Part of the explanation may be that the other Danish islands were significantly smaller than Bornholm. Denmark did recognize that these islands might constitute a special

---

[7] Note; Shelf relation in the Kattegat and the Baltic Sea dated 24 May 1967 [DNA/20], p. 3.
[8] *Ibid.*, pp. 3–4.   [9] See further Chapters 8.4 and 8.6.

circumstance. Offering this as a compromise solution to Sweden would allow Denmark to insist on full weight for Bornholm. Politics may also offer part of the explanation. The generally cordial bilateral relationship and close ties of the two Nordic countries contrasted with the, at times, tense relations between Venezuela and the Netherlands. It can easily be understood that in that light the Netherlands was unwilling to bring up the issue of islands as a special circumstance in the *North Sea continental shelf* cases, and instead preferred to focus solely on the geography between itself and Germany and in that context emphasized the equal treatment of islands.[10] Finally, Denmark and the Netherlands had different views in respect of one significant legal point. In both cases, a large part of the area concerned was only within 12 nautical miles of the islands.[11] Denmark held that all of the islands in any case were entitled to a 12-nautical-mile territorial sea, although Denmark at the time was only claiming a 3-nautical-mile territorial sea. This implied that Sweden could not claim most of the area beyond the latter limit as part of its continental shelf. To espouse this point, Denmark explicitly made a reference to this entitlement of islands in its pleadings in the *North Sea continental shelf* cases.[12] The Netherlands did not take into account this point when it considered the delimitation between Venezuela and Aruba in the 1960s, but even entertained the possibility that if Venezuela were to agree to the median line as a continental shelf boundary, it might be accorded certain rights in the area between 3 and 12 nautical miles from Aruba.[13]

Denmark only began to further consider negotiations with Canada concerning the delimitation of the continental shelf of Greenland in September 1969, after the Court had rendered its judgment in the *North Sea continental shelf* cases. In this connection, Denmark considered whether the concept of natural prolongation and the judgment's reference to the Norwegian Trough would allow Denmark to argue for

---

[10] See further Chapter 8.7, especially text at footnote 331.
[11] In the case of the Netherlands Antilles, there also was a large area beyond 12 nautical miles to the north of the islands. However, in the 1960s the focus of the Netherlands was on the delimitation of the area to the south of Aruba, i.e. between the island and the Venezuelan mainland coast.
[12] Letter of B. Jacobsen to J. Paludan of the MFA dated 31 October 1967 [DNA/20], pp. 2–3; see also Report on Continental Shelf-Meetings held at Fredensborg, November 17 and 18, 1967 (KSU/R.25) dated 19 December 1967 [DNA/53] (Report Danish CSC November 1967 [DNA/53]), p. 12.
[13] See further Chapter 7.3.3.

another boundary than the equidistance line.[14] The agreement on the delimitation of the continental shelf between Denmark and Canada of 1973 does not take into account these considerations.[15]

7.3 The Netherlands

7.3.1 Introduction

At the end of the 1950s, all parts of the Kingdom of the Netherlands seemed firmly committed to the equidistance method to delimit the continental shelf. The Netherlands had no interest in varying this method in relation to its neighbors in the North Sea. The Netherlands Antilles was faced with Venezuela's rejection of the equidistance method and the possibility of a Venezuelan claim to enclave the Netherlands Antilles. The government of Suriname had accepted the suggestion to approach its eastern and western neighbor with a proposal to delimit the continental shelf boundary by the equidistance method.[16] The harmony in the Kingdom was not to last, and the coordination of the delimitation policy of Suriname and the other parts of the Kingdom and the definition of the legal position of the Kingdom became a factor that played a not uninteresting role in the background in the negotiations and later litigation concerning the North Sea. After looking at that matter and its consequences in Chapter 7.3.2, Chapter 7.3.3 briefly looks at the developments in relation to the Netherlands Antilles in the 1960s.

7.3.2 The Netherlands and Suriname – consistency at all costs?

In August 1958, the Netherlands proposed Suriname's neighbors, France and the United Kingdom, to delimit the continental shelf in accordance

---

[14] See e.g. note to P.J. IV dated 3 September 1969 [DNA/34]; note; Danish-Canadian negotiations on the delimitation of the continental shelf between Greenland and Canada of 4 November 1969 [DNA/34]; note Delimitation of the continental shelf between Greenland and Canada dated 15 January 1970 [DNA/34].

[15] See L.M. Alexander, "Canada-Denmark (Greenland); Report Number 1-1" in J.I. Charney and L.M. Alexander (eds), International Maritime Boundaries, Vol. I (Dordrecht: Martinus Nijhoff Publishers, 1993), pp. 371–378, at pp. 374–376.

[16] The position of the territory of Netherlands New Guinea, which had played a significant role in the first stage of the formulation of the continental shelf policy of the Kingdom of the Netherlands in the 1950s (see further Chapter 3.2.3) did not have any significant impact in the period under consideration. Tensions between the Netherlands and Indonesia over the territory increased in the second half of the 1950s. The Netherlands agreed to transfer sovereignty over Netherlands New Guinea under international pressure in 1962.

with the equidistance method.[17] The United Kingdom shortly afterwards replied favorably to this proposal.[18] Notwithstanding this agreement in principle, no agreement materialized. For one thing, the delimitation of the continental shelf between Suriname and British Guiana raised two technical problems. First of all, for the territorial sea a delimitation line along an azimuth of 10° had been applied.[19] That azimuth, including its seaward terminus, deviated from an equidistance line. Consequently, it would have to be agreed how to link an equidistant continental shelf boundary to that territorial sea boundary.[20] Second, there existed uncertainty about the location of the baselines to define the equidistance line. It had been suggested that Suriname could establish a closing line across the mouth of the Corantijn River.[21] The boundary between the two countries was located on the western bank of the river. Moreover, part of the low-water line contributing to the relevant baseline had been poorly charted, and it was even suggested that before the equidistance line could be calculated a survey should be carried out to establish the relevant basepoints.[22]

These technical problems receded into the background once the more fundamental questions concerning an overall boundary agreement had come to the fore. The United Kingdom, in its acceptance of the Dutch proposal, had indicated that it intended to submit a draft treaty to the Netherlands on the entire boundary between Suriname and British

---

[17] See aide-mémoire presented by the Dutch Embassy to the FO dated 6 August 1958 [NNA/18]; note of the Dutch Embassy in Paris to the French MFA dated 20 August 1958 [NNA/18].

[18] See letter of the FO to the Dutch ambassador dated 13 January 1959 [NNA/44]. France responded favorably to the Netherlands proposal in 1963 (see note of the French MFA to the Dutch Embassy in Paris dated 11 May 1963 [NNA/18]). As the delimitation of the continental shelf boundary between Suriname and French Guiana did not affect the delimitation of continental shelf boundaries in the North Sea in a similar fashion as the boundary between Suriname and British Guiana (and Guyana after independence from the United Kingdom in 1966) the former delimitation will not be further considered in the present study. Suriname and France are yet to conclude an agreement on their maritime boundary.

[19] See also Chapter 3.2.3.

[20] See e.g. letter of R. H. Kennedy of the Hydrographic Department of the Admiralty to E. W. A. Scarlett of the Colonial Office dated 15 January 1959 (MG, Annex 24), p. 1.

[21] See e.g. note of the head of the legal department of the Ministry of the Navy to the head of Hydrography dated 15 December 1958 (MS, Annex 13).

[22] See e.g. letter of the MiFA to the Plenipotentiary Minister of Suriname dated 3 March 1959 [NNA/18], p. 1; letter of R. H. Kennedy of the Hydrographic Department of the Admiralty to E. W. A. Scarlett of the Colonial Office dated 15 January 1959 (MG, Annex 24), p. 1.

Guiana. The delimitation of the continental shelf would be just one aspect of this overall settlement.[23] The Government of Suriname had also indicated that it desired the land boundary and the delimitation of the continental shelf to be treated in a single agreement.[24] The main point of contention in respect of the land boundary concerns the upper course of the boundary river. This dispute exists till this day. Suriname considers that the New River (Upper Corantijn) constitutes the boundary and Guyana holds that the boundary is formed by the Cutari and Curuni rivers. This results in a disputed territory of some 15,000 square kilometers.

The United Kingdom submitted a draft boundary treaty at the end of 1961.[25] The government of Suriname rejected this draft because it identified the Cutari and Curuni rivers as the upper course of the Corantijn River. Subsequently, the government of Suriname prepared an alternative draft treaty, containing a compromise proposal on the land boundary.[26] The proposal located the land boundary in the Corantijn River, instead of on its western bank. In exchange, the draft treaty provided that the New River constituted the land boundary in the south. This draft provided the basis for a draft treaty that was submitted by the Netherlands to the United Kingdom in September 1962.[27]

The Dutch draft treaty on the basis of the proposal of Suriname also envisaged a new approach to the delimitation of the continental shelf. Instead of using the equidistance line, it was provided that the continental shelf would be delimited by an extension of the 10° azimuth that was employed to delimit the territorial sea. That approach implied a significant departure from the equidistance line to the advantage of Suriname.[28] The United Kingdom and British Guiana at first accepted

---

[23] See letter of the FO to the Dutch ambassador in London dated 13 January 1959 [NNA/44].

[24] See received message in cipher (Van Tilburg no. 37) dated 15 October 1958 [NNA/44]. The Dutch MiFA agreed to that approach (ibid.).

[25] The text of this draft treaty is reproduced at MG, Annex 90.

[26] See letter of the Plenipotentiary Minister of Suriname to the MiFA dated 22 May 1962 [NNA/14]; letter of the Dutch ambassador in London to the Foreign Secretary dated 10 August 1962 (CMS, Annex 4).

[27] The letter of the Dutch Embassy dated 17 September 1962 submitting the draft treaty to the FO and the text of this draft treaty are reproduced at MG, Annex 91.

[28] For a comparison of the equidistance line and the 10° azimuth line see RS, p. 88, figure 3. The government of Suriname also adopted a different position in respect of the continental shelf boundary with France, in which case Suriname opted for a 30° azimuth starting from the common land boundary, which was located to the east of the equidistance line.

to negotiate on the basis of both draft treaties. After they reneged on this offer and only wished to negotiate on the basis of the British draft treaty, negotiations for the time being did not materialize.[29]

In early 1964, the government of Suriname had a renewed interest in a proclamation on its continental shelf in the light of activities on the continental shelf and the fact that British Guiana had issued a proclamation on its continental shelf.[30] The claim of British Guiana overlapped with the continental shelf claimed by Suriname and the government of Suriname did not want to wait too long with a response to avoid that British Guiana might rely on acquired rights.[31] Because a proclamation on the continental shelf of Suriname might also affect the interests of the Netherlands and the Netherlands Antilles, the propriety of issuing a proclamation and its content were considered in a commission consisting of representatives of the three countries of the Kingdom of the Netherlands. It was agreed that there did not exist objections to a proclamation as such. The Dutch side did suggest that Suriname might also consider the option of adopting mining legislation that could also address the definition of the continental shelf, as the Netherlands presently was in the process of doing. There was interest on the part of Suriname for this suggestion, but a proclamation was preferred because it would take less time to adopt.[32] Suriname wanted a proclamation to contain a specific definition of the continental shelf boundaries with its two neighbors, French and British Guiana. Initially, there was considerable debate on this point. The Dutch side at first argued against including a reference to specific lines, while for Suriname one of the main purposes of a proclamation was to indicate Suriname's position on the boundary with British Guiana.[33] The text of the draft proclamation that was adopted by the commission did refer to specific lines, but at the same time indicated that these were provisional until the boundaries would have been established by agreement with the neighboring States. For British Guiana this concerned the 10° line starting from the western

[29] See note 30/64 dated 11 March 1964 [NNA/14], p. 3.
[30] For the earlier interest of Suriname in such a proclamation see Chapter 3, text at footnote 138 and following.
[31] Letter of the Prime Minister of Suriname to the Plenipotentiary Minister of Suriname dated 12 February 1964 [NNA/14], p. 3.
[32] Brief report of the meetings on the proclamation on the continental shelf of Suriname held at the MFA on 2, 16 and 23 April 1964 dated 1 May 1964 [NNA/18; NNA/38] (Brief report April 1964 [NNA/18; NNA/38]), p. 3.
[33] See Draft report on three meetings of the Commission Van Boetzelaer (not dated) [NNA/18; NNA/38].

bank of the Corantijn River.[34] Special circumstances in this case justified a deviation from the equidistance line.[35] The report on the meeting in this connection referred to the fact that the 10° line had to be considered as the prolongation at sea of the line which divided the territorial sea of Suriname and British Guiana.[36] During the meeting, it was also accepted that adopting the course of the Corantijn in determining the boundary of the continental shelf constituted a special circumstance in the sense of the Convention on the continental shelf.[37]

In the case of French Guiana the proclamation defined the boundary as an equidistance line. During the discussions on the boundary with French Guiana, the Suriname delegation had indicated that it preferred that the boundary with French Guiana would be defined as an azimuth of 30° E of true north. It was submitted that this boundary more or less coincided with the equidistance line, but that it was more practical because it was a straight line, whereas the equidistance line consisted of a number of segments with different bearings.[38] Suriname subsequently applied the 30° line in connection with the definition of its continental shelf in its legislation in October 1964.[39] Interestingly, the 30° line actually deviates from the equidistance line and further seaward is clearly closer to the coast of French Guiana than the equidistance line. It is not clear why this point was overlooked in the discussions on Suriname's continental shelf proclamation or later on, when Suriname's rejection of the equidistance line with British Guiana led to a conflict between Suriname and the Dutch Ministry of Foreign Affairs.[40]

The Netherlands Antilles initially had objections against a proclamation on the continental shelf of Suriname because it was thought that the reliance on special circumstances might strengthen the hand of Venezuela. These objections were dropped after the Suriname side had submitted that the specific special circumstances which existed between Suriname and British Guiana were not present between the Netherlands

---

[34] Draft for a proclamation on the continental shelf of Suriname (text agreed at the meeting of 23.4.1964) [NNA/38], article 4.

[35] Brief report April 1964 [NNA/18; NNA/38], p. 5.   [36] *Ibid.*

[37] Report on the combined meeting on the continental shelf of Suriname and the boundary between Suriname and French Guiana held at the Ministry of Foreign Affairs on 16 April 1964 [NNA/38], p. 3.

[38] Brief report April 1964 [NNA/18; NNA/38], p. 5.   [39] See text at footnote 46.

[40] Apart from the fact that this point may have gone unnoticed, another explanation may be that the proclamation referred to the equidistance line, not the 30° line.

Antilles and Venezuela.[41] The Dutch side considered that the proclamation might prejudice the interests of the Netherlands in respect of its continental shelf boundary with Germany because certain arguments Suriname was advancing in respect of British Guiana were similar to the arguments Germany had been using to reject the equidistance line. The Dutch side requested that Suriname wait with the promulgation of the proclamation until Germany and the Netherlands would have reached an agreement in principle. It was said that this would imply a delay of about two months. Following this Dutch request, Suriname remarked that the opposite might also be the case. The rejection of German arguments by the Netherlands might negatively impact on the position of Suriname. The Suriname side argued that more coordination between the negotiations on delimitation of the continental shelf of the different parts of the Kingdom was desirable. The Dutch side promised to raise this issue with Riphagen, the legal advisor of the Ministry of Foreign Affairs, who also was in charge of the negotiations with Germany.[42]

The consultations on a draft proclamation on the continental shelf of Suriname had a remarkable outcome. It is difficult, if not impossible, to square its provision on the delimitation of the continental shelf of Suriname with the position of the Netherlands in relation to Germany. This divergence can be explained by the fact that the consultations focused on the presence of specific special circumstances. As the participants observed, even in that sense there was a possible conflict of interest between Suriname and the Netherlands. The consultations failed to address the relationship between the equidistance principle and special circumstances in article 6 of the Convention on the continental shelf. If that matter would have been addressed it would have been clear that the interests of the Netherlands and the Netherlands Antilles on the one hand and Suriname on the other were diametrically opposed. While the Netherlands and the Netherlands Antilles had an interest in limiting the effect of special circumstances, the opposite was the case for Suriname. It would have been possible to construct a common position for the three countries of the Kingdom in this respect while seeking a middle ground, but this was something that was never attempted.[43] As will be further

---

[41] Note of Mr. S. D. Emanuels to the Prime Minister of Suriname dated 27 April 1964 [NNA/43], p. 2.
[42] Brief report April 1964 [NNA/18; NNA/38], pp. 5–6.
[43] Such a common position could have been built on distinguishing the three cases. The geography of the three cases differs considerably. For the Netherlands, the main issue was whether Germany's concave coast and location between two other states constituted

discussed below, the Netherlands instead sought to complete its continental shelf boundary with Germany while relying on its narrow construction of the special circumstances clause of article 6 of the Convention and keeping the involvement of the other countries of the Kingdom to a minimum. The Dutch Ministry of Foreign Affairs had to deal with the consequences of this approach once Suriname had reaffirmed its legal position in relation to Guyana in 1966.

Initially, the positions of the Netherlands and Suriname co-existed without raising any questions. After further requests for deferral of the promulgation of the proclamation,[44] Suriname was given the green light to go ahead with it in April 1965 because the agreement on the landward part of the continental shelf between Germany and the Netherlands had been signed.[45] This indicates that the ministry still did not appreciate that the fundamental opposition between the positions of Suriname and the Netherlands was also relevant for the more seaward part of the continental shelf boundary between Germany and the Netherlands.

Suriname at this time was apparently no longer interested in a proclamation, which was never promulgated. This is likely explained by the fact that Suriname in October 1964 had amended its oil and gas legislation and in that connection had clarified that the boundary of the continental shelf in relation to British Guiana was constituted by the 10° line.[46] The adoption of this legislation did not require prior consultation with the other countries of the Kingdom.

The Dutch Ministry of Foreign Affairs sought to allay the fears on the part of Suriname that the delimitation agreement with Germany might prejudice the position of Suriname. Riphagen, the legal advisor of the ministry, who had been asked to look into this matter, considered that there were insufficient grounds to assume that the envisaged agreement

---

a special circumstance or led to an inequitable result. In the case of the Netherlands Antilles the main question was whether the coastal length of Aruba in relation to the length of the coast of Venezuela required it to be treated as a special circumstance or not. In the case of Suriname, it could have been argued that the protruding coast of British Guiana constituted a special circumstance as it was a promontory, which led to a deviation of the equidistance line in front of the coast of Suriname.

[44] See e.g. send message in cipher (ref. no. 2831) of the MFA to the Governor in Paramaribo dated 19 June 1964 [NNA/18].

[45] Letter of the MiFA to the Plenipotentiary Minister of Suriname dated 13 April 1965 [NNA/43]. The agreement actually had been signed on 1 December 1964. The lapse of more than four months between signature and this notification may be explained by the fact Suriname's approval of the agreement, which had been requested in accordance with the procedure of the Statute of the Kingdom, had not yet been finalized (see further below).

[46] See Decree dated 13 October 1964, Appendix, article 1.

would negatively impact on the position of Suriname in relation to British Guiana.[47] This was a surprising conclusion. During the consultations with Suriname on a continental shelf proclamation, the Dutch side had pointed to a possible prejudice of the Dutch position in relation to Germany and, after consultations with Riphagen, had argued for delaying the promulgation of the proclamation by Suriname.[48] Riphagen's position had been adopted by the Minister of Foreign Affairs in informing the other parts of the Kingdom about the delimitation with Germany.[49]

When the approval of the agreement between Germany and the Netherlands was submitted for consideration to the Kingdom Council of Ministers in the second half of 1964, the government of Suriname indicated that it did not object to the ratification of the treaty as long as it did not have any consequences for the delimitation of the continental shelf of Suriname.[50] The treaty department of the Ministry of Foreign Affairs wondered whether such a guarantee could actually be given as ratification:

> obviously has consequences for Suriname to the extent that it confirms that the Kingdom of the Netherlands is "dedicated" to the equidistance principle and is not willingly prepared to accept a plea of special circumstances.[51]

In reply to this argument, it was pointed out that the draft agreement with Germany in principle was irrelevant for Suriname, as the reasons for a deviation from the equidistance method for Suriname were of course different from those in the North Sea.[52] Another comment recognized that the United Kingdom might seek to invoke the draft agreement but that in

---

[47] Note no. 59/64 dated 28 April 1964 [NNA/18]. A similar justification was provided later on for the fact that the Netherlands Antilles and Suriname had only been involved to a very limited extent in the delimitation of the Dutch continental shelf in the North Sea (see note no. 91 dated 17 August 1966 [NNA/40], p. 2).

[48] Note no. 59/64 dated 28 April 1964 [NNA/18].

[49] See letter of the MiFA to the Plenipotentiary Minister of the Netherlands Antilles dated 5 May 1964 [NNA/13]; letter of the MiFA to the Plenipotentiary Minister of Suriname dated 5 May 1964 [NNA/21].

[50] Letter of the Plenipotentiary Minister of Suriname to Mr. J. H. Kremer of the MFA dated 22 January 1965 [NMFA/2].

[51] Note no. 53/65 dated 29 January 1965 [NMFA/2]. Translation by the author. The original text reads "heeft uiteraard wel consequenties voor Suriname in zoverre zij bevestigt dat het Koninkrijk der Nederlanden het equidistantiebeginsel is 'toegedaan' en niet gaarne bereid is een beroep op bijzondere omstandigheden te honoreren."

[52] Note no. 9 dated 9 February 1965 [NMFA/2].

respect of British Guiana there were special circumstances, which had to be recognized internationally.[53] The government of Suriname was subsequently informed that there was no reason to assume that the ratification of the agreement with Germany would prejudice the position of Suriname. Consequently, the process of ratification would be continued.[54] It should be clear from what was said previously that the treaty department had correctly assessed the implications for Suriname.[55]

Until the middle of 1966 the continental shelf positions of the Netherlands and Suriname remained on diverging tracks without leading to difficulties. In the North Sea, the Netherlands continued its policy of concluding delimitation agreements based on equidistance with little involvement of Suriname and the Netherlands Antilles.[56] A telling example in this respect is the approval of the Special Agreement to submit the dispute with Germany to the ICJ. The Ministry of Foreign Affairs considered the question of whether this constituted an agreement which required approval from parliament or not. In that connection it was also considered whether the agreement constituted a matter of the Kingdom. It was to be expected that in particular Suriname might consider that its interests might be affected. In that light it might be advisable to submit the agreement for approval to the Kingdom Council of Ministers. The adoption of the Kingdom procedure constituted all the more reason to avoid the delay that would result from a parliamentary procedure in which Suriname and the Netherlands Antilles would also be involved.[57] Approval of the agreement by the Kingdom Council of Ministers should be handled with care. It should be avoided that the governments of the Netherlands Antilles or Suriname would feel entitled to demand a voice in the way in which the case with Germany had to be handled. Any attempt to this effect should be immediately rejected.[58]

---

[53] *Ibid.*, hand-written note on the back side of the note.

[54] Letter of J. H. Kremer of the MFA to the Plenipotentiary Minister of Suriname dated 6 April 1965 [NNA/45].

[55] See text after footnote 42.

[56] For instance, the strategy in respect of the delimitation of the continental shelf in the North Sea was considered in the Dutch Council of Ministers and not in the Council of Ministers of the Kingdom (see e.g. letter of the MiFA to the Prime Minister dated 28 January 1965 [NNA/24]); see also note no. 91 dated 17 August 1966 [NNA/40], p. 2, which observes that Suriname and the Netherlands Antilles up to that point had only been informed in some individual cases, when this had been considered to be useful.

[57] Note 46/66 of 23 May 1966 [NNA/27], especially at p. 3.

[58] Note no. 91 dated 17 August 1966 [NNA/40], p. 3. In this connection reference was made to a letter of the MiFA to the Plenipotentiary Minister of Suriname dated 28 July

Representatives of Suriname continued to discuss Suriname's land, territorial sea and continental shelf boundary with the Ministry of Foreign Affairs. In this connection, the possible interaction with the North Sea was not discussed.[59] In the light of the approaching independence of British Guiana, the Netherlands submitted a diplomatic note to the United Kingdom in which it set out the position of the Kingdom of the Netherlands in respect of all pending boundary issues. The note reconfirmed that the Kingdom considered that the maritime boundary between Suriname and British Guiana should be the 10° line.[60] At the time, the Ministry of Foreign Affairs possibly had already reached a different view on the propriety of such a boundary. Foreign Minister Luns used a meeting with Prime Minister Pengel of Suriname on 31 March 1966 to emphasize that if Suriname wanted to reach an overall settlement with British Guiana there should be no deviation from the equidistance principle.[61]

Shortly after Guyana's independence, Suriname and Guyana discussed all their pending boundary issues during a bilateral meeting.[62] A report from the Dutch Embassy in London that Suriname, during the meeting, had rejected the application of the equidistance method led to an almost immediate reaction of the Ministry of Foreign Affairs' legal advisor.[63] The fact that Suriname had taken the position that the continental shelf boundary should deviate from the equidistance line did not come as a surprise. However, as Riphagen had already pointed out in a previous meeting with the minister, it was excluded that a delegation of the Kingdom could take the legal position that special circumstances justified this deviation.[64] The suggestion that the latter argument was new is difficult to reconcile with the facts. In the contacts between Suriname and the Netherlands it had been clear from the outset that

1966 [NNA/27; NNA/40]. Interestingly, this letter indicates that the continental shelf boundary between Suriname and Guyana concerned a Kingdom Affair, which Suriname could not handle on its own. By the same token the handling of the case with Germany would be a Kingdom Affair. For a further discussion of this letter see text at footnote 68.

[59] See e.g. note no. 31 dated 30 November 1965 [NNA/35]; note no. 32 dated 1 December 1965 [NNA/35].

[60] Note verbale dated 3 February 1966 [NNA/36], para. 8.

[61] Note 64/66 dated 26 April 1966 [NNA/27].

[62] See e.g. Main points of the report of the discussions between Suriname and Guyana held at the Marlborough House in London on 23 June 1966 [NNA/36]; received message in cipher (ref. no. 6713) of the Dutch Embassy in London to the MFA dated 24 June 1966 [NNA/36].

[63] Note 37/66 dated 27 June 1966 ([NNA/27]; RG, Annex R39).   [64] Ibid., p. 1.

Suriname's position was based on a specific interpretation of article 6 of the Convention on the continental shelf.[65]

Riphagen's concern over Suriname's position was obviously caused by the fact that the Netherlands was in the process of submitting its dispute with Germany to the ICJ. As Riphagen observed, if it became known that the Netherlands in these two cases had taken a different legal position, this would be fatal for the Dutch position before the Court.[66] At this point it might have been preferable to come back on the decision to submit this matter to Court, but this was excluded in view of the advanced stage of the negotiations with Denmark and Germany.[67]

Upon a proposal of Riphagen, the government of Suriname was informed that in future negotiations with the government of Guyana, Suriname in no case should take the position that there were legal grounds to deviate from the equidistance line.[68] The Minister of Foreign Affairs observed that Suriname had been authorized to conduct these negotiations on behalf of the Kingdom. This implied that instructions for such negotiations had to be approved by the Council of Ministers of the Kingdom and that a delegation should also include Dutch experts. If the Suriname Government did not make suggestions in this respect, the minister would nominate these experts himself.[69] The minister's letter indicates that Dutch interests in the North Sea were put above Suriname's interests in relation to Guyana. It observed that the German position that equidistance could not be applied in the North Sea affected major Dutch interests. If the special circumstances on which Suriname relied would legally justify a deviation from the equidistance line, a deviation in favor of Germany certainly would also be justified.[70] Apart from the Dutch interests, Riphagen also argued that the arguments of Suriname were weaker than those of Germany.[71] However, that conclusion very much depended on one's interpretation of the meaning and scope of the special circumstances provision. The Netherlands and Suriname never attempted to reach a common position on the latter point.

[65] See text after footnote 31.

[66] Note 37/66 dated 27 June 1966 ([NNA/27]; RG, Annex R39), p. 2.   [67] Ibid., p. 1.

[68] Ibid., p. 2; letter of the MiFA to the Plenipotentiary Minister of Suriname dated 28 July 1966 [NNA/27; NNA/40], p. 2.

[69] Letter of the MiFA to the Plenipotentiary Minister of Suriname dated 28 July 1966 [NNA/27; NNA/40], p. 2.

[70] Ibid., p. 1.   [71] Note 37/66 dated 27 June 1966 ([NNA/27] RG, Annex R39), p. 1.

Following this virtual dictate of the Netherlands as regards the conduct of negotiations with Guyana, there was still an attempt to agree on the text of diplomatic notes to set out Suriname's position on its boundaries with Guyana and French Guiana. Suriname's proposed text on the continental shelf boundary with Guyana proved unacceptable to Riphagen and the notes were never finalized.[72] Suriname and Guyana only discussed their continental shelf boundary again after the judgment of the ICJ in the *North Sea continental shelf* cases had been handed down.

The interaction between the delimitations of the Netherlands and Suriname became, so to speak, part of the *North Sea continental shelf* cases in March 1968, when the dispute on the continental shelf boundary between Suriname and Guyana came to the attention of the German Embassy in The Hague.[73] In an article published in a Dutch legal journal Werners, the legal advisor of Suriname's Plenipotentiary Minister in The Hague, had observed that Suriname rejected the application of the equidistance method to delimit its continental shelf boundary with Guyana. At the same time, the Kingdom of the Netherlands before the ICJ was relying on the equidistance principle in relation to Germany. In a statement not devoid of irony, Werners submitted that if Suriname, that is the Kingdom of the Netherlands, would also submit the dispute with Guyana to the ICJ, a collision between the parts of the Kingdom would almost be inevitable.[74] Considering the position of the author and the timing of the publication one is tempted to assume that the article was intended to draw the attention of the German Embassy in The Hague. If the matter would be picked up by Germany, there would be some chance that Suriname's position would be brought to the attention of the Court. For obvious reasons, the Netherlands could not be expected to say anything about Suriname's position.

The German Embassy in The Hague immediately reported to Bonn on Werners's article, observing that the fact that two parts of the Kingdom of the Netherlands took a completely different position could be of considerable importance for Germany's case. The Embassy was going to find out if the Netherlands or Suriname had made declarations on the

---

[72] See note no. 67/66 dated 18 October 1966 [RG, Annex R40]; note 91/66 dated 4 November 1966 [NNA/36]; note dated 20 December 1967 [NNA/36].

[73] Letter of the GEH to the FO dated 7 March 1968 [AA/62].

[74] S. E. Werners, "Complicaties bij een grensgeschil" (Complications in relation to a boundary dispute), 1968 *Nederlands Juristenblad*, Issue 9 pp. 224–225, at p. 225. An English translation of the article is available at ICJ Pleadings, North Sea continental shelf cases, Vol. II, pp. 299–300.

delimitation with Guyana.[75] Upon receiving word on Suriname's position, the Foreign Office immediately instructed the German Embassy in Guyana to look into the matter.[76] The Embassy reported back that Werners had given a correct picture of the boundary dispute. The Guyanese Ministry of Foreign Affairs indicated that it would attempt to provide the Embassy with the diplomatic notes related to the boundary dispute.[77] As far as is known, the Embassy never received this documentation.[78]

Jaenicke, Germany's agent before the Court, used the information on Suriname in his first round of oral pleadings after he had argued that there was no basis for the narrow interpretation of the special circumstances clause of article 6 of the Convention on the continental shelf that Denmark and the Netherlands were offering. Their narrow interpretation would, in effect, make the equidistance method the only rule. This would seriously hamper the development of the law of the continental shelf and States would become even more hesitant to become a party to the Convention. Jaenicke expressed his disbelief that Denmark and the Netherlands were really convinced of this narrow interpretation. As a matter of fact, Germany had received information that the Kingdom of the Netherlands indeed seemed to take another view on the matter.[79] Jaenicke did not want to pass judgment on the equitability of the claims of Suriname and Guyana, but observed that he:

> just wanted to show that the Kingdom of the Netherlands does not apply such a narrow interpretation to the special circumstances clause in this case, and it is interesting that this case of the lateral boundary where the projecting coast of Guyana diverts the equidistance line before the coast of Surinam [sic]. This is just an example of the "diversion effect" which might be a justification for invoking a special circumstances clause.[80]

The message was clear. The Netherlands was applying double standards. It wanted to impose the equidistance rule on Germany, but rejected its application in a situation in which it went against its interest.

---

[75] Letter of the GEH to the FO dated 7 March 1968 [AA/62], p. 2.

[76] Letter of the FO to the German Embassy in Georgetown dated 12 March 1968 [AA/62].

[77] Letter of the German Embassy in Georgetown to the FO dated 5 April 1968 [AA/62], p. 1.

[78] From a Guyanese perspective it would be quite understandable not to provide this information. Guyana had an interest in the ICJ upholding the equidistance method, which would not be served by strengthening Germany's hand.

[79] ICJ Pleadings, North Sea continental shelf cases, Vol. I, p. 47.   [80] *Ibid.*, p. 49.

Riphagen, the Dutch agent, countered Jaenicke's suggestion that the Netherlands was applying the rules in accordance with its convenience. Riphagen pointed out that fortunately the Kingdom of the Netherlands, although responsible for the foreign relations of Suriname, was not responsible for everything that was written within its territory. Mr. Werners, who had written about the topic, was in no way a spokesperson for the government of the Kingdom.[81] Riphagen also sought to downplay the significance of the information on the 10° line. In Suriname it was sometimes advocated that this line should be established as a convenient continental shelf boundary in the framework of the settlement of various other boundary questions, including questions relating to the land boundary, by agreement.[82] In other words, Riphagen was claiming that the 10° line was not based on legal considerations. To substantiate the latter assertion, Riphagen concluded on this point by observing that:

> the Kingdom of the Netherlands, responsible for the foreign relations of Surinam, has never laid a legal claim to such boundary line, nor ever has it as yet made any proposal to the Government of Guyana relating to the establishment by agreement of a convenient boundary line on the continental shelf adjacent to Surinam.
>
> Mr. President, I do not think that I need to spend more of the time of the Court to comment on Mr. Werners' article, which is clearly so totally irrelevant to the present disputes.[83]

These were indeed the last words which were spoken on this matter during the proceedings. In the present context some further clarification is appropriate. As regards laying legal claim to the boundary line, Riphagen had taken a different view in an internal memorandum of 27 June 1966. He had written this memorandum after the meeting between representatives of Suriname and Guyana earlier that month. Riphagen pointed out that during the meeting Suriname had taken the view that the continental shelf boundary should not be the equidistance line because of the presence of special circumstances in the sense of article 6 of the Convention on the continental shelf.[84] Riphagen then continued that he had already observed during a meeting chaired by Minister of Foreign Affairs Luns:

> that there was admittedly *no* objection to Suriname trying *to agree with* Guyana that Suriname would get a larger share of the [continental shelf], but that a delegation of the Kingdom can*not* take the *legal* position that

---

[81] *Ibid.*, p. 73.   [82] *Ibid.*   [83] *Ibid.*, pp. 73 and 75.
[84] Note 37/66 dated 27 June 1966 ([NNA/27]; RG, Annex R39, p. 1).

> there is a matter of "special circumstances" in the sense of the Geneva Convention.
> Let me repeat this remark again emphatically.[85]

Riphagen was candid about the reasons for this insistence:

> When it would become apparent that the same Kingdom in another part of the world has taken a different *legal* position than in the Netherlands *the chance of winning this case* [*i.e.* the case with Germany] *becomes extremely small* ...
> If Suriname in relation to third States continues to take a legal position which *deviates* from the equidistance principle, this will be *fatal* for the Dutch position before the Court.[86]

These statements contradict Riphagen's assertion before the Court that the Kingdom of the Netherlands had never laid claim to a boundary line different from the equidistance line.

The second part of Riphagen's concluding remark before the Court is only true if a literal interpretation would be adopted. The Kingdom of the Netherlands had never made a proposal to the government of Guyana on a continental shelf boundary after Guyana's independence in May 1966. The Kingdom had made such a proposal to the government of the United Kingdom in September 1962.[87] In legal terms, the government of Guyana was the successor to the government of the United Kingdom as far as boundary questions were concerned.

This whole episode indicates the importance that is attached to consistency in legal discourse. The legal advisor of the Ministry of Foreign Affairs of the Netherlands was prepared to tell what, in everyday language cannot be termed differently than lies and half-truths in front of the highest judicial organ of the United Nations to maintain the image that the Netherlands had never deviated from the equidistance principle.

---

[85] *Ibid.* (translation by the author; emphasis in the original). The original text reads: "dat er weliswaar geen bezwaar tegen bestond dat Suriname een groter deel van het CP verkrijgt, doch dat een Koninkrijksdelegatie niet het rechtsstandpunt kan innemen, dat hier van bijzonder omstandigheden sprake is. Ik moge deze opmerking nogmaals met de meeste nadruk herhalen."

[86] *Ibid.*, pp. 1–2. Translation by the author; emphasis in the original. The original text reads: "Wanneer het dan zou blijken, dat hetzelfde Koninkrijk in een ander deel van de wereld een ander rechtsstandpunt heeft ingenomen dan in Nederland wordt de kans op het winnen van dit proces wel uiterst gering" "Wanneer echter Suriname voortgaat met het tegenover derde Staten innemen van een rechtsstandpunt dat afwijkt van het equidistantie-beginsel, is dat voor de Nederlandse positie voor het Hof fataal."

[87] The letter of the Dutch Embassy dated 17 September 1962 submitting the draft treaty to the Foreign Office and the text of this draft treaty are reproduced at MG, Annex 91.

After the Netherlands had concluded the negotiations with Germany to delimit the remainder of their continental shelf boundary on the basis of the judgment of the Court of 1969, Suriname and the Netherlands Antilles were informed that the resulting agreement had been submitted to the Dutch Council of Ministers.[88] As was set out in an internal memorandum of the Ministry of Foreign Affairs, the option of submitting the agreement to the Kingdom Council of Ministers had not been chosen to avoid giving the governments of the Netherlands Antilles and Suriname reason to enter into a detailed discussion of the negotiations with Germany. This might have led to an undesirable delay in the signature of the agreement with Germany.[89] The Netherlands Antilles and Suriname were assured that the agreement with Germany in itself did not prejudice the position of the Kingdom in the negotiations on the continental shelf boundaries of the two countries.[90] It was admitted that the judgment of the Court might have some impact on the delimitation of the continental shelf of the Netherlands Antilles and Suriname, although this was a matter on which it was not easy to provide an answer.[91] It is understandable that the Minister of Foreign Affairs sought to downplay the importance of the judgment to avoid discussions with the Netherlands Antilles and Suriname. In the light of the Court's clear rejection of the Dutch position in respect of equidistance and the Court's views on the law, the effect of the judgment could not be but significant.

7.3.3   The common interests of the Netherlands and the Netherlands Antilles

The interests of the Netherlands and the Netherlands Antilles in respect of the delimitation of the continental shelf matched perfectly. They were both faced with a neighbor, respectively Germany and Venezuela, which rejected the application of the equidistance method.[92] The delimitation of the continental shelf between the Netherlands Antilles and Venezuela

---

[88] See letter of the MiFA to the Plenipotentiary Minister of the Netherlands Antilles dated 4 January 1971 [NMFA/1]. A similar letter was sent to the Plenipotentiary Minister of Suriname on the same date (note no. 11/71 dated 13 January 1971 [NMFA/1], p. 2).

[89] *Ibid.*, p. 1.

[90] Letter of the MiFA to the Plenipotentiary Minister of the Netherlands Antilles dated 4 January 1971 [NMFA/1], p. 1.

[91] *Ibid.*

[92] As will be further discussed in Chapters 8.4.1 and 8.6.1, the interests of the Netherlands Antilles also played a minor role in determining the Dutch position during the proceedings before the ICJ.

became an issue in the fall of 1967 after Mobil and the Signal Oil Company had shown an interest in the continental shelf of Aruba.[93] The companies were only interested in a concession to the south of Aruba if the Netherlands and Venezuela would first have reached an agreement on the delimitation of the continental shelf.[94] In a discussion between Riphagen and two ministers of the Netherlands Antilles the options for an agreement with Venezuela were considered. The possibility of a deviation from the median line was immediately rejected, but concessions to Venezuela on other points might be an option. For instance, if there would be transboundary deposits, it might be possible to conclude an agreement on transboundary cooperation. To allay possible fears of Venezuela that a median line would limit Venezuela's possibilities to protect its interests, the Netherlands Antilles might offer its assistance in protecting legitimate Venezuelan interests on its side of the boundary.[95] The interest of the Netherland Antilles in a median line boundary was further confirmed in a consultation with Mobil and the Signal Oil Company. The companies considered that any deviation from the median line in favor of Venezuela would make the concession area completely uninteresting.[96]

On the basis of the outcome of the consultations with the Netherlands, the Netherlands Antilles wanted to sound out Venezuela as soon as possible to explore the possibilities of an agreement. Riphagen could agree to this wholeheartedly. If the negotiations would remain within the framework of a full acceptance of the equidistance principle for the delimitation of the continental shelf there could be no objection on account of the pending procedure involving the continental shelf in the North Sea before the ICJ.[97] The delimitation between the Netherlands Antilles and Venezuela was subsequently considered in the Dutch Council of Ministers. The Council supported the line of conduct which had been discussed between Riphagen and the ministers of the Netherlands Antilles.[98] In the meantime, the Netherlands had also

---

[93] Prior to that development, the government of the Kingdom did not consider it expedient to start consultations with the government of Venezuela because there did not exist any actual difficulties resulting from different viewpoints concerning the delimitation of the continental shelf (see K. II (BH) (1964–1965), 7800 IV, no. 8, p. 4, item 15.c).

[94] Note 54/67 dated 25 September 1967 [NNA/37], p. 1.   [95] Ibid., pp. 1–2.

[96] Ibid., p. 2.   [97] Ibid., p. 3.

[98] See Minutes of the Council of Ministers 10 November 1967 [NNA/37], item 4.c; letter of the MiFA to the Prime Minister dated 29 February 1968 [NNA/37].

become aware that Venezuela had issued concessions to a number of international companies well beyond the median line, extending up to the outer limit of Aruba's 3-nautical-mile territorial sea. This made the likelihood of a dispute with Venezuela imminent and the Dutch informed the American government accordingly.[99] In the Council of Ministers, the Minister of Foreign Affairs stressed that if a dispute did arise, it would be necessary to act with circumspection both because of the position of the Netherlands in a similar dispute with Germany and the interests of the Netherlands Antilles in the exploitation of the continental shelf and its dependent position in relation to Venezuela.[100]

However, the dispute with Venezuela did not become acute and negotiations with Venezuela did not take place before the judgment of the ICJ in the *North Sea continental shelf* cases was handed down. Following a proposal of the Netherlands Antilles to start negotiations, Venezuela indicated that it was still studying the matter. Venezuela was, moreover, negotiating with Colombia on their continental shelf boundary and would only be able to turn its attention to the Netherlands Antilles at a later stage. Venezuela was also awaiting the outcome of the judgment in the *North Sea continental shelf* cases, as it was considered to be an important precedent.[101]

The Netherlands and Venezuela eventually concluded a delimitation agreement in 1978. The changed legal context of the negotiations – in the 1950s and 1960s the focus was on the areas of continental shelf within the 200-meter isobath, whereas in the second half of the 1970s the regime of the 200-nautical-mile exclusive economic zone was generally accepted – is reflected in this agreement. The agreement not only delimits the area between Aruba and the Venezuelan mainland, but is also concerned with all areas within 200 nautical miles of Aruba, Bonaire and Curaçao. To the south of the islands the boundary differs to a limited extent from the equidistance line to the advantage of Venezuela. To the north of the islands, the boundary diverges sharply from the equidistance line and gives the islands less than half of the area they would have been attributed by equidistance lines. The agreement, in addition, gives

---

[99] See received message in cipher (ref. no. 11039) of the Governor in Paramaribo to the MFA dated 3 November 1967 [NNA/48]; Minutes of the Council of Ministers 10 November 1967 [NNA/37], item 4.c.

[100] Letter of the MiFA to the Prime Minister dated 29 February 1968 [NNA/37].

[101] Letter of the Dutch ambassador in Caracas to the MiFA dated 23 August 1968 [NNA/37].

full weight to the small islet of Aves of Venezuela in relation to the islands of Saba and Saint Eustace.[102]

7.4 Denmark and the Netherlands compared

Probably the most striking aspect of the practice of Denmark and the Netherlands discussed in this chapter is the importance both States attached to the consistency of their legal positions in respect of different boundaries. Both States did not want to give the Court, and Germany, the impression that they were applying different standards in the North Sea than to their other delimitations. The task of Denmark in this respect was simpler than that of the Netherlands. The main potential for inconsistency that Denmark was facing was the weight to be attached to a number of small islands in relation to Sweden. Denmark was not unwilling, in principle, to accommodate the Swedish position to a certain extent after it would have settled its dispute with Germany. Moreover, Denmark considered that the cases of Sweden and Germany were sufficiently distinct. In front of the Court, Denmark took the position that small islands constituted an important example of special circumstances and argued that this type of special circumstances had nothing to do with the delimitation between itself and Germany in the North Sea.

The Netherlands was facing a larger challenge than Denmark as it was more difficult to distinguish the delimitation it was facing in the North Sea from that of Suriname with Guyana. Still, as was argued above, this would not have been an impossible task. However, this would have implied that the Netherlands could not have adopted the strict view on the interpretation of equidistance it actually did in front of the Court. In view of the Dutch position in relation to Germany up to that point, this would undoubtedly have led to the impression that the Netherlands was having second thoughts about the legal arguments it had been advancing. The fact that the Netherlands, before the Court, tried to cover up the discrepancies in the position of the Kingdom is further proof of the importance that was attached to presenting a consistent view. Obviously, at this point there was little else to do besides admitting a major

---

[102] For a review of the agreement see e.g. K. G. Nweihed, "The Netherlands (Antilles)-Venezuela; Report Number 2–12" in J. I. Charney and L. M. Alexander (eds), *International Maritime Boundaries*, Vol. I (Dordrecht: Martinus Nijhoff Publishers, 1993), pp. 615–629. A comparison of the equidistance line and the boundary is provided in the figure on *ibid.*, p. 630.

inconsistency. That it got this far first of all seems to be explained by fact that the Dutch Ministry of Foreign Affairs paid insufficient attention to the implications of the reformulation of the position of Suriname in relation to British Guiana in the first half of the 1960s. Whether this was intentional or not is not completely clear. It probably would have been difficult if not impossible to arrive at a position which would have been acceptable to all three parts of the Kingdom and there was no attempt to reconcile the differences to start with. Instead, the Ministry of Foreign Affairs took measures to prevent that Suriname would continue to actively advocate a broader role for special circumstances than the Netherlands was prepared to accept in relation to Germany. In addition, the involvement of the Netherlands Antilles and Suriname in the conduct of the case with Germany was kept to a minimum. The argument to support this approach, that the delimitation primarily concerned the Dutch position in the North Sea, is difficult to accept and is clearly contradicted by the Dutch opposition against Suriname advocating its view on the role of special circumstances in relation to Guyana. In view of the common interest of the Netherlands and the Netherlands Antilles in the strict application of the equidistance principle, Suriname had nothing to gain from a discussion of this matter through the formal channels for coordinating the foreign policy interests of the countries of the Kingdom.

The pleadings of Denmark, Germany and the Netherlands before the ICJ

8.1 Introduction

The pleadings of parties to cases before the ICJ are divided into a written and oral phase. In general, both these phases consist of two rounds of pleadings. In their pleadings the parties will set out their view on the applicable law, the facts they consider to be relevant, how the applicable law should be interpreted and applied, and formal submissions indicate the way in which the case should, in their view, be decided. This process is intended to bring out the issues on which the parties remain divided and on which the Court will have to take a decision.

The Court was notified of the two agreements to submit the disputes on the continental shelf delimitations of Germany with Denmark and the Netherlands on 16 February 1967. Both bilateral agreements provided that Germany had six months to submit the Memorial, its first round of written pleadings. Denmark and the Netherlands were accorded the same time period to submit their Counter-Memorials. The Court fixed 21 August 1967 as the time limit for the filing of the German Memorial and 20 February 1968 as the time limit for the filing of the Counter-Memorials of Denmark and the Netherlands. The Court was requested to establish the time limits for the second round of written pleadings of the parties. After parties have submitted the last round of written pleadings, the Court will fix a date for the oral hearings, which are held at the Peace Palace in The Hague. The present chapter reviews the successive stages of the pleadings in chronological order.

The detail in which the process of drafting the pleadings by the parties is documented differs considerably. Whereas the Danish files contain numerous detailed reports on the drafting process, in particular the Dutch files contain little information. A part of the explanation may be that in the Netherlands and Germany a more limited number of persons was involved in the drafting process and it took place in a more informal setting. The files of the Danish Ministry of Foreign Affairs, moreover, point out that the files in the Dutch Ministry of Foreign Affairs are incomplete. Correspondence between the two ministries on the handling of the case, which are present in the Danish files, with a few exceptions are missing from the Dutch files.

In Denmark, a continental shelf committee, in which representatives of different ministries, the Danish concessionaire and a number of professors participated, met regularly to discuss the case and the deliberations in this committee and other meetings were recorded painstakingly. The introductory remarks to a discussion on the draft Danish Counter-Memorial form a telling recognition of this inclination for completeness. After listing a number of examples of editorial and linguistic changes, the record observes:

> During the proceedings the secretariat exhaustively noted all points of this nature, which were subsequently taken into account in preparing a revised draft. Neither principles nor the least disagreement being involved, it has not been found warranted to burden the historical record with a complete enumeration of such details.[1]

8.2 The timing of the proceedings

The three parties were interested in an early judgment of the Court. The timing of the proceedings was discussed between the agents of the three parties and the Registrar of the Court at the end of May 1967.[2] It was concluded that the cases could be ready for the oral hearings at the end of the fall of 1968.[3] At the time, the only other case before the Court was the *Barcelona Traction* case between Belgium and Spain. This case would be ready for hearings at about the same time. The Registrar observed that the hearings and the preparation of the judgment in the *Barcelona Traction* case would take considerable time. If the *Barcelona Traction* case would go first, it seemed quite sure that hearings in the *North Sea continental shelf* cases would only be scheduled later in 1969. This would be unfortunate as the Court assumed that the parties were interested in as early a decision as possible.[4] According to the Danish agent, although

---

[1] Report Danish CSC November 1967 [DNA/53], p. 3.

[2] The agent is the representative of a party before the Court (see Statute of the Court, article 42).

[3] See Report of Bent Jacobsen to the Danish MFA with date of receipt of 4 June 1967 [DNA/58], p. 1.

[4] See *ibid.*

it was not said in as many words, it was evident that the Court would be glad to assist the parties in moving the hearings in the *North Sea continental shelf* cases ahead of the hearings in the *Barcelona Traction* case.[5] The three agents readily declared that they were interested in an early decision. As the time limits for the first round of written pleadings had already been fixed in the Special Agreements and could not be changed, it should be assessed whether the time limits for the second round of written pleadings could be shortened. Jacobsen and Riphagen also observed that the hearings might be relatively short and they expressed the view that the Registrar would leave the question open and in any case would not plan the hearings for the *Barcelona Traction* case unless it had become clear that the *North Sea continental shelf* cases would not be ready for hearings beforehand.[6]

The Danish agent had a further meeting with the Registrar in September 1967. Jacobsen at this point explained that the Danish concessionaire had been carrying out exploratory drilling. Until now Denmark and Germany had been able to manage this matter. However, if Denmark would start exploitation activities at a successful exploratory well, it could not be excluded that Germany would request the Court for interim measures of protection.[7] This would no doubt lead to a reaction in Danish public opinion and the whole case would change in character. Neither the Court nor the parties were interested in this.[8] In Jacobsen's view the parties were interested in an early decision and he considered that it would be possible for him to convince the DUC – if there would be no delay in the proceedings of the *North Sea continental shelf* cases – to only start with the commercial exploitation of any possible finds around November 1968.[9] There is no indication that the DUC had actually discussed the option of starting exploitation with the Danish Ministry of Foreign Affairs.[10]

The Registrar agreed with Jacobsen that souring of the relations of the parties would be unfortunate. There were many reasons to prefer that the

---

[5] See *ibid.*   [6] See *ibid.*

[7] Article 61 of the Rules of Court indicates that a request for the indication of interim measures may be filed at any time during the proceedings in the case in which it is made. Article 41 of the Statute of the Court provides that the Court has "the power to indicate, if it considers that the circumstances so require, any provisional measures which ought to be taken to preserve the respective rights of either party."

[8] Report on the meeting of 12 September 1967 between the Registrar and the Danish Agent, in which the reporter also took part (not dated) [DNA/58], p. 1.

[9] *Ibid.*, pp. 1–2.   [10] See also text at footnote 196.

North Sea continental shelf cases would be heard before the *Barcelona Traction* case. After the *South West Africa* cases, which had been a political burden for the Court,[11] it would presumably be interested in a case in which the legal element had a central place.[12] The Court's handling of the case, with a relatively quick judgment, would no doubt strengthen its prestige. The Registrar also provided some suggestions as to how the parties could contribute to an expeditious treatment of the case.[13]

At two subsequent meetings of staff of the Danish Ministry of Foreign Affairs with the Registrar in January 1968, the Registrar suggested that the parties might refrain from a second round of written pleadings. The Registrar indicated that in this case there would be no problem to hear the *North Sea continental shelf* cases first, and the hearings in this case might even take place in April 1968.[14] This was an option Denmark was not willing to entertain,[15] and it was not pursued further. This approach in any case would have required the collaboration of the other two parties. Waldock, who had been consulted as to whether Germany might itself indicate that it was not interested in a second round of written pleadings, observed in this respect:

---

[11] The *South West Africa* cases were concerned with the presence of South Africa in the former German colony of South West Africa. After the First World War, South Africa had been appointed as Mandatory Power by the League of Nations to administer the territory. After the Second World War, a dispute arose between South Africa and the United Nations over the continued presence of South Africa in South West Africa. This dispute was related to the policy of racial discrimination South Africa had been implementing in the territory. In 1960, two members of the United Nations, Ethiopia and Liberia, instituted proceedings against South Africa before the Court. Although the Court in the preliminary phase of the case apparently seemed to consider that it could reach a decision on the merits, it subsequently ruled otherwise at the merits phase. As one commentary on this judgment has observed: "The inability of the Court to meet this problem is due above all to institutional causes and the fragile basis on which modern international litigation rests. This, and the legalistic nature if not pure sophistry which many purport to see in some of the Court's pronouncements, gave in turn rise to profound disappointments and misgivings ... With the 1966 judgment, the standing of the Court in political circles of the United Nations fell to its lowest point" (T. D. Gill (ed.), *Rosenne's The World Court; What it is and How it Works* (The Hague: Martinus Nijhoff Publishers, 2003), p. 142).

[12] Report on the meeting of 12 September 1967 between the Registrar and the Danish Agent, in which the reporter also took part (not dated) [DNA/58], p. 3.

[13] *Ibid.*, pp. 3–4.

[14] Report dated 12 January 1968 [DNA/49], p. 2; letter of J. Paludan to H. Waldock dated 6 February 1968 [DNA/60], p. 2.

[15] See e.g. letter of J. Paludan to H. Waldock dated 6 February 1968 [DNA/60], p. 2.

I shall be rather surprised if the Federal Republic feels that it can safely go into the Court without a second written pleading. If it does, it will deprive itself of the opportunity of bringing in fresh material from practice, and leave itself to begin the oral hearings in a somewhat defensive position.

If on the other hand the Federal Republic does waive its right to a second pleading, I really cannot see why we should have any cause to try and insist upon taking the rather extraordinary course of demanding a second pleading ourselves notwithstanding. Seeing the Court's anxiety to get on with this case, it might be a little annoyed.[16]

In the end, the *Barcelona Traction* case was ready for the hearings before the *North Sea continental shelf* cases,[17] but the Court nonetheless fixed the hearings of the latter cases first. The reasons for this are not known. Apart from the Court's apparent interest in a quick resolution of the *North Sea continental shelf* cases, Belgium, one of the parties to the *Barcelona Traction* case, reportedly was very unwilling to start the hearings in September 1968 as it wanted a longer period to prepare itself for them.[18]

8.3   The Memorials of Germany

8.3.1   The preparation of the German case and the Memorials

It would not be much of an overstatement to say that Germany had to elaborate its view on the law applicable to the delimitation of the continental shelf from scratch. The Convention on the continental shelf provided that delimitation between neighboring States had to be effected by agreement and referred to the equidistance principle and the concept of special circumstances, which could lead to an adjustment of the equidistance line. There was little guidance on how these rules had to be applied in practice. The rules applicable to the delimitation of the continental shelf had been discussed in the ILC and at the 1958 Geneva Conference on the law of the sea, which had adopted the Convention on the continental shelf. That debate remained limited to listing a number of special circumstances in general terms, without specifying what effect these circumstances, if present, should have on the equidistance line. Of even more fundamental importance was the question of whether the

---

[16] Letter of H. Waldock to J. Paludan dated 13 February 1968 [DNA/60], pp. 3–4.

[17] The Rejoinder of Spain in the *Barcelona Traction* case was submitted on 30 June 1968 and the Common Rejoinder of Denmark and the Netherlands on 30 August 1968.

[18] Report dated 12 January 1968 [DNA/49], p. 1.

Convention on the continental shelf reflected customary international law. Germany had not become a party to the Convention and was not bound by it as a rule of treaty law. If the rule contained in article 6 did not reflect customary law, it remained to be answered what the content of customary law was. In that case, it could be argued that the law left more leeway for applying different methods of delimitation than the Convention.

Germany started to consider the preparation of a detailed legal analysis in December 1965, when it was suggested to commission a report from Professor Rudolf Bernhardt of the University of Frankfurt. This analysis was first of all intended to be used in further negotiations with Denmark and the Netherlands, but might also be used in legal proceedings.[19] Bernhardt was asked to prepare a report on the legal aspects of the dispute in April 1966 and submitted his report in October 1966.[20] In this period Germany also started to organize a team for the two cases. The legal department of the Foreign Office had a central role in this respect. A special section under the direction of Truckenbrodt, the deputy head of the department, was formed to manage the case.[21]

Professor Günther Jaenicke of the University of Frankfurt was appointed as agent of Germany for the cases. Apart from representing a party before the Court, the agent in general is responsible for coordinating the work of the legal team and deciding on the organization of the pleadings. Jaenicke also had an important role in preparing the pleadings. Jaenicke closely collaborated with the legal department of the Foreign Office in preparing drafts of the German Memorials and Replies and also discussed these drafts with a number of German international law professors.[22] The Ministry for the Economy also participated in discussions on the written pleadings and was represented at the oral pleadings in The Hague. Germany also contracted foreign legal counsel. This is a common practice in cases before the Court. There may be various reasons to employ foreign counsel. An obvious reason is to tap into the knowledge of counsel with experience of appearing before the Court. Most States or their nationals will not be regularly involved in cases before the Court. Another reason may be the wish to have English

---

[19] See note dated 16 December 1965 [B102/260173], p. 2.

[20] Letter of the head of the legal department of the FO to R. Bernhardt dated 7 April 1966 [B102/260030]; Bernhardt, *Rechtsgutachten* [B102/260031].

[21] See note dated 7 September 1966 [AA/45].

[22] On the latter point see for instance letter of G. Jaenicke to the legal department of the FO dated 3 April 1968 [AA/62].

and French counsel to address the Court in its two official languages during the oral proceedings.

The legal department of the German Foreign Office discussed possible counsel with Professor Hermann Mosler, the director of the Max Planck Institute for Comparative Public Law and International Law, who had accepted to act as the German judge ad hoc in the two cases. Mosler considered that it most of all seemed appropriate to contract French counsel. France had made a reservation to article 6 of the Convention on the continental shelf and Mosler considered it possible that France would be facing similar difficulties as Germany was currently having in the North Sea.[23] Mosler expressed a preference for Paul Reuter, whom the Foreign Office had already considered. Reuter had acted as agent of France in a number of cases before the ICJ and also advised other States that had appeared before the ICJ. Other possible candidates mentioned by Mosler were Charles Rousseau, Suzanne Bastid, Sir Humphrey Waldock, James Fawcett and Hans Blix. If the Ministry wanted to look beyond Western Europe, in view of the universality of the Court, it might consider Jimenez de Aréchaga or Andrassy.[24] The Ministry for the Economy and Transport of Lower Saxony had suggested engaging Elihu Lauterpacht,[25] who in the past had written an opinion for the North Sea consortium, Germany's concessionaire.[26] Lauterpacht was instead hired by the DUC, which acted quickly after it got wind that the German government was interested in Lauterpacht.[27]

Truckenbrodt, of the legal department of the German Foreign Office, met with Reuter at the beginning of December 1966. Reuter showed himself to be very well-informed about the Court, talking amusingly about the internal affairs of the Court, with an eye for mischievous asides.[28] Already at this first meeting Reuter made a large number of constructive suggestions concerning the case and Truckenbrodt concluded that Reuter was exactly the expert the Foreign Office was looking for.[29] During a second meeting with Reuter in Paris in the second half of December 1966, Truckenbrodt, Fleischhauer and Treviranus of the Foreign Office's legal department discussed the substance of

---

[23] Note dated 5 May 1966 [AA/45], p. 4.   [24] *Ibid.*, p. 5.

[25] Letter of the Ministry for the Economy and Transport of Lower Saxony to the Minister for the Economy dated 24 May 1966 [AA/46], p. 3.

[26] E. Lauterpacht, Opinion "The German Continental Shelf" dated 6 November 1963 [B102/260126].

[27] Report Danish CSC May 1967 [DNA/39], p. 3.

[28] See note dated 8 December 1966 [AA/51], p. 1.   [29] See *ibid.*

Germany's case for three hours, looking at the report that had been prepared by Bernhardt. Reuter agreed with Bernhardt's conclusion on the applicable law and considered the report to be a good piece of work.[30] At a later stage of the proceedings Reuter provided Jaenicke with a thorough analysis of the Counter-Memorials of Denmark and the Netherlands. Jaenicke found Reuter's ideas to be very useful and essentially corresponding with his own thinking.[31] Reuter did not act as counsel for Germany during the oral pleadings, but he was informed about their course.[32] During the oral proceedings, Germany did make use of the services of Shigeru Oda, who was later to become a long time judge in the ICJ (1976–2003) and the American lawyer Henry Hermann. Oda had been a member of the Japanese delegation to the 1958 Geneva Conference on the law of the sea, which had negotiated the Convention on the continental shelf, and had published a book critically appraising the continental shelf regime.[33]

Professor Bernhardt's report for the Foreign Office, which was finalized in October 1966, provided a detailed assessment of the legal aspects of the delimitation of Germany's continental shelf. Working on the report and thinking the matter over, Bernhardt's conviction that there existed very important reasons for the German legal position had become rather stronger than weaker.[34] Bernhardt's report confirmed that there was ground for optimism. Based on an analysis of the drafting history of article 6 of the Convention and State practice, Bernhardt among others concluded that the equidistance principle had not acquired the status of a rule of customary law. If it were to be concluded that article 6 reflected customary law, it in any case had to be adjusted if special circumstances were present. This was the case for the North Sea. It was, however, most likely that customary law required States to reach agreement on the basis of equitable considerations. Criteria for a delimitation on the basis of equity in particular concerned the configuration of the coasts, their relation to the continental shelf and traditional and

---

[30] Note dated 27 December 1966 [AA/51].

[31] Letter of G. Jaenicke to the legal department of the FO dated 3 April 1968 [AA/62]. A copy of the comments of Reuter is attached to this letter.

[32] See letter of G. Jaenicke to P. Reuter dated 29 October 1968 [AA/64].

[33] S. Oda, *International Control of Sea Resources* (A.W. Sythoff, Leyden: 1963, pp. 147–177). Oda became involved in the cases in April 1968 (see letter of G. Jaenicke to the FO dated 25 April 1968 [AA/62]), but he did not contribute to Germany's written pleadings.

[34] Letter of R. Bernhardt to the deputy head of the legal department of the FO dated 21 July 1966 [AA/47].

actual use of the shelf. Bernhardt did recognize that the equidistance method in particular cases could result in a suitable and equitable solution.[35]

Bernhardt's report no doubt laid the groundwork for the German Memorial in the two cases. A large part of the analysis of the drafting history of article 6 of the Convention on the continental shelf and State practice is similar and a number of specific ideas and suggestions contained in the report were also included in the Memorials. Both the report and the Memorials did not reject the equidistance method outright, but acknowledged that it could lead to an equitable result in individual cases.[36] To mention just one other example, Bernhardt had suggested that it could be pointed out that if the equidistance method in the future would be applied to divide ocean basins – the international community in the middle of the 1960s had started to entertain the idea that it would become possible to mine the deep seabed – it would lead to absurd results. Bernhardt suggested illustrating this idea with a reference to the island territories in the Pacific Ocean.[37] The Memorial picked up this idea but illustrated it by referring to the North Atlantic Ocean.[38]

There are also obvious differences between the Memorials and Bernhardt's report. The report provides an objective assessment of the strengths and weaknesses of the German case, while the Memorials are a legal brief positively stating the German case. One example illustrating this point is the scope of the special circumstances provision of article 6 of the Convention on the continental shelf. Bernhardt only observed that these circumstance are limited to circumstances related to the coast or the sea and do not include unrelated factors such as a country's population or surface area.[39] The Memorials adopt the same stance, but make it into an argument to support the reasonableness of the German case:

> Despite its need and capacity to exploit the continental shelf, Germany does not wish to base its claim on these considerations. All the more, therefore, the Federal Republic of Germany is of the opinion that the apportionment of the submarine areas of the North Sea should be made primarily according to the geographical criterion described above.[40]

---

[35] Bernhardt, *Rechtsgutachten* [B102/260031], pp. 74–76.
[36] See *ibid.*, p. 75; GM, para. 39.   [37] Bernhardt, *Rechtsgutachten* [B102/260031], p. 55.
[38] GM, para. 67 and figure 15.
[39] Bernhardt, *Rechtsgutachten* [B102/260031], pp. 62–63.   [40] GM, para. 79.

Another example of the difference between the documents is the treatment of publications. Bernhardt concluded that legal doctrine gave even less to go by than State practice.[41] The Memorials repeatedly refer to legal doctrine to support the German arguments.[42]

The Memorials, on a number of points, considerably elaborate on Bernhardt's report. After the legal department of the Foreign Office had discussed the report with Reuter in December 1966, it identified questions which needed further research and clarification.[43] A number of these points were argued in the Memorials. This, for instance, concerned the idea of coastal façades and frontal projections and the suggestion to present equidistance as a purely geometrical delimitation method, which as such was not inherently linked to the concept of the continental shelf and the sovereign equality of States.[44]

8.3.2 The Memorials

The Memorial of Germany in the case with Denmark was identical to the Memorial submitted in the case with the Netherlands, apart from the introductory chapter, which in each case treated the diplomatic history of the dispute with the State concerned. There is no indication that it was ever seriously considered to distinguish between the two cases on other points. Germany's case was built on the fact that the geography of its coast and its relationship to the coast of its two neighbors made delimitation by equidistance inappropriate. This made the choice for identical pleadings an obvious one.

The Memorials make the geographical situation of the North Sea as a whole a central tenet of the German case. As the Memorials observe, the North Sea is surrounded by the United Kingdom and the European continent and geologically consists of a single continental shelf.[45] On this basis, the Memorials developed a number of arguments. First, the fact that the North Sea constituted a single continental shelf implied that it was held in common between the coastal States and had to be delimited in its entirety.[46] In dividing this common good, each State should be

---

[41] Bernhardt, *Rechtsgutachten* [B102/260031], p. 44.
[42] See e.g. GM, paras 34 and 70–72.
[43] See note dated 13 January 1967 [AA/51]. This note was also forwarded to Reuter (see letter of the legal department of the FO to P. Reuter dated 13 January 1967 [AA/51]).
[44] See note dated 13 January 1967 [AA/51], pp. 2 and 6; these issues are addressed at e.g. GM, paras 30–31, 66, 70 and 78.
[45] GM, paras 7–8.   [46] *Ibid.*, paras 8, 28 and 30.

given a just and equitable share.[47] Germany emphasized that the principle of a just and equitable share for each State adjacent to a common continental shelf was a principle of law and not a principle of equity.[48] In other words, the Court was requested to arrive at a legal decision and not a decision *ex aequo et bono*. The latter would not be possible under the agreement by which the parties had submitted the dispute to the Court.[49] The justification of the claim that the principle of a just and equitable share was a principle of international law appeared to be somewhat hesitant. Germany submitted that it was a principle of international law because it derived its binding force from the conviction of the international community. It was then argued that it *"could* be regarded as an emanation of the principle of equality of States."[50] Germany acknowledged that the general principle of a just and equitable share would not allow the boundaries between the States concerned to be fixed, but added there were more precise criteria which could be applied to the special case of the North Sea.[51]

Another consequence of the fact that the entire North Sea was part of the continental shelf of the coastal States, which must be divided up between them, was that delimitations in this area were different from cases of delimitation in which the continental shelf constituted only a narrow area off the coast of States.[52] Germany submitted that the method of drawing a lateral equidistance line between adjacent States had been developed with mainly the latter situation in mind. However, even in that case the equidistance line was less likely to result in an equitable division of the continental shelf than in the case of opposite coasts, because the outcome to a much larger degree depended on the configuration of the coast and at times single points on a salient part of the coast.[53] Germany considered that the North Sea constituted a very special situation, where a continental shelf had to be divided by several coastal States. That problem could not be solved by the application of "methods developed for drawing maritime boundaries in normal geographical situations."[54] According to Germany, the fact that equidistance lines did not lead to an equitable result in its case was illustrated by a comparison of the lengths of the coasts of Germany, Denmark and the Netherlands to the equidistance area of each of them. Measured by one straight line between the islands of Borkum and Sylt, the German coast

---

[47] *Ibid.*, para. 30.   [48] *Ibid.*, para. 37.   [49] *Ibid.*   [50] *Ibid.*; emphasis provided.
[51] *Ibid.* On these criteria see further below.   [52] *Ibid.*, para. 8.
[53] *Ibid.*, paras 41 and 44.   [54] *Ibid.*, para. 41.

stood in a ratio of 6 to 9 to both the Danish and Dutch coast if they were measured in a similar fashion.[55] The equidistance area of the other two States was over 61,000 km^2, but Germany's share was only some 23,600 km^2.[56]

The equidistance method also did not take into account the principle of equality of all adjacent States. This did not mean that all States should receive an equal share of the continental shelf. As the Memorials pointed out, Germany had never proposed such a division. However, Germany considered that it was at least justified in hoping – this is again an example of the cautiousness of the Memorials – that any criterion for dividing the North Sea would not reduce the share of Germany disproportionally. If the principle of equality of the coastal States was to be interpreted in this sense, it was evident that the equidistance method was not equitable as it reduced the German share of the North Sea to 1/25 of the total area.[57]

Third, the equidistance method cut Germany off from the middle of the North Sea. In the case of an enclosed sea it could not be maintained that the middle of the sea appertained to one State, in view of the fact that all States had an equal title to the continental shelf. The Memorials cautioned that this did not necessarily mean that Germany should reach to the middle of the North Sea, but it did at least demonstrate that the equidistance method, which accorded Germany only a small corner of the North Sea, did not lead to a just and equitable solution.[58]

Another point of attack on the equidistance line by the Memorials was the distinction between the equidistance line and the median line. The latter would produce an equitable outcome in most situations involving opposite States, but this was much less so for the lateral equidistance line. As this argument indicates, the Memorial reduced equidistance to a method of delimitation and rejected that it was the overriding principle in relation to the delimitation of the continental shelf.[59] According to Germany that principle was the principle of a fair and equitable share. This conclusion was not only supported by the arguments set out above, but also by State practice in the form of unilateral acts, such as the

---

[55] *Ibid.*, para. 78. The Memorial does not provide figures for these coastal lengths. A straight line between Borkum and Sylt measures some 200 kilometers. A straight line between the Dutch boundary with Belgium and the most eastern of the Dutch Frisian Islands measures about 325 kilometers. This is also the approximate distance between Romø and the northeastern tip of the Danish mainland. These figures result in a coastal ratio of 6 to 9.8.

[56] *Ibid.*   [57] *Ibid.*, para. 80.   [58] *Ibid.*, para. 81.   [59] See e.g. *ibid.*, paras 36 and 39.

declaration of President Truman on the continental shelf of the United States.[60] The analysis of State practice also pointed out that the equidistance method was not a rule of customary law and as such was not applicable between the parties.[61]

Apart from subordinating the equidistance method to the principle of an equitable apportionment, the Memorials set out that there was no inherent link between the basis of continental shelf entitlement and the equidistance line. It was submitted that such a link did exist in the case of the territorial sea. Entitlement to the territorial sea was based on distance from the coast and as a consequence the outer limits of the territorial seas would intersect at a point at equal distance from the baselines of the States concerned. The equidistance line between both States would also reach the outer limit of their territorial seas at that same point. This "equidistant" point was the necessary consequence of the geometric construction of the outer limit of the territorial sea.[62] In the case of the continental shelf, entitlement was based on contiguity to the coast. Contiguity should however not be understood as propinquity to the nearest point on the coast, but as propinquity to the coast in general.[63] Two arguments were advanced to support this view.[64] First, for the exploitation and control of the continental shelf, the nearest coastal area or port from which exploitation could be effected was relevant, not the nearest point on the coast. This argument does not seem to be particularly convincing as international law does not ascribe any role to this factor in determining entitlement to the continental shelf. Moreover, it undermines the German argument in respect of the territorial sea. In that case actual control also does not necessarily depend on the nearest point on the coast. The second German argument did address the nature of continental shelf entitlement:

> To make the delimitation of the continental shelf more than necessary dependent on the coastal contours ... would be a departure from the geographical basis underlying the legal title of the coastal State to the continental shelf before its coast, namely the continuance of its territories into the sea. On the contrary, the tectonic-geographical connection between land and shelf should be an argument in favour of the thesis that special configurations of the coast should have no influence on the apportionment of a common continental shelf between the adjacent States.[65]

---

[60] Ibid., para. 31.   [61] Ibid., para. 62.   [62] Ibid., para. 45; see also paras 65–66.
[63] Ibid., para. 66.   [64] Ibid.   [65] Ibid.

The Memorials distinguished the provisions on continental shelf entitlement of the Convention on the continental shelf from its provision on delimitation. Articles 1 to 3 of the Convention, which according to Germany were its basic principles, were generally recognized as customary law. This raised the question of whether other provisions of the Convention, such as article 6 dealing with the delimitation between neighboring States, also reflected customary law. According to the Memorials a reply depended upon whether or not these other provisions were linked in such a way with the basic principles that the Convention would not be capable of application or implementation without them. The Memorials asserted that this was not the case. It was true that a necessary consequence of continental shelf rights was that in the case of conflicting claims these had to be apportioned between the States concerned. There was, however, no cogent reason why this apportionment should be made according to the equidistance method.[66] The Memorials furthermore pointed out that article 6 differed from articles 1 to 3 because the Convention allowed making reservations to article 6.[67]

Germany did acknowledge that the equidistance method could lead to a just and equitable apportionment.[68] The Memorials further submitted that the fact that the equidistance method did not necessarily lead to a just and equitable apportionment put the burden of proof in this respect on the party who claimed that it did.[69] In other words, in the case at hand Denmark and the Netherlands had the burden of proof in respect of the equidistance method. Although the Memorials submitted that the delimitation between Germany and its neighbors was governed by customary international law and not article 6 of the Convention on the continental shelf, the Memorials analyzed the drafting history of article 6. They concluded that:

> the so-called principle of equidistance was not laid down in the Continental Shelf Convention for the reason that it was a generally recognized rule of international law. Rather did the International Law Commission as well as the signatory States regard it as a useful method for an equitable apportionment of the continental shelf between States being opposite or adjacent to each other, insofar as circumstances permitted. Thus Article 6 is not a codification of already existing international law, but it is the outcome of an effort to develop the existing legal situation, with its demand for an equitable solution, by the establishment of a method which it was assured would, under normal geographical

---

[66] *Ibid.*, para. 61.   [67] *Ibid.*, para. 55.   [68] *Ibid.*, para. 37.   [69] *Ibid.*, para. 39.

> conditions, lead to an equitable and just apportionment of the continental shelf between the States concerned.[70]

This reasoning implied that even if the rule of article 6 were found to reflect customary international law, the burden of proving that it would be appropriate to apply the equidistance method in this case would also rest with Denmark and the Netherlands. The Memorials also took care to explain that Denmark and the Netherlands had not made any effort to discharge that obligation. They had shown "no inclination to negotiate on any other basis than that of the strict application of the equidistance principle."[71] Germany's suggestion to Denmark and the Netherlands that the best way to achieve a generally acceptable apportionment would be a multilateral conference involving all coastal States "unfortunately ... met with no response."[72] Germany sought to turn its own passivity into an advantage by suggesting that it had refrained from unilaterally defining its continental shelf boundaries in view of the friendly relations and cooperation between the North Sea States and because of the negotiations it had been conducting with Denmark and the Netherlands. Now it had not taken this step in view of the case pending before the Court.[73]

The "normal geographical conditions" under which the equidistance method would lead to an equitable solution were subsequently further defined by the Memorials. This "normal case" was:

> a more or less straight coastline, so that the areas of the shelf apportioned through the equidistance boundary more or less correspond to the shorelines (façades) of the adjacent States. Should this not be the case, and should therefore no equitable and appropriate solution result, the clause of the "special circumstances" applies.[74]

The Memorials then turned to the question of why Germany's case did not correspond to this normal case. The Memorials did take care to explain that the German situation was not unique. It was set out that if in a major indentation of the coastline one or both seaward sides belonged to a neighboring State, the problem corresponded to islands which lie before the coast. In both cases the equidistance line would cut the State at the back of the indentation off from the sea.[75] The Memorials illustrated the effect of the presence of islands and major indentations of the coastline by figures showing boundaries on the basis of the equidistance

[70] *Ibid.*, para. 53. [71] *Ibid.*, para. 27. [72] *Ibid.*, para. 75. [73] *Ibid.*, para. 94.
[74] *Ibid.*, para. 70. [75] *Ibid.*, para. 72.

principle off the West African coast, East Pakistan (present-day Bangladesh) and its neighbors, between the United States and Canada in the Gulf of Maine, between Romania and the Soviet Union (presently Ukraine), between Albania and Greece, between France and the United Kingdom in the English Channel, and between Colombia, Venezuela and the Netherlands Antilles.[76] These examples were chosen with care. They included the States of nationality of individual judges. As was remarked during the preparation of the Memorials: "[t]he idea is: each judge a good example as a warning, so that he can also clearly see the danger of a decision against Germany for his own country."[77]

The final part of the Memorials was dedicated to setting out a possible solution for the delimitation between Denmark, Germany and the Netherlands. The Memorials took the so-called sector principle as their starting point.[78] This principle has been applied in the Arctic and Antarctica in connection with claims to territory and maritime zones.[79] The sector principle involves the determination of sectors in the polar regions by drawing lines along meridians up to, respectively, the geographical North Pole or South Pole. These lines thus delimit a wedge-shaped area with its point at the pole. The Memorials recognized that there was considerable opposition to the sector principle to claim territory but submitted that apparently no serious objections had been raised against the use of the principle as an equitable method of delimiting spheres of interest in the Arctic Ocean and Antarctica. As such it was "still a valid precedent for a just and equitable division of such enclosed regions."[80] This seems too optimistic an assessment. For instance, Norway in its national legislation used the equidistance line to define the extent of its continental shelf in relation to its neighbors, including in relation to the Soviet Union in the Barents Sea. In that area the equidistance line diverges sharply from a sector line between the two States.[81] Second, a glance at figure 20 of the Memorials, which depicts

---

[76] *Ibid.*, figures 7–14.

[77] Note of Treviranus of the legal department of the FO to Witt dated 31 March 1967 [AA/ 84], p. 2. Translation by the author. The original text reads: "Der Gedanke ist: jedem Richter ein gutes Beispiel zur Warnung, damit er auch die Gefahren einer Entscheidung gegen Deutschland für sein Land deutlich sieht."

[78] GM, paras 82–83.

[79] For a brief discussion of the sector principle and further literature see R. Y. Jennings and A. Watts (eds), *Oppenheim's International Law*, 9th edn (Harlow: Longman, 1992), p. 693.

[80] GM, para. 83.

[81] For a discussion of the Norwegian position and its negotiations with the Soviet Union and subsequently the Russian Federation see e.g. Oude Elferink, *Maritime boundary*

sector lines in the Arctic Ocean, shows that the sector principle obviously would not lead to a just and equitable division of the area between all of the coastal States concerned.

The sector principle did provide Germany with a welcome precedent to assess its own situation against. For one thing it offered Germany an argument to extend its claim to the median line between the United Kingdom and the continent. Its application to the North Sea required some tweaking, as it would have been difficult to apply it to the North Sea as a whole. In the latter case, the center of the North Sea would have been located to the north of the area in dispute between Germany, Denmark and the Netherlands and the argument that there was a similarity with the polar regions, where the sectors converged on a central point, would have been difficult to maintain. By excluding the northern part of the North Sea because it concerned only the United Kingdom and Norway, Germany arrived at an area that was "roughly circular, without doing violence to the geographic realities."[82] From the center of this circular area straight lines could be drawn to the end points of Germany's boundaries with Denmark and the Netherlands. This resulted in a continental shelf area of some 36,700 km^2 for Germany, 53,900 km^2 for Denmark and 56,300 km^2 for the Netherlands. As the Memorials pointed out, the ratio of these areas roughly corresponded to the ratio of the coastal façades of the three States as previously defined by Germany.[83] On this basis the sector principle provided a well-suited standard of evaluation of what constituted a just and equitable apportionment of the continental shelf in the North Sea.[84]

Did the Memorials' sector claim actually constitute the best possible opening bid for Germany? This approach gave Germany a significantly smaller area of continental shelf than either Denmark or the Netherlands. Moreover, as was pointed out above, the German assertion that the use of sector lines in the polar regions constituted a valid precedent for the delimitation in the North Sea is far from convincing. But was there an alternative to the sector approach that would have given Germany a better starting position? An alternative would have to meet three conditions: it should give Germany a larger share than a sector claim; should not put the boundaries between the United Kingdom and the continent, and between Norway and Denmark, which were not

---

delimitation, pp. 224 and 226–246. For a figure showing the equidistance line and the sector line in this case see ibid., p. 383.

[82] GM, para. 86.   [83] Ibid.; a figure illustrating this point is provided at ibid., figure 21.

[84] Ibid., para. 87; see also ibid., para. 96.

disputed by Germany,[85] into question; and should give Germany access to the median line with the United Kingdom. The latter was achieved by the sector claim and this access was an important aspect of Germany's shelf delimitation policy.

The Memorials themselves indicate that the sector approach was not needed to justify the median line boundary between the United Kingdom and the continent. The Memorials had explained at length that median lines in general did result in an equitable solution and had set out that this was also the case in the specific geographic context of the North Sea.[86] The boundary between Denmark and Norway, which was nearly aligned with a sector line under the German proposal, might have posed more difficulties, as it was rather a lateral equidistance line than a median line. Its acceptance might at first sight have seemed to be contradictory to Germany's rejection of lateral equidistance lines with Denmark and the Netherlands. However, there is also a significant difference between the two cases. The equidistance line between Denmark and Norway extends to the median with the United Kingdom, whereas the German equidistance area does not even extend halfway to that line. In other words, Germany only needed to find an alternative basis for its claims in relation to Denmark and the Netherlands.

An alternative for a sector claim could have been an approach using coastal façades and their frontal projections. It could have been argued that these coastal projections of the three parties extended to the median line with the United Kingdom and that the overlap of these projections had to be divided between the parties.[87] As will be apparent from the preceding discussion, elements for such an approach, such as the idea of coastal façades, are already present in the Memorials.

It is not clear why the Memorials did not opt for a more assertive approach. Part of the explanation may be that it took a long time to develop the positive case for Germany. At a meeting between the German agent and representatives of the Foreign Office and the Ministry for the Economy in the middle of April 1967, just over a month before the final draft of the Memorials was scheduled to be finalized, the basic elements of Germany's positive case as contained in the Memorials were not yet in place.[88] The cautious shelf delimitation policy of the Foreign Office likely provides a further explanation for the Memorials' approach. This is confirmed by the comments of Treviranus

---

[85] Ibid., para. 23.   [86] Ibid., para. 89.   [87] See further text after footnote 360.
[88] See note dated 18 April 1967 [B102/318714], pp. 1–2.

of the Foreign Office's legal department on Germany's tactics for the proceedings in reviewing the draft of the German Replies. Treviranus assumed that Germany in the end could only count on relatively minor improvements and should aim for a limited but sure gain instead of going all or nothing.[89] Comments on the draft Memorials by the Ministry for the Economy confirm that it favored a more assertive stance. The comments query why the draft Memorials did not positively espouse the view that the continental shelf should be divided in accordance with the parity principle, which had been proposed by Germany during the negotiations with Denmark and the Netherlands.[90]

Germany, in the submissions of the Memorials, requested the Court to declare that: the delimitation between the parties was governed by the principle that each coastal State is entitled to a just and equitable share; the equidistance method was not a rule of customary international law; and could only be employed if it would achieve a just and equitable apportionment.

8.4   The Counter-Memorials of Denmark and the Netherlands

8.4.1   The preparation of the Danish and Dutch cases and the Counter-Memorials

Denmark and the Netherlands continued their existing close collaboration during the preparation of their cases. This was considered to be essential to avoid Germany taking advantage of the least discrepancy between the Danish and Dutch positions.[91] The only real difference between the two Counter-Memorials concerns the description of the factual background, which in each case focuses on the bilateral relation with Germany.

Denmark and the Netherlands approached the preparations for the proceedings completely different. Already in June 1966, the Danish government took the initiative to set up a committee, which was to

---

[89] Letter of H. Treviranus of the legal department of the FO to G. Jaenicke dated 20 May 1968 [AA/63], p. 7. Interestingly, Treviranus at one point submitted that a sector solution would not be appropriate, but only a solution giving Germany a wide opening to the middle of the North Sea (note dated 15 May 1968 [AA/63], p. 1).

[90] See note dated 22 June 1967 [B102/260032], p. 13.

[91] Report Danish CSC December 1966 [DNA/138], p. 6; Report on continental shelf-meetings held in Copenhagen September 5 and 6, 1967 (KSU/R.22) dated 21 September 1967 [DNA/138] (Report Danish CSC September 1967 [DNA/138]), p. 12.

advise the Ministry of Foreign Affairs on the case with Germany.[92] This continental shelf committee consisted of representatives of the Ministries of Foreign Affairs, Public Works and Justice, the Danish agent before the Court, the shipping company A.P. Møller, which represented the DUC, and Professors Foighel, Philip and Ross of the University of Copenhagen. Starting from August 1966 the committee met regularly. Committee members prepared preliminary reports for the Danish Counter-Memorial and contributed to drafts of the Counter-Memorial. Drafts of the Counter-Memorial were discussed in the committee and it was involved in decisions concerning the overall approach to the Counter-Memorial.

The preparation of the Dutch case was mainly carried out by the legal department of the Ministry of Foreign Affairs. Riphagen, the legal advisor of the Ministry, acted as the Dutch agent and was involved in drafting the written pleadings. Apart from Riphagen, a limited number of staff of the legal department was directly involved in the case. Whereas Denmark had already actively started to prepare texts for the Danish Counter-Memorial well before Germany had submitted the Memorial, the Netherlands was waiting for the Memorial.[93]

Denmark and the Netherlands did start to discuss the coordination of their work at an early stage. Riphagen met with Sørensen, the legal advisor of the Danish Ministry of Foreign Affairs, to discuss organizational matters in December 1966. Sørensen had been nominated as the Danish agent, and Riphagen was nominated as the Dutch agent shortly after this meeting.[94] Riphagen and Sørensen discussed the possibility of hiring foreign counsel, the appointment of the judge ad hoc in the two cases and the collection of State practice concerning the delimitation of the continental shelf. Sørensen wanted to hire Sir Humphrey Waldock and was wondering whether the Danish and Dutch government might jointly contact Sir Humphrey. This proposal was supported wholeheartedly by Riphagen.[95] Waldock, among others, had served as counsel in various cases before the ICJ and was an influential member of the ILC.[96]

---

[92] Note dated 28 June 1966 [DNA/141]; Report on the meeting of the advisory committee for the case with Germany on the continental shelf (Continental shelf committee) held on Monday 29 August 1966 (KSU/MR.1) [DNA/138], p. 1.

[93] Report Danish CSC May 1967 [DNA/39], p. 2.

[94] Note 85/66 dated 23 December 1966 [NNA/34].   [95] *Ibid.*

[96] Since 1961, Waldock had been serving as special rapporteur for the law of treaties and between 1962 and 1966 drafted five reports on this topic.

In view of the importance of State practice for determining the content of the law, Riphagen and Sørensen proposed to instruct Danish and Dutch embassies to collect material on the positions of States and bilateral agreements.[97] The embassies were requested to gather information on the boundaries between adjacent States in lakes, territorial waters and the continental shelf.[98] Where methods other than equidistance had been utilized, it was "essential to clarify the reasons, whether they be the configuration of the coast or others."[99]

Denmark and the Netherlands agreed that the Netherlands would propose a judge ad hoc. Originally, Rear Admiral Mouton had been considered for this position.[100] Mouton had written a monograph on the regime of the continental shelf[101] and was considered to be one of leading experts on the subject. Mouton's health at that time made his nomination as a judge ad hoc problematical. The resignation of Sørensen as the Danish agent led to a reconsideration of the choice of the judge ad hoc.[102] At a meeting with Riphagen, Paludan of the Danish Ministry of Foreign Affairs indicated that the Danish government realized that the Court would not be favorably impressed if Sørensen, the only Danish lawyer of international stature would not participate in the case and that it would do everything possible to get him back on board. Paludan circumspectly asked Riphagen if there would be a possibility of offering Sørensen the position of the judge ad hoc. Riphagen did not oppose this suggestion, because he had not committed himself to Mouton and he above all considered that Sørensen should play a role in the

---

[97] See note 85/66 dated 23 December 1966 [NNA/34].

[98] See circular of the MFA of March 1967 attached to a letter of the MiFA dated 3 March 1967 [NNA/23], p. 2.

[99] See ibid., p. 3.   [100] Note 85/66 dated 23 December 1966 [NNA/34].

[101] M. W. Mouton, The Continental Shelf (The Hague: Nijhoff, 1952). Mouton also dealt with this subject in his lectures at the Hague Academy of International Law (M. W. Mouton, "The Continental Shelf," 1954 (85) Recueil des Cours, pp. 347–463).

[102] Sørensen officially had resigned as agent for reasons of health, but the real cause of his resignation was the course the first meetings in the Danish continental shelf committee had taken (note 18/67 dated 17 March 1967 [NNA/29], p. 1; Report on the meeting of the advisory committee for the case with Germany on the continental shelf (Continental shelf committee) held on Tuesday 21 February 1967 (KSU/MR.3) [DNA/138], p. 1). Sørensen apparently felt that he was not given sufficient room to deal with the case as he considered appropriate (ibid). The report on the second meeting of the committee indicates that various of its members had criticized the formulation of the question to be submitted to the Court and suggested that this matter should be further discussed with Germany. This suggestion was rejected by Sørensen and Paludan of the MFA (Report Danish CSC December 1966 [DNA/138], pp. 1–3).

proceedings.[103] This did not lead to immediate agreement on Sørensen's selection as the judge ad hoc. The Dutch apparently wanted to keep open the option of nominating Mouton. Only after Paludan had asked Riphagen about Mouton's health in September 1967, and it became clear it was unlikely that he could participate in the proceedings,[104] was Sørensen eventually selected as judge ad hoc. Interestingly, article 17(2) of the Statute of the Court provides that no member of the Court may participate in the decision of any case in which he has previously taken part as agent. As is indicated by article 31(6) of the Statute, this provision also applies to judges ad hoc. Sørensen had resigned as the Danish agent before his nomination had been communicated to the Court and Denmark had only informed the Court of the nomination of Jacobsen, his replacement, as agent.[105]

A first meeting between Riphagen, Paludan, Jacobsen and Waldock, who acted as counsel for the two States, took place in March 1967. Apart from organizational matters, they also discussed the merits of the case. Waldock was very familiar with the issue of special circumstances, because he had already dealt with this matter in relation to the Persian Gulf.[106] He was convinced that the Danish position was correct and that from a geographical perspective only the equidistance principle would lead to equity. Various of the German arguments were "quite hopeless."[107] Waldock asked to be authorized to contact Elihu Lauterpacht to see if he could be of assistance. Lauterpacht had been doing a lot of work on this matter for Royal Dutch Shell. Jacobsen and Paludan granted their authorization enthusiastically, but Riphagen only agreed hesitatingly[108] finding it problematic that Lauterpacht was also advising the DUC.[109]

Waldock and the two agents would meet regularly to discuss the cases. Jacobsen and Riphagen agreed that Waldock should have a central role in preparing the two Counter-Memorials.[110] Waldock also met a number of times with the Danish continental shelf committee. Riphagen initially

---

[103] Note 18/67 dated 17 March 1967 [NNA/29], p. 1.
[104] Draft of a report attached to a letter of Jacobsen to Paludan of the MFA dated 11 September 1967, [DNA/79], p. 3.
[105] See letter of the Ambassador of Denmark to the Netherlands to the Registrar dated 22 February 1967 (ICJ Pleadings, *North Sea continental shelf* cases, Vol. II, p. 369).
[106] Note 18/67 dated 17 March 1967 [NNA/29], pp. 1–2.
[107] Report Danish CSC May 1967 [DNA/39], p. 2.
[108] Note 18/67 dated 17 March 1967 [NNA/29], p. 2.   [109] *Ibid.*
[110] Notes of Bent Jacobsen on a Meeting with Sir Humphrey Waldock on 22 and 23 May 1967 [DNA/133], p. 3.

had agreed to also participate in a meeting of the committee, but sub-sequently changed his mind. The Dutch Ministry of Foreign Affairs had a policy not to engage in consultations with Dutch concessionaires concerning the case and would be placed in an impossible position if it became known that Riphagen had participated in a meeting in which a representative of the DUC had also been present.[111] It is actually not certain that the Dutch Ministry of Foreign Affairs had instituted a policy to refrain from discussing the matter with concessionaires. The records from the archives of the Ministry do not contain any information in this respect, but there had been meetings with representatives of Shell to discuss the delimitation of the Dutch continental shelf previously. J.W. Josephus Jitta of Shell submitted a note on the continental shelf dispute a couple of months after Riphagen's participation in the meeting of the Danish continental shelf committee had been discussed and he met with Maas Geesteranus of the Ministry of Foreign Affairs in June 1968 to discuss the German Reply.[112]

When the Danish Ministry of Foreign Affairs tried to convince Riphagen to reconsider his position, by pointing out that the Danish situation, with one concessionaire instead of many, was fundamentally different from that of the Netherlands,[113] Riphagen put forward the additional argument that the two cases were formally separate and that in order to maintain his independence as the Dutch agent he had to abstain from participating in a meeting with the committee.[114] In the light of these not altogether convincing arguments, it should not be excluded that Riphagen did not want to participate in a meeting for a different reason. There existed certain differences between the Netherlands and Denmark on how to deal with the substance of the matter. This in particular concerned the question of how to deal with the special circumstances provision.[115] The participation of the Danish concessionaire in the committee may have provided a welcome pretext

---

[111] Telex of the Danish Embassy in The Hague to the MFA dated 9 August 1967 [DNA/138].
[112] See letter of J. W. Josephus Jitta to I. Foighel dated 28 November 1967 [DNA/50] and letter of J. W. Josephus Jitta to G. W. Maas Geesteranus dated 20 June 1968 [NNA/30].
[113] Telex of the Danish Embassy in The Hague to the MFA dated 14 August 1967 [DNA/138].
[114] Telex of the Danish Embassy in The Hague to the MFA dated 22 August 1967 [DNA/138].
[115] See further below.

to avoid being drawn into a debate on the substance of the matter with the committee.

At the meeting of the Danish continental shelf committee with Waldock, it was recognized that the German Memorial had been cleverly drafted and that it might prove difficult to refute certain of its arguments.[116] A similar conclusion had been reached by Lauterpacht in his preliminary assessment:

> The Memorial seems to me to be a very skilled piece of work. It is characterized by such unTeutonic features as tact, restraint and reasonableness. It could have been a heavy, elaborate and unreadable document, but it is not. It will, I think, be bound to make a good impression on the Court. The Germans having thus made a good start, the Danish case is made no easier.[117]

At the same time, the committee concluded that the Memorial was unbalanced in a number of respects. For instance, the approach to equity was tailored to meet the German interests and employed equity to argue in terms of continental shelf apportionment, rather than delimitation.[118] The argument in relation to the sector theory was the only important new idea introduced by the Memorial and it needed to be addressed. It was among others criticized because it was nothing but a variation of the equidistance method, was not a principle of delimitation but acquisition of territory, and in the particular case had not been applied to Belgium, France and Sweden.[119]

The main point of discussion during the preparations of the Counter-Memorials probably was how to deal with the reference to special circumstances in article 6 of the Convention on the continental shelf.[120] The importance of this point was obvious. The whole case could be said to hinge on it. Broadly speaking, two outcomes would be possible. Special circumstances could be viewed as being equal in status to the equidistance method or as an exception which could only be applied in exceptional cases. In the former case there would be much more scope to depart from the equidistance line than in the latter. As was observed at one point by Paludan, the deputy head of the political and

---

[116] Report Danish CSC September 1967 [DNA/138], p. 2.
[117] E. Lauterpacht, Danish continental shelf; Preliminary comments on the German Memorial dated 2 September 1967 [DNA/59], p. 1.
[118] Report Danish CSC September 1967 [DNA/138], pp. 2–3. [119] *Ibid.*, p. 3.
[120] See also Report on the meeting of the advisory committee for the case with Germany on the continental shelf (Continental shelf committee) held on Friday 22 December 1967 (KSU/MR.7) [DNA/53] (Report Danish CSC December 1967 [DNA/53]), p. 9, para. 90.

legal department of the Danish Ministry of Foreign Affairs, if the judges could not be convinced of the Danish point of view on the role of special circumstances – "the case would be lost."[121]

A first draft on the special circumstances provision for the Danish Counter-Memorial was prepared by Bent Jacobsen, the Danish agent.[122] Members of the Danish continental shelf committee were not convinced by the agent's approach. They considered that the chapter on special circumstances should start with an objective legal explanation of the concept of special circumstances, after which the rules as interpreted should be applied to the specific case. It was also considered that the draft of the agent was too polemic and "too direct in its argumentation against the German views."[123] To seek to deal with these points Professor Foighel prepared a new draft on the basis of the work of Jacobsen.[124] A further draft was prepared jointly by Jacobsen and Foighel.[125] At a meeting of the Danish continental shelf committee with Waldock in December 1967, Jacobsen had come around to the view of Foighel and other members of the committee. The Counter-Memorial should present a clear general interpretation of the special circumstances clause.[126] Waldock argued against this approach, submitting that "[o]ne should be careful not to get too involved in problems of general interpretation which might not be accepted by the Court."[127] It was not possible to be too categorical about whether specific geographical conditions constituted special circumstances. This had to be decided in the circumstances of the specific case. Finally, it would in any case be possible to further develop the argument in further pleadings after it would have become known what Germany had argued in the Rejoinder.[128] The chairman of committee proposed a compromise between the two views. Saying too much at this point was not an attractive option, but if Denmark said too little, Germany might refrain from further addressing the argument.[129] This compromise proposal was elaborated in a new draft text. The

---

[121] Report on the meeting of 12 October 1967 concerning the chapter on "Special Circumstances" in the Danish pleading dated 24 October 1967 [DNA/59], p. 4 (translation by the author. The original text reads "var sagen tabt"); see also e.g. Report Danish CSC November 1967 [DNA/53], p. 5.

[122] See letter of B. Jacobsen to Sir Humphrey Waldock dated 5 October 1967 [DNA/59], p. 1.

[123] *Ibid.*   [124] See letter of I. Foighel to J. Paludan dated 9 October 1967 [DNA/18].

[125] Report on the meeting of 12 October 1967 concerning the chapter on "Special Circumstances" in the Danish pleading of [DNA/59], p. 4.

[126] Report Danish CSC November 1967 [DNA/53], p. 5.   [127] *Ibid.*, p. 6.   [128] *Ibid.*

[129] *Ibid.*, p. 7.

meeting agreed that this text provided a good basis for a final text,[130] but this was not the end of the matter. The special circumstances clause also divided Denmark and the Netherlands.

In September 1967, during informal discussions between the Dutch and Danish agents and Waldock and the chairman of the Danish continental shelf committee, Riphagen indicated that he was in agreement with the outline of the Counter-Memorial prepared by the committee. The only major difficulty in bringing the two Counter-Memorials in line was that the Netherlands also had a problem in respect of the continental shelf delimitation involving the Netherlands Antilles and Venezuela. Riphagen was reluctant to commit himself to the argument that islands constituted a special circumstance,[131] but Jacobsen had the impression that Riphagen agreed that the issue of islands would be treated in the Danish Rejoinder for the benefit of both cases.[132] Riphagen and Jacobsen agreed that giving too much consideration to circumstances extraneous to a case would be the surest way to losing it.[133]

The draft of a chapter on special circumstances for the Danish Counter-Memorial did argue that islands constituted a category of special circumstances. The Danish agent in this connection referred to islands and peninsulas as two examples of unusual coastal circumstances, which, as they do not truly represent the coast of a State, should be disregarded for the purpose of drawing equidistance lines.[134] Stated this categorically, the argument in any case could have little appeal to the Netherlands. Denmark at first refrained from sharing the drafts on the special circumstances with the Netherlands.[135]

The chapter on special circumstances, which was prepared after the meeting of the Danish continental shelf committee of November 1967, was submitted to Waldock. He criticized it as being academic and that did not fit well with the other chapters contained in the same part of the Counter-Memorial he had drafted himself. It was, moreover, better to keep different options open and to wait and see how Germany would further develop its arguments. Finally, it would be unfortunate if the Danish and Dutch Counter-Memorials would not be identical on such

[130] *Ibid.*, pp. 14–16. [131] Report Danish CSC September 1967 [DNA/138], p. 13.
[132] Letter of B. Jacobsen to W. Riphagen dated 3 May 1968 [NNA/30], p. 2.
[133] Note dated 10 January 1968 prepared by B. Jacobsen [DNA/60], p. 3.
[134] Letter of B. Jacobsen to Sir Humphrey Waldock dated 5 October 1967 [DNA/59], p. 2.
[135] See Short report on an internal working meeting dated 7 October 1967 [DNA/138], p. 1; letter of B. Jacobsen to Sir Humphrey Waldock dated 5 October 1967 [DNA/59], p. 1.

an important point.[136] Jacobsen accepted Waldock's proposal to rewrite the chapter with the reservation that Denmark at a later stage could set out its legal views on the special circumstances clause. Jacobsen found that the text prepared by Waldock, which had also been accepted by the Dutch, was much more in line with the Danish fundamental views than could have been expected on the basis of earlier discussions.[137] Still, Jacobsen concluded that it should be acknowledged that there might be an essential difference between the next round of pleadings of Denmark and the Netherlands, but that would be a lesser disadvantage than refraining from making the Danish views known.[138]

The key passage on special circumstances of the Counter-Memorials only referred to the fact that the general rule of equidistance could be modified "by reason of some exceptional feature."[139] Jacobsen considered that this abstract reference would not be understood by someone who was not familiar with the matter. Only a direct reference to islands and promontories would make it possible to understand the implications of these considerations in practice.[140] The Counter-Memorials also strongly rejected that the continental shelf delimitations of Germany in the North Sea had nothing in common with the case of islands, as had been submitted by Germany in the Memorial.[141] Neither the Danish and Dutch governments nor the Court were "called upon in the present case to express any opinion as to what should be the solution of that particular problem [of islands] under Article 6 of the Convention" on the continental shelf.[142] This line of argument offered the best chance that the Court would indeed refrain from saying anything about islands and it meant that nothing specific had to be said about islands in the Counter-Memorials.

8.4.2   The Counter-Memorials

The two Counter-Memorials were largely identical. The major differences are contained in their Part I, which set out the facts and the history

---

[136] Report Danish CSC December 1967 [DNA/53], p. 9, para. 71.   [137] Ibid., para. 72.

[138] Note dated 10 January 1968 prepared by B. Jacobsen [DNA/60], pp. 2–3; see also letter of J. Paludan to Sir Humphrey Waldock dated 6 February 1968 [DNA/60], pp. 1 and 3.

[139] DCM, para. 156; NCM, para. 151.

[140] Note dated 10 January 1968 prepared by B. Jacobsen [DNA/60], p. 2.

[141] DCM, para. 147; NCM, para. 141; see also DCM, paras 145 and 150; NCM, paras 139, 144 and 145.

[142] DCM, para. 147; NCM, para. 141.

of the two disputes. The introductions to the Counter-Memorials immediately set the tone. The disputes had arisen because Germany "has
thought fit to lay claim to areas of the continental shelf" which lie nearer
to the coasts of Denmark and the Netherlands.[143] What is more,
Germany had "declined to acknowledge" the right of Denmark and the
Netherlands to delimit their continental shelf in accordance with the
principles recognized as applicable at the 1958 Geneva Conference and
the Convention on the continental shelf.[144] Whereas Denmark and
the Netherlands were asking for a delimitation in accordance with the
international law, Germany was asking for an apportionment "according
to the Federal Government's own notion of what is due to the Federal
Republic *ex aequo et bono*."[145]

Denmark and the Netherlands had a logical explanation for
Germany's disdain for the law, but they were careful not to say so in so
many words. Chapter I of Part I of the Danish Counter-Memorial on
geology and geography was most of all concerned with the activities of
the DUC on the continental shelf in the North Sea. This information was
put up front for tactical purposes. It should:

> make it clear to the Court at an early stage, that Germany was claiming
> exactly that part of the continental shelf which offered the best chances of
> finding valuable resources, and in which the concessionaire had, in fact,
> undertaken drillings.[146]

Chapter I stressed that Denmark's geological research on the continental
shelf only started in 1963, whereas Denmark had already expressed the
view that the continental shelf should be delimited by equidistance in
1952.[147] In other words, Denmark's choice for the equidistance principle
was not motivated by knowledge about the shelf's resources. The Danish
criticism of Germany actually was quite beside the point. As was set out
in detail in Chapter 2.3.3, Germany considered the equidistance method
inappropriate for the delimitation of its continental shelf in the North
Sea in preparation for the 1958 Conference – well before Denmark had
started exploratory activities in 1963. Chapter 2 of the Dutch Counter-
Memorial detailed that activities on the Dutch continental shelf had been
taking place up to the median line with Germany.[148]

---

[143] DCM, para. 2; NCM, para. 2.   [144] DCM, para. 2; NCM, para. 2.
[145] DCM, para. 2; NCM, para. 2.
[146] Report Danish CSC November 1967 [DNA/53], p. 3, para. 10.   [147] DCM, para. 7.
[148] NCM, para. 11.

Denmark and the Netherlands took a different approach to the history of their negotiations with Germany. The Dutch Counter-Memorial limited itself mostly to giving a factual description and concluded that the negotiations had resulted in a partial boundary and an agreement to submit the dispute over the remainder of the boundary to the Court.[149] Denmark explicitly sought to refute the impression which might have been conveyed by the Memorial "that only a formal presentation of official viewpoints took place, and 'the other Party' was intransigent in upholding the Continental Shelf Convention and the equidistance principle."[150] Whether the Danish argument was brought across effectively is questionable. At various points it was admitted that Denmark (and the Netherlands) had had no inclination to abandon the position that the law mandated the application of the equidistance principle. In addition, even a possible regime for joint utilization, which disregarded the legal positions as was proposed by Germany, was only acceptable to the other two States if delimitation lines would be agreed upon. It was also acknowledged that Germany had made various proposals on delimitation principles different from equidistance, albeit without presenting any concrete proposal or elaboration.[151]

The almost identical Part II of both Counter-Memorials is dedicated to a discussion of the applicable law and its implications for the North Sea. A first chapter criticizes the German submission that the fundamental goal of the delimitation of the continental shelf is to achieve a just and equitable apportionment between the States concerned. According to the Counter-Memorials the claim of Germany amounted to a request to the Court to settle the delimitation *ex aequo et bono*:

> Without a framework of legal criteria to determine what is "just and equitable", the concept of a "just and equitable apportionment" lacks any legal content.[152]

Because the Memorials allegedly did not provide this framework, the claim of Germany did not fall within the terms of the questions put to the Court, which had been requested to provide "directions regarding the applicable framework of legal criteria."[153] This submission is wrong on both counts. The Memorials did elaborate on the implications its fundamental notion of continental shelf delimitation and the questions put

---

[149] *Ibid.*, paras 28–31.   [150] DCM, para. 27.   [151] *Ibid.*, paras 33–36.
[152] *Ibid.*, para. 38; NCM, para. 33.   [153] DCM, para. 38; NCM, para. 33.

to the Court in any case did not require the kind of specificity that the Counter-Memorials suggested.

Another criticism of the Memorials' fundamental notion of continental shelf delimitation was more to the point. The Counter-Memorials argued that there was no question of sharing out a common continental shelf between the coastal States of the North Sea. Instead, what was required was the delimitation of a continental shelf boundary between neighboring coastal States.[154] Two arguments were advanced to reject the German view. First, that view confused the geological and legal concept of the continental shelf. There perhaps were reasons to consider the continental shelf as a "unit" from the former perspective, but from a legal perspective the continental shelf was primarily a "space" like land, sea and air and as such was *a priori* susceptible to any limitation or division. Second, the mere fact that more than one State laid claim to the same space did not make it a *common* space to be divided between them. The normal legal situation in respect of, for instance, disputed territory was not that the territory was divided but that the better claim prevailed.[155] The view that the continental shelf is not a common area, but that instead in certain cases there exist areas of overlapping claims to the same area, is in line with the notion that the continental shelf is an area adjacent to, and an extension of, land territory. At the same time, the Danish-Dutch argument is not wholly convincing as it divorces the legal concept of the continental shelf completely from the geological concept. As is evidenced by article 1 of the Convention on the continental shelf and its drafting history such a complete separation of the two concepts actually did not exist.[156]

The opposition of the geological and legal continental shelf could be said to be essential to the Danish-Dutch case, as it allowed viewing the continental shelf as a featureless "space." Under that view, it could be argued that adjacency to the coast, which provided the basis for continental shelf entitlement, constituted a purely geometrical concept. If distance from the coast provided the basis of continental shelf entitlement it would follow logically that equidistance should provide the central rule for the delimitation of the continental shelf between

---

[154] DCM, paras 40 and 49–50; NCM, paras 35 and 43–44.

[155] DCM, para. 49; NCM, para. 43.

[156] For a detailed review of the drafting history of article 1 see Oxman, "The Preparation of Article 1," pp. 245–305, 445–472 and 683–723. This review illustrates the significance of geological concepts for understanding the implications of article 1 of the Convention.

neighboring States. This idea was already introduced in the criticism of
the German notion of apportioning a common continental shelf between
neighboring States:

> Nor have any of the other North Sea States sought to treat the continental
> shelf beneath that sea as legally a unity. On the contrary, every single one
> of them – with the exception of the Federal Republic of Germany – has
> demonstrably regarded its claim as limited to that part of the continental
> shelf every point of which is nearer to its coast than to that of any other
> State.[157]

Subsequently, this concept of the shelf was further elaborated and firmly
linked article 6 of the Convention to its articles 1 and 2:

> there were cogent reasons why Article 6 should state the equidistance
> principle as the general rule – reasons which are linked to the *ratio legis* of
> Articles 1 and 2 . . . Inherent in the concept of a coastal State's title *ipso*
> *jure* to the areas adjacent to its coast is the principle that areas nearer to
> one State than to any other State are to be presumed to fall within its
> boundaries rather than within those of a more distant State . . . In other
> words, this principle establishes a direct and essential link between the
> provisions of Article 6 regarding the equidistance principle and the basic
> concept of the continental shelf recognized in Articles 1 and 2 of the
> Geneva Convention of 1958.[158]

Under articles 1, 2 and 6 a State laying claim to an area closer to the coast
of another State had to establish the legal grounds on which its claim
should be given precedence.[159]

According the Counter-Memorials, Germany's notion of apportioning a just and equitable share of a common continental shelf to each of
the coastal States of the North Sea was also flawed for another reason. In
both cases, the compromis between the parties requested the Court to
determine the boundary between Germany and the other party to the
case. A delimitation between Germany and the other party would not by
itself determine the total area appertaining to each of them, since that
would be dependent upon their other boundaries with States which were
not a party to the dispute before the Court. Thus the principle formulated by Germany could not constitute a rule of international law as
between the parties before the Court.[160] This indeed was a fundamental,
if not the fundamental, aspect of the cases. As Germany had consistently
argued, the equidistance method did not lead to an equitable result for

---

[157] DCM, para. 49; NCM, para. 43.   [158] DCM, para. 115; NCM, para. 109.
[159] DCM, para. 116; NCM, para. 110.   [160] DCM, para. 51; NCM, para. 45.

itself because it was squeezed in between Denmark and the Netherlands. On the other hand, if these two delimitations were viewed in isolation, they could be argued to be equitable. The equidistance line in both cases was more or less a perpendicular to the coast of Germany and its neighbor, suggesting that both were treated in an equal fashion.

A second chapter of Part II of the Counter-Memorials dealt with the proposition that a delimitation of a maritime area in accordance with generally recognized rules of international law is *prima facie* valid and opposable to other States. Or, as had been suggested in the introduction of the Counter-Memorials, Germany declined Denmark and the Netherlands the right to delimit their boundaries unilaterally in accordance with these rules. To support this proposition, the Counter-Memorials invoked the judgment of the Court in the *Anglo-Norwegian fisheries* case, which observed that:

> The delimitation of sea areas has always an international aspect; it cannot be dependent merely upon the will of the coastal State as expressed in its municipal law. Although it is true that the act of delimitation is necessarily a unilateral act, because only the coastal State is competent to undertake it, the validity of the delimitation with regard to other States depends upon international law.[161]

The reliance on this finding in the *Anglo-Norwegian fisheries* case at first sight comes across as somewhat peculiar. The Court in this case had to rule on the legality of the straight baselines of Norway. The establishment of baselines indeed can only be effected unilaterally by the coastal State. Only after that determination by the coastal State, can other States object if they consider that the coastal State has not acted in accordance with international law. On the other hand, it is generally accepted that a maritime boundary with another State cannot be determined unilaterally. The debate in the ILC and at the 1958 Conference indicated that the need for agreement between neighboring States was a fundamental aspect of continental shelf delimitation.

It can be asked why the Counter-Memorials so emphatically argued that Denmark and the Netherlands were entitled to delimit their continental shelf boundary with Germany unilaterally. This stance may have contributed to the impression that Denmark and the Netherlands were not willing to seriously consider any alternative to their own position and wanted to impose it on Germany. A convincing explanation is not

---

[161] [1951] ICJ Reports, p. 116, at p. 132; quoted at DCM, para. 58; NCM, para. 52.

readily available, but it seems that there was a genuine belief that this argument had to be made.[162] It did justify the unilateralism of Denmark and the Netherlands and shifted the burden of proving the presence of special circumstances on Germany. It would, however, also have been possible to disregard the first point and the second point could have been made on the basis of a specific interpretation of article 6 of the Convention on the continental shelf.

The Counter-Memorials were diametrically opposed to the German Memorials on all major issues in respect of article 6 of the Convention on the continental shelf. Equidistance was not just one method of delimitation, which could only be applied if it was shown that it led to an equitable result. The Counter-Memorials did not deny that "equitable principles" had been mentioned in State practice prior to the equidistance principle, but argued that this rule had been transformed into the rule contained in article 6.[163] Equidistance constituted the general rule unless special circumstances justified another boundary and equidistance was inherent to a coastal State's title to the continental shelf.[164] Equidistance was, so to speak, the equitable delimitation method *par excellence*.

Like the Memorials, the Counter-Memorials discussed the work of the ILC, the 1958 Geneva Conference and State practice. The 1953 session of the ILC was considered to be a turning point in the development of the delimitation provision for the continental shelf. Especially many smaller States had raised strong objections to the ILC's proposal that disputes concerning delimitation should be settled *ex aequo et bono*.[165] The Counter-Memorials implicitly suggested that smaller States were looking for protection by the law – and the Court – from the demands of more powerful States.

Following the commentaries of governments on the ILC's work, the equidistance principle had been introduced as the general rule and the major principle in the draft delimitation provision.[166] The drafting history showed that this recognition of equidistance as the general rule in the absence of agreement was equally applicable to States with opposite and adjacent coasts.[167] The commentary on the 1958 Conference focused on the attitude of Germany. Germany's proposal for an

---

[162] Report Danish CSC September 1967 [DNA/138], p. 5, para. 24.
[163] DCM, para. 61; NCM, para. 55.   [164] DCM, paras 114–115; NCM, paras 108–109.
[165] DCM, para. 64; NCM, para. 58.   [166] DCM, para. 67; NCM, para. 61.
[167] DCM, para. 71; NCM, para. 65.

international regime for the continental shelf had failed, after which it fully participated in the discussion of the ILC's draft articles.[168] Germany did make some comments on the delimitation provision, but these were unrelated to the issues in dispute with Denmark and the Netherlands and the negative vote of Germany on the text of the Convention was not motivated by opposition to its article 6.[169] That rejection in any case was short-lived as Germany signed the Convention on 30 October 1958.[170]

Before looking at State practice in respect of article 6 of the Convention on the continental shelf, the Counter-Memorials discussed practice in respect of other maritime boundaries and boundaries in lakes and rivers.[171] The purpose of this discussion was twofold. First, it was intended to show that the equidistance method had already been quite widely recognized prior to the 1950s. Second, this practice illustrated that the equidistance method had been applied not only in the case of oppositeness, but had also been used to determine lateral boundaries.

State practice showed a wide acceptance of the Convention on the continental shelf. The German argument that only a minority of States had accepted the Convention was "a little surprising."[172] The number was impressive by any standard if compared to the past practice of States in accepting multilateral conventions.[173] The value of this large number of ratifications was not materially weakened by the "so-called 'reservations' to Article 6."[174] The statements of four States on article 6 in accepting the Convention only served to confirm the generally recognized rules of law governing continental shelf boundaries:

> By invoking the exception of "special circumstances" included in Article 6, the four States concerned expressly recognized the validity, and claimed the benefits, of the provisions of that article.[175]

What is meant by this rather obscure formulation is not completely clear. The reservations of France and Venezuela referred to the presence of special circumstances and that this led to the inapplicability of the equidistance method. If this had been intended to merely confirm the general validity of article 6, there would have been no need to do so, as the article itself already referred to special circumstances.

---

[168] DCM, para. 75; NCM, para. 69.
[169] DCM, paras 77 and 79; NCM, paras 71 and 73.
[170] DCM, para. 79; NCM, para. 73.   [171] DCM, paras 84–90; NCM, paras 78–84.
[172] DCM, para. 92; NCM, para. 86.   [173] DCM, para. 92; NCM, para. 86.
[174] DCM, para. 93; NCM, para. 87.   [175] DCM, para. 98; NCM, para. 92.

The Counter-Memorials paid considerable attention to the agreed continental shelf delimitations in the North Sea, which Germany "evidently [found] somewhat embarrassing."[176] All of this practice, including the two partial boundaries concluded by Germany, reflected the principles of article 6 of the Convention.[177] This practice might seem impressive, but if it could not be shown that these cases were similar to that of the pending delimitations between Germany and Denmark and the Netherlands, it would be of little value. This practice did not indicate how the States concerned viewed the relationship between equidistance and special circumstances. The Counter-Memorials were silent on both counts.

The last part of Part III of the Counter-Memorials criticized Germany's application of the law to its delimitations with Denmark and the Netherlands. Germany had inflated the scope of special circumstances in such a way that the equidistance principle almost became the exception, rather than the rule.[178] In any case, it was not clear what Germany was driving at. The partial boundaries with Denmark and the Netherlands were based on equidistance. How could it be that an equidistance line was perfectly appropriate near the coast but ceased to be so "further out to sea when the coastline is 'more or less straight' *and no geographical factor other than that coast-line influences materially the course of the equidistance line?*"[179]

Of course, Germany would contend that this conclusion ignored the wider geographical context. To counter such an argument, the Counter-Memorials advanced a number of points. Germany did not make any real attempt to examine the actual configuration of the coastlines of itself and Denmark and the Netherlands.[180] First of all, the bend in the German coast could not be equated to a bay or gulf as was submitted by Germany.[181] Second, the coasts on both sides of this bend were not only open but more or less straight and the German coast in both directions extended for more than 100 kilometers. Finally, there were no offshore islands influencing the delimitation. In short, the coastal configuration was quite unremarkable and could hardly be less exceptional.[182]

---

[176] DCM, para. 104; NCM, para. 98.   [177] DCM, para. 110; NCM, para. 104.
[178] DCM, para. 131; NCM, para. 125.
[179] DCM, para. 132; NCM, para. 126 (emphasis in the original).
[180] DCM, para. 140; NCM, para. 134.   [181] DCM, para. 142; NCM, para. 136.
[182] DCM, paras 143–146; NCM, paras 137–140.

Even if the bend in the German coast would be found to constitute a special circumstance, it still would not justify another boundary line. The continental shelf of Denmark and the Netherlands delimited by the equidistance method was perfectly normal and neither State gained anything at the expense of Germany because of anything unusual about their own coasts.[183] Only if a State's share would be made abnormally large by reason of some unusual feature and another State's continental shelf was made abnormally small by the same feature would an adjustment of the equidistance line be justified. Only in such a case would a modification be equitable and just to both States.[184] As is readily apparent from this reasoning, it entails that a completely different treatment of States with more or less similar coast would still be just.

Germany's sector claim and coastal frontage theory were dismissed as respectively "highly artificial and arbitrary," "the Federal Republic's magic circle" and "nothing but an artificial construction."[185] Neither the sector claim nor the coastal façade were credibly linked to the actual geography, were not supported by State practice and were not justified by the applicable law.[186] Much of this criticism is to the point and exposed weaknesses in the specific elaboration of the positive German case. However, that would not be sufficient if the Court would not also be convinced that the premises of the Danish-Dutch affirmative case were more convincing than those of Germany. This basically concerned the question what importance the need for an equitable outcome should have on the selection of delimitation methods. As had been pointed out by Lauterpacht in his initial comment on the German Memorial:

> it is important to counterbalance the skilful emphasis in the Memorial on the principle of the "just and equitable share." Although it may well be possible to argue that in so far as the debates in the International Law Commission and elsewhere refer to "just and equitable" they refer to a boundary line rather than to an area, the fact remains that it will be difficult effectively to negate the essential substance of Chapter I and II of the Memorial.[187]

In their submissions Denmark and the Netherlands asked the Court to determine that their bilateral delimitations with Germany were governed

---

[183] DCM, para. 152; NCM, para. 147.   [184] DCM, para. 156; NCM, para. 151.
[185] DCM, paras 163 and 167; NCM, paras 158 and 162.
[186] DCM, paras 157–174; NCM, paras 152–167.
[187] E. Lauterpacht, Danish continental shelf; Preliminary comments on the German Memorial dated 2 September 1967 [DNA/59], p. 2.

by the principles of international law reflected in article 6(2) of the Convention on the continental shelf. Since there were no special circumstances justifying another boundary, these boundaries had to be determined by application of the equidistance principle.

8.5   Germany's Replies

8.5.1   The preparation of the Replies

The Counter-Memorials of Denmark and the Netherlands did not make a profound impression in Germany. Blomeyer of the legal department of the Foreign Office found the reasoning of the Counter-Memorials surprisingly weak and was very optimistic about the outcome of the proceedings.[188] Jaenicke, the German agent, was largely in agreement with Reuter's detailed criticism of the Counter-Memorials.[189] Reuter's general impression was that the Counter-Memorials were not of uniform quality and a bit disorganized. Their main strength lay in the criticism of the German approach, but the positive case of the Counter-Memorials was not very convincing. The tone of the Counter-Memorials was rather unpleasant and at times a little aggressive.[190]

The commentary of Reuter and other observations on the Counter-Memorials identified some of the key issues for Germany to address in its Replies. For one thing, it would be essential for Germany to clarify its position on the process leading up to the conclusion of the Convention on the continental shelf and its position in the negotiations with Denmark and the Netherlands.[191] As Reuter observed, the Counter-Memorials contained a two-pronged attack on Germany: on the legal and the moral plane. It would be fairly easy to refute the legal arguments,

---

[188] Note dated 10 April 1968 [B102/260033], p. 1.

[189] Letter of G. Jaenicke to H. Blomeyer of the legal department of the FO dated 3 April 1968 [AA/62].

[190] See P. Reuter, *Brèves observations sur le Counter Memorial du Danemark* (not dated), enclosure to a letter of G. Jaenicke to H. Blomeyer of the legal department of the FO dated 30 April 1968 [AA/62] (Reuter, *Brèves observations* [AA/62]). Reuter's comments concern the Danish Counter-Memorial, but in view of the fact that both Counter-Memorials apart from the introduction are almost identical, they equally apply to the Dutch Counter-Memorial.

[191] Letter of G. Jaenicke to H. Blomeyer of the legal department of the FO dated 30 April 1968 [AA/62]; Reuter, *Brèves observations* [AA/62], pp. 7–8.

which were concerned with among others the significance of the signature of a treaty and the absence of reservations.[192] The attack was much more dangerous on the moral plane. Denmark in its Counter-Memorial had attempted to demonstrate – and Reuter found that it had succeeded to give this a certain appearance of credibility – that Germany had changed its position after it had become aware of the existence of an oil deposit in a clearly determined area.[193] To address this issue, the German Embassy in Copenhagen was instructed to gather additional information on the activities of the DUC.[194] The available information indicated that the results of the successful Danish drilling had only become known in the winter of 1967; one month after Germany had submitted its Memorials.[195] The assertions in the Danish Counter-Memorial concerning the significance of the finds were certainly exaggerated and only served a tactical purpose. Much more exploratory activity would be required to establish what constituted the most promising areas in the North Sea. In the disputed area there was only a question of expectations and chances.[196]

Another important point to address in the Reply concerned the status of the Convention on the continental shelf.[197] Treviranus considered that Denmark and the Netherlands would only be able to reinforce their legal position if they would be able to demonstrate that the Convention reflected customary international law.[198] Other points raised by Reuter among others concerned how specific Germany would like to be as regards its positive claim, the fact that delimitation could not be viewed purely in a bilateral context as was asserted by the Counter-Memorials, their assertion that the equidistance method applied *ipso jure*, which negated the significance of the reference to agreement in article 6 of the Convention on the continental shelf, and the claim that Denmark and the Netherlands were entitled to unilaterally delimit their boundaries in accordance with international law.[199] These were indeed important points and they would continue to play a part throughout the proceedings.

---

[192] *Ibid.*, p. 7.   [193] *Ibid.*
[194] Letter of the FO to the German Embassy in Copenhagen dated 14 March 1968 [AA/62].
[195] Note dated 15 May 1968 [AA/63], p. 1.   [196] *Ibid.*, pp. 1–2.
[197] Reuter, *Brèves observations* [AA/62], pp. 6–7; Letter of H. Treviranus of the legal department of the FO to G. Jaenicke dated 1 April 1968 [AA/62], pp. 1–2.
[198] *Ibid.*, p. 1.   [199] Reuter, *Brèves observations* [AA/62], pp. 1–6.

A first draft of the Replies was prepared by Jaenicke, the German agent.[200] Jaenicke generally agreed with Reuter's views on the main arguments of the Counter-Memorials and the options to refute them.[201] Jaenicke also profited from a discussion with a number of German professors in international law.[202] The Foreign Office also provided detailed comments.[203] Otherwise, the Foreign Office left Jaenicke a free hand in preparing the first draft of the Replies.[204] It was originally envisaged that this draft would be ready by the end of April 1968 and could be discussed between the government agencies involved in the case, at the latest, in the beginning of May.[205] Apparently, this meeting never took place and there is no indication that apart from the Foreign Office other government agencies had an opportunity to comment on the draft of the Replies. The Foreign Office only received the entire draft by the middle of May, just two weeks before the Replies had to be submitted to the Court.[206]

8.5.2   The Replies

The German Reply in its case with Denmark was completely identical with the Reply in its case with the Netherlands. In a brief introduction, Germany noted that the dispute submitted to the Court in essence was a dispute about the applicable law and as a consequence the Replies refrained from providing detailed comments on the way in which the Counter-Memorials had presented the facts and the history of the disputes.[207] That being said, the introduction did provide two examples to illustrate that the Counter-Memorials in certain instances were misleading. These examples were used to bring home that Germany had not, as

---

[200] Note dated 10 April 1968 [B102/260033], p. 1.
[201] Letter of G. Jaenicke to the legal department of the FO dated 3 April 1968 [AA/62].
[202] *Ibid.*
[203] See the letter of A. Blomeyer of the legal department of the FO to G. Jaenicke dated 13 March 1968 [AA/62], which among other refers to a 77-page paper *Ideas for Drafting the German Reply*. This paper is not included in the file in which this letter is contained. Additional suggestions are contained in a letter of H. Treviranus of the legal department of the FO to G. Jaenicke dated 1 April 1968 [AA/62].
[204] Note dated 10 April 1968 [B102/260033], p. 1.   [205] *Ibid.*
[206] On this latter point see letter of H. Blomeyer of the legal department of the FO to G. Jaenicke dated 15 May 1968 [AA/63], p. 1. The commentary on the draft by the legal department of the FO is contained in two letters of H. Treviranus to G. Jaenicke of respectively 17 and 20 May 1968 [AA/63].
[207] GR, paras 2–3.

the Counter-Memorials intimated, acknowledged that the Convention on the continental shelf constituted an expression of customary international law.[208] The introduction was also used to point out that Denmark and the Netherlands had proceeded with the ratification of their bilateral delimitation agreement, which established an equidistance boundary in the area in dispute with Germany. Germany prior to ratification had notified both States that this agreement could not have any effect on Germany's delimitations with Denmark and the Netherlands.[209] Without saying in so many words, the Replies informed the Court that Denmark and the Netherlands continued their policy of presenting Germany with *faits accomplis* even now that the dispute was before the Court.

The Replies are divided in three main chapters. A short first chapter contains a refutation of the Danish-Dutch assertion that the principle of a just and equitable share proposed by Germany was not a rule of law but an attempt to settle the disputes *ex aequo et bono*. Chapters II and III deal respectively with the role of equidistance in the delimitation of the continental shelf and how an equitable delimitation might be achieved in the North Sea.

The Replies explain that the principle of a just and equitable share, unlike the principle of *ex aequo et bono*, in no way constituted a deviation from the existing law.[210] The principle of a just and equitable share constituted a general principle of law, which, in accordance with article 38(1) of the Statute of the Court, is one of the sources of law the Court is required to apply to settle disputes submitted to it.[211]

The Replies also attempted to provide the principle of a just and equitable share with more specific content. First of all, it was observed that the principle prescribed that:

> the share to be allotted to each State should be measured out "equitably", i.e., with impartial reason and fairness according to the weight of all factors pertinent to the right of the State over the submarine areas before its coast.[212]

The Replies further submitted that it was generally accepted that certain general principles of law had an inherent self-evident and necessary validity. The principle that each State may claim a just and equitable share in resources to which more than one State has an equally valid title

[208] *Ibid.*, para. 3.   [209] *Ibid.*, para. 4.   [210] *Ibid.*, para. 9.   [211] *Ibid.*, paras 7 and 10.
[212] *Ibid.*, para. 9.

could be said to be of this character.[213] The function of the principle of a just and equitable share was intended to supplement the emerging law on the continental shelf. Whereas there was a general recognition concerning the entitlement of coastal States to the continental shelf, a generally accepted rule on the delimitation of the continental shelf between neighboring States was still lacking.[214] Article 6(2) of the Convention on the continental shelf was only "one cautious step in the attempt to find a formula which might lead to an equitable solution of the boundary problem."[215] It was not sufficient because it offered no criteria to assess when the equidistance line should not be applied because of the presence of special circumstances.[216]

Finally, the distinction the Counter-Memorials made between the "delimitation" and "sharing out" of areas of continental shelf was "rather artificial."[217] By arguing that the Special Agreements did not request the Court to decide which rules and principles governed the sharing out of areas of continental shelf, but instead asked the Court to pronounce itself on the rules of delimitation, the Counter-Memorials would effectively deny the Court the power to assess whether a delimitation in accordance with the equidistance method led to an equitable result.[218] The argument that the Court had to look at the share of continental shelf of each State to assess whether a specific method led to an equitable result implied that it had to look beyond the bilateral delimitation context, something that had been strongly rejected by the Counter-Memorials.

Chapter II of the Replies, dealing with the applicability of the equidistance line between the parties, pointed out that Denmark and the Netherlands had convincingly demonstrated the merits of the equidistance method. Germany indeed had already done the same in the Memorials and also pointed out the shortcomings of the method.[219] What the Counter-Memorials had failed to do was show that there existed an obligation for Germany to accept the equidistance principle as an obligatory rule of international law.[220] The Counter-Memorials had suggested that Germany's position at the 1958 Conference, its signature of the Convention without a reservation to its article 6, and the German reliance on the Convention in issuing its proclamation on the continental shelf in 1964 were all relevant to determining the German position in relation to the equidistance method. The Replies

---

[213] *Ibid.*, para. 11.   [214] *Ibid.*, para. 12.   [215] *Ibid.*, para. 13.   [216] *Ibid.*
[217] *Ibid.*, para. 14.   [218] *Ibid.*   [219] *Ibid.*, para. 19.   [220] *Ibid.*; see also *ibid.*, para. 20.

countered that Germany at the 1958 Conference could in no way have known that Denmark and the Netherlands would go as far as to argue that a unilateral delimitation of the continental shelf in accordance with equidistance in principle would be valid against other States. Germany had voted in favor of article 6 because it constituted a workable solution, provided that sufficient attention would be paid to its purpose, i.e. achieving an equitable solution. Moreover, Germany's main concern in 1958 was not the delimitation of the continental shelf, but the possible negative impacts of the regime of the continental shelf on the freedom of the high seas.[221] This account of the German position may not be completely accurate – opposition to the regime of the continental shelf was at least in part a result of Germany's limited equidistance area in the North Sea[222] – but Germany's attitude at the Conference certainly had not led to acceptance of the equidistance principle in relation to Denmark and the Netherlands.

The Replies' argument on the signature of the Convention was somewhat defensive as it, among others, acknowledged that Germany had taken a passive attitude.[223] It might have been sufficient to point out, as the Replies also did, that Germany's signature did not make the Convention binding law for Germany.[224] The Replies also pointed out that upon signature, Germany could still expect that it could come to a settlement along equitable lines, as article 6 of the Convention expressly referred to settlement by agreement. The Counter-Memorials' question as to why Germany had not ratified the Convention was easily explained. There was no question of a change in the position of the German government. Rather, it had become clear that Denmark and the Netherlands insisted on equidistance as the only valid rule for the delimitation of the continental shelf while relying on article 6 of the Convention. This new fact led the German government to reconsider ratification as long as the interpretation of article 6 remained uncertain.[225] Germany's 1964 continental shelf proclamation and legislation also did not point to acceptance of article 6 of the Convention, as these instruments only relied on the Convention's articles 1 and 2.[226] The Replies also took care to bring across that the Counter-Memorials' argument that the partial continental shelf boundaries of Germany with Denmark and the Netherlands showed that it had accepted the

---

[221] *Ibid.*, para. 26.   [222] See Chapters 3.2.2 and 4.3.   [223] GR, para. 25.   [224] *Ibid.*
[225] *Ibid.*, para. 27.   [226] *Ibid.*, para. 28.

equidistance principle was disingenuous.[227] Germany had explicitly reserved its rights in this respect.[228]

To wind up the argument on Germany's attitude, the Replies addressed the point that Germany allegedly had changed its position after further information on the oil and gas potential of specific areas had become known. The Replies observed that the kind of survey work Germany had carried out in the early 1960s in any case could not have provided it with reliable information on the presence of oil and gas deposits in the area. Only actual drilling, as undertaken by Denmark in 1967, might have resulted in such information.[229] It should be noted that this may be true as far as actual deposits is concerned, to which the Replies refer. Germany already knew at least as early as the middle of 1965 that the area of overlapping claims offered the best chance of finding natural gas.[230] This is not to say that Germany changed its position because of this information – indeed there never was such a change in position – but the potentiality of the area no doubt contributed to shaping Germany's policy. Obviously, Germany found that the Court did not need to be bothered with this kind of detail.

In discussing exploratory activities, the opportunity to once more stress Germany's reasonableness and its opponents' intransigence was not neglected. The Replies pointed out that German exploration in the Danish equidistance area stopped on the request of Denmark. To the contrary, the Danish government had allowed drilling in the disputed area, which was in line with the unfounded claim that Denmark and the Netherlands could unilaterally delimit their continental shelf boundary by equidistance in relation to other States.[231]

After disposing of the contention that Germany had accepted the equidistance principle, the Replies turned to the question of whether the rules contained in article 6(2) of the Convention had become customary international law. Practice prior to the 1958 Convention mostly concerned the delimitation of lakes, rivers and the territorial sea. This often concerned cases in which the equidistance method gave States an equal area. Moreover, it was in any case difficult to conclude that the practice in respect of relatively narrow belts of waters could be transposed to continental shelf areas further offshore. Finally, practice was far from uniform. In conclusion, it was difficult to maintain that there was a

---

[227] *Ibid.*, para. 29; see also *ibid.*, para. 30.
[228] See also the discussion in Chapters 5.2.4 and 5.3.   [229] GR, para. 31.
[230] See note dated 13 August 1965 [B102/318713], p. 1.   [231] GR, para. 31.

rule of customary law giving preeminence to the equidistance principle, which was consolidated in article 6(2).[232] The Replies also pointed out that the argument that the reservations in respect of article 6 of the Convention by a number of States merely invoked the existence of special circumstances was flawed. That situation was already covered by the article itself. Rather, the reservations were intended to clarify that these States did not accept the interpretation that equidistance could be automatically invoked if a party rejected the presence of special circumstances.[233]

As far as State practice after 1958 was concerned, the Replies noted that there only had been concluded a limited number of treaties. The Counter-Memorials had failed to show that the rules of article 6 had been applied because it was considered that equidistance constituted the "general rule."[234] Moreover, there was no proof that States, which had used the equidistance method in the case of opposite coasts, accepted that this method also had to be applied in other geographic situations.[235] The Replies dealt separately with the practice of the North Sea coastal States, observing that "[i]n the last resort" the Counter-Memorials relied on this practice. Germany had already explained why its bilateral treaties with Denmark and the Netherlands could not be invoked as precedents.[236] All other treaties were concerned with opposite coasts in which case the equidistance line produced a broadly equal division, giving each party a just and equitable share.[237]

The Replies rejected the thesis of the Counter-Memorials that entitlement to the continental shelf was based on propinquity in the sense that a State should have jurisdiction over areas nearer to its coast than the coasts of other States.[238] If it had been intended to attribute this specific meaning to the term "adjacent" in article 1 of the Convention on the continental shelf, the principle of equidistance was no more than its logical consequence and there would have been no need for article 6 of the Convention. According to the Replies, article 1 of the Convention recognized the rights of the coastal State to the shelf, but did not imply a rule for deciding conflicting claims.[239] The Replies also proposed a different understanding of the term propinquity. The recognition of the concept of the continental shelf and the rights of the coastal State

---

[232] *Ibid.*, paras 34–39; see also *ibid.*, para. 44.
[233] *Ibid.*, para. 50; see also *ibid.*, paras 45–46.   [234] *Ibid.*, para. 49.   [235] *Ibid.*, para. 51.
[236] *Ibid.*, para. 52; and see text at footnote 227.   [237] *Ibid.*, para. 53.
[238] *Ibid.*, para. 56.   [239] *Ibid.*, para. 58.

over the shelf were based on the generally accepted fact that the continental shelf was a "natural continuation of the State's territory into the sea."[240] This idea, which would be picked up by the Court in its judgment, was not a novelty. It had already been advanced in the proclamation of President Truman to claim rights for the United States over the continental shelf.[241]

Chapter III of the Replies further elaborated on the Memorials' argument that the North Sea constituted a special case and responded to the Counter-Memorials. The Replies, while stressing that Germany was not bound by article 6 of the Convention on the continental shelf, pointed out that this article "if interpreted in harmony with its real purpose" in essence was similar to the rules advanced by Germany.[242] Article 6 had to be understood as a procedure to arrive at an equitable solution in the specific case. Special circumstances should not be applied in exceptional cases, but rather their absence was "a necessary precondition for the application of the equidistance line."[243]

In setting out Germany's solution to the disputes, the Replies focused on three arguments advanced in the Counter-Memorials, namely that:

a) the North Sea was no special case which justified a different method than equidistance;
b) the two bilateral delimitations had to be viewed as individual problems, without looking at the impact of other delimitations; and
c) the breadth of the coastal frontage of each party facing the North Sea was not a relevant criterion to assess the equitableness of an equidistance boundary.

All three contentions had to be rejected.[244] In addressing the first point, the Replies no longer started from the premise that the North Sea consisted of a single area of continental shelf as had been done in the Memorials, but focused on coastal geography. In areas like the North Sea, with more than two States involved, equity would more often require departures from the equidistance line because of the coastal configuration.[245] The Replies repeated the argument of the Memorials

---

[240] *Ibid.*, para. 60.

[241] The proclamation provided as one justification for the United States' claim that "the continental shelf may be regarded as an extension of the land-mass of the coastal nation and thus naturally appurtenant to it" (Presidential Proclamation no. 2667, concerning the Policy of the United States with respect to the Natural Resources of the Subsoil and Sea Bed of the Continental Shelf of 28 September 1945).

[242] GR, para. 67.   [243] *Ibid.*, para. 70.   [244] *Ibid.*, para. 77.   [245] *Ibid.*, para. 78.

that the boundaries of Germany could not be viewed in isolation as this did not allow making an assessment of the impact of the application of the equidistance method. The Replies submitted that the contention of the Counter-Memorials that Germany's limited share of the continental shelf resulted solely from its own coast was mistaken. What counted was the outcome of the delimitation, not whether this had to be attributed to the "gaining" or the "losing" State.[246] The Counter-Memorials had tried to disguise the inequitable result of the equidistance method in presenting a figure of the bilateral delimitation with one State while omitting the bilateral delimitation of Germany with the other State. Those figures suggested that the equidistance method gave Denmark and the Netherlands no more than their "normal" share – the Counter-Memorials had argued that only if the equidistance line gave a State an "abnormal" share it was inequitably. The Reply submitted that an addition of the other equidistance line to each figure showed that the shares of Denmark and the Netherlands were not as "normal" as they claimed if compared to Germany's share.[247]

The defense of Germany's position, that the coastal front of each State facing the delimitation area was relevant for determining an appropriate delimitation method, was developed along two lines. On the one hand, the Replies tried to demonstrate the weaknesses in the critique of the Counter-Memorials that focused on arguments related to distance from the coast. By submitting that distance from the coast was the standard to evaluate a specific method of delimitation the Counter-Memorials were making "the principle of equidistance its own standard for the equitableness of its application."[248] Moreover, the German coast was only slightly further from the center of the North Sea than the coast of the other North Sea States, but equidistance gave it much less area than Germany's proposed sector solution.[249] On the other hand, the Replies provided a more detailed explanation of Germany's own theory concerning the role of coastal frontages in determining continental shelf boundaries between adjacent States:

> The configuration of the coast should be irrelevant in this respect: the breadth of the coastal front should be measured on the basis of the general direction of the coast, thereby eliminating the effect of indentures as well as promontories. If such configurations [i.e. indentures or promontories] would have the effect to apportion parts of the continental shelf

[246] *Ibid.*, para. 84. [247] *Ibid.*, para. 85. [248] *Ibid.*, para. 97; see also *ibid.*, paras 93–96.
[249] *Ibid.*, para. 95.

which appear to an unbiased observer as a continuation of one State's territory, to another State such an effect has to be regarded as a circumstance ... which excludes the application of the equidistance method for the determination of the boundary between these States as inequitable.[250]

In explaining the implications of this general point, the Replies first provided a number of examples involving adjacent coasts. The Replies argued that in these cases an equitable delimitation should allow each neighboring State a continental shelf extending seaward as a corridor with a uniform breadth.[251] In other words, in the case involving three adjacent States, the breadth of the continental shelf of State lying in the middle did not become less moving seaward.

After the Replies discussed the case of three adjacent States, they moved to the case where the coasts of three States lay in a bend or an almost circular line. According to the Replies this was a more difficult case because the continental shelf of the States concerned converged to the middle of the area to be delimited.[252] As a consequence, it was appropriate to divide the area in sectors between the coastal States.[253] In the hypothetical cases presented by Germany the boundaries of all of the States concerned centered on a single point in the middle of the area of continental shelf. However, these models, like the similar solution presented in the Memorial, neglect a critical point. In the North Sea the application of this approach would only give Germany a sector-shaped area of continental shelf, i.e. one point of contact with the median line with the United Kingdom, whereas Denmark, and in particular the Netherlands, would have a much longer boundary along that median line. As was argued above in respect of the Memorial,[254] and was also suggested by Germany's model involving adjacent States, it would have been more advantageous for Germany to develop a model based on a frontal projection of the coast of each State extending seaward as a corridor of uniform breadth.[255] The equal division of the area of overlap of these corridors would result in giving the middle State a larger opening on the center of the area concerned than a sector centering on a single point.[256]

---

[250] *Ibid.*, para. 88.   [251] *Ibid.*, para. 89 and figures 1–3.   [252] *Ibid.*, para. 90.
[253] *Ibid.*, paras 90–93 and figures 4–5.   [254] See text after footnote 84.
[255] Interestingly, this approach was adopted in the judgment of the Court in the two cases (see further Chapter 9.3).
[256] Admittedly, the Replies hinted at the fact that a "pure" sector approach might not be appropriate (see e.g. GR, para. 92), but they certainly did no more than that.

The Replies repeated the submissions of the Memorials and explicitly added a request to the Court to declare that article 6(2) of the Convention on the continental shelf had not become customary international law and that the delimitation involving the parties had to be settled by agreement and should apportion each of the parties an equitable share in the light of all relevant factors.

8.6 The Common Rejoinder of Denmark and the Netherlands

8.6.1 The preparation of the Common Rejoinder

After the first round of written pleadings had been completed, the date for the submission of the German Replies and the Danish and Dutch Rejoinders had to be established. An Order of the Court of 26 April 1968 held that Denmark and the Netherlands were parties in the same interest and joined the proceedings in the two cases. The order enjoined Denmark and the Netherlands to submit a "common Rejoinder."[257]

The Danish Embassy in The Hague immediately questioned the Registry of the Court concerning the exact meaning of the wording "common rejoinder." As was explained by the Registrar, what was expected was "one document, one text."[258] This required Denmark and the Netherlands to reconsider their approach to their second round of written pleadings. The Netherlands had initially indicated that it considered that close coordination would not be possible because of the question of how to deal with islands as a category of special circumstances.[259] The Netherlands considered that no reference to islands as special circumstances should be made in the Rejoinder in view of the Dutch difficulties with Venezuela over the boundary between the island of Aruba and the Venezuelan mainland. On the other hand, Denmark considered it essential to refer to islands in discussing special circumstances. Jacobsen held this matter to be of such importance that if a satisfactory solution would not be found he would refuse to sign the Common Rejoinder and he informed Riphagen that in this case the matter would have to be further discussed between the Ministers of Foreign Affairs.[260] Discussions involving the two agents and Waldock

---

[257] For a discussion of the Order see also Chapter 6.5.
[258] Letter of B. Jacobsen to W. Riphagen dated 3 May 1968 [NNA/30], p. 1.
[259] Telex dated 29 April 1968 [DNA/142], p. 2.
[260] Note concerning the division of competence in the preparation of The Hague process on the North Sea shelf dated 25 June 1968 [DNA/137], p. 2.

led to the conclusion that, although the Court was expecting "one document, one text" it would be possible to include sections which would represent the views of one of the parties only.[261] In the end, the Netherlands accepted almost all of the arguments in respect of special circumstances which had been prepared by Denmark. Riphagen only was unable to agree to the section which detailed in which instances special circumstances justified to deviate from the equidistance method.[262] Jacobsen did not consider it possible to expect that Riphagen would further accommodate Denmark on this point.[263] Consequently, the Common Rejoinder contains a 4-page section with individual observations of Denmark and the Netherlands.[264] In the final analysis, Chapter 3 and the separate comments of Denmark seem to add little to the Rejoinder. Paludan found that the chapter on special circumstances was not very strong, but the best that could be accomplished. The two separate parts with individual comments did not function well and it was difficult to see the added value of Jacobsen's separate remarks.[265]

First drafts of the three substantive chapters of the Common Rejoinder dealing with the question as to the essence of the dispute, the applicability of the principles contained in article 6(2) of the Convention on the continental shelf and the interpretation of the special circumstances clause were prepared by respectively Riphagen, Jacobsen and Waldock.[266] Because of the need for close collaboration between the agents and Waldock in preparing the Common Rejoinder, the role of the Danish continental shelf committee was expected to become less pronounced.[267] The limited time for preparing the Rejoinder also implied that there would be less time for discussions. The committee, as a matter of fact, only met once to discuss the drafts prepared by Jacobsen, Riphagen and Waldock after they had already been discussed between their authors and Paludan and Krog-Meyer of the Danish Ministry of Foreign Affairs, who also were respectively the chairman and deputy

---

[261] Report on the Continental Shelf meetings held in Copenhagen; June 11, 1968 (KSU/MR.9) dated 21 August 1968 [DNA/39] (Report Danish CSC June 1968 [DNA/39]), p. 2, paras 6–7; note concerning the division of competence in the preparation of The Hague process on the North Sea shelf dated 25 June 1968 [DNA/137], p. 2.

[262] Report on the meeting of the advisory committee for the case with Germany on the continental shelf (Continental shelf committee) held on Tuesday 16 July 1968 (KSU/MR.10) dated 2 August 1968 [DNA/39] (Report Danish CSC July 1968 [DNA/39]), p. 2, para. 3.

[263] Ibid., p. 8, paras 46–47.   [264] CR, paras 142–143.

[265] Letter of J. Paludan to E. Krog-Meyer dated 25 September 1968 [DNA/61].

[266] Report Danish CSC June 1968 [DNA/39], p. 3, para. 10.   [267] Ibid., para. 8.

chairman of the committee.[268] The committee does not seem to have had any significant impact on the content of the Rejoinder.[269] The more limited role of the Danish continental shelf committee may have been a not unwelcome side-effect of the ICJ's instruction to submit a Common Rejoinder. The meetings of the committee in preparation of the Danish Counter-Memorial had led to much discussion and numerous drafts. In commenting on the payment of Waldock by Denmark and the Netherlands, Riphagen could agree to a 50–50 division because the interest of the case for both sides was comparable, although:

> it could not be denied that Sir Humphrey has had to spend considerably more time and effort on the Danes (because of the fact that in Denmark a commission of about 15 men "assists" the Danish agent and from that side also many useless ideas had been submitted!).[270]

The basic approach on how to answer the German Replies was probably decided during a meeting between Jacobsen, Riphagen, Waldock and Paludan and Krog-Meyer of the Danish Ministry of Foreign Affairs.[271] There is little information on that meeting. It is safe to say that the German Replies required Denmark and the Netherlands to revisit all the major points of the Counter-Memorials. The meeting decided on the main points to be made in the introduction of the Rejoinder and the content of its main chapters.[272] That basic approach is reflected in the Common Rejoinder as submitted to the Court.

8.6.2 The Common Rejoinder

The Common Rejoinder consists of a short introduction, three substantive chapters and formal submissions. Of the three substantive chapters, dealing respectively with the essence of the issues before the Court, the applicability of the principles contained in article 6(2) of the Convention on the continental shelf and the interpretation of the special circumstances, the second chapter is by far the longest – some fifty pages as

---

[268] See Report Danish CSC July 1968 [DNA/39], p. 2, para. 3.
[269] See also *ibid.*, p. 6, para. 32.
[270] Note 37/68 dated 17 June 1968 [NNA/34]. Translation by the author. The original text reads "niet te ontkennen is, dat Sir Humphrey aanzienlijk meer tijd en moeite heeft moeten besteden aan de Denen (vanwege het feit dat in Denemarken een commissie van ongeveer 15 man de Deense Agent 'bijstaat' en uit die hoek ook veel onbruikbare ideeën zijn gekomen!)."
[271] See Report Danish CSC June 1968 [DNA/39], p. 2, para. 7 and p. 3, paras 9–10.
[272] See *ibid.*, p. 3, paras 9–10.

compared to the twenty-four and nine pages of chapters 1 and 3 respectively. Chapter 2, among others, contains a detailed analysis of State practice.

The introduction to the Common Rejoinder focuses on two substantive issues. First of all, Denmark and the Netherlands sought to keep their cases with Germany separate. For Germany's case it was essential to look at the joint effect of the two delimitations and the opposite was true for its opponents. The joining of the cases by the Court might contribute to accepting the German view more easily. Denmark and the Netherlands appreciated the convenience for the Court of joining the cases and having a Common Rejoinder.[273] At the same time, they emphasized that their respective cases concerned different boundaries in different parts of the North Sea. The Counter-Memorials had stressed the entirely separate character of the disputes before the Court. Their partial boundaries with Germany had been concluded through separate negotiations and been based exclusively on the coasts of the respective parties.[274] Of course, these separate negotiations had also pointed out that there had only been agreement on a partial boundary because the parties differed about the impact of the presence of a third State.

The second point of the introduction concerned the submissions of Germany. According to the Rejoinder the new final submission of Germany, which provided that a delimitation should be effected by agreement to apportion a just and equitable share to each of the parties in the light of the relevant factors, was extraordinary because it seemed to question the very basis of the Special Agreements by which the cases had been submitted to the Court.[275] Basically, the argument of the Rejoinder was that if the Court were to accept this submission of Germany, the parties would not have any indication on how they should go ahead with the negotiation of an agreement.[276] This assertion might, in itself, have already given the impression that Denmark and the Netherlands were pushing it beyond the limit. The Replies obviously had given the Court sufficient arguments to further specify the relevant factors. The Rejoinder even went as far as suggesting that Germany was asking the Court to pronounce a *non liquet* and send the parties back to the negotiating table without sufficient legal criteria.[277] Moreover, Denmark and the Netherlands were entitled to know:

---

[273] CR, para. 2.　[274] *Ibid.*, para. 3.　[275] *Ibid.*, para. 5.　[276] *Ibid.*, paras 6–7.
[277] *Ibid.*, para. 8; see also *ibid.*, paras 25–26.

> before they are asked in negotiations to yield a single metre of continental
> shelf which naturally appertains to them under the principles contained
> in the Continental Shelf Convention ... upon precisely what *legal* basis
> that metre ought to be regarded as appertaining to the Federal Republic
> rather than to Denmark or, as the case may be, the Netherlands.[278]

If this were the standard to be adopted by the Court, it would probably have little other choice than accepting the Danish-Dutch argument concerning the equidistance method.

Chapter 1 of the Rejoinder discusses the essence of the issues before the Court. It looked at the rules and principles of international law advanced by the parties and criticized the German concepts of coastal frontage, sector principle and special circumstances. The Rejoinder submitted as a starting point that the principles and rules of delimitation presuppose the coexistence of various States, each already having a more or less defined space in which it exclusively exercises sovereignty and sovereign rights. Taken this starting point, international law did not pretend to "distribute" the total available space between States, but rather accepted that each State determined its own exclusive sphere of activities in space and delimitation was only concerned with determining the boundary line between States, that is, the exact points where the extension in space of sovereign rights of two States meet each other.[279] This conception of delimitation relied heavily on the situation in respect of land boundaries.[280] In that case there are indeed many instances in which historical developments have led to the acceptance that a specific river basin or mountain range divides the territory of two States. Delimitation provides technical rules to determine the exact location of a boundary in such instances.

After discussing the situation of land boundaries, the Rejoinder submitted that it was particularly apparent in the case of fresh water and sea areas that the extension of sovereign rights in space constituted the starting point of the rules and principles of international law.[281] Of course, in the case of the continental shelf the question existed how the spatial extent of the rights of coastal States had to be determined. The Rejoinder submitted that the concept of natural continuation of the land territory into the water and consequently the concept of propinquity were at the basis of judicial settlement and State practice in respect of delimitation. As had already been argued in the Counter-Memorials, that concept of natural continuation and propinquity naturally led to the

---

[278] *Ibid.*, para. 8.   [279] *Ibid.*, para. 15.   [280] See *ibid.*, para. 16.   [281] *Ibid.*, para. 17.

application of the equidistance principle as the starting point to determine where the overlapping entitlements of two States meet.[282] The problem with this conception is that it ignores that entitlement in the case of the continental shelf may lead to an area of overlapping claims which is in no way related to distance/equidistance lines. As a matter of fact, the Rejoinder's argument does not even consider the potential extent of these entitlements, but immediately looks at the point at which they meet by using the concept of propinquity defined in nearness to the coast to arrive at the starting point for delimitation.

The Rejoinder criticized the German conception of overlapping entitlements, emphasizing that the principle of a just and equitable share implied that the whole continental shelf of the North Sea was shared between the coastal States.[283] On that basis it was not clear why Denmark and the Netherlands should provide Germany with an additional continental shelf area.[284] This representation of the German argument was not completely to the point. Even in the Memorials, the idea that the continental shelf of the North Sea constituted a whole had not been fundamental for setting out the German position and the Replies had given it even less emphasis. The focus of the German argument as regards overlapping entitlement throughout was on the situation involving the parties to the proceedings.

The Rejoinder's criticism of the German concept of coastal fronts relied on a number of propositions. Neither the Geneva Conventions nor their *travaux préparatoires* or State practice contained any indication of the concept of coastal fronts Germany had propagated.[285] Second, Germany had failed to specify the coastal fronts of Denmark and the Netherlands. The Rejoinder suggested that Germany's argument implied that a large part of the coasts of Denmark and the Netherlands would be ignored.[286] Third, Denmark and the Netherlands submitted that the Replies were not clear on whether the general direction or the breadth of the coastal front was determinative. In addition, the Replies overlooked that the total coastline of a single State in many cases did not have one general direction.[287]

The Rejoinder's criticism of the sector concept was less to the point, as it had already been largely abandoned in the Replies, as was also admitted in so many words by the Rejoinder.[288] Germany's concept of special circumstances risked to completely sever the relationship between the

---

[282] *Ibid.*   [283] *Ibid.*, para. 23.   [284] *Ibid.*, para. 24.   [285] *Ibid.*, para. 26.
[286] *Ibid.* and figure A.   [287] *Ibid.*, para. 26.   [288] *Ibid.*, paras 27–29.

location of the boundary line and coastal geography.[289] If an equidistance line had to be adjusted because it did not give a coastal State an equitable share, that adjustment would not be based on the relationship of the boundary to the coast in a specific location.[290]

The Rejoinder's argument on Germany's concept of coastal fronts and special circumstances both hinge on the centrality of propinquity/equidistance in relation to entitlement and the delimitation process. If those principles are not the starting point in either of these cases the relationship between the actual coast and the boundary is much more loosely defined than the Rejoinder posits.

The Rejoinder also rejected that the principle of a just and equitable share advanced by Germany could be applied under article 38(1)(c) of the Statute of the Court, which refers to general principles of law. General principles of law could only be applied if there did not exist principles of international law which allowed the Court to decide the dispute submitted to it.[291] In the view of Denmark and the Netherlands these principles were reflected in article 6 of the Convention on the continental shelf. If article 6 was found to be inapplicable, article 1 and 2 of the Convention expressed generally binding principles, a matter on which the parties agreed. These articles led to an unqualified application of the equidistance method.[292] That this rule did not provide for special circumstances was no fault of Denmark and the Netherlands. After all, Germany rejected that the equidistance-special circumstances rule was applicable between the parties.[293] Germany's principle of a just and equitable share also could not be applied because it was incompatible with the principles on which the positive law on the matter was based.[294] This entire argument in the end again hinged on the nature of continental shelf entitlement and its relationship to delimitation.

Chapter 2 of the Rejoinder focused on the applicability of the principles of delimitation contained in article 6 of the Convention to the situation involving Denmark and Germany and the Netherlands and Germany. Chapter 2 first of all countered the assertion of the Replies that the Counter-Memorials did not distinguish between equidistance as a method and the rule of law which prescribed that Germany had become bound to accept this method of delimitation vis-à-vis Denmark and the Netherlands.[295] The Rejoinder submitted that this distinction was unjustified in the context of the Convention on the continental shelf.[296] Article

---

[289] *Ibid.*, para. 30. [290] *Ibid.* [291] *Ibid.*, para. 118. [292] *Ibid.*, para. 119. [293] *Ibid.*
[294] *Ibid.*, para. 117. [295] *Ibid.*, para. 32. [296] *Ibid.*, para. 36.

6 recognized that equidistance was both a method and a principle of delimitation.[297] The remainder of chapter 2 was dedicated to proving the status of the equidistance principle as a generally recognized rule of international law. After reconfirming the arguments set out in chapter 3 of the Counter-Memorials,[298] the Rejoinder elaborated a number of points. The Rejoinder rejected that there was a distinction between opposite and adjacent coasts as was argued by Germany.[299] Obviously, if there was no distinction between the two cases, the significant amount of State practice in respect of opposite coasts was also relevant for cases involving adjacent coasts.[300] In this respect the Rejoinder focused on explaining why the case of opposite and adjacent coasts was treated in separate paragraphs of article 6 of the Convention on the continental shelf.[301] This point was convincingly argued and as the Rejoinder also pointed out, Germany had failed to ask itself why the provisions in both paragraphs of article 6 were exactly the same.[302] However, the Rejoinder failed to address the German argument that coastal geography in cases of adjacency impacted differently on the equidistance line than in cases of oppositeness. This was a much more fundamental issue than the structure of article 6.

The Rejoinder reiterated that the role of the equidistance-special circumstances rule was essentially the same in the delimitation of the territorial sea and the continental shelf.[303] This perspective implied that the rules for the delimitation of the continental shelf were solidly grounded in the law applicable to the delimitation of maritime boundaries existing prior to the adoption of the Convention on the continental shelf.[304] The Rejoinder submitted that Germany's argument to the contrary only served "to underline the embarrassment which the Federal Republic feels when it finds itself confronted by so clear and general an acceptance of the equidistance-special circumstances rule within the international community."[305] This was not the only instance in which the Rejoinder suggested that Germany's views conflicted with the whole of the international community. For instance, the Rejoinder submitted that Germany provided the only example of "a State's resisting the normal application of the principle of Article 6."[306] This kind of

[297] *Ibid.*, para. 33; see also *ibid.*, paras 32–37.   [298] *Ibid.*, paras 39–40.
[299] *Ibid.*, para. 41.   [300] See also *ibid.*, paras 42–44.   [301] *Ibid.*, paras 45–47.
[302] *Ibid.*, para. 46.   [303] *Ibid.*, paras 48–53.
[304] See also *ibid.*, para. 56, to which reference is also made below.   [305] *Ibid.*, para. 53.
[306] *Ibid.*, para. 39.

hyperbole may have conveyed the impression that Denmark and the Netherlands as a matter of fact were not that sure that Germany's position differed fundamentally from that of other States.

The Rejoinder's detailed discussion of State practice first considered delimitations in fresh water bodies and the territorial sea. The preponderance of the equidistance method in those cases showed that the authors of the Convention on the continental shelf framed article 6 "after paying regard to the experience made with such methods in State practice and that they chose the method which they considered most suitable for the purpose."[307] In discussing State practice in relation to the continental shelf, the Rejoinder first looked at the North Sea. The Replies had suggested that reliance on the practice involving the parties was somewhat inappropriate and that other agreements concerned opposite coasts, which distinguished them from the cases involving Germany. According to the Rejoinder, the Counter-Memorials had cited the treaty between the Netherlands and Denmark merely as part of the general evidence of the conviction of States that article 6 of the Convention on the continental shelf expressed the applicable principles and rules of international law. This similarly applied to practice of the other North Sea States.[308] The Rejoinder remained silent on Germany's argument on its partial boundaries with Denmark and the Netherlands. Instead, it was argued that the Replies showed that Germany in this case was in reality relying on – however unjustified – the special circumstances clause which formed an integral part of article 6 of the Convention on the continental shelf.[309] The Rejoinder discussed the case of Belgium in most detail. The significance of this case was explained by the fact that Belgium had a limited continental shelf due to the proximity of neighboring States.[310] Unlike Germany, Belgium did not believe this to be a reason for displacing the equidistance principle as the applicable method of delimitation.[311] Whether this comparison was really helpful to the Danish-Dutch case is doubtful. There is little similarity between the two cases. The coast of Belgium and its adjacent neighbors do not form a concavity and the equidistance method gives Belgium a continental shelf boundary of some 40 kilometers with the opposite coast of the United Kingdom, whereas the Belgian coast measures some 60 kilometers. This contrasts starkly with the case of Germany. Germany has a much longer coast, but if equidistance were to be applied its continental shelf is some 140 kilometers distant from the

---

[307] Ibid., para. 56.   [308] Ibid., para. 59.   [309] Ibid.   [310] Ibid., para. 62.   [311] Ibid.

median line with the United Kingdom. According to the Rejoinder, the significance of the Belgian position was increased by its complete consistency with the position adopted by Iraq.[312] Again, the comparison with Germany is problematic. The coast of Iraq is much shorter than that of its two neighbors, Iran and Kuwait.[313]

The Rejoinder submitted that its review of State practice demonstrated that Germany stood alone in making a sharp distinction between States with opposite and adjacent coasts. Moreover, in both instances equidistance provided an objective criterion for determining what constituted an equitable delimitation and certain geographical circumstances, such as an insignificant offshore island, might amount to special circumstances.[314] Denmark and the Netherlands reaffirmed that article 6(2) of the Convention on the continental shelf reflected general international law.[315]

The Rejoinder paid considerable attention to the position of Germany in relation to the rule contained in article 6 of the Convention on the continental shelf.[316] In the Replies, Germany had argued that Denmark and the Netherlands had failed to show that Germany had become bound to regard the equidistance rule as an obligatory rule of international law. The Rejoinder largely maintained the arguments advanced in the Counter-Memorials to the effect that Denmark and the Netherlands had *ipso jure* rights to the continental shelf based on adjacency, inherent in which was the concept that the continental shelf nearer to the coast of one State than another State had to be considered to be adjacent to the former State. This was further reinforced by the position of the equidistance rule in article 6 of the Convention.[317] The Rejoinder also continued to rely on the *Anglo-Norwegian fisheries* case to justify the Danish and Dutch unilateralism in defining their continental shelf boundaries.[318]

A review of the Rejoinder suggests that Denmark and the Netherlands were hard pressed to convincingly argue that Germany's conduct evidenced that it had accepted the equidistance method. For instance, to counter the German argument that Germany in 1958 could not possibly know that Denmark and the Netherlands would later interpret the equidistance-special circumstances rule in the way they did, the Rejoinder

---

[312] *Ibid.*, para. 63; see also *ibid.*, paras 71–72.
[313] Similar observation in respect of Belgium and Iraq were made by Jaenicke during the oral pleadings (OP, pp. 174–175).
[314] CR, para. 74.   [315] *Ibid.*, para. 75.   [316] *Ibid.*, para. 76–107.   [317] *Ibid.*, para. 83.
[318] *Ibid.*, para. 81. For a discussion of this argument see text at footnotes 161 and following.

submitted that this "is precisely the way in which all the other North Sea States, other than the Federal Republic, have automatically proceeded to interpret it."[319] The Rejoinder placed a similar reliance on clairvoyance in rejecting the German assertion that it could not have foreseen that Denmark and the Netherlands would subsequently claim that article 6 allowed them to unilaterally determine their continental shelf boundaries.[320] The Rejoinder's rejection of Germany's argument on its 1964 Proclamation on its continental shelf is equally unconvincing. The Rejoinder submitted that the Proclamation did not make "the slightest difference between the different parts of the Geneva Convention."[321] Actually, the Proclamation paraphrases articles 1 and 2(1) of the Convention and then adds:

> The detailed delimitation of the German continental shelf vis-à-vis the continental shelves of other States will remain the subject of international agreements with those States.[322]

There is no reference to the equidistance principle or any other rule of delimitation in the Proclamation.[323]

Chapter 3 of the Rejoinder briefly elaborated the concept of special circumstances. There was no basis for the German assertion that the special circumstances clause had to be understood broadly. Among others the drafting history of article 6 of the Convention on the continental shelf pointed out that the clause was intended to correct the principle of equidistance in cases in which its strict application would lead to an unacceptable result from a legal point of view.[324] Specifically, the term special circumstances could be invoked:

> against a State whose continental shelf boundary under the equidistance principle reflects projecting geographical features (primarily certain islands and peninsula) whereas it cannot be applied against a State whose continental shelf has a solid geographical connection with the territory of that State thereby constituting a natural continuation of the State in conformity with the general geographical configuration.[325]

According to Rejoinder this interpretation "will always offer a criterion providing objective directions for the determination of the line of the

---

[319] *Ibid.*, para. 105 (emphasis suppressed).   [320] *Ibid.*   [321] *Ibid.*, para. 92.
[322] Proclamation of the Federal Government on the exploration and exploitation of the German continental shelf of 22 January 1964, para. 1.
[323] See also Von Schenk, "Die Festlandsockel-Proklamation," pp. 492–493.
[324] CR, para. 123.   [325] *Ibid.*, para. 124 (emphasis suppressed).

boundary."[326] Provided of course, that one would be able to determine what constituted a projecting feature and when there was a solid geographical connection.

In the case of Germany, Denmark and the Netherlands, there was nothing special about their coastlines. An adjustment of the equidistance line would lead to an encroachment on the fully legitimate continental shelf of another State and a redistribution of continental shelf areas, a result which was not contemplated by the special circumstances clause.[327] Germany's concept of coastal frontage was totally divorced from the concept of special circumstances because it led to a delimitation which was not related to specific basepoints while ignoring other basepoints.[328]

Chapter 3 concluded with a number of individual observations of Denmark and the Netherlands. Denmark illustrated the application of special circumstances by a number of concrete examples involving islands.[329] This added little to the argument contained in chapter 3, but these examples certainly were a better illustration than the example used in the preceding paragraphs of the chapter. The latter example referred to a hypothetical case, which was said to show a similarity with Germany, although it is difficult to detect any resemblance.[330] Whereas the coast of the parties to the cases could all be said to face the middle of the North Sea, in the example the middle State was located directly in front of the coast of an opposite State and the two neighboring States faced open waters. The Netherlands, in its separate comments, limited itself to observing that it did not consider it necessary to comment on whether coasts other than those in the North Sea warranted a deviation from the equidistance line.[331]

The Rejoinder confirmed the submissions of the Counter-Memorials. In addition, the Court was requested to determine, if it rejected those submissions, that the boundary was determined on the basis of nearness to the coast. This submission was based on Rejoinder's interpretation of articles 1 and 2 of the Convention on the continental shelf set out above, which implied an unconditional application of the equidistance method.

8.7   The oral pleadings

The hearings in the *North Sea continental shelf* cases took place at the Peace Palace in The Hague, the seat of the ICJ, from 23 October to 11

---

[326] *Ibid.*, para. 129.   [327] *Ibid.*, para. 135.   [328] *Ibid.*, para. 141.   [329] *Ibid.*, para. 142.
[330] For the argument in this respect see *ibid.*, paras 124–125 and 135.   [331] *Ibid.*, para. 143.

November 1968. Like the written pleadings, the oral pleadings consisted of two rounds, in which Germany pleaded first. The German pleadings were mainly presented by Jaenicke, Germany's agent. In both rounds of pleadings, Shigeru Oda also made an appearance. The pleadings of Denmark and the Netherlands were presented by the two agents, Riphagen and Jacobsen, and Waldock. Riphagen and Jacobsen introduced the two cases separately and stressed that they had to be viewed in isolation. Because the legal argument in the two cases to a large extent was identical, the agents and Sir Humphrey had arranged for a presentation that avoided repetition and did not waste time as far as possible.[332]

The two rounds of written pleadings had put the most important issues which divided the parties into focus. Denmark and the Netherlands argued that continental shelf entitlement was based on nearness to the coast. This linked entitlement inextricably to the equidistance method and implied a very limited role for special circumstances. Delimitation of continental shelf boundaries between neighboring States was a bilateral process, which did not require looking at the impact of third States and the share of continental shelf a State was attributed as a result of its bilateral delimitations. The equidistance-special circumstances rule of article 6 of the Convention on the continental shelf reflected customary international law. State practice before the adoption of the Convention already showed the preeminence of equidistance in the delimitation of all bodies of water. Practice in respect of the delimitation of the continental shelf after 1958 confirmed the significance of this rule.

Germany differed with Denmark and the Netherlands on all these points. Continental shelf entitlement was not based on nearness to the coast but rather on the fact that the continental shelf constituted the natural prolongation of the land territory. In the case of the North Sea, the continental shelf of the parties understood in this sense overlapped. This view on entitlement implied that there was no inherent linkage with the equidistance principle. Delimitation was to result in a just and equitable share for the States concerned. This implied that it was always necessary to look beyond the bilateral delimitation between the States concerned. Germany contended that the equidistance-special circumstances rule contained in the Convention on the continental shelf had not become customary international law. State practice might show that equidistance had been applied in a considerable number of cases, but

---

[332] OP, p. 70.

because of the differences between opposite and adjacent coasts, it was not justified to lump all this practice together. Moreover, the fact that States had used the equidistance method in a particular case did not prove that they considered that this method had to be applied in all cases. Finally, Germany considered that the special circumstances clause had a broad range of application.

In view of the entrenched positions of the parties it had to be expected that the oral pleadings would not lead to any substantial changes. One of the members of the German team present at the oral pleadings in The Hague in fact concluded that they "had brought little news."[333] Although this is no doubt a fair summary of the substance of the oral proceedings – both sides revisited their written pleading at great length – the oral pleadings at the same time did contribute to bringing into relief a number of significant points.

Denmark and the Netherlands put further emphasis on the bilateral nature of their disputes with Germany. In their first round introduction, Jacobsen and Riphagen reminded the Court that their partial boundaries with Germany had been concluded on the basis of bilateral discussions and quite independently from the other partial delimitation.[334] The boundary dispute between Germany and the Netherlands had already fully matured before Germany had invited Denmark to negotiate a bilateral boundary.[335] It was true that Germany had informed the Netherlands of its intention to invite all of the North Sea States to a conference, but the Netherlands had simply taken note of this and Germany later gave up this intention. Germany had then instigated tripartite talks, but these had been concerned with the coordinated discussion of the two bilateral disputes.[336] As a matter of fact the tripartite talks ended in two separate Special Agreements.[337] Riphagen submitted that all of this did not constitute mere formalities but reflected:

> the very root of the question of substance now submitted to the Court . . .
> [T]his bilateral approach is not only in conformity with the whole philosophy of the rules of international law relating to boundaries . . .
> but is also more particularly in conformity with the wording of Article 6, paragraph 2, of the Geneva Convention on the Continental Shelf.[338]

---

[333] Letter of H. D. Treviranus to B. Rüster dated 26 November 1968 [AA/64]. Translation by the author. The original text reads "brachten . . . wenig Neues."

[334] OP, pp. 69 and 71.   [335] Ibid., p. 71.   [336] Ibid.; see also ibid., p. 72.   [337] Ibid.

[338] Ibid.

This obviously was a partisan reading of the negotiating history. Denmark and the Netherlands did not want to look beyond their bilateral relationship with Germany, but the latter had made it clear from the outset that an assessment of the reasonableness of a bilateral delimitation required looking at the regional context.

Waldock again explained that Denmark and the Netherlands considered that the two Special Agreements only allowed the Court to determine the rules applicable to the bilateral delimitation of Germany with Denmark and the Netherlands respectively. The Special Agreements did not allow the Court to determine the rules applicable to the sharing out of the continental shelf between three States.[339] The case law of the Court and its predecessor indicated that the question submitted to the Court could not be amplified unilaterally by one of the parties in its pleadings.[340]

Jaenicke succinctly disposed of the Danish-Dutch arguments in his second round pleading. Germany did not deny that the Court was being faced with two separate cases.[341] However, it would be impossible for the Court:

> to pass judgment on the equitableness of a continental shelf boundary without considering the whole geographical situation and its effect on the share it apportions to the one or the other State.[342]

Riphagen, in his second round argument, attempted to ground the bilateral nature of delimitations in the decentralized nature of power and authority in the present state of international law.[343] The delimitation of boundaries was not effected by a centralized world authority that handed out a just and equitable share to individual States. To the contrary, the determination of a boundary line was always a bilateral affair. There was no indication that under the applicable rules of international law the boundary between one pair of States had to be determined or was influenced by the boundary between one of these States and a third State. As a matter of fact, this was also confirmed by the bilateral delimitations which had been effected by the other North Sea States, and which Germany had accepted as determining the legal situation.[344]

---

[339] *Ibid.*, p. 81. [340] *Ibid.*, p. 82. [341] *Ibid.*, p. 165. [342] *Ibid.*, p. 166.

[343] This argument is similar to an argument presented in the Common Rejoinder, but in the latter case less emphasis was placed on the nature of the international legal system (see CR, p. 460).

[344] OP, pp. 214–215.

Waldock, in his second round argument, invited the Court to consider what would have been the situation if either Denmark or the Netherlands would have insisted on bringing their case separately to the Court. In that case the Court's determination in the *Monetary gold* case, that it could not rule on a matter which would require it to take a position on the rights of third States, would apply.[345] How could the Court in that case for instance:

> have listened to an argument that it should determine the principles for delimiting the Danish-German boundary on the basis of particular assumptions regarding the determination of the Netherlands-German and also the Netherlands-Belgian boundaries?[346]

This might be seen as an additional argument for the Court to reject the German view on the applicable rules of delimitation. However, it could also be seen as a last minute attempt to build in a safeguard against a finding of the Court in favor of Germany. What Denmark and the Netherlands were saying was that the Court could not take a decision in favor of Germany in the two individual cases because it would require a ruling on the rights of a third party. Denmark and the Netherlands apparently preferred the absence of a decision on the basis of procedural grounds over a decision favorable to Germany. In that case it would be clear that the Court did not accept their views on the delimitation, but it would be prevented from clearly spelling out the applicable rules. That would have made it more difficult for Germany to rely on the law, and would leave Denmark and the Netherlands more leeway to rely on non-legal considerations.

Germany further elaborated its positive case during the oral pleadings. As was observed in Chapter 8.5.2, in its second round of written pleadings Germany had distanced itself to some extent from the sector principle. Instead, more emphasis had been placed on the idea of coastal façades and their frontal projections. This approach had the potential to go beyond a sector-based claim. The latter approach would lead to an area for Germany of some 36,000 km^2.[347] During Germany's first round

---

[345] The *Monetary gold* case had been brought by Italy against France, the United Kingdom and the United States. The Court held that it could not rule on the claim of Italy because it would require determining questions relating to the lawful or unlawful character of certain acts of Albania in relation to Italy. As the Court observed: "[t]o go into the merits of such questions would be to decide a dispute between Italy and Albania" (Judgment of 15 June 1954; [1954] ICJ Reports, p. 32).

[346] OP, p. 270.

[347] During the oral pleadings the figure was said to be around 36,700 km^2 (*ibid.*, pp. 180–181 and figure 6, at p. 182).

of oral pleadings, both Jaenicke and Oda presented arguments related to the idea of coastal façades, without however becoming very specific about the exact implications for the delimitation of the continental shelf. According to Jaenicke, the coastal façade of Denmark was represented by a straight line running due north from the end point of the land boundary with Germany. Germany's coastal front could be best represented by a line running between the latter point and the end point of the land boundary with the Netherlands. This line was also referred to as the line between Borkum and Sylt. For the Netherlands, Jaenicke proposed a line running from the latter end point in an approximately west southwestern direction. This latter choice was explained by the fact that further south the North Sea got narrower and gradually passed into the Channel.[348] These coastal front lines are identified in Figure 8.1. Although this was a defendable position and could be explained on the basis that this coastal front took into account the entire coast of the Netherlands facing north and northwest and was visually attractive as it suggested that the coastal façade eliminated the convexity of this part of the Dutch coast, Jaenicke did not commit himself to this position, as he told the Court:

> you may take what you like. I don't mind whether you take some other line as the coastal front, that would be more favourable to us. I have taken as the coastal front that which is the least favourable to us.[349]

Lines perpendicular to these coastal fronts converged more or less at the center of the North Sea where the equidistance lines between the United Kingdom and the continent and between Denmark and Norway met. Jaenicke specified that these were not boundary lines. Rather he was:

> just trying to say what could be approximately regarded as to be the natural continuation of these States' territories into the sea, how they converge into each other and what part would be regarded as belonging to the continental shelf of one or the other, as being the extension of their territories into the sea.[350]

He concluded that a sector-like part of the shelf better could be regarded as the natural continuation of Germany into the continental shelf than the equidistance area.[351] Jaenicke did not commit himself on the size, exact shape or location of this sector-like part.

[348] *Ibid.*, p. 41.   [349] *Ibid.*   [350] *Ibid.*   [351] *Ibid.*

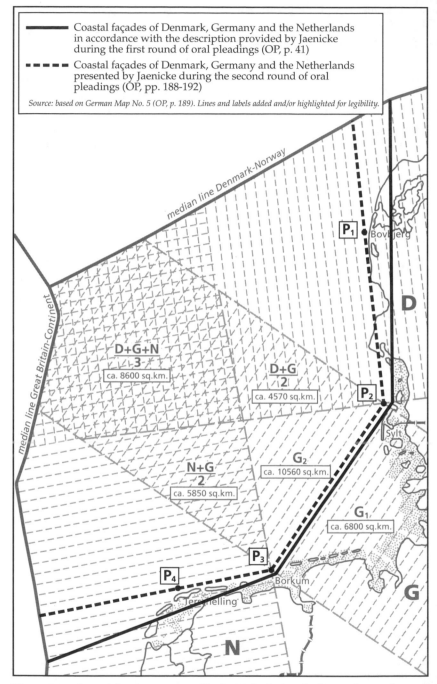

Figure 8.1 Coastal façades mentioned by Jaenicke during the first and second round of oral pleadings before the ICJ

Oda, in his first round pleadings, suggested that the German coastal front represented by the line between Borkum and Sylt provided a starting point for further evaluation and discussion. He felt that it had significance:

> because it avoids deriving from the coastal configuration such an *a priori* predominance of one coastal State over the adjacent coastal State as is inherent in the equidistance method.[352]

Oda did not say anything about the Danish and Dutch coasts or possible delimitation lines.

The Court apparently also had not gotten a clear picture as to the precise implications of the proposals of Jaenicke and Oda. Judge Fitzmaurice asked whether either Jaenicke or Oda could indicate how they would draw the line of demarcation on the basis of the assumption that a straight baseline would be drawn between the islands of Borkum and Sylt.[353]

Denmark and the Netherlands, in their first round pleadings, did not embark on a detailed discussion of the German argument related to coastal façades, but limited themselves to criticizing it as being unclear and without basis in the law.[354]

Germany did finally get more specific in its last round of oral pleadings. Jaenicke first explained the effect of disregarding a limited part of the coasts of Denmark and the Netherlands on equidistance lines. He had already alluded to this argument in his first round pleadings, but now also showed a figure to illustrate his point.[355] Although this argument might suggest that Germany would finally move beyond the sector claim it had previously presented in its written pleadings, the opposite was actually true. Jaenicke indicated that the present German claim was the sector claim it had already presented in the Memorials. This sector comprised an area of about 36,700 km^2.[356] Jaenicke also further elaborated the coastal front concept and its implications for the delimitation of the continental shelf. In the case of Germany, the line between Borkum and Sylt was maintained as the representation of the coastal front, but for Denmark and the Netherlands, the coastal fronts presented in the first round of oral pleadings were readjusted to Germany's disadvantage.[357] This was a remarkable retreat, as Jaenicke in his first round

---

[352] *Ibid.*, p. 63. [353] *Ibid.*, p. 64. See *ibid.*, pp. 64–65 for the initial reply of Jaenicke.
[354] See e.g. *ibid.*, pp. 76 and 132. [355] See *ibid.*, pp. 170 and 172 and figure 7, at p. 171.
[356] *Ibid.*, pp. 180–181 and figure 6, at p. 182.
[357] These coastal fronts were presented on a figure during the pleadings. This figure is reproduced as Figure 8.1. Jaenicke's first round position has been added to Figure 8.1 to allow making a comparison.

presentation had indicated that he had taken the coastal fronts that were least favorable to Germany.[358] In the case of Denmark, the shift was minimal, but it was much more substantial in the case of the Netherlands. Figure 5, presented to Court, showed the resulting frontal projections of these second round coastal fronts. The figure did not include delimitation lines, but did indicate how the area of overlap should be divided between the three States. For Germany this would result in an area of about 36,380 km^2. As was also suggested by Jacobsen in his rejoinder, it seems that Germany had shifted the coastal fronts of Denmark and the Netherlands it had defined previously to allow that this figure would be approximately the same as the areas of Germany's sector claim.[359] As a curious aside, it can be noted that Germany's insistency on the sector claim apparently left the Court with the impression that Germany had already presented it during the negotiations between the parties prior to the proceedings,[360] something which was patently not the case.

It remains somewhat of a mystery why Germany retreated from its first round position and focused to such an extent on the sector claim. Menzel, who acted as counsel for Germany during the oral pleadings, suggested after the case that the sector theory was Germany's main argument and that idea of coastal façades was mainly intended to provide support for the outcome arrived at by the sector theory.[361] Menzel's analysis also indicates that Germany only fully appreciated the advantages of looking at coastal façades during the oral pleadings. The use of coastal fronts allowed mathematical determination of the area of continental shelf of each coast. This implied that the equidistance method "had lost its monopoly of arithmetic precision."[362] This late realization of the full potential of using coastal façades, coupled with the German wish not to overclaim, may well explain that this idea was not further developed in Germany's pleadings.

---

[358] *Ibid.*, p. 41.   [359] *Ibid.*, p. 232.
[360] See Judgment of 20 February 1969, [1969] ICJ Reports, p. 3, at p. 14, para. 5 and Map 3.
[361] Menzel, "Der Festlandsockel der Bundesrepublik," pp. 84–86. On the other hand, Blomeyer, who had been present at the hearings, observed that the sector approach had only been presented as one possible solution (letter of Blomeyer of the FO to the Ministry for the Economy and Transport of Lower Saxony dated 26 November 1968 [AA/48]).
[362] Menzel, "Der Festlandsockel der Bundesrepublik," p. 84. Translation by the author. The original text reads "das Monopol der rechnerischen Präzision war verloren gegangen."

The coastal fronts Germany had presented in its first round pleadings would have given Germany a larger share of continental shelf.[363] If these coastal fronts are used to determine the share of continental shelf in accordance with the methodology Germany applied in figure 5 presented to the Court its area would have been in the range of 40,500 to 41,000 km^2. An equally defendable representation of the Dutch coastal front would have put this figure in the range of 42,000 to 42,500 km^2. This alternative coastal façade is identified in Figure 8.2. Figure 8.2 also identifies bisector lines between these coastal fronts. Unlike Germany's second round position,[364] its first round coastal front resulted in the bisector lines extending approximately up to the median line with the United Kingdom. As Figure 8.2 also indicates, using the alternative coastal front for the Netherlands, bisector lines would have given Germany a common boundary with the United Kingdom.

It does not seem that the above approach carried more of a risk for Germany than the course it actually took during its second round. Germany could, in any case, expect to be attacked on this point. If the Court would accept Germany's submissions concerning delimitation principles, the Court at worst could have rejected the German view on the direction of the coasts of the Netherlands and Denmark. In that case, it was unlikely that the Court would have adopted a more disadvantageous coastal front for Denmark and the Netherlands than Germany had now done itself. Moreover, the coastal façades Germany used in its second round argument contained a drawback, which it would have avoided if it would have stuck to its original definition of the coastal façades. In his second round argument, Jacobsen pointed out that if the German coastal fronts included in its figure 5 were used as the basis for drawing equidistance lines, these lines would converge before they reached the equidistance line with the United Kingdom.[365] That Jacobsen's argument had some appeal is suggested by the opinion of judge Ammoun, who voted with the majority of the Court, appended to the judgment. The opinion contains a figure showing his preferred outcome of the two delimitations. This result seems to be identical to the

---

[363] See Figure 8.1 for a depiction of these coastal fronts.
[364] See further text at footnote 365.
[365] OP, pp. 230 and 232 and figure A on p. 231. This figure is included as Figure 10.3 here. The lines to which Jacobsen referred start at points P_2 and P_3 and converge at the label D+G+N/3.

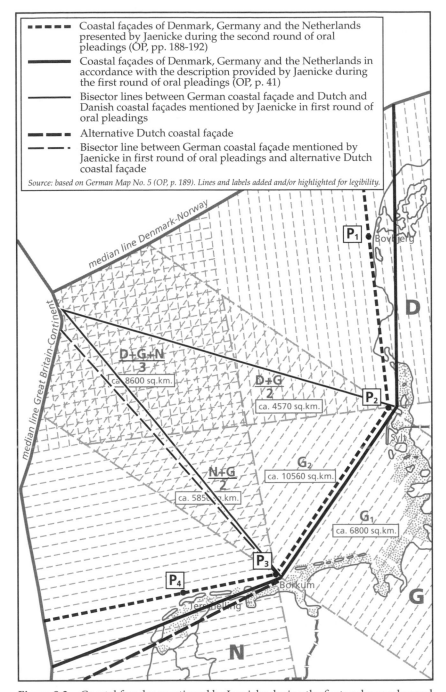

- - - - - Coastal façades of Denmark, Germany and the Netherlands
            presented by Jaenicke during the second round of oral
            pleadings (OP, pp. 188-192)
———————   Coastal façades of Denmark, Germany and the Netherlands in
            accordance with the description provided by Jaenicke during
            the first round of oral pleadings (OP, p. 41)
———————— Bisector lines between German coastal façade and Dutch and
            Danish coastal façades mentioned by Jaenicke in first round of
            oral pleadings
■ ■ ■ ■   Alternative Dutch coastal façade
— — — ·   Bisector line between German coastal façade mentioned by
            Jaenicke in first round of oral pleadings and alternative Dutch
            coastal façade

Source: based on German Map No. 5 (OP, p. 189). Lines and labels added and/or highlighted for legibility.

Figure 8.2   Coastal façades mentioned by Jaenicke during the first and second round of oral pleadings before the ICJ and alternative Dutch coastal façade lines presented by Jacobsen.[366] Jacobsen could not have made his argument if Germany had remained loyal to its original views. Equidistance lines between these coastal front lines would have given Germany a point of contact with the United Kingdom. If the other alternative coastal front for the Netherlands described above would have been used, Germany would have had an opening on the median line between the United Kingdom and the continent of some 20 kilometers.

There had been some discussion on how to respond to Germany's second round argument on coastal fronts during the preparation of the second round pleadings by Denmark and the Netherlands. Jacobsen considered that it was necessary to give the Court a complete picture of the German inconsistencies. Krog-Meyer pointed out that there might be a risk that the debate in this way would become focused on the German case.[367] Waldock in particular warned against using charts and diagrams:

> [t]o do so would be to provide food for those looking for a compromise – something many judges were all too prone to do. The Court would well understand the German ideas were cock-eyed without any visual demonstration of the fact.[368]

Jacobsen, in his second round pleading, provided a detailed analysis of the inconsistencies in the German argument on the share it should actually receive as it had been developed during the pleadings.[369] It cannot but be said that Jacobsen made a convincing argument. Whether this was sufficient to discredit the whole German case was quite a different matter. Germany's case was not based on particular geometrical methods and Germany had always made it clear that there were no hard and fast rules to determine what constituted a just and equitable share in its particular case.

---

[366] See Judgment of 20 February 1969, Sep. op. Ammoun [1969] ICJ Reports, p. 100, figure at p. 153. The description of these lines by judge Ammoun suggests that they are equidistance lines between the line between Borkum and Sylt and the actual baselines of Denmark and the Netherlands (*ibid.*, p. 152). However, that is not the case. The line AC in the figure at *ibid.*, p. 153, representing the boundary between Denmark and Germany does not take into account a part of the Danish baseline. The separate opinion of judge Ammoun was invoked by the Netherlands during the negotiations after the judgment in justifying one of its compromise proposals (see further Chapter 10, text at footnote 289).

[367] Summary Record of the Meeting of the Continental Shelf Advisory Committee held on November 5, 1968 dated 6 November 1968 [DNA/67], p. 2.

[368] *Ibid.*   [369] OP, pp. 223–241.

The relationship between delimitations involving opposite and adjacent coasts was another point that was further clarified during the hearings. This was the direct result of a question judge Fitzmaurice put to Denmark and the Netherlands. Fitzmaurice queried whether their contention that there was no essential difference was really correct. As Fitzmaurice observed, apart from the distorting effects of rocks and islands, which could be met by the application of the special circumstances exception, the median line between opposite coasts in principle always gives the States concerned areas of the same size, whereas a lateral equidistance line often attributed areas of a different size to the States concerned in a way that could not be explained merely by the length of their coastlines.[370]

In answering judge Fitzmaurice's question, Waldock presented a number of hypothetical cases.[371] No doubt these examples were carefully selected. Still, they pointed out that the same geographical feature led to a more unequal division in the case of lateral coasts as compared to opposite coasts.[372] Waldock also conceded that these differences would increase if a lateral boundary would move further seaward,[373] which was exactly the point Germany had been making about lateral equidistance lines all along. Waldock's final argument, that his examples showed that equidistance also did not necessarily lead to equality in the case of opposite coasts, probably also did not help to improve the Danish-Dutch case.[374] The failure of Denmark and the Netherlands to refute the German argument on this point, after they had been explicitly invited to do so, most probably was a severe blow to their case. Apart from showing that equidistance did not lead to similar outcomes in the case of opposite and lateral coasts, it implied that it was less likely that the Court would take into account State practice involving opposite coasts in looking at the rules involving lateral coasts. That judge Fitzmaurice's question made an impact is also suggested by the fact that paragraph 57 of the judgment of the Court is strikingly similar to the proposition he had presented to Denmark and the Netherlands.

Other questions by judges may also have had some impact on the course of the proceedings. Judge Jessup asked Germany to produce a map showing median lines between Germany and the United Kingdom and Germany and Norway.[375] Jaenicke in his second round argument used these maps to explain that they provided a telling illustration that

---

[370] *Ibid.*, p. 163.   [371] *Ibid.*, pp. 248–252.   [372] *Ibid.*, p. 252.   [373] *Ibid.*   [374] *Ibid.*
[375] *Ibid.*, pp. 65–66.

Denmark and the Netherlands were clearly mistaken in arguing that there was no ascertainable center in the North Sea to take into account in determining the extent of a sector claim of Germany.[376] Apart from the median lines judge Jessup wanted to see on paper, Germany had also determined a hypothetical equidistance line between the Netherlands and Norway. These lines and existing equidistance lines were presented in one figure, which showed that all these lines converged near the tip of the German sector claim.[377] Jaenicke concluded that there was a real center in that part of the North Sea on which the continental shelf of the parties and Norway and the United Kingdom converged.[378] Jaenicke's presentation was attacked by Jacobsen. Germany had failed to include the equidistance lines between the parties before the Court. These lines would not have intersected that central area identified by Germany. The same was no doubt true for Belgium and France.[379] Jacobsen was actually right on both counts. None of these equidistance lines intersected the center identified by Germany. However, Jacobsen's argument probably only highlighted that Germany's continental shelf was unduly cut-off by the equidistance lines Denmark and the Netherlands were advocating and that Belgium and France were irrelevant to the broader picture as they were located too far south. The Court used the map presented by Germany as the basis for a figure in connection with its description of the general background of the cases in its judgment. Although the judgment indicated that the hypothetical equidistance lines might be of little significance in practice,[380] the reliance on the map suggests that the idea it expressed had struck a chord with the majority of the judges.

Denmark and the Netherlands also took into account cues from the Court. The President of the Court had asked whether the Court could be provided with all relevant information on the work of the committee of experts, which had advised the special rapporteur of the ILC on possible methods to delimit the territorial sea and the continental shelf in 1953.[381] Although Germany in its second round had paid scant attention to the committee's work, Waldock embarked on a lengthy analysis of its significance in his rejoinder.[382] Interestingly, in stressing the significance of

---

[376] *Ibid.*, p. 181.
[377] See the discussion at *ibid.*, pp. 181 and 186 and the figures at pp. 182–185 and 187.
[378] *Ibid.*, p. 186.   [379] *Ibid.*, p. 226.
[380] Judgment of 20 February 1969, [1969] ICJ Reports, p. 3, at p. 14, para. 4 and p.15, figure 2.
[381] OP, p. 162.   [382] *Ibid.*, pp. 259–265.

the committee of experts and the work on article 6 generally, Waldock conceded that State practice outside of this framework might have shown an obligation to delimit boundaries on the basis of equitable principles, although it was indefinite as to the basis for determining what constituted a delimitation on equitable principles.[383] Previously, Denmark and the Netherlands had stressed the significance of equidistance during this period.

The pleadings indicate that all three parties considered consistency to be of great importance and that they tried to convince the Court that the other party was being inconsistent. Chapter 7.3.2 set out the episode involving the differences between the position of the Netherlands and Suriname. Jaenicke, during the first round of written pleadings, referred to the case of Suriname to show that the Netherlands in that case did not adhere to a narrow interpretation of the special circumstances clause.[384] On their part, Denmark and the Netherlands also tried to press home the inconsistencies in Germany's position. Jacobsen, Riphagen and Waldock, in their first round pleadings, revisited the partial continental shelf boundaries of Denmark and the Netherlands with Germany. In their written pleadings, Denmark and the Netherlands had argued that these agreements showed that Germany had accepted the equidistance-special circumstances rule. Jacobsen, in his introduction of the Danish case, pointed out that the Danish government could not find any reason why the remainder of the boundary in the North Sea, which was determined by points on the same coast as the first stretch of the boundary, should not be determined on the basis of the same principles as the partial boundary.[385] Riphagen in his introductory statement stressed another aspect of the partial boundary between Germany and the Netherlands. The Netherlands had not only accepted the German low-tide elevation of Hohe Riff, but also the German island of Borkum in determining the location of that boundary. As was illustrated on a map, the boundary would have been located much further to the east if this offshore island had not been taken into account. The Court would no doubt appreciate the resemblance of this map to the map of Haiti and the Dominican Republic, which had been discussed by Germany as a situation involving a coastal configuration which made equidistance inequitable.[386] Lest the Court had not been paying attention, the point was brought home by Waldock. Germany had "had no scruples or

---

[383] *Ibid.*, p. 258.   [384] *Ibid.*, p. 47; see further Chapter 7.3.2.   [385] *Ibid.*, pp. 69–70.
[386] *Ibid.*, p. 72–73.

hesitations about using the German island of Borkum and the low-tide elevation of Hohe Riff, both of which lie off the mainland of the Netherlands, as base points for delimiting the partial boundary."[387] In discussing the Danish-German partial boundary, Waldock pointed out that the German baseline in this case comprised the island of Sylt, about half of which stretched in front of the Danish coast.[388] In other words, in this case Germany had also taken a restrictive view on the operation of the special circumstances clause. A similar point was made later on in respect of the median line delimitation of the continental shelf in the Baltic Sea between Denmark and Germany, which allegedly had led to an unequal division of the area concerned:

> We have not heard the Federal Republic complain of that, nor has Denmark complained of it, because she thought it was the Federal Republic's legal right. But Denmark also thinks that she is entitled to her rights off her North Sea coasts.[389]

Whether Germany had actually profited from the application of the equidistance principle in these cases is debatable. Borkum and Sylt are not isolated islands but part of a chain of islands which are close together and in this sense jointly are similar to a mainland coast. The reference to the location of the land boundary behind the islands ignores that this land boundary continues between the islands and the Danish and Dutch coastal islands. In the Baltic Sea a median line is not more advantageous to Germany than to Denmark.

At the end of their pleadings the parties maintained the submissions of, respectively, the Replies and the Common Rejoinder. Germany made a clarification in respect of one of its submissions.[390]

8.8 The approach of the parties to the pleadings

Germany and Denmark and the Netherlands both had to set out their own positive case and to discredit the other side's positive case. The most striking aspect of Germany's pleadings is no doubt its cautious approach to its positive case. One could easily conclude that Germany was too cautious. As was demonstrated during the second round of oral pleadings by Jacobsen, the Danish agent, the scheme Germany had presented during its last round of oral pleadings logically led to a delimitation which would not accord Germany access to the median line with the

---

[387] *Ibid.*, p. 86.   [388] *Ibid.*   [389] *Ibid.*, p. 252.   [390] *Ibid.*, pp. 210–211 and 284.

United Kingdom, even though this was one of the central aims of Germany's shelf policy. The exact reasons for Germany taking this approach are not known. The explanation of the German agent of this scheme indicated that one important reason may have been that Germany was looking for an outcome that would not differ too much from the sector claim it had presented previously during its pleadings.[391] Whether the wish to appear consistent completely explains the German pleading is, however, doubtful. As was set out above, Germany during its first round of oral pleadings had already submitted a different view on the Dutch coastal front and retracting from that position, which had been explicitly identified as being the least advantageous for Germany, also led to an inconsistency in the German case. All in all, it seems more likely that Germany did not want to be seen as making an exaggerated claim. It can be doubted that there really was a risk in this respect. If the Court would accept that the equidistance principle was not a mandatory or guiding principle, a rejection of the German claim would not leave Germany empty-handed. Germany, at worst, risked being told that the outcome should be less than what it had been claiming before the Court.[392]

As was argued above, Denmark and the Netherlands were successful in setting out certain weaknesses in Germany's positive case. They were less so in arguing certain aspects of their own positive case. First, it was difficult to square their attempts to show that Germany had accepted the equidistance principle with the facts. Second, Denmark and the Netherlands may have put too much emphasis on the centrality of the equidistance principle in the rule on the delimitation on the continental shelf. This approach did not seem to do justice to the drafting history of the Convention on the continental shelf. If the role of equidistance was that obvious, why was the principle not adopted at the outset, and why was there so much uncertainty about the meaning and scope of special circumstances? The drafting history also did not provide support for the contention that absolute proximity was inherent to the basis of continental shelf entitlement. Admittedly, it would in any case have been difficult to strike the right balance in this respect and it would have been

---

[391] See *ibid.*, p. 191.

[392] To the contrary, Münch shortly after the judgment concluded that Germany probably did well in only discussing the most moderate possibility that had been suggested in a 1964 report by Menzel (Münch, "Das Urteil," pp. 473–474). Münch does not provide any arguments for this conclusion.

difficult to admit that special circumstances had a broader scope of application without risking that Germany's case would also be covered by it. Another weakness of the positive case of Denmark and the Netherlands was that, when prodded by the bench, they were not able to convincingly demonstrate that the equidistance principle had similar implications in delimitations involving opposite and adjacent coasts. This had a number of consequences. It implied that State practice involving opposite coasts could be considered to be irrelevant for the disputes before the Court and, equally important, it reconfirmed that Germany had a point in making a distinction between opposite and adjacent coasts.

9

The judgment of the Court

9.1   Introduction

The pleadings of the parties had presented the Court with two contrasting views of the law concerning the entitlement to, and delimitation of, the continental shelf. It fell on the Court to sort out to what extent either view actually reflected the law. More than anything else, the pleadings of the parties confirmed that it would be difficult for the Court to determine the content of the law if it would find that the equidistance principle was not applicable. That the Court was, by no means, confronted with an easy task is confirmed by the German Foreign Office's assessment of the cases after the conclusion of the oral proceedings. It was acknowledged that the outcome was and remained completely open. This was hardly surprising in view of the composition of the Court and the rapid but still unconsolidated development of the law concerning the continental shelf.[1]

The judgment of the Court was handed down on 20 February 1969. Eleven of the judges voted in favor of the judgment and six judges voted against. Although the vote was far from unanimous, this was a clear majority.[2] Scheuner, one of the advisors to the German team in The Hague, concluded that this should facilitate the negotiations as Denmark and the Netherlands would know that they had nothing to expect from the Court if the parties were to request a further judgment.[3]

The judgment implied a clear rejection of the position of Denmark and the Netherlands. The first finding of the judgment's operative paragraph was that the use of the equidistance method was not obligatory between the parties.[4] In itself, this represented a clear victory for Germany. However, as the judgment did not provide the parties with

[1] Circular letter of the FO dated 19 December 1968 [B102/260033], p. 3.
[2] See also text at footnote 94.
[3] Letter of U. Scheuner to H. Blomeyer of the FO dated 21 February 1969 [AA/64].
[4] Judgment of 20 February 1969, [1969] ICJ Reports, p. 3, at p. 33, para. 101(A).

detailed guidance for their further negotiations, it remained to be seen what gains Germany would eventually be able to reap from this victory during new negotiations.

A first section of this chapter looks at how the Court dealt with the arguments of the parties. It then looks at the kind of guidance the judgment provided to the parties for negotiating the remainder of their boundaries on the basis of the judgment. The judgment and the views of individual judges that are appended to the judgment do allow establishing what the Court considered the general characteristics of an outcome should be. This matter is considered in Chapter 9.4.

9.2 The Court's consideration of the arguments of the parties

Before looking at the kind of guidance the Court offered the parties for their further negotiations, this section considers the judgment's treatment of the arguments contained in the pleadings of the parties. An important aspect of the case of Denmark and the Netherlands was that the two disputes had to be viewed separately. In order to substantiate that point they had advanced procedural grounds – the cases had been brought separately and remained so even though they had been joined – and arguments of substance – delimitation was bilateral in nature and did not require consideration of the consequences of the presence of a third State.

The Court dealt with the Danish and Dutch argument on the relation of the two cases in one slightly enigmatic paragraph.[5] The Court acknowledged that although the cases had been joined, they remained separate "at least in the sense that they relate to different areas of the continental shelf of the North Sea, and that there was no *a priori* reason why the Court must reach identical conclusions in regard to them." The Court then observed that Denmark and the Netherlands had acted in close cooperation, making that "[t]o this extent" the two cases could be treated as one and even though either case might have been brought separately, that did not alter the character of the problem with which the Court was actually faced, having regard to the manner in which the parties had brought the matter before the Court.[6] The latter observation especially might suggest that the way in which Denmark, Germany and the Netherlands had brought the cases had a decisive impact on its outcome. This view has, for instance, been advanced by Evans:

---

[5] *Ibid.*, p. 19, para. 11.   [6] *Ibid.*

> If two States want a broader regional perspective to feature in the reasoning underlying their maritime boundary delimitation then the options are clear: they can negotiate a regional solution or submit a case (or cases) involving the other relevant regional actors. That, of course, is what happened in the *North Sea* cases and explains why the regional dimension was of relevance and could be taken into account.[7]

At the same time, paragraph 11 also contains indications that the way in which the cases were brought did not significantly impact on their outcome. The Court acknowledged that the two cases remained separate. The rider it added to this observation confirms that it had dealt with the two delimitations separately and could also have done this if only one of the cases would have been brought to it. The last sentence of paragraph 11 also suggests that a separate filing of one of the disputes would not have led to a different outcome, although it could also be argued that the Court was saying that this was a situation with which it was not faced and which it consequently did not have to consider. Finally, the Court observed that the two delimitations in fact gave rise to a single situation. This observation suggests that the approach of the Court was not the result of the joining of the cases, but a result of the geography of each individual case. In other words, in the view of the Court the applicable law required it to look for each of the two cases at the wider geographical context.[8] The Court would also have had to base itself on the applicable law if one of the cases had been brought separately.[9]

---

[7] M. D. Evans, "Maritime Boundary Delimitation: Where Do We Go From Here?" in D. Freestone, R. Barnes and D. Ong (eds), *The Law of the Sea; Progress and Prospects* (Oxford University Press, 2006), pp. 137–160, at p. 151 (footnote omitted).

[8] This point is also addressed by the Court in paras 57 and 89 of its judgment.

[9] The separate opinion of judge Padilla Nervo [1969] ICJ Reports, pp. 89–90 and the dissenting opinions of judges Tanaka (*ibid.*, pp. 191–192) and Morelli (*ibid.*, pp. 209–212) also address this point. Judges Padilla Nervo and Morelli observed that the combined effect of the two bilateral delimitations led to an inequitable result. Judge Padilla Nervo expressed the belief that the Parties by submitting the matter to the Court in the way they did, recognized that the matter constituted an integral whole (*ibid.*, p. 90). Judge Morelli observed that it would have been quite possible for the Court to reach the conclusions it did if it would only have had two of the three States concerned before it (*ibid.*, p. 211). Judge Tanaka concluded that the two cases had to be considered as separate and independent from a substantive viewpoint and suggested that the Court could not have reached the same judgment if one of the cases would have been brought separately (*ibid.*, p. 192). Judge Tanaka's conclusion is explained by a different view on the applicable law. His opinion stresses the bilateral nature of the delimitation process and the centrality of equidistance. Judges Padilla Nervo and Morelli indicate – as does the judgment of the Court – that the equitable nature of a bilateral delimitation has to be assessed in a broader context.

The Court, as a matter of fact, has repeatedly taken the view that it can take into account the coast of a third State in deciding the bilateral delimitation of the parties appearing before it. As the Court reaffirmed in its recent judgment of 4 May 2011 on the application of Costa Rica for permission to intervene in the *Territorial and Maritime Dispute (Nicaragua v. Colombia)* the taking into account of all coasts and coastal relationships in no way signifies that the legal interests of a third State may be affected.[10] Although the latter point seems to be problematic,[11] this finding confirms that Denmark and the Netherlands probably would not have fared differently if they had rejected the joining of their cases. The dynamics of the negotiations in relation to the submission of the disputes to the Court in any case would have made it difficult to keep the two cases completely separate.

Even if the Court had relied on its jurisprudence in the *Monetary gold* case, which had been invoked by Waldock during the oral pleadings, to rule that it could not have decided a separately brought case because a necessary third party was lacking, this would not have assisted Denmark and the Netherlands. The finding that a necessary third party was lacking would have implied that the Court considered that the applicable law required it to look beyond the bilateral relation of the parties. That is, the Court would have found that the equidistance method was not applicable in the instant case because of the single situation resulting from two separate delimitations.

The judgment of the Court summarily dismissed the argument of Germany concerning a just and equitable share.[12] The Court agreed with Denmark and the Netherlands that this notion was foreign to:

> the basic concept of continental shelf entitlement, according to which the process of delimitation is essentially one of drawing a boundary line between areas which already appertain to one or other of the States affected.[13]

The judgment dealt at much more length with the contentions of Denmark and the Netherlands. Leaving aside the introductory part of

---

[10] Judgment of 4 May 2011, para. 85.

[11] See A. G. Oude Elferink, "Third States in Maritime Delimitation Cases: Too Big a Role, Too Small a Role or Both?" in A. Chircop, T. L. McDorman, S. J. Rolston (eds), *The Future of Ocean Regime-building; Essays in Tribute to Douglas M. Johnston* (Leiden: Martinus Nijhoff Publishers, 2009), pp. 611–641, at pp. 633–638.

[12] Judgment of 20 February 1969, [1969] ICJ Reports, p. 3, at pp. 21–23, paras 18–20.

[13] *Ibid.*, p. 22, para. 20.

the judgment, which among others set out the submissions of the parties and the text of the Special Agreements, more than half of the judgment is dedicated to dismissing the Danish-Dutch arguments. The judgment dealt with these contentions under three headings. It first considered whether Germany had become bound by the rules contained in the Convention on the continental shelf and subsequently considered whether the equidistance method was binding on Germany as a matter of customary law. In the latter respect the Court both considered whether equidistance was inherent to the concept of the continental shelf and whether the method had become a binding rule through State practice.

The Convention on the continental shelf as such was not binding on Germany, because it had not become a party to it. It was, however, possible that Germany by its conduct had become bound by the Convention.[14] The judgment submitted that this could only be the case if there had been "a very definite, very consistent course of conduct" on the part of Germany.[15] The judgment concluded that all of the arguments advanced by Denmark and the Netherlands were either ultimately negative or inconclusive and were all capable of varying interpretations or explanations.[16] In addition, any inferences about Germany's conduct would be nullified by the fact that as soon as the parties began to consider the delimitations in the North Sea, Germany immediately reserved its position with respect to those delimitations based on equidistance which might be prejudicial to its position.[17]

The judgment found that there was a certain appeal to the Danish-Dutch argument that the equidistance principle was inherent in the continental shelf concept. As a matter of normal topography the greater part of the continental shelf of a coastal State would be nearer to its coasts than any other coast. However, this did not imply that any part of the continental shelf must be placed in this way.[18] The Court moreover rejected that there was a complete identity between the notions of adjacency and absolute proximity, as had been submitted by Denmark and the Netherlands. Adjacency and other similar terms all had a somewhat imprecise character and specific areas often could properly be said to be adjacent to more than one coast.[19] According to the judgment, the notion of natural prolongation was more fundamental to continental shelf entitlement than proximity. Title to the continental shelf was

---

[14] *Ibid.*, p. 25, para. 27.   [15] *Ibid.*, para. 28.   [16] *Ibid.*, pp. 26–27, para. 32.
[17] *Ibid.*, p. 27, para. 33.   [18] *Ibid.*, pp. 29–30, para. 40.   [19] *Ibid.*, p. 30, paras 41–42.

attributed to the coastal State because a given area was a prolongation under water of the land territory of a State. From this it followed that:

> whenever a given submarine area does not constitute a natural – or the most natural – extension of the land territory of a coastal State, even though that area may be closer to it than it is to the territory of any other State, it cannot be regarded as appertaining to that State; – or at least it cannot be so regarded in the face of a competing claim by a State of whose land territory the submarine area concerned is to be regarded as a natural extension, even if it is less close to it.[20]

The judgment subsequently picked up this concept of natural prolongation in providing guidance to the parties for their future negotiations. What is immediately evident is that the concept in itself offers little guidance in determining a boundary. A natural prolongation can be common to two or more States, in which case it apparently is "most natural" to one of the States concerned. The latter notion is difficult to square with the notion of natural prolongation and continental shelf entitlement, which either exists or does not exist.

The Court's judgment also found that the genesis and development of the equidistance method only served to confirm the conclusion that it was not inherent to the concept of the continental shelf.[21] The ILC had never entertained the idea that it had to adopt the equidistance method because it gave expression to a principle of proximity inherent in the basic concept of the continental shelf.[22] To the contrary, the whole history of the continental shelf regime indicated that two concepts were more fundamental – delimitation by agreement and in accordance with equitable principles.[23]

Before looking at the question of whether the equidistance method had become a rule of customary law, the judgment briefly considered the relationship between the application of the equidistance method in the case of opposite and adjacent coasts in continental shelf delimitations and the use of the equidistance method to delimit lateral boundaries in the territorial sea. In the case of opposite coasts natural prolongations in general would meet and overlap and the median line would result in an equal division of the area involved. On the other hand, in the case of adjacent coasts, the equidistance line did not necessarily divide areas equally between States and often left an area which was the natural prolongation of one State to the other State. The case of the delimitation

---

[20] *Ibid.*, p. 31, para. 43. [21] *Ibid.*, pp. 32–36, paras 47–56. [22] *Ibid.*, p. 33, para. 48.
[23] *Ibid.*, para. 47.

of the territorial sea between adjacent coasts was different from the delimitation of the continental shelf between such coasts. The territorial sea was relatively close to the coasts and the distorting effect caused by certain coastal configurations would only become marked further seaward.[24] These findings were important for the judgment's subsequent analysis of State practice. In determining whether State practice had converted the equidistance method into a rule of law between the parties, practice involving opposite coasts and territorial sea delimitations would have to be disregarded because these differed from the delimitation of the continental shelf between adjacent States. This implied that there was only a limited amount of relevant State practice, which made it unlikely that it had led to the creation of a rule of customary law. This was indeed the approach the judgment took.[25] As far as continental shelf delimitations between States with adjacent coasts were concerned, the parties discussed only one situation which to some extent resembled their own situation.[26] In short, State practice was simply insufficient to prove that the use of equidistance was obligatory for the delimitation between the parties.[27] In that light, the judgment concluded that it was unnecessary for the Court to determine whether or not the configuration of the German coast constituted a special circumstance. Since the equidistance method was not obligatory, it ceased to be legally necessary to "prove the existence of special circumstances in order to justify not using that method."[28] This reasoning of the Court misrepresents the argument of Denmark and the Netherlands, which had argued that it was not the equidistance principle as such which reflected customary law, but the combined equidistance-special circumstances rule.[29]

The Court's reluctance to look into the combined equidistance-special circumstances rule may be explained by the fact that there was no agreement among the majority of the judges on the content of special circumstances. The judgment indeed refers to "the very considerable, still unresolved controversies as to the exact meaning and scope of this

---

[24] *Ibid.*, pp. 36–37, paras. 57–58.

[25] *Ibid.*, p. 45, paras 79–80. The approach of the Court on this point has been criticized as "geographically inaccurate" because certain of the coastal relationships which the Court classified as opposite rather should be viewed as adjacent (Antunes, *Conceptualisation*, p. 53). Apart from the fact that Antunes's view is debatable, it can be noted that the parties in the proceedings accepted that the examples he mentions constituted cases involving opposite coasts (see e.g. the references in Annex 13 to the DCM).

[26] Judgment of 20 February 1969, [1969] ICJ Reports, p. 3, at p. 45, para. 79.

[27] *Ibid.*, pp. 45–46, para. 82.   [28] *Ibid.*, p. 46, para. 82.

[29] See also Antunes, *Conceptualisation*, p. 56.

notion."[30] The drafting history of article 6 sheds little light on this matter and the same can be said about State practice. As the Court itself concluded, there was only one case which was somewhat similar to the case before it. This hardly constituted a basis for resolving the meaning and scope of special circumstances. Interestingly, the Court in setting out its views on the applicable rules and the content of equitable principles also had to face this problem. The Court has been criticized that it, in this case, did not proceed to a similar analysis of State practice, which would have revealed that there was "[n]o settled, extensive and virtually uniform practice using equitable principles."[31] One answer to this criticism could be that the fact that there was no State practice which had led to the formation of specific rules implied that the Court had to deduce such rules from the fundamental concepts underlying the continental shelf regime. Having to choose between two problematic approaches, from the perspective of the Court it could be said to be more attractive to present an authoritative statement on the fundamental notions of the continental shelf regime than delving into the question of how the concept of special circumstances had to be interpreted and applied. The question of the definition of the continental shelf had been on the agenda of the United Nations since the middle of the 1960s. The debate indicated that a different approach to the definition of the continental shelf was called for and a judgment looking into this issue could provide a contribution. The judgment in the *North Sea continental shelf* cases indeed would have a significant impact on the development of the law.[32] The approach the Court took offered it better opportunities to reestablish itself in the aftermath of the *South West Africa* cases than a narrow focus on the special circumstances clause.[33]

In looking at the customary law status of the equidistance method, the judgment also considered the significance of article 6 of the Convention on the continental shelf. The judgment did not accept the Danish-Dutch argument that article 6 of the Convention was directly linked with articles 1 and 2 of the Convention or their arguments concerning the significance of the reservations in respect of article 6. Articles 1 and 2 expressed established principles of international law, but this was not the case for article 6.[34]

---

[30] Judgment of 20 February 1969, [1969] ICJ Reports, p. 3, at p. 42, para. 72.

[31] Antunes, *Conceptualisation*, p. 54.

[32] See e.g. Oude Elferink, "Article 76," pp. 272–273.

[33] On the latter point see also Chapter 8, text at footnotes 11 and following.

[34] Judgment of 20 February 1969, [1969] ICJ Reports, p. 3, at pp. 38–41, paras 63–69.

9.3   The Court's guidance to the parties

The findings of the Court on the applicable law are contained in paragraph 101 of the judgment. After concluding that the equidistance method is not obligatory between the parties and that there is no other single method which is obligatory in all circumstances, paragraph 101 provides specific guidance to the parties. As this part of paragraph 101 is essential for a discussion of where the judgment left the parties, it is useful to quote it in full:

> (C) the principles and rules of international law applicable to the delimitation as between the Parties of the areas of the continental shelf in the North Sea which appertain to each of them beyond the partial boundary determined by the agreements of 1 December 1964 and 9 June 1965, respectively, are as follows:
>
> (1) delimitation is to be effected by agreement in accordance with equitable principles, and taking account of all the relevant circumstances, in such a way as to leave as much as possible to each Party all those parts of the continental shelf that constitute a natural prolongation of its land territory into and under the sea, without encroachment on the natural prolongation of the land territory of the other;
> (2) if, in the application of the preceding sub-paragraph, the delimitation leaves to the Parties areas that overlap, these are to be divided between them in agreed proportions or, failing agreement, equally, unless they decide on a régime of joint jurisdiction, user, or exploitation for the zones of overlap or any part of them;
>
> (D) in the course of the negotiations, the factors to be taken into account are to include:
>
> (1) the general configuration of the coasts of the Parties, as well as the presence of any special or unusual features;
> (2) so far as known or readily ascertainable, the physical and geological structure, and natural resources, of the continental shelf areas involved;
> (3) the element of a reasonable degree of proportionality, which a delimitation carried out in accordance with equitable principles ought to bring about between the extent of the continental shelf areas appertaining to the coastal State and the length of its Coast measured in the general direction of the coastline, account being taken for this purpose of the effects, actual or prospective, of any other continental shelf delimitations between adjacent States in the same region.

Paragraph 101(C)(1) first of all confirms that delimitation of the continental shelf is to be effected by agreement. This implies a rejection of the Danish-Dutch thesis that they were entitled to unilaterally determine their continental shelf boundaries in relation to Germany in accordance with international law. The implications of the duty to delimit continental shelf boundaries by agreement are set out in the reasoning of the Court preceding the dispositif. The reasoning of a judgment can be used to construct the meaning of the dispositif.[35]

According to the Court, Denmark, Germany and the Netherlands were:

> under an obligation to enter into negotiations with a view to arriving at an agreement, and not merely to go through a formal process of negotiation as a sort of prior condition for the automatic application of a certain method of delimitation in the absence of agreement; they are under an obligation so to conduct themselves that the negotiations are meaningful, which will not be the case when either of them insists upon its own position without contemplating any modification of it.[36]

As the Court also indicated, the obligation to enter into negotiations did not imply an obligation to reach agreement.[37] The Court further considered that the negotiations between the parties in 1965 and 1966 had not been meaningful, as was required, because Denmark and the Netherlands, which were convinced that the equidistance principle alone was applicable, had not wanted to depart from that rule. Equally, Germany could not accept the situation resulting from the application of that rule.[38] The Court concluded that fresh negotiations had to take place on the basis of its judgment.[39]

The reasoning of the Court was a slap on the wrist for Denmark and the Netherlands. As is indicated in paragraph 85(a) of the judgment, the obligation to engage in meaningful negotiations implied that a party

---

[35] The findings of the Court contained in the dispositif of a judgment are *res judicata* for the parties to a case. It is binding on the parties and must be implemented by them in good faith. *Res judicata* also implies that an issue which has been decided by a judgment cannot be reopened. The doctrine of *res judicata* not only applies to the dispositif of a judgment but also to the reasoning which led to the adoption of the operative part of the judgment (see e.g. *Territorial and Maritime Dispute (Nicaragua v. Colombia)* Application by Honduras for permission to intervene, judgment dated 4 May 2011, para. 70).

[36] Judgment of 20 February 1969, [1969] ICJ Reports, p. 3, at p. 47, para. 85(a).

[37] *Ibid.*, pp. 47–48, para. 87.

[38] The Court had asked the parties for additional information on these negotiations. This additional information was submitted to the Court after the closure of the hearings (see ICJ Pleadings, *North Sea Continental Shelf* cases, Vol. II, pp. 303–363).

[39] Judgment of 20 February 1969, [1969] ICJ Reports, p. 3, at p. 48, para. 87.

could not insist on its own position without contemplating any modification of it. The Court pointed out that this is what Denmark and the Netherlands had been doing in their negotiations with Germany in 1965 and 1966. This conclusion was mitigated by the Court's acknowledgement that Denmark and the Netherlands had acted in this way because of their conviction that their position was in accordance with the applicable law. The reasoning of the Court could be said to put Denmark and the Netherlands on warning. If the parties would not be able to reach agreement on the basis of the judgment of the Court and would return to it for a further decision on the location of their boundaries with Germany, they could expect that the Court would be looking at their track record. The Court in that case could have found that Denmark and the Netherlands had not complied with their obligation to conduct themselves in such a way that negotiations had been meaningful. Such a finding might reflect negatively on the international reliability of both States. To the extent that Denmark and the Netherlands wanted to avoid such negative consequences, the Court's finding in its judgment could be expected to set certain limits on the way in which both States could conduct the further negotiations.

Paragraph 101(C)1 of the judgment also set out the substantive rules applicable to the delimitation of the continental shelf between the parties. Agreement had to be effected in accordance with equitable principles and taking into account all the relevant circumstances. The Court in its judgment explained that the requirement that agreement had to be arrived at in accordance with equitable principles was one of the basic notions which from the beginning of the continental shelf regime had reflected the legal thinking on the matter.[40] The implications of the requirement to delimit continental shelf boundaries in accordance with equitable principles was first explained negatively, when the judgment commented on the sometimes inequitable nature of the equidistance method. The Court observed that the slightest irregularity in a coastline impacting on an equidistance line would have more unreasonable consequences the further from shore one would get.[41] In the case of the North Sea, the claims of several States converged, met and intercrossed at a considerable distance from the coast. A delimitation based on equidistance would ignore these geographical circumstances.[42] This reasoning of the Court indicates that the application of the equidistance method was considered to be inequitable because it resulted in an

---

[40] Ibid., p. 46, para. 85.   [41] Ibid., p. 49, para. 89(a).   [42] Ibid., para. 89(b).

unjustified unequal treatment. This is confirmed by the Court's further explanation of the implications of applying equity. It did not necessarily imply equality and did not require that the situation of a State with an extensive coastline should be rendered similar to that of a State with a restricted coastline. However, States in a comparable situation should be given broadly equal treatment.[43] The judgment of the Court does not explicitly elaborate further on the content of equitable principles, but a number of specific principles can be inferred from the Court's reasoning. For instance, the judgment indicates that: "the relative weight to be accorded to different considerations naturally varies with the circumstances of the case."[44] This reflects an equitable principle that due respect has to be given to all relevant circumstances, which has been explicitly formulated by the Court subsequently.[45]

Paragraph 101(C)1 also indicates that the parties have to take into account all relevant circumstances. The judgment does not define what it understood by relevant circumstances. In discussing the equidistance method, the judgment refers to the "certain geographical circumstances"[46] and it is clear from the judgment that coastal geography has an important role to play in the delimitation process. However, the judgment also indicates that circumstances unrelated to coastal geography may be relevant. Reference is, for instance, made to geology and the idea of the unity of any deposits.[47] The lack of clarity of the judgment on this point is reinforced by the use of a number of other terms without clearly specifying their relationship or their relationship to relevant circumstances. For instance, at one point the judgment refers to the considerations States may take into account in applying equitable procedures. The relative weight of these considerations varies with the circumstances of the case.[48] The judgment then refers to the factors in question, presumably referring to the considerations to which it referred previously and observes that in balancing these factors it would assume that "various aspects must be taken into account. Some are related to the geographical aspect of the situation, others again to the idea of the unity of any deposits."[49] Immediately after the quoted text, the judgment indicates that these criteria can provide an adequate basis for a decision

---

[43] *Ibid.*, pp. 49–50, para. 91.   [44] *Ibid.*, p. 50, para. 93.
[45] *Continental Shelf (Libyan Arab Jamahiriya/Malta)* case, Judgment of 3 June 1985, [1985] ICJ Reports, p. 39, para. 46.
[46] Judgment of 20 February 1969, [1969] ICJ Reports, p. 3, at p. 49, para. 89.
[47] *Ibid.*, pp. 50–51, para. 94.   [48] *Ibid.*, p. 50, para. 93.   [49] *Ibid.*, pp. 50–51, para. 94.

adapted to the factual situation.[50] In itself, this lack of clarity did not have to be a problem. The judgment contained more detailed information about considerations/factors/aspects that had to be taken into account. However, uncertainty concerning the relationship between various terms contained in the judgment might complicate negotiations if the parties would seek to focus on determining their content and relationship.

Paragraph 101(C)(1) of the judgment indicates the result a delimitation in accordance with equitable principles is to achieve. The delimitation should leave each party as much as possible all those parts of the continental shelf that constitute a natural prolongation of its land territory, without encroachment of the natural prolongation of the land territory of the other party. The second requirement is the mirror image of the first. It achieves that the delimitation also leaves as much as possible of the natural prolongation to the other party. This finding implies a recognition that the continental shelves of States may overlap. Without an overlap there would not be a need for a delimitation to start with. Paragraph 101(C)(1) does not indicate how the parties are to implement these findings on natural prolongation. In general, the reasoning of the Court indicates that it considers that coastal geography is central in considering natural prolongation. For instance, the judgment observes that in the case of opposite coasts, the natural prolongations of two States will generally meet halfway between them.[51] The rejection of the appropriateness of the equidistance method in the cases of lateral coasts which are convex or concave[52] implies that the frontal projections of the coasts should be used in looking at the delimitation of overlapping natural prolongations.

In respect of the North Sea, the judgment indicates that coastal geography implies that the natural prolongations of the parties "converge, meet and intercross."[53] The fact that the judgment in this connection indicates that there is no outer limit of the continental shelf in the North Sea[54] implies that the natural prolongations of the three parties extend to the equidistance line with the United Kingdom. The question was how to determine how to leave as much as possible of the natural prolongation of each State to that State, without encroaching on the natural prolongation of the other State. In view of the Court's reasoning on opposite and lateral coasts generally, it would seem logical to expect that the Court would refer to the frontal projections of the

---

[50] Ibid.   [51] Ibid., p. 36, para. 57.   [52] Ibid., pp. 17–18, para. 8 and pp. 31–32, para. 44.
[53] Ibid., p. 49, para. 89.   [54] Ibid.

coasts as being the primary element for carrying out this exercise. The Court did not make this specific point, although the judgment's reference to avoiding encroachment on the natural prolongation of another State implicitly endorses it. The judgment in any case did not indicate how the coastal front in that connection had to be determined. The Court did discuss methods to determine coastal fronts in a different context.[55] It could be said that this discussion implicitly pronounces itself on natural prolongation. Paragraph 98 of the judgment is in particular concerned with identifying an approach to allow the application of the equidistance method. If this approach is read in the context of paragraph 101(C)(1), it is clear that the Court in paragraph 98 was setting out how it would be possible to employ the equidistance method while leaving each party as much as possible its natural prolongation without encroachment on the natural prolongation of the other party.

The judgment indicates that apart from geography, geology might be relevant in determining whether a specific area should form part of the natural prolongation of one State rather than of another State in effecting a delimitation.[56] This pronouncement is problematic. The reasoning of the Court indicates that it was talking about areas which were part of the continental shelf of both States. The Court did not clarify how this geological criterion should be taken into account if it was contradicted by the coastal geography. For instance, would it be possible to delimit the continental shelf on the basis of geological criteria if this would lead to an encroachment on the natural prolongation of the other party? Moreover, the North Sea has a complex geological history and different layers of sedimentary rock have a widely diverging extent. The judgment of the Court does not even start to address that problem.[57]

Paragraph 101(C)(1) pronounces itself on the principles and rules of international law applicable to the delimitation of the continental shelf. Delimitation concerns the establishment of a boundary between two States. This implies that the application of the rules and principles concerned should not leave an area of overlap. However, paragraph 101(C)(2) envisages the possibility that the application of paragraph (C)(1) leaves areas that overlap. Paragraph (C)(2) envisages that such areas are to be divided between the parties in agreed proportions or, failing agreement, equally, unless the parties agree on a regime of joint jurisdiction or exploitation. Although conceptually debatable, it perhaps

---

[55] *Ibid.*, p. 52, para. 98; see also text at footnote 66.   [56] *Ibid.*, p. 51, para. 95.
[57] For a further discussion of this point see also text at footnote 68.

should not come as a surprise that the Court concluded that the guidance it had provided would not easily result in agreement on the location of specific boundaries. That being said, it is probably more striking that the Court seems to have been confident that the parties would be able to agree without difficulty on an approach which would only leave areas of overlap "in certain localities."[58]

Paragraph 101(D) lists the factors which are to be taken into account in the negotiations between the parties. Paragraph 101(D) indicates that this is an open-ended list. The freedom of the parties in this respect was stressed by the Court, as it observed that there was "no legal limit to the considerations which States may take account of for the purpose of making sure that they apply equitable procedures."[59] This statement indicates that a State may also introduce considerations which are unrelated to the regime of the continental shelf. This, however, does not imply that such considerations have to be accepted by another State. Otherwise, the guidance on the applicable law provided by the Court would be without meaning. The statement in paragraph 93 rather seems to be intended to stress the freedom of the parties to negotiations to also take into account extra-legal considerations if they agree that this is conducive to reaching agreement. As the judgment also indicates, the delimitation of the continental shelf in accordance with the substantive delimitation law would require the parties to base themselves on considerations related to the regime of the continental shelf.[60]

Paragraph 101(D) does not indicate how the different factors to be taken into consideration are to be balanced up. The judgment does indicate that in most cases different factors indeed have to be balanced up, instead of the reliance on one of them to the exclusion of all others.[61]

---

[58] *Ibid.*, p. 52, para. 99. In rejecting the German claim to a just and equitable share, the Court similarly had suggested that the area of overlapping claims only involved "a disputed marginal or fringe area" (*ibid.*, p. 22, para. 20).

[59] *Ibid.*, p. 50, para. 93. The implications of paragraph 93 were also considered by Denmark and the Netherlands (see Chapter 10, text at footnotes 36 and 220).

[60] This is for instance reflected in paragraph 101(C)(1) of the operative part of the judgment. This point was clarified by the Court in its judgment in the *Continental Shelf (Libyan Arab Jamahiriya/Malta)* case of 3 June 1985, in which it observed: "Yet although there may be no legal limit to the considerations which States may take account of, this can hardly be true for a court applying equitable procedures. For a court, although there is assuredly no closed list of considerations, it is evident that only those that are pertinent to the institution of the continental shelf as it has developed within the law, and to the application of equitable principles to its delimitation, will qualify for inclusion" ([1985] ICJ Reports, p. 40, para. 48).

[61] Judgment of 20 February 1969, [1969] ICJ Reports, p. 3, at p. 50, para. 93.

The relative weight to be accorded to different considerations according to the judgment varies with the circumstances of the case.[62] These indications point to the difficulties parties may encounter in seeking to agree on the outcome of the balancing up of different factors. This difficulty is reinforced by the fact that the different factors identified by the Court are of a different nature. In balancing up the different factors, the parties would have to act in accordance with paragraph 101 (C)(1) of the judgment.

A first factor to be taken into account in the negotiations is the general configuration of the coasts of the parties and the presence of special or unusual features.[63] The general configuration of the coasts of parties is described in some detail in paragraph 91 of the judgment. The Court first of all pointed out that equity did not necessarily imply equality. In the present case there was a situation of broad equality. According to the Court the coasts of the parties were comparable in length. For the Court this implied that the parties had been given broadly equal treatment by nature. However, due to the concave nature of the German coast, application of the equidistance method would deny Germany equal or comparable treatment. In a situation like that between the parties, it would be "unacceptable that a State should enjoy continental shelf rights considerably different from those of its neighbours."[64] In looking at another aspect of the cases, the Court indicated how the general configuration of the coast of the parties might be represented by straight lines.[65] In that connection, the Court referred to the discussions of the parties on coastal fronts during the proceedings. Especially to eliminate or diminish the distortions of the equidistance method, a straight baseline between the extreme points at either end of the coast concerned could be used, or in some cases a series of such lines.[66]

As regards the second factor mentioned in paragraph 101(D)(1), the judgment does not contain any further reference to "unusual features," but in two instances refers to, respectively, a "special feature" and an

---

[62] *Ibid.*   [63] *Ibid.*, pp. 53–54, para. 101(D)(1).   [64] *Ibid.*

[65] See also text after footnote 73.

[66] Judgment of 20 February 1969, [1969] ICJ Reports, p. 3, at p. 52, para. 98. The use of the term "straight baseline" in this context seems unfortunate and confusing. The term is used to refer to a straight line between points on the low-water line, which is part of the baseline to measure the breadth of the territorial sea and other maritime zones (see e.g. Convention on the territorial sea and the contiguous zone, article 4). These straight baselines will always be seaward of the low-water line. The reasoning of the Court indicates that the straight lines to which it refers may also be landward of the low-water line.

"incidental special feature."[67] Paragraph 13 points out that a special feature, such as an islet or small protuberance, is minor in itself, but because of its location has a disproportionally distorting effect on an otherwise acceptable boundary. On the other hand, paragraph 91 indicates that the Court considers that the concavity of the German coast is an incidental special feature, the effects of which have to be abated because otherwise it could result in an unjustified difference in treatment. If the Court's reference to a special feature in paragraph 101(D)(1) was intended to refer to this second case it would seem to be superfluous as it is already covered by that paragraph's reference to the general coastal configuration of the parties. If the Court in paragraph 101(D)(1) only intended to refer to minor features, such as islets and rocks, its relevance for the delimitation between Germany and it neighbors is also far from clear. The islands in front of the coasts of the three States can hardly be considered to be islets and the reasoning of the Court in fact indicates that these islands are considered to be an integral part of the coast.[68] Only the German offshore island of Heligoland can readily be classified as an islet, but it did not have any effect on the equidistance lines between the parties, and Germany had never argued that it should be used in determining its continental shelf boundaries.[69] The promontory of Blaavands Huk on the Danish coast might be classified as a "small protuberance," but if the parties would take into account the general coastal configuration this feature would in all likelihood be disregarded under that heading.

Paragraph 101(D)(2) enjoins the parties to take the physical and geological structure and natural resources of the continental shelf into account, to the extent that information is available or readily ascertainable. The reference in paragraph (D)(2) to the physical and geological structure of the continental shelf suggests that the Court considered that this concerned two distinct terms. The judgment does not contain a clear definition of either term or make a clear distinction between them. To the contrary, the reasoning in the judgment rather suggests that these terms were used interchangeably.[70] The reasoning of the Court further indicates that when it is talking about the physical and geological shelf it

---

[67] Judgment of 20 February 1969, [1969] ICJ Reports, p. 3, at p. 20, para. 13 and p. 50, para. 91.

[68] See e.g. *ibid.*, p. 52, para. 98.

[69] Heligoland consists of two small islands, which together measure some 1.4 km². They are between 40 and 45 kilometers from the German coast.

[70] See Judgment of 20 February 1969, [1969] ICJ Reports, p. 3, at p. 51, para. 95.

is only considering the surface configuration of the shelf, not the structure of the subsoil. For instance, paragraph 95 in discussing geological aspects refers in particular to definitions of the continental shelf listed in the Yearbook of the ILC. All these definitions exclusively refer to the surface configuration of the shelf and do not make any reference to the structure of the subsoil.[71] This focus is curious as the surface configuration of the continental shelf in the southeastern part of the North Sea is unremarkable and there is nothing to suggest that this factor could be relevant to the delimitation of the shelf, as is suggested by the judgment.[72]

The judgment indicates that the presence of natural resources, which is also mentioned in paragraph 101(D)(2), is not a factor that has to be taken into account in determining the course of the boundary. Natural resources may frequently lie on both sides of a continental shelf boundary. This could lead to the wasteful exploitation of resources. The Court considered that the parties were aware of this problem and also of possible ways of solving it, such as an undertaking between the States concerned to guarantee the most efficient exploitation of the resources concerned.[73]

Paragraph 101(D)(3) of the judgment indicated that there has to be a reasonable degree of proportionality between the continental shelf areas of the coastal State resulting from delimitations and the length of its coast. Read in conjunction with paragraph 98 of the judgment it is clear that paragraph 101(D)(3) requires the comparison of the respective coast of one party to the respective coast of the other party and make a similar comparison for their continental shelf areas. A delimitation has to ensure that there is a reasonable proportionality between these two ratios. Paragraphs 98 and 101(D)(3) also indicate a number of other considerations to be taken into account in applying this proportionality test. First, the judgment indicates that the entire area of continental shelf and the entire coastline had to be taken into account. In the case of Denmark and Germany, this requires taking into account the (prospective) delimitation with the Netherlands and for Germany and the Netherlands the (prospective) delimitation with Denmark. Second, the judgment

---

[71] This concerns the definitions contained in Yearbook ILC, 1956, Vol. I, p. 131.

[72] Judgment of 20 February 1969, [1969] ICJ Reports, p. 3, at p. 51, para. 95.

[73] *Ibid.*, pp. 51–52, para. 97. This interpretation is confirmed by the introductory part of the judgment, in which the Court set out that it had received limited information on the location of natural resources and then explained that the Court: "has not thought it necessary to pursue the matter, since the question of natural resources is less one of delimitation than of eventual exploitation" (*ibid.*, p. 12, para. 17).

indicates that one or more straight lines may be used to determine the general directions of the coastlines.[74] These straight lines could also be used if the parties wished to employ the equidistance line.[75] The only criterion the judgment provides in connection with the drawing of these straight lines is that they can establish the necessary balance between States with straight, and those with markedly concave or convex, coasts.[76] This implies that in the case of a markedly concave coast, one or more straight lines could be drawn seaward of the coast and, in the case of a markedly convex coast, one or more straight lines could be drawn landward of the actual coast. Interestingly, the Court did not consider that the coastlines of Denmark and the Netherlands were markedly convex. Instead, it referred to them as moderately and roughly convex.[77] The Court did not indicate how the effect of such coasts might be abated to achieve an equitable outcome. It stands to reason that where straight coasts need not be adjusted at all, moderately concave or convex coasts should be adjusted to some extent, but to a lesser degree than markedly concave or convex coasts. In that light, the judgment can be said to imply that for Denmark and the Netherlands straight lines landward of their coasts would have to be used if the equidistance method were to be applied. This implied using lines which were less advantageous to Denmark and the Netherlands than those Germany had presented in its last round of pleading.[78] This approach finds support in paragraph 91 of the judgment.[79] Although that paragraph first refers to the fact that the coastline of *one* of the States concerned would deny that State equal or comparable treatment, it subsequently indicates that the difference in treatment is due to the combined effect of a markedly concave coast and a roughly convex coast. This indicates that all coasts would require some form of adjustment.

---

[74] *Ibid.*, p. 52, para. 98.   [75] *Ibid.*   [76] *Ibid.*

[77] See *ibid.*, at p. 17, para. 8 and p. 50, para. 91. Paragraph 8 refers to the Danish and Dutch coasts as convex to a moderate extent. Paragraph 91 refers to these coasts as roughly convex. The French text of paragraph 91 refers to the coast of Denmark and the Netherlands as "plutôt convexe" (fairly or rather convex).

[78] See Chapter 8.7. For the Netherlands, Germany had presented a line which ran parallel to the northward facing coast of the Netherlands. In the case of Denmark, Germany had presented a line which cut off a very limited part of the Danish coast. It would seem difficult to argue that this line already abated the effect of the entire Danish coast facing the North Sea.

[79] See also text at footnotes 92 and following.

9.4 The judgment's and individual judges' views on the possible location of continental shelf boundaries between the parties

As the preceding analysis indicates, various aspects of the judgment of the Court are open to different interpretations. Before looking at the parties' assessment of the judgment and their preparation for further negotiations, it is interesting to pause for a moment to determine what the judgment actually has to tell as far as the location of the boundaries between the parties is concerned.

As will be apparent from the preceding analysis, the basic task facing the parties was to ensure that a delimitation was to leave, as much as possible, to each party the parts of the continental shelf which constituted a natural prolongation of its land territory. The judgment provides little guidance on the Court's views on the location of these areas. The most concrete indication in this respect is contained in paragraph 89 of the judgment. As was set out above, this paragraph read in conjunction with other parts of the judgment indicates that the natural prolongations of the three States extend to the median line with the United Kingdom and that a natural prolongation is most directly linked to the coast to which it is directly in front.

The judgment also pronounces itself on the extent of the continental shelf areas of the parties resulting from the bilateral delimitations. Paragraph 91 of the judgment concludes that the coasts of Denmark, Germany and the Netherlands are broadly equal. For the Court the application of the equidistance method, which gave Germany a much smaller area of continental shelf than its neighbors, resulted in an unacceptable difference in treatment. Paragraph 91 also indicates the kind of treatment Germany should get, namely treatment "equal or comparable to" that given to Denmark and the Netherlands. A "considerably different" treatment would be unacceptable. It is of course difficult to put an exact figure on these terms. The only thing which can be concluded with certainty is that the Court considered that application of the equidistance method led to an unacceptable difference in treatment. According to the German Memorials, the equidistance method gave Germany an area of approximately 23,600 km^2, Denmark an area of about 61,500 km^2 and the Netherlands an area of about 61,800 km^2.[80] Calculations by the Dutch Hydrographic Department from July 1969 put

---

[80] GM, para. 78.

these figures at respectively 52,000, 23,400 and 61,600 km^2 for Denmark, Germany and the Netherlands.[81] During the pleadings, Germany had claimed an area of 36,700 km^2. 7,600 km^2 was part of the Danish equidistance area and 5,500 km^2 of the Dutch equidistance area.[82] If the figures of the German Memorials would be applied, the German approach would leave Denmark with some 53,900 km^2 and the Netherlands with some 56,300 km^2 and if the figures of the Dutch Hydrographic Department would be used, this would be respectively 44,400 and 56,100 km^2. These figures imply that under the German claim presented during the pleadings, the German area would be respectively 32 or 17.5 percent less than the Danish share and 35 or 34.5 percent less than the Dutch share. It seems reasonable to conclude that, with the possible exception of the lower figure in relation to Denmark, this still would not result in a treatment "equal or comparable" to the other two States.

Another indication on the kind of outcome the Court envisaged is contained in paragraph 98 of the judgment. The Court suggested that the parties could use straight lines instead of their actual baselines if they wished to employ the equidistance method. As was discussed previously, the reasoning of the Court implies that for Germany such a straight line is seaward of its baseline. The only feasible option is a line between Borkum and Sylt.[83] For Denmark and the Netherlands different options to represent their coasts by straight lines might be employed. As Figure 8.2 indicates, even straight lines which discount a limited part of the Danish and Dutch coast would result in a point of contact between the continental shelf of Germany and the median line with the United Kingdom. As was set out above, paragraph 98 leaves room to argue that straight lines along the Danish and Dutch coasts should result in discounting a larger part of their actual coast lines. In that case, the equidistance approach proposed by the Court would result in a longer boundary between Germany and the United Kingdom.

Apart from the judgment, a number of the individual opinions of the judges also consider the possible location of the continental shelf

---

[81] See note; North Sea Continental Shelf dated 11 August 1969 [NNA/31]. These figures excluded the territorial sea and the Skagerrak.

[82] See OP, p. 210.

[83] Paragraph 98 of the judgment indicated that a straight line between the extreme points of the coast would be an appropriate straight line. Alternatively more than one straight line could be employed. In the case of Germany the latter approach would not make it possible to draw lines significantly seaward from the actual coast.

boundaries between Germany and its two neighbors. The president of the Court, Bustamante y Rivero, in his separate opinion concurred with the judgment's view that the natural prolongations of the parties converged in the center of the North Sea in the area of the median line between the United Kingdom and the continent. Bustamante y Rivero also considered that these natural prolongations projected frontally.[84] Because the three natural prolongations overlapped, it would not be possible for the German boundaries to run parallel, but the boundaries would progressively near each other, resulting in a triangular or trapezoidal area for Germany.[85] Bustamante y Rivero offered two specific solutions to the delimitation. The two boundaries of Germany with its neighbors could either extend to a point or segment on the median line between the United Kingdom and the continent or to the median line between the United Kingdom and Germany, leaving a small corridor belonging to Denmark and the Netherlands.[86]

Judge Jessup in his separate opinion considered certain questions related to the sector claim that Germany had advanced in its second round of oral pleadings.[87] Jessup's opinion indicates that he considered that Germany might also claim areas located beyond this sector claim as he mentions that a German triangular sector could not extend beyond the median line between Germany and the United Kingdom and the median line between Germany and Norway. The latter line is to the north of the sector claim Germany presented at the hearings. The views of Jessup implied that a Danish-Dutch corridor would remain between the continental shelf of Germany and the United Kingdom. The separate opinion of Jessup is also interesting for another reason. The opinion focused on the location of areas where exploitation had already produced promising results.[88] Jessup indicated that the Danish activities had not resulted in a Danish title to the area concerned, but he held that the private interests which were in question had to be taken into account.[89] Jessup's opinion leaves room to argue that these interests should either be taken into account in determining the course of the boundary or

---

[84] Judgment of 20 February 1969, Sep. op. Bustamante y Rivero, [1969] ICJ Reports, p. 57, at p. 61, para. 6.

[85] Ibid. [86] Ibid., p. 64, para. 6.

[87] Ibid., Sep. op. Jessup, [1969] ICJ Reports, p. 66, at p. 81.

[88] Ibid. The opinion refers to exploitation but actually is concerned with the area in which the Danish consortium had carried out successful exploratory drilling.

[89] Ibid., pp. 79–81.

alternatively should be respected by Germany if the area concerned would be located on the German side of a continental shelf boundary. Jessup's discussion of this point was not based on all relevant information. The parties had not informed the Court of the agreement on activities on the continental shelf they had reached in August 1966. In implementing that arrangement Germany had reserved its rights in respect of the activities Jessup considered in his opinion both in respect of Denmark and the Danish concessionaire.[90] If that information would have been available to Jessup, it stands to reason that he would have reached a different conclusion.[91]

Judge Ammoun, in his separate opinion, indicated that he considered that the equidistance-special circumstances rule had to be applied.[92] Because the German coast constituted a special circumstance, it should be replaced by the straight line between Borkum and Sylt. For Denmark and the Netherlands their actual coast lines could be employed. Judge Ammoun's approach resulted in a triangular area extending from the Borkum-Sylt line, which remained well short from the median line between the United Kingdom and the continent.[93] It should be noted that judge Ammoun's solution is not in accordance with the approach set out in the judgment if the parties would wish to apply the equidistance method as judge Ammoun himself acknowledged.[94] As was argued above, the judgment's approach should lead to a larger area for Germany, which would likely extend up to, or be close to, the equidistance line with the United Kingdom.

Judge Morelli had voted against the judgment of the Court because he had a different view on the applicable law. However, Morelli did agree with the Court that the equidistance method led to a gravely inequitable result. According to Morelli, this was evident from the remarkable disproportion between the area of continental shelf of each State and the length of their respective coastlines. This would even be the case if the actual German coastline would be substituted by the straight line

---

[90] Note verbale dated 16 August 1966 [AA/45].

[91] It can furthermore be observed that the issue of the possible rights of the concessionaire had not been put in front of the Court.

[92] Judgment of 20 February 1969, Sep. op. Ammoun, [1969] ICJ Reports, p. 100, at p. 148, para. 52.

[93] *Ibid.*, pp. 151–152, paras 55–56. A map illustrating judge Ammoun's proposed approach is included at *ibid.*, p. 153. For a further discussion of judge Ammoun's proposed methodology see Chapter 9, text at footnote 366.

[94] *Ibid.*, p. 152, para. 56.

between Borkum and Sylt.[95] Morelli did not further specify what would constitute an equitable result, but his unequivocal assertion that the solution proposed by judge Ammoun still did not lead to an equitable solution is telling enough. Judge Morelli's opinion indicates that if the parties would return to the Court if they could not agree on boundaries themselves, Denmark and the Netherlands would not be faced by a 11 to 6 majority which supported a significantly larger German share but a majority of 12 to 5.

Most of the above opinions have in common that they consider that a just outcome of the negotiations between the parties would accord Germany an area of continental shelf which would extend in a roughly triangular or trapezoidal shape to the equidistance line between the United Kingdom and the continent, or at least close to that line. Most of the opinions also indicate that the delimitation could result in more advantageous boundaries for Germany than it had claimed during the oral pleadings, something which is also suggested by the judgment itself.

---

[95] *Ibid.*, Dis. op. Morelli, [1969] ICJ Reports, p. 197 at, p. 209, para. 15.

10

The negotiations following the judgment

10.1 Introduction

The judgment of the Court changed the legal framework for the negotiations between Denmark, Germany and the Netherlands dramatically. Prior to the judgment, Denmark and the Netherlands could stick to their interpretation of article 6 of the Convention on the continental shelf and argue that from a legal perspective there was no scope for adjusting the equidistance line. Now that they had been told by the Court that the equidistance principle was not applicable, they would have to come up with arguments justifying as small as possible a deviation from the equidistance line. They had also been told by the Court that they should negotiate seriously. This required them to engage with the arguments set out in the judgment. At the same time, the less than clear implications of the judgment left them sufficient room for maneuver. In addition, as the further negotiations would show, the political context would continue to impact on the negotiations at least to the same extent as the redefined legal framework. Denmark and the Netherlands would also have to consider the nature of their further collaboration. Before the judgment, they were united by their insistence on the equidistance method. Now that they would have to make concessions to Germany there was a larger risk of disagreement. A specific interpretation of the judgment might work to the advantage of one party but against the other. The German sector claim provides an apt illustration. Some 58 percent of the additional German area was on the Danish side of the equidistance lines as compared to only 42 percent on the Dutch side.

Germany would have to figure out how to use the judgment to its advantage. Although there could be no doubt that the judgment implied a significantly larger share for Germany than it would have gotten under the equidistance principle, it left a large margin of appreciation. This also indicated that Germany could not sit back and let the judgment work for it, so to speak. Getting a positive result would likely require a commitment to back up the legal arguments with political resolve. Another part of Germany's strategy might be to seek to play Denmark and the Netherlands apart and win over one of the other parties to the German side.

This chapter first provides an overview of the preparations of the parties to the new negotiations and then discusses these negotiations.

10.2   Germany's assessment of the judgment and preparations for further negotiations

It should not come as a surprise that the Court's judgment was positively received in Germany.[1] Apart from the judgment's favorable decision on the applicable law, the political aspects of the outcome were welcomed. The judgment had significantly improved Germany's chances of getting an appropriate share of the continental shelf without burdening the relationship with Denmark and the Netherlands.[2] The proceedings had not led to polemics in Denmark and the Netherlands and although public opinion in both countries was disappointed about the outcome, reactions did not show emotional overtones. It was, moreover, stressed that the case had been conducted in a businesslike manner. The fear that the composition of the ICJ might negatively impact on Germany had been proven to be unjustified. The dissenting opinions of the judges voting against the judgment did not reveal any political motives.[3] The continued sensitivity of Germany's relationship with its neighbors was emphasized in the concluding remark in the briefing for the Foreign Minister. The fact that Germany, although it was not a member of the United Nations, had been able to defend its legal views and its interests in such a successful manner against two small neighbors, who had been occupied during the Second World War, had to be judged positively from a political perspective. This gave reason to also use good legal positions in a moderate but staunch way in the future, where political considerations in other respects imposed considerable restraints.[4]

---

[1] See e.g. telex from the GEH to the FO dated 20 February 1969 [AA/64]; letter of U. Scheuner to H. Blomeyer of the FO dated 21 February 1969 [AA/64]; letter of H. Blomeyer of the FO to E. Menzel dated 28 February 1969 [AA/64]; note for the FM dated 10 March 1969 [AA/73], p. 1.

[2] Note for the FM dated 10 March 1969 [AA/73], p. 1.

[3] *Ibid.*, p. 2. For an account of the German reservations to submit the dispute to the Court see Chapter 6, text at footnote 3.

[4] *Ibid.*, p. 3.

Although this assessment stressed the positive political aspects of the outcome of the cases, it also indicated that Germany might be unwilling to stand by its interpretation of the judgment of the Court if there would be a risk that the further negotiations with Denmark and the Netherlands would lead to political complications.

The preliminary German evaluation of the situation after the judgment indicated that there were certain doubts that Germany would be able to profit fully from its victory in The Hague. Not surprisingly, it was recognized that many points of the judgment, including the key passage of the dispositif referring to leaving each of the parties as much as possible of its natural prolongation, seemed to be open to different interpretations.[5] Blomeyer, who was positive about the outcome of the case, still did not want to make any prediction in view of the tenacity of Germany's negotiating partners.[6] Another evaluation of the judgment concluded that the Danish discoveries of hydrocarbons, just to the north of the equidistance line between Denmark and Germany, possibly constituted the major difficulty for the upcoming negotiations.[7] It was almost inevitable that these discoveries would be located in the area that would be claimed by Germany on the basis of the judgment of the Court.

The German position for the further negotiations was determined in consultations between the Foreign Office, other ministries and the four coastal states, on whose behalf the state of Lower Saxony acted. At a first meeting in early March, it was agreed that, as a minimum, the negotiations should give Germany a 15-nautical-mile boundary with the United Kingdom.[8] Prior to the meeting a 10-nautical-mile boundary had been suggested as this would leave a German corridor of sufficient width to carry out oil and gas activities without immediately reaching the boundaries with neighboring States.[9] This approach was based on a wedge-shaped area of the continental shelf of Germany.[10]

An important consideration in determining a German opening bid for the next round of negotiations was that Germany should acquire as much area as possible to the west and northwest of its equidistance

---

[5] Note (draft) dated 5 March 1969 [AA/73], p. 1; see also note dated 3 March 1969 [B102/260033], p. 1.
[6] Letter of H. Blomeyer of the FO to E. Menzel dated 28 February 1969 [AA/64].
[7] Note dated 26 February 1969 [AA/84; AA/64], p. 3.
[8] Circular letter of the Minister of Defense dated 10 March 1969 [B102/260033], p. 2.
[9] Note dated 26 February 1969 [AA/84; AA/64], p. 2.   [10] Ibid.

area.[11] According to the North Sea consortium only that area would eventually give further chances of finding commercially interesting deposits. Only after this area would have been secured, would it be possible to decide if exploratory activities on the German shelf could be continued.[12] Specific proposals focused on the determination of the natural prolongations of Germany, Denmark and the Netherlands and their overlap.[13] In general, the proposals employ a straight line coastal front for each of the three States and define the natural prolongation as the frontal projection of these straight lines. For Germany most proposals used the straight line between Borkum and Sylt. For Denmark and the Netherlands different lines were proposed. For Denmark this ranged from a line between the terminus of the land boundary with Germany and the northeastern tip of Jutland to a line along the Danish North Sea coast running approximately north to south. For the Netherlands this ranged from a line between the two termini of the land boundaries of the Netherlands with Belgium and Germany and a line running along the north-facing coast of the Netherlands.

The Foreign Office and the Ministry for the Economy considered that a number of the proposals, which had been submitted by the Minister for the Economy and Transport of Lower Saxony, were problematical.[14] It was, among others, considered that it was difficult to reconcile those proposals with the tenor of the ICJ's judgment.[15] Some of the proposals actually gave Germany a larger share of the shelf than Denmark and the Netherlands. The Foreign Office concluded that it did seem opportune to involve a representative of the states in the German delegation, although the Ministry for the Economy had argued against this. Such a representative could be expected to report objectively on the negotiations, which could prevent or lessen possible differences of opinion.[16] The German

---

[11] Note dated 3 March 1969 [B102/260033], p. 2; Circular letter of the Minister for the Economy and Transport of Lower Saxony dated 24 March 1969 [AA/73], p. 5; letter of Preussag AG to the FO dated 8 April 1969 [AA/73], p. 7.

[12] Ibid.

[13] For the positions of different agencies see circular letter of the Minister of Defense dated 10 March 1969 [B102/260033]; circular letter of the Minister for the Economy dated 12 March 1969 [AA/73]; letter of the Ministry of Transport to the FO dated 18 March 1969 [AA/73]; note dated 24 March 1969 [AA/73]; circular letter of the Minister for the Economy and Transport of Lower Saxony dated 24 March 1969 [AA/73]; circular letter of the Minister of Transport dated 25 March 1969 [AA/73].

[14] Note (undated) [AA/73]; note dated 8 May 1969 [B102/260034].

[15] Ibid., p. 3; note (undated) [AA/73], p. 3.   [16] Ibid., p. 4.

delegation included a representative from Lower Saxony from the outset.[17]

The proposals for a German opening bid were reviewed at an inter-departmental meeting.[18] The meeting agreed that the best approach was contained in a proposal of the Minister for the Economy, which for the Netherlands employed the straight line between the termini of its two land boundaries and for Denmark a straight line between the endpoint of the boundary between Germany and Denmark and Cape Hanstholm. The latter line runs almost south to north. The choice for Cape Hanstholm was justified by the fact that it marked the eastern limit of the North Sea in an 1882 Agreement on the North Sea and represented the point at which the coast changed direction. To the south of Cape Hanstholm the coast faced west and to the north it faced west north-west.[19] If the area of overlap of the natural prolongations of these coastal façades would be divided equally, Germany would get about 48,000 km^2 or about a third of the continental shelf to be divided between the three countries. It was considered that the choice for the endpoints of the Danish and Dutch coastal façades was based on the geography of the North Sea. In short, the proposal was presentable and could be justified and as such constituted a useful German starting position.[20] Interestingly, in an internal review of the proposals, the Foreign Office had criticized the Dutch coastal façade, which was now agreed upon by the meeting, because it was not in accordance with the judgment as it did not give the Netherlands access to the center of the North Sea.[21] The use for Denmark of a straight line running to the northeastern tip of Jutland, which was not adopted in the interdepartmental meeting, had been rejected for similar reasons.[22] At the same time, the review had acknowl-edged that for tactical reasons Germany would have to make a high opening bid.[23]

Germany also reconsidered how to deal with Danish or Dutch activ-ities on the North Sea shelf. The Ministry for the Economy and the Ministry for the Economy and Transport of Lower Saxony had urged the Foreign Office that the practice of reserving Germany's rights should be

[17] See also Von Schenk, "Die vertragliche Abgrenzung," p. 378, footnote 25.
[18] Report dated 12 May 1969 [AA/73]. The meeting was attended by representatives of the FO, the ME, the Ministry of the Interior and the Ministry for the Economy and Transport of Lower Saxony (ibid., p. 1).
[19] Ibid., p. 2.   [20] Ibid.
[21] Note dated 24 March 1969 [AA/48], p. 2; note (undated) [AA/73], p. 3.   [22] Ibid.
[23] Ibid., p. 4.

continued.[24] The Foreign Office initially considered whether Germany should accept drilling in two specific locations at all,[25] but concluded that it seemed to be safe to wait with a written reaction until the negotiations would have been resumed.[26] Subsequently, the Foreign Office presented a note to the Dutch Embassy to safeguard Germany's rights because it wanted to avoid that the Netherlands might get the impression that these locations were not in the area Germany considered to be in dispute.[27]

10.3 Denmark's assessment of the judgment and preparations for further negotiations

Immediately after the judgment, Denmark and the Netherlands had a preliminary exchange of views on its implications.[28] The discussions indicated that the exact implications of the judgment were not clear and that the Court failed to give guidelines on how the parties should implement its findings.[29] Apart from certain technical and procedural points, the meeting did not reach any conclusions.[30] It was agreed that Denmark and the Netherlands would each further look into the implications of the judgment and meet again at a later stage.[31]

In a statement to the press, the Danish Minister of Foreign Affairs admitted that the judgment naturally was a disappointment.[32] The judgment went against what Denmark considered to be right and it could easily have a negative impact on the efforts, which had been made in the framework of the United Nations since the 1950s, to further develop an international legal order. This last aspect no doubt affected small and new States.[33] At the same time, the minister observed that it was

---

[24] Note dated 24 March 1969 [AA/48], p. 1; letter of the Minister for the Economy to the FO dated 28 March 1969 [AA/48].

[25] Note dated 24 March 1969 [AA/48], p. 3.   [26] Note dated 23 April 1969 [AA/48].

[27] See circular letter of the FO dated 5 May 1969 [AA/48], p. 2. The note as presented to the Netherlands likely is identical to the draft note dated 25 April 1969 attached to the note dated 23 April 1969 [AA/48] (see also Protocol on the continental shelf negotiations at the Foreign Office in Bonn on 18/19 June 1969 dated 21 June 1969 [AA/73] (German Protocol June 1969 [AA/73]), p. 6).

[28] Summary record of the Danish-Dutch meeting held on February 21, 1969 dated 26 February 1969 [DNA/39].

[29] See e.g. ibid., pp. 5–6.   [30] Ibid., p. 1.   [31] Ibid., p. 11.

[32] Statement of the MiFA concerning the shelf case dated 11 March 1969 [DNA/147], p. 2.

[33] Ibid.

important to emphasize that the judgment did not lay down precise rules, which the parties were required to follow. The parties were required to enter into negotiations, but were free to decide on the considerations they would like to take into account in concluding an agreement. The judgment required that the outcome had to be just and equitable for both parties.[34] As the briefing for the minister in connection with his statement observed, Denmark might be able to use this latter point to its advantage.[35] The briefing also submitted that the possibility to introduce a range of further considerations implied that the negotiations would now be more in the nature of traditional foreign policy negotiations, in which the overall political relationship with Germany would also be a factor.[36] Interestingly, Waldock's analysis of the judgment submitted to ambassador Sandager Jeppesen, who was to head the Danish negotiating team, considered that the judgment curtailed the liberty of the parties to advance specific arguments. Waldock recognized the contractual freedom of the parties, but for him the essential question was if there was "a *legal* limit to the considerations which a Party may insist upon in good faith to the point of bringing the negotiations to a deadlock?"[37] Waldock concluded that if the parties were to resubmit the matter to the Court, it would be likely to regard itself competent to decide the criteria, which were "*legally* relevant within the meaning of its Judgment."[38]

The judgment of the Court was the subject of detailed analysis. Next to its own analysis, the Ministry of Foreign Affairs received written observations from Waldock, Foighel, Philip and Lauterpacht,[39] who had all been closely involved in the proceedings. Geological aspects of the case were discussed with the Danish Geological Survey and Professor Sorgenfrei, who acted as an advisor to the Danish concessionaire, while geographical aspects were considered by the Danish Hydrographic Institute.[40] The written opinions on the judgment provide a detailed review of many of its aspects, as is illustrated by Waldock's argument on the contractual freedom of the parties, which was discussed in the

---

[34] *Ibid.*, p. 3.

[35] Note for the MiFA concerning the delimitation of the continental shelf in the North Sea dated 10 March 1969 [DNA/39], p. 3.

[36] *Ibid.*, p. 2.

[37] H. Waldock, Note on the law laid down by the International Court of Justice in the North Sea Cases dated 19 March 1969 [DNA/82], para. 24 (emphasis in the original).

[38] *Ibid.* (emphasis in the original).

[39] Note; Shelf case. Position Report dated 3 June 1969 [DNA/39], p. 2.   [40] *Ibid.*

previous paragraph. The review of documents related to the preparation of the negotiations with Germany indicates that the focus was not on such legal niceties. This is aptly illustrated by a statement of Sandager Jeppesen to the Danish continental shelf committee. The Ministry of Foreign Affairs was interested in hearing further views on the interpretation of the judgment, but was not interested in losing itself in a *post mortem* of the judgment. The focus should be on developing a reasonable framework for further work on the matter.[41]

Denmark had two fundamental interests in the further negotiations with Germany. Denmark should lose as little as possible of its equidistance area and should secure the four locations where drilling by the Danish concessionaire had shown the presence of oil and gas.[42] As Figure 10.1 indicates, one of these locations was squarely inside the sector Germany had claimed during the oral pleadings and two were just to the north of that sector. For reasons of domestic policy it would be extremely difficult for Denmark to give up this area in whole or in part. It was assumed that Germany would be sympathetic to this point of view, which was also supported by Jessup's separate opinion.[43] The geology of the continental shelf could also be used to support Denmark's case. About 25 percent of the Danish shelf consisted of the Fyn-Grindsted High, which by all accounts did not contain any oil or gas deposits.[44] It would be equitable to take this fact into account in assessing Denmark's share of the continental shelf.[45] In addition, the promising locations from a geological perspective seemed a separate oil province, which was similar to the northern part of the North Sea, but different from its southern part.[46] The Danish interests, as formulated above, implied that there was little chance of a conflict of interest with the Danish concessionaire and contacts had pointed out that no such conflict indeed existed.[47] Nonetheless, the Ministry of Foreign Affairs consulted with the

---

[41] Report on the meeting of the advisory committee for the case with Germany on the continental shelf (Continental shelf committee) held on Thursday 6 March 1969 (KSU/MR.11) [DNA/142], p. 2.

[42] See e.g. note; Some main points of the shelf case problem dated 3 June 1969 [DNA/39], p. 3.

[43] Points of conversation to be used in the discussions with the German head of delegation dated 22 April 1969 [DNA/143], p. 2. For a review of Jessup's opinion in this respect see Chapter 9, text at footnotes 88 and following.

[44] See Figure 10.1 for the location of the Fyn-Grindsted High.

[45] Note; Some main points of the shelf case problem dated 3 June 1969 [DNA/39], p. 3.

[46] *Ibid.*, pp. 3–4.

[47] Note; Shelf case; Report on positions dated 3 June 1969 [DNA/39], p. 3.

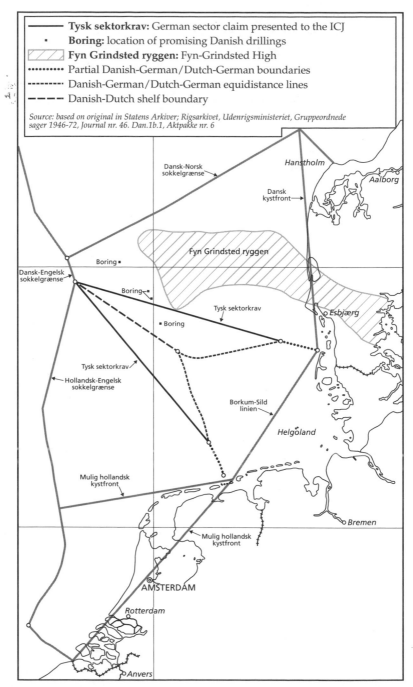

Figure 10.1   Location of the German sector claim in relation to promising Danish drillings and the Fyn-Grindsted High

Ministries of Public Works and Justice to assess whether Denmark's freedom to contract might be limited by the nature of the concessionaire's concession, which applied to the whole Danish shelf, and a number of reports on this matter were prepared by, among others, the legal adviser of the Ministry of Foreign Affairs.[48] The Danish State would be bound by the concession during new negotiations, but at the same time the public interest might be more important than the rights of the concessionaire. A further delimitation of these respective rights would only be possible in the light of a specific negotiating result.[49]

The Danish authorities also considered the impact of coastal geography. Like Germany and the Netherlands, they focused on straight lines to represent the coasts of the parties. For itself, Denmark opted for the straight line which had already been used by Germany in the oral pleadings.[50] This line was the most advantageous for Denmark if it would be used for equidistance calculations. However, it was appreciated that if this line would be used in a proportionality test between coastal lengths and continental shelf areas it might be disadvantageous to Denmark. The interests of Denmark and the Netherlands might not run parallel in this respect as the Netherlands might argue that its whole coast was relevant in this connection.[51] Germany could be expected to insist on the line between Borkum and Sylt. Denmark could argue that these islands were in front of the coast of respectively Denmark and the Netherlands. As a consequence, it could be submitted that Germany should use a shorter straight line.[52]

Denmark did not yet develop specific proposals for its boundary with Germany. As the preceding analysis points out, Denmark could be expected to rely on a mix of arguments to try to secure its interests. Denmark did not entertain the possibility that Germany might advance a significantly larger claim than it had advanced during the oral proceedings. Denmark possibly would only be able to secure the promising locations of its concessionaire if Germany would get an even smaller share of the shelf.[53]

---

[48] *Ibid.*   [49] *Ibid.*, p. 4.
[50] Note; Some main points of the shelf case problem dated 3 June 1969 [DNA/39], p. 2.
[51] *Ibid.*, pp. 2–3.   [52] *Ibid.*, p. 2.
[53] See also Points of conversation to be used in the discussions with the German head of delegation dated 22 April 1969 [DNA/143], p. 1.

10.4   The Dutch assessment of the judgment and preparations for further negotiations

Luns, the Dutch Minister of Foreign Affairs, regretted what he qualified as the disappointing judgment of the ICJ.[54] The Dutch Government regretted that the Court had only been willing to indicate some vague guidelines, without further clarification, as factors to be taken into consideration.[55] A first assessment of Maas Geesteranus, the deputy legal adviser of the Ministry of Foreign Affairs, on the outcome was more dispassionate. He observed that the Court had rejected both the German and Danish and Dutch positions. The judgment provided the parties some guidelines for the further negotiations. It looked like these guidelines would lead to a more generous part of the continental shelf for Germany than the equidistance method.[56]

There is relatively little information on the analysis of the Court's judgment prior to the negotiations with Germany in the archives of the Dutch Ministry of Foreign Affairs. One memorandum prepared by Maas Geesteranus considered the concept of natural prolongation from a geomorphological and geological perspective.[57] The interest in these aspects of natural prolongation, which had a secondary importance in the judgment of the Court, is readily explained by the fact that natural prolongation defined in terms of coastal geography had little to offer to the Netherlands. Geomorphology and geology might provide arguments to undermine Germany's reliance of coastal geography. To prepare his memorandum Maas Geesteranus had consulted with the Dutch Geological Service and the BIPM, a subsidiary of Shell, which possessed much more detailed information than the Geological Service.[58] The BIPM had indicated that it would only assist the Netherlands on the basis of strict confidentiality to prevent damage to the company's German interests.[59] Only if it would be required to refute German geological maps, would the BIPM consider putting one of its maps at the

---

[54] Minutes of the Kingdom Council of Ministers dated 21 February 1969 [AA/32], item 4.d. According to Jansen, Riphagen, the Dutch agent, was touched "even on the emotional plane" that he had not been able to convince the Court that "his position was the right one" (P. E. L. Jansen, *Willem Riphagen 1919–1994* (Den Haag: T.M.C. Asser Instituut, 1998), p. 7).

[55] National Budget for the year 1970; Chapter V; Foreign Affairs; Explanatory Memorandum no. 2 (Parliamentary paper 10300 V (1969–1970), no. 2), p. 32.

[56] Note 20/69 dated 20 February 1969 [NNA/31].

[57] Note 30/69 dated 11 March 1969 [NNA/31].   [58] *Ibid.*, pp. 2–3.   [59] *Ibid.*, p. 2.

disposal of the Netherlands.[60] The available information indicated that geology did not point in one direction. Strata of different geological eras were linked in different ways to the continent and pointed to different possible boundaries.[61] Maas Geesteranus's preliminary conclusion was that the geology of the shelf could not be used to the advantage of the Netherlands. It would be better to wait and see what Germany knew about the geology of the shelf.[62] A complication of geological arguments was that it led to different results in the case of Germany and Denmark. Arguments favoring the Netherlands almost certainly would work to the detriment of Denmark.[63] The former channel of the Ems River on the bed of the North Sea possibly provided a further argument to the Netherlands. The channel generally ran in a northern direction and lay to the east of the partial boundary with Germany. A choice for this channel could be justified by the fact that it constituted a logical continuation of the land boundary, which has been located in the Ems for centuries. Maas Geesteranus considered that reliance on this argument would only be feasible to counter a German claim based on geology.[64] The Dutch Geological Service provided an assessment of the prospectivity of the part of the Dutch equidistance area that might be claimed by Germany. The prospectivity of the area west of the meridian of 5° E was much better than that of the nearshore area east of this meridian. The Geological Service considered that the Netherlands should only give up half, or if need be all, of the latter area.[65] Maas Geesteranus held that it would not be possible to completely comply with the proposal of the Geological Service. It could only be justified on the basis of structural geological arguments, which the Service itself had concluded to be unattractive for the Netherlands.[66]

A possible Dutch opening bid to Germany was discussed between the Ministry of Foreign Affairs and the Ministry of Economic Affairs.[67] The latter was interested in an early settlement to allow exploratory activities on the northern part of the Dutch shelf to continue and proposed that the Netherlands would take the initiative in the negotiations. The fact that the Netherlands at present probably knew more about the prospectivity of the area also justified making an early start with the negotiations.[68] The Ministry of Economic Affairs could agree to a

---

[60] *Ibid.*, p. 3.   [61] *Ibid.*, pp. 3–4.   [62] *Ibid.*, p. 5.   [63] *Ibid.*   [64] *Ibid.*, p. 4.
[65] Note 40/69 dated 24 March 1969 [NNA/33], p. 1. These two areas are identified in Figure 10.2.
[66] *Ibid.*, pp. 1–2.   [67] Note 49/69 dated 3 April 1969 [NNA/31].
[68] Note 30/69 dated 11 March 1969 [NNA/31], p. 5.

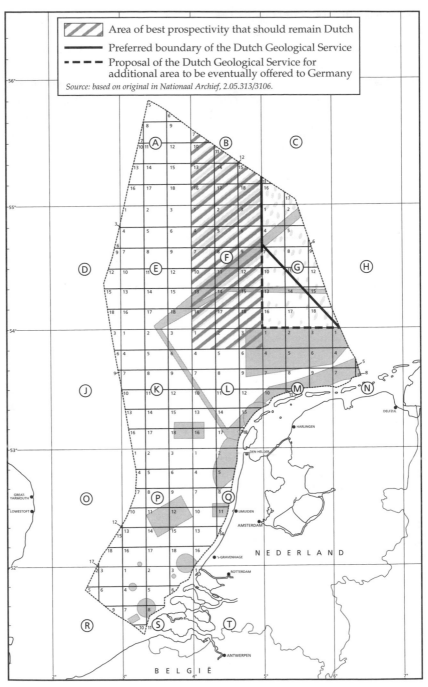

Figure 10.2   Preferred outcome of the Dutch Geological Service based on prospectivity of the Dutch equidistance area

proposal on a boundary which would run from the endpoint of the partial boundary between Germany and the Netherlands and the point of intersection of a German-Dutch coastal front bisector and a Danish-German coastal front bisector.[69] This proposal resulted in a boundary line that would only give a small part of the area with the best prospectivity to Germany.[70] The bisector lines were prepared on the basis of the coastal fronts Germany had presented during the hearings in front of the Court. These were the most advantageous coastal fronts for Denmark and the Netherlands one could think of and Germany was likely to come up with a different view in the upcoming negotiations. This apparently was also the view of Maas Geesteranus, who observed that this opening bid represented the area which the Netherlands could, at least, be expected to have to give up to Germany in the light of the judgment of the Court.[71] He considered it more likely that Germany would aim for a boundary with the Netherlands starting from the terminus of the partial boundary and ending on the trijunction point between Denmark, the Netherlands and the United Kingdom.[72] Maas Geesteranus had a number of reasons to assume that Germany would be looking for that outcome. The German agent had indicated this boundary during the oral pleadings before the Court.[73] Second, Maas Geesteranus considered that the Court in paragraph 5 of its judgment and in particular judge Jessup in his separate opinion had more or less established that this was the maximum Germany could claim.[74] Actually, paragraph 5 mistakenly referred to this line as representing the boundary which Germany had wanted to obtain during the negotiations with the Netherlands and Jessup's opinion did not refer to this line and left open other outcomes.[75] Other parts of the judgment and some of the other separate opinions also indicate that Germany was not prevented from making a claim to areas to the south of the line identified by Maas Geesteranus.[76] Finally, a

---

[69] Note 49/69 dated 3 April 1969 [NNA/31].

[70] This proposal is depicted in Figure 10.3. For a comparison of this proposal to the equidistance line see Figure 10.4.

[71] Note 49/69 dated 3 April 1969 [NNA/31].

[72] *Ibid.*; Note 34/69 dated 17 March 1969 [NNA/31], p. 4. This line is depicted in Figure 10.3.

[73] *Ibid.*, p. 4. The German agent had referred to this line in discussing what constituted the German claim (see OP, p. 180; see also *ibid.*, p. 182, figure 6).

[74] Note 34/69 dated 17 March 1969 [NNA/31], p. 4.

[75] For the former point see also Chapter 8, text at footnote 360 and for the latter point see also Chapter 9, text at footnote 87 and following.

[76] See further Chapter 9.4.

manager from British Petroleum had informed Maas Geesteranus in early March that he had been told by a contact in the German branch of British Petroleum that the German authorities were intent on using this line.[77] There is no indication that this was the case and at a meeting on 4 March 1969 the German authorities had already agreed that, at a minimum, Germany should have a 15-nautical-mile boundary with the United Kingdom,[78] and not a single point of contact, which would be the consequence of maintaining the German claim before the Court. The one-sided reading of the judgment and the incorrect information on Germany's current position led to an overly optimistic assessment of Germany's opening bid. Maas Geesteranus even considered that Germany might be modest enough to use the claim it had presented as its opening bid for the further negotiations. In that case, the Netherlands might even get a better result in the western part of the shelf, possibly by giving up some additional area further to the east.[79] However, Maas Geesteranus also did not exclude that Germany for tactical reasons might present a rather more expansive claim.[80]

The judgment of the Court led the Netherlands to reconsider its leasing policy for the continental shelf. Just before the judgment had been issued, the Netherlands had already made one adjustment. Oil companies which had informed the Ministry of Economic Affairs that they were planning to drill in one of their license areas had always been informed that there was no objection. In consultation with the Ministry of Foreign Affairs, it was decided to now also point out to British Petroleum, which planned to start drilling on 15 February 1969, that "perhaps it is needless to say that there is an ongoing dispute between the Neth.-Germany and Denmark-Germany concerning the [continental shelf] boundary, on which the ICJ has not yet rendered a decision."[81] The specific formulation was probably carefully chosen to convey that the Netherlands considered that this dispute was common knowledge and that companies had been acting on their own risk all along. It was furthermore agreed between the Ministry of Foreign Affairs

[77] Note 34/69 dated 17 March 1969 [NNA/31], p. 4.   [78] See text at footnote 8.
[79] Note 34/69 dated 17 March 1969 [NNA/31], p. 4.   [80] Ibid., p. 3.
[81] Hand-written note on send message in cipher (ref. no. 777) of the MFA to the DEB dated 4 February 1969 [NNA/31]. Translation by the author. The original text reads "Wellicht ten overvloede wijs ik er op dat over de CP-grens een geschil Ned.-Duitsland en Denemarken-Duitsl. gaande is, waarover het Int. Gerechtshof nog geen uitspraak heeft gedaan."

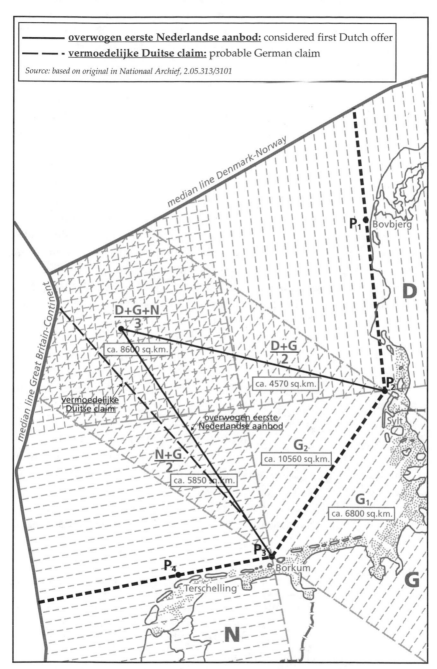

Figure 10.3 First proposal considered by the Netherlands and German claim expected by the Dutch

and the Ministry of Economic Affairs that no license for exploration would be granted for the moment and that incidental exploratory licenses would not be granted to the north of the parallel of 54° N.[82] The endpoint of the partial continental shelf boundary with Germany was located at this parallel.

The Ministry of Foreign Affairs also considered how to deal with existing exploratory licenses. A number of licenses had already been applied for by 15 November 1967 and been granted on 4 March 1968. A second round of applications had started in the fall of 1968 and had been closed in February 1969. For the first group, the Netherlands in the negotiations with Germany should attempt an arrangement to protect the acquired rights of concessionaires in the area that would go to Germany. At least two companies informed the Minister of Economic Affairs that they considered it of the utmost importance that their rights and interests in the areas concerned would not be impaired and requested that these rights and interests be taken into account in their full extent in the eventual conclusion of an agreement with Germany. It was also emphasized that the companies had complied with all their obligations under Dutch legislation and had made considerable investments.[83] It was submitted that in support of its attempt to safeguard these rights, the Netherlands could maintain that it had applied the equidistance method in good faith. Moreover, Germany had refrained from defining its claim. This had made it impossible for the Netherlands to determine the area in which a certain restraint might be expected from it. It would not have been reasonable to expect from the Netherlands that it would itself limit its claim or refrain from activities on its entire shelf.[84] An arrangement to respect acquired right might be supported by reference to the separate opinion of judge Jessup and the fact that German companies had been granted a Dutch license for a couple of the blocks concerned.[85]

It was considered unlikely that Germany would be prepared to consider that blocks, which would have been granted after the judgment of

---

[82] Note 34/69 dated 17 March 1969 [NNA/31], p. 5.

[83] See letter of the Director of British Petroleum Exploratie Maatschappij Nederland N.V. to the Minister of Economic Affairs dated 14 August 1969 [NMEA/2]. The German sister company of the Dutch British Petroleum affiliate informed the German FM of this request and asked to take into account the interest of the preservation of the concessionary rights (letter of BP Benzin und Petroleum Aktiengesellschaft dated 19 September 1969 [AA/48]).

[84] Note 34/69 dated 17 March 1969 [NNA/31], p. 1.   [85] Ibid., p. 2.

20 February 1969, had been granted in good faith. Looking at various options, it was concluded that it seemed to be preferable to ask Germany its ideas about the location of the boundary. The Netherlands could then issue licenses to the south of this line.[86] Because the oil companies were interested in an early decision on the second licensing round, as this would allow them to continue their activities on the Dutch shelf, a quick initiative by the Netherlands would be required.[87]

10.5 The negotiations – the deconstruction of the judgment

The following subsections provide a review of the 9 negotiating rounds during which an agreement between Denmark, Germany and the Netherlands was worked out. Each subsection will also look at the preparations and review of the negotiations as appropriate. This also concerns the bilateral consultations between Denmark and the Netherlands, which they held prior to each of their meetings with Germany.

10.5.1 Cracks in the Danish-Dutch front

Both Germany and the Netherlands were interested in an early start of further negotiations.[88] Germany considered that a settlement would allow turning the page on the judgment, which had led to disappointment in Denmark and the Netherlands.[89] The Netherlands was interested in getting to know the German claim to be able to resume exploratory activities to the south of it.[90] On the other hand, Denmark considered that a thorough review of the judgment was called for.[91]

Ambassador Fack, who had replaced Riphagen as the head of the Dutch negotiating team,[92] paid a visit to the German Foreign Office on 15 April 1969, where he discussed matters with Truckenbrodt. Fack

---

[86] Ibid., pp. 3 and 5.

[87] Ibid., pp. 2–3 and 5; see also note 49/69 dated 3 April 1969 [NNA/31].

[88] See e.g. ibid.; note (draft) dated 5 March 1969 [AA/73], p. 1.

[89] Telex of the FO to the Embassies in The Hague and Copenhagen dated 26 February 1969 [AA/64].

[90] Note 49/69 dated 3 April 1969 [NNA/31].

[91] Summary record of the Danish-Dutch meeting held on February 21, 1969 dated 26 February 1969 [DNA/39], p. 2; see also the note dated 14 March 1969 prepared by Truckenbrodt [AA/73].

[92] Fack replaced Riphagen because the latter was expected to have to spend much time on the Barcelona Traction case, in which he acted as the judge ad hoc for Belgium (note

explained that the Netherlands was interested in learning the German position because of its interest in allowing further exploratory activities.[93] Truckenbrodt showed understanding, but explained that it would only be possible to inform the Netherlands after a further round of internal deliberations.[94] Fack was told that the slow pace of these deliberations was most of all caused by the relationship between the Federal authorities and the states.[95] A review of the German internal deliberations indicates that the delay was not, in particular, attributable to this relationship, but Truckenbrodt's statement was bound to reconfirm the Dutch perception that the German coastal states were pressing the Federal authorities to maximize Germany's claim. Fack and Truckenbrodt agreed that they would be in contact again by the end of May.[96] They further agreed that a tripartite handling of the negotiations would be unavoidable in view of ICJ's judgment, which required that the delimitation between all parties would be based on the same principles, although bilateral contacts might also be useful from time to time.[97]

Fack also presented the Dutch view on the delimitation: an equidistance line between the straight lines P_2-P_3 and P_3-P_4 contained in the German map 5 presented during the oral pleadings.[98] The area which this approach would give to Germany would depend on how this equidistance line would be connected to the existing boundary and the end point of the new boundary. The latter point would depend on the boundary between Denmark and Germany. If the German-Dutch boundary would stop at the equidistance line between the Netherlands and Denmark, which the two States had accepted as their common boundary in 1966, the Dutch offer to Germany comprised some 2,750 km^2.[99] Fack explained that this delimitation was a synthesis between three elements:

40/69 dated 24 March 1969 [NNA/33], p. 1, hand-written note; note; Shelf case; Report on positions dated 3 June 1969 [DNA/39], p. 3). Oral hearings in the *Barcelona Traction* case took place in June and July 1969 and a judgment was rendered in February 1970. Fack had had some involvement with the negotiations during the first phase of the negotiations with Germany, when he served in the DEB (Fack, *Gedane Zaken*, pp. 73–74).

[93] Note dated 16 April 1969 prepared by R. Fack [NNA/40], p. 1; note dated 16 April 1969 [AA/73], p. 1.

[94] Note dated 16 April 1969 prepared by R. Fack [NNA/40], p. 1.

[95] Note dated 16 April 1969 [AA/73], p. 1.

[96] There is no indication that this meeting actually took place.

[97] Note dated 16 April 1969 prepared by R. Fack [NNA/40], p. 2.

[98] *Ibid.*, p. 1; see also note dated 16 April 1969 [AA/73], p. 1. For a depiction of the proposal see Figure 10.3.

[99] This figure is provided in the documents mentioned in footnote 231.

the Dutch preference for some form of equidistance, the German wish to rectify the unfavorable configuration of its coast and the exhortation of the Court to correct certain distortions because of equity. Fack had the impression that Truckenbrodt had readily taken his hint.[100] The German account of the encounter observed that Fack's suggestion only met Germany's views to a limited extent.[101]

Sandager Jeppesen, who was to conduct the further negotiations on behalf of Denmark, paid a visit to Truckenbrodt on 12 May 1969. They did not discuss the substance of the matter, but agreed that it would stand to reason to negotiate in a trilateral setting.[102] Sandager Jeppesen and Truckenbrodt agreed that a first meeting in June should seek to establish a common interpretation of the judgment and to sort out the principles and factors which had to be taken into consideration.[103] Sandager Jeppesen did not refer to the possibility of further bilateral contacts with Germany, as Fack had done.[104] The Danish Ministry of Foreign Affairs had also been interested in arranging a meeting with Fack to discuss the possibilities of further coordinating the positions of Denmark and the Netherlands in the upcoming meeting in June,[105] but this meeting did not materialize. The Danish ambassador in The Hague expressed his surprise about Fack's "passivity" in relation to the Danish continental shelf delegation to Maas Geesteranus.[106] Although Fack had been continuously absent from The Hague, it was not excluded that the Dutch reserve was tactical in nature.[107] This may well have been the case.[108] Fack had paid a visit to the Danish ambassador after his talk with Truckenbrodt. Fack indicated that the Dutch had wanted to learn more about the stage of Germany's preparations and the time schedule

---

[100] Note dated 16 April 1969 prepared by R. Fack [NNA/40], pp. 1–2.

[101] Note dated 16 April 1969 [AA/73], p. 1.

[102] See letter of the FO to the German Embassy in Copenhagen dated 19 May 1969 [AA/73]; note; Shelf case; Report on positions dated 3 June 1969 [DNA/39], pp. 2–3.

[103] Report on positions dated 3 June 1969 [DNA/39], p. 3.

[104] Letter of the FO to the German Embassy in Copenhagen dated 19 May 1969 [AA/73].

[105] Note; Shelf case; Report on positions dated 3 June 1969 [DNA/39], p. 3.

[106] Letter of the Danish ambassador in The Hague to Krog-Meyer of the MFA dated 30 May 1969 [DNA/143].

[107] Note; Shelf case; Report on positions dated 3 June 1969 [DNA/39], p. 3.

[108] This is also suggested by a report of the Dutch ambassador in Bonn on a meeting with Sandager Jeppesen. The latter repeated to be prepared to meet with Fack and hinted at the desirability of a common Danish-Dutch position. The ambassador did not respond to these suggestions (received message in cipher (ref. no. 6246) of the DEB to the MFA dated 13 May 1969 [NNA/31]).

for negotiations, but remained silent on the fact that he had informed Truckenbrodt about the Dutch views on its boundary with Germany.[109] The coordination of positions with Denmark might have frustrated the option of further exploratory talks with Germany at the bilateral level.

The Danish and Dutch delegations met on 17 June 1969, the day before the first trilateral discussion with Germany had been scheduled in Bonn. The discussion revealed that there existed considerable differences.[110] First of all, ambassador Fack indicated that it might be an idea for Denmark and the Netherlands to assess which issues were bilateral and which issues were trilateral. Such an assessment would allow establishing whether the negotiations later had to be pursued on a trilateral or a bilateral basis. Sandager Jeppesen immediately indicated that Denmark wanted to continue the close cooperation with the Netherlands and that Denmark in principle considered that the negotiations had to continue on a trilateral basis.[111]

When Sandager Jeppesen raised Denmark's interest in arguing that the German coastal front should be represented by a shorter line than the Borkum-Sylt line, Fack submitted that the Netherlands had problems with its own coastal front, whether it was defined as one line representing its entire coast or a line along its northern coast. The Netherlands were only prepared to accept the Dutch coastal front advanced by Germany during the oral hearings for the limited purpose of drawing an equidistance line between that line and the Borkum-Sylt line. Fack wondered whether Denmark would also be able to accept that idea, using the Danish coastal front Germany had presented during the oral hearings. The Netherlands had prepared a sketch map to illustrate this idea.[112] Fack admitted that this solution was not perfect for Denmark as the tip of the German sector was on the Danish side of the equidistance line and Denmark would be giving a larger area to Germany.[113] Sandager Jeppesen replied that the proposal was unacceptable for Denmark.

---

[109] Note dated 16 April 1969 prepared by R. Fack [NNA/40], p. 2.

[110] Information on the meeting is included in Report meetings on delimitation of the continental shelf in Bonn on 17, 18 and 19 June 1969 [NNA/31] (Dutch Report June 1969 [NNA/31]); Report on the Danish-Dutch preparatory meeting in Bonn on 17 June 1969 concerning the delimitation of the continental shelf in the North Sea [DNA/40].

[111] Report on the Danish-Dutch preparatory meeting in Bonn on 17 June 1969 [DNA/40], p. 2.

[112] A Danish representation of the sketch map attached to ibid. is reproduced as Figure 10.4.

[113] Report on the Danish-Dutch preparatory meeting in Bonn on 17 June 1969 [DNA/40], pp. 3–4.

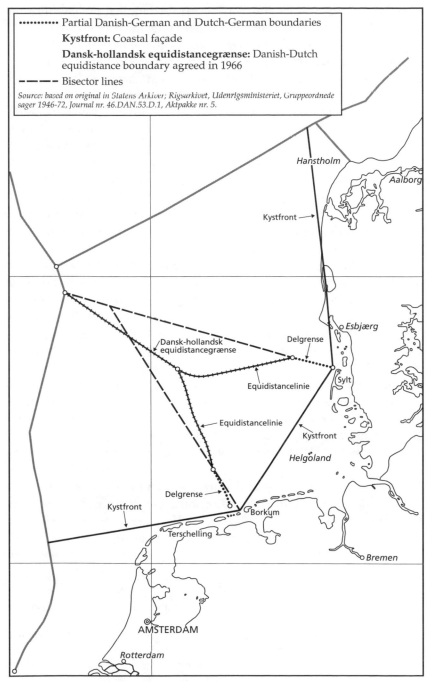

Figure 10.4 Dutch view on coastal façades and bisector line boundaries (Danish reconstruction of Dutch figure presented during a bilateral meeting of 17 June 1969 and the trilateral meeting of 18 and 19 June 1969)

First of all, it was inequitable that Denmark would have to give a larger area to Germany than the Netherlands. Second, the proposal resulted in a Dutch-German boundary in Denmark's equidistance area. It was not acceptable that two of the parties would present the third party with a *fait accompli* of this nature. The Danish-Dutch equidistance line remained relevant in relation to Denmark and the Netherlands. Finally, the location of the successful drillings of the DUC implied that Denmark could not accept a simple geometrical solution consisting of a triangle, which in all likelihood would cut across the area of interest to Denmark.[114]

Denmark also had a more fundamental problem with the Dutch approach. The negotiations with Germany should not start with the discussion of specific geometrical solutions, but should first identify the relevant criteria contained in the judgment. These criteria then could be used for arriving at a solution, which would be acceptable for all three parties. This approach had the additional advantage that if the parties would eventually return to the Court, they could show that they had had meaningful negotiations.[115] Fack understood that Denmark might be interested in using as many of the judgment's criteria as possible, but this was not the case for the Netherlands. The idea of proportionality might harm the interest of both Denmark and the Netherlands and the only geological argument the Netherlands was interested in was the location of the former bed of the Ems.[116] Sandager Jeppesen's reaction indicated that he was not prepared to give up the possibility of using geological arguments. He also hinted that other equitable considerations might even be worse for the Netherlands: the Netherlands already had large proven reserves because of the gas field of Slochteren and the Netherlands and Germany were collaborating on energy policy in the framework of the European Economic Community.[117] The disagreement on these points also spilled over to the question of a possible joint development regime with Germany. The Netherlands indicated itself squarely against this option and Denmark indicated that it would not work against such a solution, but also not in favor, implying that the Netherlands was not sure of Danish support.[118] Internally, Denmark had previously concluded that if Germany were to raise the issue, Denmark would reserve its position as there were a lot of complex aspects to joint development

---

[114] *Ibid.*, pp. 4–5.   [115] *Ibid.*, p. 5; see also *ibid.*, p. 4.
[116] *Ibid.*, pp. 4–5. On the later issue see text at footnote 64.   [117] *Ibid.*, p. 5.
[118] *Ibid.*, pp. 5–6.

schemes.[119] At the end of the meeting, Sandager Jeppesen took care to thank the Dutch delegation for its frankness.[120]

At the start of the trilateral meeting in Bonn on 18 and 19 June 1969, Germany presented its views on natural prolongation. The German delegation explained that it had based itself on figure A of the Common Rejoinder, but for Denmark had selected a more advantageous coastal front.[121] The map in addition identified the extent to which the natural prolongations overlapped.[122] Germany invited the other delegations to comment on this map.[123] By mistake the German delegation handed Fack, the head of the Dutch delegation, a different figure, which used the same coastal fronts for Denmark and Germany, but a much more advantageous coastal front for the Netherlands.[124] At the same time, this coastal front was less advantageous than the one Jaenicke had presented to the Court. Fack used this mistake to brush aside the initial German map. The Netherlands would rather like to start immediately with map 5 or 6.[125] The second German map did make it clear to Denmark and the Netherlands that Germany's opening bid was open to significant adjustments,[126] but even this second map implied a much larger German claim than the Netherlands had been counting on and it would probably result in situating three of the promising Danish locations on the German continental shelf.[127]

---

[119] Note; Some main points of the shelf case problem dated 3 June 1969 [DNA/39], p. 4.

[120] Report on the Danish-Dutch preparatory meeting in Bonn on 17 June 1969 [DNA/40], p. 6.

[121] Instead of a line running from the northeastern tip of Jutland to a point at or near the Danish-German boundary in the territorial sea, the latter point was connected to Cape Hanstholm.

[122] The map is reproduced in Figure 10.5.

[123] Report on the Danish-German-Dutch meeting in Bonn on 18 and 19 June 1969 concerning the delimitation of the continental shelf in the North Sea dated 31 June 1969 [DNA/142] (Danish Report June 1969 [DNA/142]), p. 2; Dutch Report June 1969 [NNA/31], p. 2; German Protocol June 1969 [AA/73], p. 1.

[124] See Dutch Report June 1969 [NNA/31], p. 2; copy of the second German map with a hand-written note [B102/260034]. This map is reproduced in Figure 10.6.

[125] Dutch Report June 1969 [NNA/31], p. 2.

[126] See also note dated 4 September 1969 [AA/73], p. 4.

[127] A bisector between the lateral limit of the natural prolongations of Germany and Denmark (and Germany and the Netherlands) would not divide the area of overlap equally. A bisector line thus would not be a logical choice to divide the area of overlap and there would not be one other specific method readily available. This makes it difficult to assess with certainty how this German proposal would have affected the four promising Danish locations.

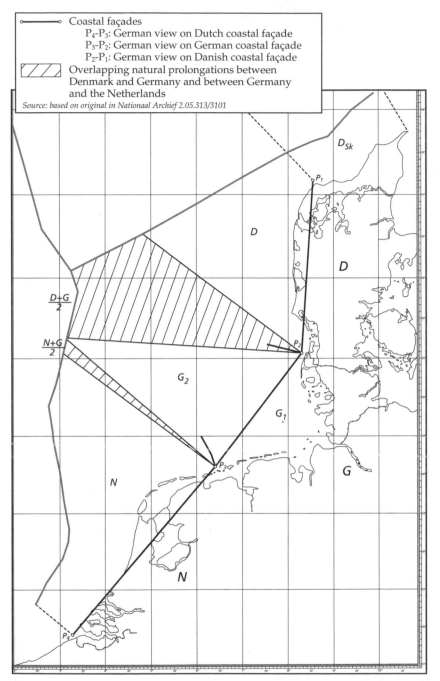

○─────○ Coastal façades
P₄-P₃: German view on Dutch coastal façade
P₃-P₂: German view on German coastal façade
P₂-P₁: German view on Danish coastal façade
Overlapping natural prolongations between Denmark and Germany and between Germany and the Netherlands
Source: based on original in Nationaal Archief 2.05.313/3101

Figure 10.5 German view on natural prolongation presented during the trilateral meeting of 18 and 19 June 1969

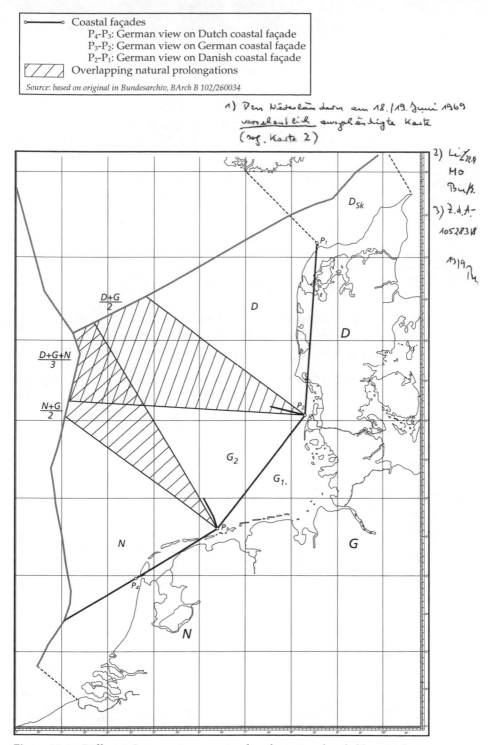

Source: based on original in Bundesarchiv, BArch B 102/260034

Figure 10.6 Different German view on natural prolongation; handed by mistake to ambassador Fack during the trilateral meeting of 18 and 19 June 1969

Denmark and the Netherlands rejected the initial German map as a basis for discussion.[128] Apart from his quip of immediately using German map 5 or 6, Fack submitted that the German views had been overtaken by the judgment. The judgment was very vague on natural prolongation and only one thing was certain: the outcome of the negotiations had to lie between the views of the parties as presented to the Court,[129] that is, the German share would lie between 23,000 km^2 and 36,700 km^2. Fack then referred to the solution he had already outlined at the German Foreign Office and during the bilateral talks with Denmark on 18 June 1969.[130]

Instead of focusing on the issue of natural prolongation, Fack introduced 5 points for the further negotiations:

1. determination of the Dutch-German boundary;
2. as a consequence, adjustment of the Dutch-Danish boundary agreement;
3. possible area of joint control/exploitation;
4. protection of established rights of private interests; and
5. treatment of new requests for concessions in the "disputed" area.[131]

Fack also submitted that a solution should be more political than legal in nature and that delay would burden Germany's relations with the Netherlands.[132] Sandager Jeppesen considered that the Dutch proposal was too concrete. The parties should start their negotiations with a map with no lines on it. As he had also indicated in the bilateral talks with the Dutch, he favored to first focus on the content of the judgment and identify the concrete directives and factors, which could be taken into account by the parties, such as coastal fronts, proportionality, special geographical circumstances, geological factors, energy resources in general and other possible factors.[133]

Truckenbrodt, the head of the German delegation, tried to use the Dutch and Danish arguments in another attempt to focus the discussion on the issue of natural prolongation. Fack's first point required

---

[128] See also note 61/69 dated 21 June 1969 [NNA/31], p. 1.
[129] German Protocol June 1969 [AA/73], pp. 2–3.
[130] Danish Report June 1969 [DNA/142], p. 4, footnote.
[131] Dutch Report June 1969 [NNA/31], p. 2; see also German Protocol June 1969 [AA/73], p. 2; Danish Report June 1969 [DNA/142], p. 3.
[132] German Protocol June 1969 [AA/73], pp. 2–3; Dutch Report June 1969 [NNA/31], p. 2.
[133] Danish Report June 1969 [DNA/142], p. 3.

considering this issue and it was important to follow the judgment closely. Before focusing on the factors listed in paragraph 101(D) of the judgment, the implications of paragraph 101(C), which mentioned natural prolongation, had to be considered.[134] Treviranus submitted that, contrary to what Fack claimed, the judgment was far from vague. It was systematic and had a logical structure and had committed the parties to a considerable extent. It had after all been the intention of the Court to render a judgment which could be implemented.[135]

In reply, Sandager Jeppesen submitted that the idea of coastal fronts – which Germany has used to determine the natural prolongations of the three States – was new, and not everything was clear. The judgment had to be understood as a compromise and proportionality also had to be taken into consideration. He was for a step-by-step approach and against using points and lines at this stage.[136] Fack argued that the Netherlands did not see a need to change its coastal front. According to Fack, the judgment only required the parties to replace their actual coasts by straight lines if they wished to do this.[137] As was set out in Chapter 9.3, the judgment had actually suggested that if equidistance would be employed, something the Netherlands still preferred, a straight line should be employed to avoid the undue effect on that line of the convex Dutch coast. Fack asked the German delegation to explain how the present German position could be reconciled with the binding statements of the German agent on the Dutch coastal front during the oral proceedings.[138] Treviranus replied that Germany could only feel itself bound by the arguments that had been confirmed by the judgment. The argument of the German agent that Fack had invoked stemmed from the sector theory and the idea that the natural prolongations met at the center of the North Sea. These ideas had not been adopted by the judgment. The idea of natural prolongation had been clearly recognized by the judgment and the present German map was elaborated on that basis.[139]

---

[134] German Protocol June 1969 [AA/73], p. 2; Danish Report June 1969 [DNA/142], p. 3.

[135] Dutch Report June 1969 [NNA/31], p. 3.

[136] German Protocol June 1969 [AA/73], p. 3; see also Danish Report June 1969 [DNA/142], p. 5.

[137] German Protocol June 1969 [AA/73], p. 3; see also Danish Report June 1969 [DNA/142], p. 4.

[138] German Protocol June 1969 [AA/73], p. 3; see also Danish Report June 1969 [DNA/142], p. 4.

[139] German Protocol June 1969 [AA/73], p. 4; Danish Report June 1969 [DNA/142], pp. 4–5. For a discussion of the legal merits of the Dutch and German positions

Now that Germany had explained its views on natural prolongation, Truckenbrodt and Treviranus once more invited Denmark and the Netherlands to set out their views on natural prolongation. Sandager Jeppesen suggested that this issue could be included in a list of issues to be considered in further negotiations and Fack replied that one could and should not yet go into so much detail.[140] After some further discussion, Germany took back the map it had proposed as a basis for discussing natural prolongation and the parties worked out a list of topics for discussion at the next meeting:[141]

1. what is the total area to be divided between the parties?;
2. what constitutes the natural prolongation of the parties (paragraph 101(C)(1) of the judgment)?;
3. how should the overlapping areas resulting from point 2 be divided (paragraph 101(C)(2) of the judgment)?;
4. what is the significance of coastal length mentioned in paragraph 101 (D)(3) of the judgment?;
5. what is the significance of the factors mentioned in paragraph 101(D) (2) of the judgment, such as the physical and geological structure of the shelf and natural resources?;
6. should the coastal configurations mentioned in paragraph 101(D)(1) of the judgment be taken into consideration?;
7. are there other factors to be taken into consideration?; and
8. established rights.[142]

The final point was included on the insistence of Denmark. Germany considered that this was not an issue of relevance for the delimitation of the continental shelf.[143] Apparently, Denmark was seeking to create an opening for an argument to be allocated the four promising drilling locations of the Danish concessionaire.[144] At first sight, the agreed list

concerning the binding nature of statements during pleadings see further text at footnote 345 and following.

[140] German Protocol June 1969 [AA/73], p. 4.
[141] Danish Report June 1969 [DNA/142], pp. 5–6; see also German Protocol June 1969 [AA/73], p. 3.
[142] This list is based on the lists contained in the reports of the Danish, Dutch and German delegations, which differ to a limited extent (Report of the German-Danish-Dutch continental shelf negotiations in Bonn on 18/19 June 1969 [AA/73], pp. 2–3; Dutch Report June 1969 [NNA/31], p. 4; Danish Report June 1969 [DNA/142], pp. 6–7).
[143] Danish Report June 1969 [DNA/142], pp. 6–7; Dutch Report June 1969 [NNA/31], p. 4.
[144] See also note 61/69 dated 21 June 1969 [NNA/31], p. 1.

of topics might suggest a considerable fidelity to the judgment as all the relevant paragraphs of the judgment's dispositif were included. However, the list left all parties ample room to introduce argument they thought fit. As the preceding discussion shows and as was also concluded by the German delegation, whereas Germany had advanced the view that the judgment set considerable limits to the freedom of the parties, Denmark and the Netherlands had stressed the freedom of the parties and considered that the judgment had to be taken into account, but this did not concern the application of geometrical concepts or purely legal considerations, but decisions of a political nature.[145]

The meeting also shortly considered joint development. Germany indicated that it did not have a strong preference for this option and the Netherlands spoke out against it. Denmark also preferred a delimitation, but did not want to exclude joint development as a fallback position. The parties agreed that they would not come up with specific proposals for joint development at the next meeting.[146] Germany did not meet the Dutch request to define its claim by a specific line in order to allow the Netherlands to determine in which area it could allow drilling activities without problems. On the other hand, the Netherlands could not accept a German proposal that Germany would only object to drilling if it considered that this concerned the area claimed by it, as this might lead to uncertainty about the exact extent of the disputed area for years. As a consequence, it was agreed to continue the existing practice in which all planned drilling was notified to Germany.[147]

At the trilateral meeting the differences between Denmark and the Netherlands, which had revealed themselves during their preparatory meeting of 17 June 1969, also became clear to Germany.[148] While the Netherlands expressed its preference to immediately start discussing specific proposals, Denmark continued to insist on the need to look at all factors mentioned in the dispositif of the judgment. Fack also rejected

---

[145] Report of the German-Danish-Dutch continental shelf negotiations in Bonn on 18/19 June 1969 [AA/73], p. 1.

[146] *Ibid.*, p. 5; see also Dutch Report June 1969 [NNA/31], p. 3.

[147] German Protocol June 1969 [AA/73], pp. 6–7; Dutch Report June 1969 [NNA/31], p. 4.

[148] At the same time, there was also a large measure of agreement between Denmark and the Netherlands. Apart from their agreement on the kind of guidance the judgment provided to the parties, they also supported each other on specific points. For instance, Fack seconded the Danish wish to work with a map without additional lines and the Danish request to include acquired rights in the list of topics to be discussed (see Danish Report June 1969 [DNA/142], p. 3; German Protocol June 1969 [AA/73], p. 6).

the relevance of specific factors mentioned in the judgment. He for instance considered that the proportionality test of paragraph 101(D)(3) of the judgment only came into play after paragraphs 101(C)(1) and (2) had been applied. If there were no difficulties in the latter respect, proportionality might not need to be applied, as was also observed by Bustamante in his separate opinion.[149] After the first day of the trilateral talks, Maas Geesteranus had spoken with his Danish counterpart and told him that continued Danish support for the proportionality test might lead him to advise his delegation that the Netherlands should negotiate separately with Germany. After a two hour talk between the two delegations it was decided to continue the collaboration.[150] During the second day of the trilateral meeting, after the parties had already agreed on the list of topics for a next meeting, Fack also put the option of breaking with Denmark on the table. He indicated that the negotiations would not necessarily result in one trilateral agreement, but could also lead to three bilateral agreements. It might be considered to structure the negotiations accordingly.[151] Joint development also need not necessarily be discussed between all three partners.[152] Fack not only raised the possibility of going it alone because Denmark was considering a different approach to the delimitation, but also because Denmark seemed to envisage a slower pace of the negotiations than was in the interest of the Netherlands.[153] The Netherlands had proposed that the next meeting should take place in The Hague in July 1969 and was prepared to make specific proposals. After Denmark had indicated that it would only be available in the second half of August, the delegations agreed on 19 and 20 August 1969.[154]

10.5.2   Natural prolongation as a geographical concept?

The report of the German Foreign Office on the June 1969 meeting recognized that it had revealed conflicting views on the role of the

---

[149] German Protocol June 1969 [AA/73], p. 5. Bustamante had argued that the judgment of the Court had not conferred the proportionality test "the character of an obligatory principle" (judgment of 20 February 1969, Sep. op. Bustamante y Rivero, [1969] ICJ Reports, p. 57, at p. 59, para. 4).

[150] Note concerning the Danish-Dutch relationship in the shelf case dated 16 August 1969 [DNA/40], p. 3.

[151] Dutch Report June 1969 [NNA/31], p. 3; German Protocol June 1969 [AA/73], p. 6.

[152] German Protocol June 1969 [AA/73], p. 6.

[153] Note 61/69 dated 21 June 1969 [NNA/31], p. 2.

[154] German Protocol June 1969 [AA/73], p. 7.

judgment in the negotiations. Germany considered that the judgment limited the freedom of the parties, whereas Denmark and the Netherlands downplayed the significance of the judgment in this respect.[155] Another account of the meeting found that it had already become clear that the differences of views with, in particular, the Netherlands would be substantial. The fact that the Netherlands was picked out in this respect probably is explained by the fact that Denmark, unlike the Netherlands, had not yet expressed a specific position. On the upside, the talks were characterized by a friendly atmosphere and the fact that the Netherlands was interested in a quick settlement coincided with the German interests.[156]

As was set out in the previous section, the meeting between Denmark and the Netherlands and the trilateral talks had revealed discrepancies between the two States. A number of reports in the Danish press on the trilateral meeting led the Dutch Ministry of Foreign Affairs to doubt whether Denmark was laying all its cards on the table.[157] When asked about one of the reports, the deputy head of the political and legal department of the Danish Ministry of Foreign Affairs considered it embarrassing that the Netherlands had given any credence to the report, which obviously was not in accordance with what had transpired at the meeting. Maybe the report was a canard of German origin. He assured the Dutch Embassy that Denmark would not deviate from its declared policy to negotiate hard.[158]

A note on a proposal for the Danish negotiating mandate, which was prepared between the negotiating rounds of June and August 1969, does indicate that it was clear that the interests of Denmark did not run parallel to those of the Netherlands in respect of many possible

---

[155] Report of the German-Danish-Dutch continental shelf negotiations in Bonn on 18/19 June 1969 [AA/73], p. 1.

[156] Note of section III to the Deputy Minister for the Economy dated 20 June 1969 [B102/260034], p. 1.

[157] See telex of the DEB to the MFA dated 19 June 1969 [NNA/31]; received message in cipher (ref. no. 8114) of the DEC to the MFA dated 30 June 1969 [NNA/31]; letter of the DEC to the MFA dated 16 July 1969 [NNA/22]; memorandum 71/69 dated 29 July 1969 [NNA/31].

[158] Received message in cipher (ref. no. 8114) of the DEC to the MFA dated 30 June 1969 [NNA/31]. Another account of this episode is contained in note concerning the Danish-Dutch relationship in the shelf case dated 16 August 1969 [DNA/40], pp. 3–4. According to the note, the Dutch Embassy delivered an aide-mémoire in which it was indicated that the Dutch side "deeply deplored" that the Danish delegation had given detailed statements on the negotiations to the press, contrary to what had been agreed upon by the parties.

arguments.[159] Nevertheless, Denmark wanted to continue the close collaboration with the Netherlands. Just after the June negotiations Sandager Jeppesen, the head of the Danish delegation, stressed that Denmark had an interest in the best possible relations with the Netherlands.[160] The above-mentioned note on a proposal for the Danish negotiating mandate took into account the possible impact Danish arguments might have on the collaboration with the Netherlands. For instance, it was advised to point to the differences in the geology of the area in front of the Danish coast as compared to the German and Dutch coasts. The note added that in view of the cooperation with the Netherlands, this was an extremely delicate subject.[161] That cooperation also indicated that Denmark, because of the problem the Netherlands had with the definition of the coastal front along its coast, should be cautious about taking too firm a view on that matter.[162]

Before the meeting of August 1969 the Danish authorities had a further look at the arguments Denmark might employ. A number of these arguments had already been discussed previously and it is not necessary to look at them again. An example in this respect is the argument in relation to the geology of the shelf.[163] Further consideration of the factor of proportionality, which the Dutch had opposed from the start, led to the conclusion that it did not seem to be of much advantage to either Denmark or the Netherlands.[164] If the Netherlands would seek to employ this argument to its advantage, Denmark could make the same argument as Germany to the effect that proportionality worked to the advantage of the Netherlands because of its convex coast. Reliance on this convexity was not in accordance with the judgment.[165]

At the trilateral meeting in June 1969, Denmark had requested that the issue of acquired rights would be included in the list of topics to be discussed at the next meeting. Denmark considered that this concept might be used to ensure that the promising locations of the Danish concessionaire would remain on the Danish side of the continental

---

[159] Note; Continental shelf dated 26 June 1969 [DNA/39].

[160] Report on the meeting of the advisory committee for the case with Germany on the continental shelf (Continental shelf committee) held on Wednesday 25 June 1969 (KSU/MR.12) dated 31 July 1969, p. 5.

[161] Note; Continental shelf dated 26 June 1969 [DNA/39], p. 5.   [162] *Ibid.*, pp. 4–5.

[163] See text at footnote 45 and following. These arguments are set out at pp. 12–14 in an untitled and undated note, which on the basis of its content and the place in the folder in which it is included [DNA/143] can be dated at around 20 July 1969.

[164] *Ibid.*, p. 20.   [165] *Ibid.*, p. 22.

shelf boundary. Further consideration led to the conclusion that if it could be shown that Germany had given Denmark reason to believe that it had acted in good faith, Denmark perhaps could request some form of compensation if it would have to give up the locations, but the reference to the concept of acquired rights would be problematic.[166] It would be difficult to maintain that Germany had given any indication that the drilling activities were free of risk.[167] A review by the Ministry of Foreign Affairs concluded that Germany was likely to reject the argument of acquired rights by referring to the diplomatic note in which it had reserved its rights.[168] It was also concluded that the judgment gave little support to argue that the drilling sites had to be taken into account in the delimitation. Only judge Jessup's separate opinion gave some support to this argument.[169]

The Danish Ministry of Foreign Affairs was given the mandate to discuss the continental shelf delimitation with Germany and the Netherlands.[170] The Danish approach to the negotiations should leave Denmark the possibility to introduce all criteria mentioned in the judgment and to avoid that the discussion would focus on specific geometrically constructed lines, which would harm the Danish interests in the area.[171] What was meant by the latter point is that straight lines, like the ones Germany had proposed in June, would in all likelihood place the promising locations of the Danish concessionaire on the German side of a boundary. It was considered too early for Denmark to make a delimitation proposal itself.[172] One analysis of possible outcomes indicated that Denmark did not yet have a clear idea of the kind of outcome it

---

[166] Note on the significance of the concept "acquired rights" in relation to the delimitation of the continental shelf dated 15 August 1969 [DNA/143]. The note had been prepared by O. Espersen of the Ministry of Justice on the request of Sandager Jeppesen (see Report on the meeting of the advisory committee for the case with Germany on the continental shelf (Continental shelf committee) held on Wednesday 25 June 1969 (KSU/MR.12) [DNA/142] dated 31 July 1969, p. 6).

[167] Note on the significance of the concept "acquired rights" in relation to the delimitation of the continental shelf dated 15 August 1969 [DNA/143], p. 1.

[168] Note (untitled, undated) [DNA/143], pp. 16–17.

[169] Note; Significance of natural resources dated 6 August 1969 [DNA/143]. The ICJ had a possibility to revisit this matter a couple of years later in the *Aegean Sea Continental Shelf (Greece v. Turkey)* and clearly rejected that unilateral activities in a disputed area could have any consequences for a delimitation (*Aegean Sea Continental Shelf*, Interim Protection, Order dated 11 September 1976, [1976] ICJ Reports, p. 3, at p. 11, para. 28).

[170] Note; Ministerial meeting of 11 July 1969 [DNA/147].

[171] Note; Continental shelf dated 26 June 1969 [DNA/39], pp. 5–6.   [172] *Ibid.*, p. 4.

should aim for, apart from safeguarding the promising locations and minimizing the German share.[173]

Before the trilateral meeting of 19 and 20 August 1969, Denmark and the Netherlands again held a prior consultation.[174] The proposal that the Netherlands intended to present to Germany formed the most important point of discussion. This proposal was based on the idea of using Dutch and German coastal fronts to determine an equidistance line.[175] The proposal met the Danish objections in one respect – it stopped at the equidistance line between Denmark and the Netherlands. All other Danish objections were brushed aside by the Netherlands. The Dutch arguments indicated that the Netherlands in no case would be prepared to reconsider its position and the Dutch were basically telling the Danes that it was not their problem that Denmark wanted to hold on to the promising locations of its concessionaire.[176] Denmark also pointed out that the size of the area – a mere 2,500 km^2, whereas Germany during the pleadings had claimed more than 13,000 km^2 in addition to its equidistance area – meant that it was hardly imaginable that Germany would find it acceptable. Fack considered that Germany nonetheless would find it difficult to reject the Dutch proposal as it would put Germany in a bad light if the parties were to go back to the Court.[177] In view of the content of the Dutch proposal this was hardly a credible argument.

The bilateral talks did not go down well with the Danes. The day after the meeting Krog-Meyer, the head of the legal and political department of the Ministry of Foreign Affairs, immediately wrote a note in which he set out all the instances in which the Netherlands had complicated the

---

[173] Note (untitled; undated) [DNA/143], pp. 17–19a.

[174] Accounts of the meeting are contained in memorandum 85–69 dated 26 August 1969 [NNA/31]; Report on the Danish-Dutch preparatory meeting in Copenhagen on Friday 15 August 1969 concerning the delimitation of the continental shelf in the North Sea dated 16 August 1969 [DNA/143].

[175] See the illustration of the proposal contained in Figure 10.7. This proposal was based on the possible solution for the German-Dutch delimitation the Netherlands had previously set out to Denmark and Germany. The present proposal further elaborated this idea into a proposal on a specific boundary. To this end, the lines representing the coastal fronts of the Netherlands and Germany were shifted seawards to the terminus of their partial continental shelf boundary and the prolongation of the partial boundary in the form of a bisector the coastal front lines started from the terminus of the partial boundary.

[176] See Report on the Danish-Dutch preparatory meeting in Copenhagen on Friday 15 August 1969 concerning the delimitation of the continental shelf in the North Sea dated 16 August 1969 [DNA/143], pp. 2 and 4.

[177] Ibid., p. 4.

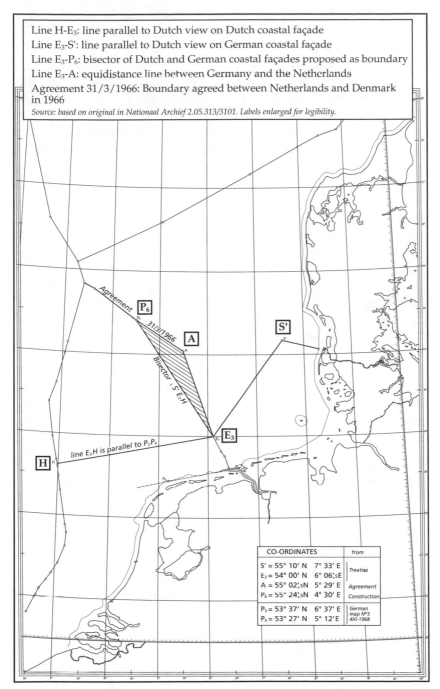

Line H-E$_3$: line parallel to Dutch view on Dutch coastal façade
Line E$_3$-S': line parallel to Dutch view on German coastal façade
Line E$_3$-P$_6$: bisector of Dutch and German coastal façades proposed as boundary
Line E$_3$-A: equidistance line between Germany and the Netherlands
Agreement 31/3/1966: Boundary agreed between Netherlands and Denmark in 1966
Source: based on original in Nationaal Archief 2.05.313/3101. Labels enlarged for legibility.

CO-ORDINATES	from
S' = 55° 10' N 7° 33' E	Treaties
E$_3$ = 54° 00' N 6° 06',5E	
A = 55° 02',5N 5° 29' E	Agreement
P$_6$ = 55° 24',5N 4° 30' E	Construction
P$_3$ = 53° 37' N 6° 37' E	German map N°5 4XI-1968
P$_4$ = 53° 27' N 5° 12'E	

Figure 10.7   Dutch proposal on a boundary with Germany presented during the trilateral meeting of 23 and 24 September 1969

continental shelf negotiations in the past. The Dutch had frustrated meaningful negotiations in 1965/66 by their insistence that the law indicated that the boundaries had to be delimited by equidistance.[178] During the preparation of the written pleadings, the Dutch had not taken any initiative for meetings, discussions or exchanges of views and had categorically rejected that the decisive issue of islands could be addressed, notwithstanding the views of Waldock. Immediately after the judgment, the Netherlands had made it clear that it wanted to back out of the cooperation with Denmark.[179] Subsequently, Fack had tried to arrange a bilateral meeting with Germany before the trilateral negotiations and had avoided a meeting with Denmark. During the first round of trilateral talks, the Netherlands had threatened going it alone if Denmark would not modify its position.[180] The meeting of 15 August had shown how deep the conflict of interest between Denmark and the Netherlands actually was. The Dutch were seeking to force a break, but apparently wanted Denmark to take the initiative. Krog-Meyer expected that if the Netherlands, during the next trilateral round of negotiations, would get the impression that the negotiations would drag on, it would itself cause a break. The Danish negotiating tactic would be much more difficult to implement if the Dutch would reach a separate agreement with Germany.[181] The Dutch Government in August 1969 was considering informing Denmark that it was intent on arriving at an early agreement with Germany in bilateral talks without discussing the significance of the judgment.[182] This idea was not pursued because during the trilateral talks in August 1969 it was agreed that the Netherlands would make a proposal to Germany on their bilateral boundary.[183]

The trilateral meeting of 19 and 20 August 1969 was mostly devoted to a discussion on natural prolongation.[184] In general, the meeting did not

---

[178] Note concerning the Danish-Dutch relationship in the shelf case dated 16 August 1969 [DNA/40], p. 1.

[179] Ibid., p. 2.   [180] Ibid., p. 3.   [181] Ibid., p. 5.

[182] See letter of the Dutch ambassador in Copenhagen to Maas Geesteranus of the MFA of 19 August 1969 and attached concept of an aide-mémoire [NNA/33].

[183] Letter of the Dutch ambassador in Copenhagen to Maas Geesteranus of the MFA of 19 August 1969 [NNA/33], hand-written note.

[184] An account of this meeting is contained in Report of the German-Danish-Dutch continental shelf negotiations in The Hague on 19/20 August 1969 [AA/73]; Protocol of the negotiations in The Hague on 19/20 August 1969 dated 21 August 1969 [AA/73] (German Protocol August 1969 [AA/73]); Report on the second round of negotiations between Federal Republic of Germany on the one side and Denmark and the Netherlands on the other side on the delimitation of the continental shelf of the North Sea held in The Hague on 19 and 20 August 1969 dated 27 August 1969

advance much beyond a presentation of the respective views and it did not bring the parties closer together. At the start of the meeting, Germany invited Denmark and the Netherlands to comment on the map Germany had introduced during the first meeting in an apparent attempt to force them to make their views on natural prolongation more specific.[185] Denmark indicated that the use of straight lines to determine the natural prolongations of the parties was not in accordance with the judgment, which indicated that apart from geometrical and geographical considerations, the geology of the shelf was also relevant to determining natural prolongation. In this connection, Sandager Jeppesen also referred to the presence of the Fyn-Grindsted High. Denmark did not want to indicate the specific consequences it attached to geological considerations for the determination of the natural prolongations of the parties.[186] More in general, Denmark before reaching specific conclusions on natural prolongation wanted to discuss all the elements of the judgment and did not want to commit itself to specific lines.[187]

Germany considered that the judgment in any case accorded geology, which was mentioned in paragraph 101(D), a secondary role. Geology should only come into play after the parties had determined the natural prolongations in accordance with paragraph 101(C).[188] Treviranus submitted that the judgment had referred to the geophysical shelf, but that there were no features in the southeastern part of the North Sea, such as for instance the Norwegian Trough, which could impact on the delimitation.[189] The implication of this observation was obvious. Denmark would not be acting consistently if it would seek to introduce geological arguments, whereas Denmark and Norway had completely ignored the Norwegian Trough in determining their common boundary. The Netherlands agreed with Germany that geology only had a secondary role to play for their bilateral delimitation, but recognized that Denmark did attach significance to this element.[190] Denmark also introduced a map on the geology of the North Sea, which had been prepared by a

---

[NNA/31] (Dutch Report August 1969 [NNA/31]); Report on the Danish-German-Dutch meeting in The Hague on 19 and 20 August 1969 concerning the delimitation of the continental shelf in the North Sea dated 27 August 1969 [DNA/147] (Danish Report August 1969 [DNA/147]).

[185] See e.g. Danish Report August 1969 [DNA/147], p. 1.   [186] See e.g. ibid., pp. 3–4.
[187] See e.g. ibid., p. 2; Dutch Report August 1969 [NNA/31], p. 11.
[188] See e.g. Dutch Report August 1969 [NNA/31], pp. 4–5.
[189] Danish Report August 1969 [DNA/147], p. 4.   [190] Ibid., p. 5.

Source: based on original in P. Heybroek, U. Haanstra, D.A. Erdman "Observations on the geology of the North Sea area" *in Origin of oil, geology and geophysics - Seventh world petroleum congress, Mexico, 1967; proceedings,* Vol. 2, pp. 905-916 figure 3 at p. 908

Figure 10.8   North Sea Area: The Zechstein Basin

Dutch geologist.[191] The German and Dutch delegations were not familiar with the map and Fack added that the significance of the map was doubtful as Erdman had acted in a private capacity.[192] The Netherlands considered that, as far as natural prolongation was concerned, only its north-facing coast could be taken into consideration.[193]

The Netherlands again criticized Germany for seeking a larger share in the continental shelf than it had done during the pleadings. Apart from the pleadings, Fack now also invoked an article by Jaenicke, who had been the German agent in the two cases, in the *Neue Zürcher Zeitung*. Jaenicke submitted that although the judgment left it to the parties to draw practical conclusions from the criteria it had formulated, the sector claim Germany had formulated seemed most consistent with the Court's criteria.[194] Fack added that Jaenicke had made this statement being fully aware of the judgment.[195] Groepper, the German head of delegation,[196] pointed out that the judgment had created a new situation and Germany was no longer bound by the positions it had adopted during the pleadings.[197] Jaenicke had written his article in a private capacity and could not commit the German government.[198] Although the latter observation certainly is right, Jaenicke's remarks all the same did not help to boost the credibility of Germany's current negotiating position. It is in any case

---

[191] The map on which Denmark relied during the negotiations is contained in P. Heybroek, U. Haanstra, D. A. Erdman, "Observations on the geology of the North Sea Area," in *Origin of Oil, Geology and Geophysics – Seventh World Petroleum Congress, Mexico, 1967; Proceedings*, Vol. 2, pp. 905–916, at p. 908. The map is included as Figure 10.8.

[192] Danish Report August 1969 [DNA/147], p. 5. This map was further discussed during the subsequent trilateral meeting in September 1969 (see further text at footnotes 240 and following).

[193] See e.g. Dutch Report August 1969 [NNA/31], p. 2.

[194] G. Jaenicke, "Der Streit über den Nordsee-Festlandsockel; Grundsätzliches Urteil des Internationalen Gerichtshofs," *Neue Zürcher Zeitung*, 27 March 1969, p. 3. A similar view had been expressed by Menzel in an interview ("Wie Weit Reicht Deutschland unter Wasser?") with the German weekly *Der Stern* of 31 March 1969 [NNA/22]. Menzel had acted as adviser to the German delegation during the pleadings before the ICJ at the end of 1968.

[195] German Protocol August 1969 [AA/73], p. 6.

[196] Groepper had replaced Truckenbrodt as the head of the German delegation after the first round of trilateral discussions in June 1969.

[197] On this point see further text at footnote 345 and following.

[198] Dutch Report August 1969 [NNA/31], p. 12; German Protocol August 1969 [AA/73], p. 6. According to Von Schenk the judgment of the Court did not provide any basis to assume that Germany was limited by Jaenicke's claim before the Court. Paragraph 18 of the Court's judgment explicitly rejected a division of the continental shelf into converging sectors and its further reasoning was limited to formulating general principles and rules (Von Schenk, "Die vertragliche Abgrenzung," p. 379, note 26).

remarkable that Jaenicke felt the need to make his views public while the negotiations had not yet been concluded. Fack also indicated that if the case would again be presented to the Court, Germany could not credibly present a claim which differed from what it had previously pleaded before the Court.[199]

At a later point in the discussion, Germany again solicited the views of the other parties in respect of natural prolongation. Germany conceded that the judgment did not explicitly support its approach to determining natural prolongations – i.e. using perpendiculars to a straight line coastal front – but learning the views of Denmark and the Netherlands was important in achieving progress.[200]

Denmark and the Netherlands clashed a couple of times during the meeting. Fack at one point suggested that each delegation must have some ideas concerning natural prolongation and that it might be useful to compare these views as it might offer an avenue to reaching some positive conclusions. Sandager Jeppesen retorted that there could not be question of three views, but that the parties together had to come to a common understanding of all the elements of the judgment. When Fack prodded Sandager Jeppesen and said that six months after the judgment he should have some idea as to natural prolongation, Sandager Jeppesen simply stated that this was not the case.[201] During the second day of the negotiations, ambassador Fack sought to create an opening for separate negotiations with Germany. Fack submitted that the judgment of the Court had also been intended for other delimitation disputes. This explained why certain of its elements might appear to be contradictory. It might well be that some elements might be applied to one case, whereas other elements might be decisive in another one. As a consequence, he could imagine that different solutions would be applied for Denmark and the Netherlands without prejudice to the other party. Sandager Jeppesen rejected this suggestion outright. It was not possible to commit oneself to a specific solution without a proper understanding of the implications of the different elements of the judgment for Denmark and the Netherlands.[202] The Netherlands also indicated that it was prepared to make a specific proposal of its own.[203] Denmark had

---

[199] German Protocol August 1969 [AA/73], p. 7.
[200] Danish Report August 1969 [DNA/147], p. 7.
[201] Dutch Report August 1969 [NNA/31], pp. 6–7.
[202] See German Protocol August 1969 [AA/73], p. 5; Dutch Report August 1969 [NNA/31], p. 10.
[203] Dutch Report August 1969 [NNA/31], p. 8.

raised serious objections to this proposal during its bilateral meeting with the Netherlands.[204]

Although Denmark in general avoided becoming specific, in a number of cases it introduced detailed argument. For instance, Denmark criticized the line Germany was using to represent its coastal front because it took into account the German island of Sylt, which was lying in front of the Danish coast. According to Denmark, the judgment did not indicate that the north tip of Sylt could be used. The land boundary between Denmark and Germany ended further south and that point had to be taken into account to ensure that the continental shelf of Germany would be sufficiently linked to its land territory. The German side pointed out that the judgment did not support this view. The coastline provided the starting point for continental shelf delimitation. The East Frisian Islands formed part of the German coast and were closely connected to the mainland coast – most of the intermediate internal waters were uncovered at low-tide. In any case, even if the Danish view as regards Sylt was correct, the straight line defining the lateral extent of the German natural prolongation also intersected the terminus of the land boundary.[205]

10.5.3 Different approaches of Denmark and the Netherlands: unity in diversity?

Reporting on the meeting of August 1969 to the German Foreign Minister, the Foreign Office's legal department was cautiously optimistic. It noted that in particular Denmark showed little willingness to take into account aspects of the judgment which were unfavorable to it. Progress was also made difficult because Denmark and the Netherlands clearly had difficulty in coordinating their positions. It was expected that the negotiations would be a long and time-consuming struggle, but Germany, supported by the judgment, did not have a weak position. At the moment there was no reason to abandon the German maximum position.[206] A more nuanced and detailed assessment of the situation was given in an internal note of the legal department, which was intended to set out the German options.[207] Since the judgment consisted of a number of different, and at times not necessarily

---

[204] See further text after footnote 175.
[205] See German Protocol August 1969 [AA/73], p. 3.
[206] Note dated 27 August 1969 [AA/73].   [207] Note dated 4 September 1969 [AA/73].

compatible, elements, which did not automatically lead to a specific solution, the parties had a broad range of possible solutions. This allowed the other side to engage in protracted hermeneutic expositions, without risking the reproach that they were stalling – although subjectively such a reproach was already deserved.[208]

The legal department held that the current German position could be deduced from the tenor of the judgment, but did not follow obligatorily from it.[209] It was also unlikely that the German position would be upheld by the Court if the parties were to return to it. The amalgam of different factors and elements of the present judgment suggested that the majority of the Court was likely to look for a middle ground between the positions of the parties.[210] The legal department also did not expect that reference of the matter to the Ministers of Foreign Affairs – an option Denmark and the Netherlands had already suggested – would strengthen the German position.[211] Denmark and the Netherlands had already repeatedly successfully used the argument that an undesirable outcome might burden the bilateral relations with Germany.[212] Denmark and the Netherlands were obviously aware of the German sensitivities on this point and were seeking to exploit them to get out from under the negative consequences of the judgment.[213]

Although the judgment gave Germany a sound basis to claim a larger area than it had done before the Court, the legal department concluded that the likeliness that Germany would have to accept a compromise solution implied that the German share in the end probably would be hardly larger than the sector claim it had presented to the Court.[214] Because the delimitation might be referred back to the Court, the German delegation should maintain the initiative and take a positive stance. In the longer run it would not be possible to simply maintain the original German position, even though the other side had not given a serious reply to it, but Germany would have to indicate that it was willing to compromise. In relation to the Netherlands, Germany should explore whether its acceptance of the Dutch north-facing coast to determine the natural prolongations would not be used by the Netherlands to extract further concessions. Denmark should be probed as to how it would like to resolve the issue of the promising drilling sites, if these would come located in part, or in their entirety, on the German continental shelf.[215]

---

[208] *Ibid.*, pp. 1–2.   [209] *Ibid.*, p. 2.   [210] *Ibid.*, p. 3.   [211] *Ibid.*, pp. 10–11.   [212] *Ibid.*
[213] *Ibid.*, p. 6.   [214] *Ibid.*, p. 20.   [215] *Ibid.*

Reviewing the second round of negotiations, Maas Geesteranus, the deputy legal advisor of the Dutch Ministry of Foreign Affairs, did not consider it likely that the parties would reach agreement on a common understanding of the implications of the concept of natural prolongation. Whether the tactic of undermining the German "certainties" about the judgment, which the Netherlands and Denmark were currently pursuing on the Danish initiative, was working was hard to tell. Maas Geesteranus furthermore doubted whether Denmark had already formed a clear picture of what the result of this tactic should be. The Netherlands delegation considered that it did not make sense to continue to beat around the bush and aimed to prepare a couple of specific proposals on the Dutch-German boundary.[216]

For its part, Denmark had the impression that its tactics were working. The German side seemed to accept that it was not in the interest of the parties to look for schematic solutions, and that it would be necessary to take a pragmatic approach.[217] The Danish arguments seemed to have put Germany on the defensive[218] and Sandager Jeppesen considered that there was no need to change the current Danish tactic. In the next round of negotiations Denmark should seek to establish how far it could get with arguments concerning the geology of the shelf.[219]

In their bilateral meeting preceding their discussions with Germany, Denmark and the Netherlands both tried to get the other party to accept its approach for the trilateral talks.[220] Denmark wanted to undermine the German contention that the judgment was sufficiently clear and allowed the parties to execute it in an unequivocal manner and wanted to stress the importance of paragraph 93 of the judgment, where the Court had said there was no legal limit to the considerations which States may take into account. Denmark considered that the Netherlands had a similar interest in addressing this point of principle and also sought

---

[216] Note no. 86/69 dated 26 August 1969 [NNA/31].

[217] Note dated 22 August 1969 [DNA/147].

[218] Note dated 2 September 1969 [DNA/39], p. 3. Interestingly, a German assessment of negotiations held that the Danish head of delegation had acted in an uncertain and anxious manner (note of 4 September 1969 [AA/73], pp. 8–9).

[219] Report on the meeting of the advisory committee for the case with Germany on the continental shelf (Continental shelf committee) held on Monday 1 September 1969 (KSU/MR.13) dated 9 September 1969 [DNA/39], p. 5.

[220] An account of this meeting is contained in Report on the Danish-Dutch preparatory meeting in Copenhagen on 22 September 1969 dated 5 October 1969 [DNA/144].

Dutch support on other points.[221] Sandager Jeppesen reiterated Denmark's commitment to the significance of geology for determining natural prolongation.[222]

Fack appreciated that Denmark had brought up the point of negotiating tactics. The Netherlands considered that the negotiations thus far had not resulted in any tangible results and the Dutch delegation had been instructed to present a specific proposal to Germany during the upcoming meeting. There might be an advantage to Denmark and the Netherlands taking different approaches to the negotiations.[223] In further discussing these points, both parties basically repeated the views they had expressed during their previous bilateral meeting. Denmark wanted to limit Germany's room for maneuver by seeking agreement on points of principle and did not want to associate itself in any way with the Dutch proposal, whereas the Netherlands wanted to move forward with its proposal and did not show any enthusiasm for the Danish tactics and did not want to get involved in the Danish arguments concerning geology.[224]

The trilateral meeting of 23 and 24 September 1969 could, to a large extent, be said to consist of two sets of bilateral talks.[225] The Netherlands upon the invitation of Germany introduced its proposal concerning a continental shelf boundary between the two States. Denmark and Germany had detailed discussions on whether the German island of Sylt should be taken into account as part of the German coast and the significance of geology for the determination of natural prolongation.[226]

At the start of the negotiations Denmark raised two preliminary points. One concerned the question to what extent the judgment left room for different interpretations. Denmark used this argument to set out its view that paragraph 93 left the parties considerable freedom.

---

[221] *Ibid.*, pp. 1–2. For a discussion on the meaning and implications of paragraph 93 of the judgment see Chapter 9, text at footnote 59 and following.

[222] *Ibid.*, p. 2.   [223] *Ibid.*   [224] *Ibid.*, pp. 2–5.

[225] See also note 97/69 dated 26 September 1969 [NNA/31], p. 1.

[226] An account of this meeting is contained in Report of the German-Danish-Dutch continental shelf negotiations in Copenhagen on 23/24 September 1969 dated 5 October 1969 [AA/74] (German Report September 1969 [AA/74]); Report on the third round of negotiations between Federal Republic of Germany on the one side and Denmark and the Netherlands on the other side on the delimitation of the continental shelf of the North Sea held in Copenhagen on 23 and 24 September 1969 [NNA/31] (Dutch Report September 1969 [NNA/31]); Report on the Danish-German-Dutch meeting in Copenhagen on 23 and 24 September 1969 concerning the delimitation of the continental shelf in the North Sea dated 3 October 1969 [DNA/147] (Danish Report September 1969 [DNA/147]).

Germany doubted the wisdom of discussing this proposition in detail, but twice suggested that it would be helpful if the parties would give their views on points of principle, such as natural prolongation.[227] Denmark also raised the question of whether the judgment allowed Germany to claim a continental shelf up to the median line between the European continent and the United Kingdom, as Germany was assuming. According to Denmark, the judgment implied that Germany's shelf could only extend to the more landward median line between Germany and the United Kingdom.[228] A couple of the separate opinions of the judges had indeed referred to this latter solution, but the other option had also been mentioned and was not excluded by the judgment.[229]

The Dutch proposal for a boundary basically entailed a bisector between its northern coast and the straight line between Borkum and Sylt.[230] This proposal would give Germany some 2,750 km^2 more than the equidistance area.[231] The Netherlands went to great pains to explain that this proposal only concerned the Netherlands and was without prejudice to the position of Denmark.[232] According to the Netherlands, its proposal constituted a synthesis between the German and Dutch views, which made it fully in compliance with the judgment of the Court.[233] When the German delegation suggested that the Dutch proposal did not reflect the Dutch view on natural prolongation but was intended to resolve the bilateral delimitation, Fack submitted that the proposal was also based on his delegation's view on natural prolongation, but first of all was concerned to reach a political solution for a political problem.[234] Germany for the moment reserved its position on the Dutch proposal but indicated that this was not intended to stall the negotiations and promised to give a reaction during the next round of

---

[227] See e.g. Dutch Report September 1969 [NNA/31], pp. 1–2.

[228] See e.g. ibid., pp. 2–3; Danish Report September 1969 [DNA/147], p. 3.

[229] See further Chapter 9.4.

[230] For a more detailed explanation of this proposal see footnote 175. For a depiction of the proposal see Figure 10.7.

[231] This figure is among others mentioned in Report on the fourth round of negotiations between Federal Republic of Germany on the one side and Denmark and the Netherlands on the other side on the delimitation of the continental shelf of the North Sea held in Bonn on 4 and 5 November 1969 [NNA/31] (Dutch Report November 1969), p. 17; note 108/69 dated 7 November 1969 [NNA/31], p. 1.

[232] See Dutch Report September 1969 [NNA/31], pp. 6–7.   [233] Ibid., p. 7.

[234] Ibid., p. 10. The Netherlands presented its view on natural prolongation at a subsequent meeting (see text at footnote 296).

negotiations.[235] Outside of the meeting room the German head of delegation qualified the proposal as "minimal."[236]

Denmark further elaborated the argument that Sylt did not form part of the coastal front of Germany, which it had raised during the previous round of negotiations.[237] Denmark again argued that Sylt, in part, was lying in front of the Danish mainland coast, not the German mainland coast. This absence of "a real coast" meant that Sylt could not be used by Germany. Sylt also constituted a "special and unusual feature" in the sense of paragraph 101(D) of the judgment. Paragraph 43 of the judgment also supported the Danish point of view.[238] Germany argued that the Danish interpretation was contrary to article 1 of the Convention on the continental shelf, which treated islands equal to mainland coasts. Moreover, Sylt was part of a chain of islands, which taken together formed the coast. Sylt was also connected to the mainland and as a consequence in reality was not an island. Denmark rejected the latter point because the connection between Sylt and the mainland was an artificial dyke and not natural. Germany submitted that this linkage was natural and had only been built up and broadened. According to Germany, the idea that a coast had to be backed up by a *hinterland* was contrary to the judgment and, as a matter of fact, would be more advantageous to Germany, which had a much larger landmass. Paragraph 43 did not support the Danish view, but rather illustrated that parts of the continental shelf closer to Denmark could still be the natural prolongation of Sylt. In the context of the discussion of Sylt, Denmark also submitted that its arguments also applied to the German island of Borkum near the Dutch coast. The Netherlands refrained from commenting on this point.[239]

In introducing its arguments on geology and natural prolongation, Denmark referred to the article "Observations on the Geology of the North Sea Area" of three Dutch geologists.[240] Sandager Jeppesen immediately indicated that he did not want to draw any conclusions on the relevance of geology for natural prolongation.[241] He also took care to indicate that, although the article and attached map were of Dutch

---

[235] *Ibid.*, pp. 8 and 10.   [236] Note 97/69 dated 26 September 1969 [NNA/31], p. 2.
[237] The arguments in this respect are set out in detail in Dutch Report September 1969 [NNA/31], pp. 11–12; Danish Report September 1969 [DNA/147], pp. 6–8.
[238] Dutch Report September 1969 [NNA/31], p. 11.
[239] German Report September 1969 [AA/74], p. 2.   [240] Heybroek, "Observations".
[241] The arguments in this respect are set out in Dutch Report September 1969 [NNA/31], pp. 4–5 and 13–14; Danish Report September 1969 [DNA/147], pp. 4–5 and 9–11.

origin, the Dutch delegation did not have to feel bound by it.[242] As a matter of fact, the authors were employed by a subsidiary of Royal Dutch Shell and were not affiliated to the Dutch government. Sandager Jeppesen submitted that one of the maps attached to the article indicated a significant geological structure, which extended from the German northwestern coast in a northwesterly direction. Sandager Jeppesen submitted that this zone could be of importance in determining natural prolongation. Instead of committing himself to a specific line, he asked his German counterpart if the map could be used to draw a boundary between the German and Danish continental shelf. Groepper in first instance showed himself prepared to think about this. Prior to the meeting the German authorities had already considered the map and concluded that it represented a single geological component of the North Sea and as such did not represent the geology of the shelf in its entirety.[243] The Dutch delegation only intervened in the debate between Denmark and Germany to indicate that geology was not helpful in delimiting the continental shelf between the Netherlands and Germany. Although Fack admitted that the matter might be different for other parts of the North Sea, he strongly argued against the relevance of the data included in the map and geology in general.

During the second day of the meeting, Denmark and Germany revisited the arguments concerning natural prolongation and geology at length. Denmark rejected Germany's geometrical vision on natural prolongation because it was not supported by the judgment. The German view, moreover, would place Denmark in a disadvantageous position. Denmark's equidistant boundary with Norway implied that under this approach, it could not extend to the center of the North Sea in the same fashion as Germany. Groepper considered that the agreement between Denmark and Norway was not relevant for Germany and that Denmark possibly would not have concluded this agreement if it would have known the judgment's views on natural prolongation and the Norwegian Trough.[244] Germany repeatedly requested Denmark to indicate its view on natural prolongation on the basis of geology on a map as

---

[242] The map on which Denmark relied during the negotiations is contained in Heybroek, "Observations," p. 908. The map is included as Figure 10.8.

[243] See Position on the presentation of a map on salt pillows of the North Sea by the Danish delegation in The Hague dated 16 September 1969 [AA/73]; note dated 19 September 1969 [B102/260034].

[244] Danish Report September 1969 [DNA/147], p. 7. For the judgment's observations on the Norwegian Trough see Chapter 5, footnote 228.

this would help to further the negotiations. Denmark indicated that it was only willing to do this after the parties would have agreed on the relevant elements of the judgment. In the end, Denmark indicated certain parameters to assess the relevance of geology and that it wanted to have a further look at them with Germany.[245] As was also observed by the Dutch delegation,[246] Sandager Jeppesen's remarks hinted at a possible boundary between the geological structure off the German coast he had identified and the Fyn-Grindsted High.

The discussions between Denmark and Germany on Sylt and natural prolongation must have once more driven home the point that it was possible for the parties to debate for hours without coming any closer to a common understanding of the judgment of the Court. The report of the Dutch delegation on the meeting indicates a certain exasperation in this respect. It observed that the issue of Sylt led to a discussion without end between the other two delegations and that most of the morning of the second day of the meeting was lost on a debate on natural prolongation between the other two delegations.[247] At the same time, the discussions on geology and natural prolongation did seem to have advanced sufficiently to allow Denmark to prepare a specific proposal on its boundary with Germany.

10.5.4   The German rejection of the initial Dutch and Danish offers

In the margin of the meeting of 23 and 24 September 1969 Fack had proposed to Groepper that the Netherlands and Germany, next to the trilateral meetings, might also hold bilateral talks.[248] Germany declined this offer for tactical reasons. The urgency of the Dutch to reach an agreement might help in overcoming the stalling tactics of the Danes.[249] The Netherlands did not further insist on a bilateral meeting. As Riphagen observed in commenting on the original Dutch initiative, there was not yet a good chance of an agreement with Germany and it

---

[245] Danish Report September 1969 [DNA/147], p. 11.

[246] Dutch Report September 1969 [NNA/31], p. 14.   [247] *Ibid.*, pp. 11 and 14.

[248] Note of Lantzke dated 26 September 1969 [B102/260034], p. 2; memorandum 97/69 dated 26 September 1969 [NNA/31], p. 3; telex of the FO to the GEH dated 26 September 1969 [AA/73].

[249] Note of Lantzke dated 26 September 1969 [B102/260034], p. 2; telex of the FO to the GEH dated 26 September 1969 [AA/73].

was too early to risk offending Denmark by striking a bilateral deal with Germany.[250]

Germany considered the initial Dutch proposal to be unsatisfactory.[251] Apart from its limited size, the offer concerned an area which had little oil and gas potential.[252] The Ministry for the Economy was in favor of immediately making a counter-offer to the Dutch during the next round of trilateral negotiations, which were scheduled for 4 and 5 November 1969.[253] The counter-offer basically envisaged that the overlapping area resulting from the natural prolongations, as defined by Germany in the negotiations, would be divided in two. The proposal was considered to be attractive because it stuck to the original German position and was in accordance with the judgment of the Court.[254] In addition, a specific proposal would allow Germany, just like the Netherlands, to avoid further general discussions on the interpretation of the judgment and to focus on agreeing on a boundary. This might also push the Danes to taking a more active approach and also avoided the risk of giving the impression that Germany was not interested in a quick settlement.[255] The ministry considered that Germany remained interested in a quick settlement both because the German North Sea consortium could resume its activities and it would lead to less difficulties with rights of companies which now held concessions to these areas.[256] Two commentaries prepared in the legal department of the Foreign Office set out a number of objections to the proposal of the Ministry for the Economy.[257] The proposal could hardly be said to be in accordance with the reference to proportionality in paragraph 101(D)(3) of the judgment as it gave Germany, which had a shorter coast than the Netherlands, a larger share of the continental shelf.[258] By taking this position, Germany moreover would give the impression that it was sticking to its original position, instead of showing

---

[250] Note 97/69 dated 26 September 1969 [NNA/31], p. 3.

[251] See e.g. German Report September 1969 [AA/74], p. 2; note of Lantzke dated 26 September 1969 [B102/260034], p. 1.

[252] Circular letter of the Minister for the Economy and Transport of Lower Saxony dated 8 October 1969 [AA/74].

[253] Circular letter of the Minister for the Economy dated 23 October 1969 [AA/74].

[254] *Ibid.*, p. 2.   [255] *Ibid.*, pp. 2–3.   [256] *Ibid.*, p. 3.

[257] Note dated 24 October 1969 [AA/74]; note dated 27 October 1969 [AA/74]. This actually concerns two versions of the proposal of the ME. Due to the similarity of the proposals, the commentary on the first version applies equally to the final version of the proposal of the ME.

[258] Note dated 24 October 1969 [AA/74], pp. 1–2.

flexibility.[259] It would, moreover, be difficult to move from this first offer to subsequent offers which could also be said to be in accordance with the judgment.[260] The submission of the Ministry for the Economy that the proposal had the tactical advantage that it could lead to a compromise through gradual concessions on both parts also had a downside. In view of the initial Dutch offer, it was likely that the end result would lie somewhere between 35,000 and 37,000 km^2, that is, it would probably be less than what Germany had claimed in front of the Court.[261] As an alternative to the proposal of the Ministry for the Economy it was suggested to further explore how the Netherlands viewed the implications of paragraph 101(D)(3) of the judgment and to offer to talk about the option to give the Netherlands a larger share of the overlapping areas that would result from the application of paragraph 101(C)(2) of the judgment.[262] Whether all of the criticism on the proposal of the Ministry for the Economy was justified is questionable. For instance, the fact that the proposal did not reflect paragraph 101(D)(3) of the judgment could also be viewed as an advantage as it would allow later acceptance of a further adjustment to meet a possible criticism in that respect. In that light, the criticism that it would not be possible to make further proposals in accordance with the judgment is misguided. The objection that the proposal would suggest that Germany was not willing to compromise seemed to be more to the point. In view of paragraph 101(C)(2) of the judgment, a proposal to divide the overlapping area was already implicit in the position on natural prolongation presented in June 1969. The proposal of the Ministry for the Economy was presented with a minor modification at the trilateral meeting of 4 and 5 November 1969.[263] A significant difference was that the German delegation did not present this as a compromise proposal, but as the German view on the outcome of paragraph 101(C)(2) of the judgment.[264]

In respect of Denmark, Germany intended to stick to its original position in order to force the Danes to develop a view on how the

---

[259] *Ibid.*, p. 5; note dated 27 October 1969 [AA/74], pp. 1–2.

[260] Note dated 27 October 1969 [AA/74], p. 1; note dated 24 October 1969 [AA/74], p. 5.

[261] Note dated 24 October 1969 [AA/74], p. 3.   [262] *Ibid.*, pp. 4–5.

[263] The proposal had linked the line resulting from an equal division of the area of overlap to the terminus of the existing partial boundary between Germany and the Netherlands. The proposal as presented left the question how the line dividing the area of overlap and the partial boundary pending (see also note dated 27 October 1969 [AA/74], p. 2).

[264] For the discussion of the German proposal/view see further text at footnote 294 and following.

continental shelf should be delimited. In this way, the disputed area would be roughly defined.[265] That would allow Germany to develop further initiatives along similar lines as were being envisaged in relation to the Netherlands.

The Danish Ministry of Foreign Affairs, after the meeting of 23 and 24 September 1969, paid further attention to the arguments related to Sylt, Germany's access to the middle of the North Sea and the development of a proposal for a boundary based on geology. Staff of the political and legal department considered that putting too much emphasis on the issue of Sylt carried the risk that the discussions would be focused on coastal geography and not geology.[266] This obviously would play into Germany's hand. Second, there was a certain contradiction between Denmark's argument on Germany's access to the middle of the North Sea and its geological argument. Germany might use the latter argument to argue that its shelf could extend beyond its own median line with the United Kingdom up to the median line between the latter and Denmark and the Netherlands.[267] These concerns were addressed by Krog-Meyer, the head of the department, who recalled that the positions on Sylt and the median line were tactical. They were intended to arrive at a sensible solution, but need not be a part of it.[268] The argument on the median line had been useful for two reasons. It had given a chance to show the Dutch, who had already used this argument, that Denmark was willing to support them. In addition, the argument had had "nuisance value" during the last round of talks.[269] The issue of Sylt could be used to force the Germans to continue to talk about all the principles contained in the judgment. It probably was "the best 'red herring' in the actual negotiations."[270] As far as geology was concerned, it really only could be used by Denmark. That being said, Krog-Meyer believed that it would not be possible to transform natural prolongation into a geological criterion. The whole judgment on this point took geography as the starting point, but gave consideration to geology.[271]

---

[265] Note of Lantzke dated 26 September 1969 [B102/260034], p. 2.

[266] Note on the issues of the northern part of Sylt and the median line problem dated 2 October 1969 [DNA/144], p. 2.

[267] Ibid., p. 4.

[268] Note on the note of Engel of 2.10.1969 concerning Sylt and the median line problem dated 8 October 1969 [DNA/144], pp. 1–2.

[269] Ibid., p. 2.

[270] Ibid., p. 2. Translation by the author. The original text reads "bedste 'red herring' i selve forhandlingsgangen."

[271] Ibid., pp. 2–3.

Denmark prepared a delimitation proposal on the basis of the geological data included in the map it had presented during the trilateral meeting of 23 and 24 September 1969.[272] It was considered a tactical advantage to present this proposal while the modest Dutch proposal was still on the table. The Danish proposal concerned a similar area of just over 3,000 km^2.[273] The proposal would also show that Denmark had contributed to making the negotiations meaningful.[274] The proposal entailed that Denmark would offer Germany a part of the area with salt structures to the north of the equidistance line between the two States.[275] This area was a considerable distance from the location of the promising drilling sites.[276] If Germany found this offer unacceptable, it could be enlarged without posing the risk of losing the sites.[277] To the east and the west of the area Denmark was offering, the boundary of the salt structures was to the south of the equidistance line between Germany and Denmark. In this case, Denmark accepted the equidistance line as the boundary. This latter point would also allow Denmark to reply to a possible German contention that its position was solely based on geological considerations.[278] The Danish proposal was also discussed at a meeting with the Danish concessionaire, which had developed an alternative proposal and had also presented two further proposals of its British advisor, Elihu Lauterpacht.[279] The Ministry of Foreign Affairs considered that all these proposals had their disadvantages.[280] The concessionaire agreed that, taking into consideration the

---

[272] This proposal was first developed in note concerning the geological delimitation and options for further renouncement of areas to Germany dated 2 October 1969 [DNA/144] and is further elaborated in note; Shelf case dated 16 October 1969 [DNA/39].

[273] Note; Shelf case dated 16 October 1969 [DNA/39], pp. 4–5.   [274] Ibid., p. 5.

[275] For a depiction of the proposal see Figure 10.9.

[276] Report on the meeting of the advisory committee for the case with Germany on the continental shelf (Continental shelf committee) held on Thursday 9 October 1969 (KSU/MR.14) dated 20 October 1969 [DNA/39], p. 4.

[277] Two options in this respect were already sketched in note concerning the geological delimitation and options for further renouncement of areas to Germany dated 2 October 1969 [DNA/144], pp. 3–4.

[278] See also the remarks of Sandager Jeppesen in Report on the meeting of the advisory committee for the case with Germany on the continental shelf (Continental shelf committee) held on Thursday 9 October 1969 (KSU/MR.14) dated 20 October 1969 [DNA/39], p. 7.

[279] These three proposals are depicted on the figure attached to Report of the meeting with the representatives of the concessionaire concerning the shelf case Wednesday 15 October 1969 dated 27 October 1969 [DNA/44].

[280] See ibid., pp. 3–4.

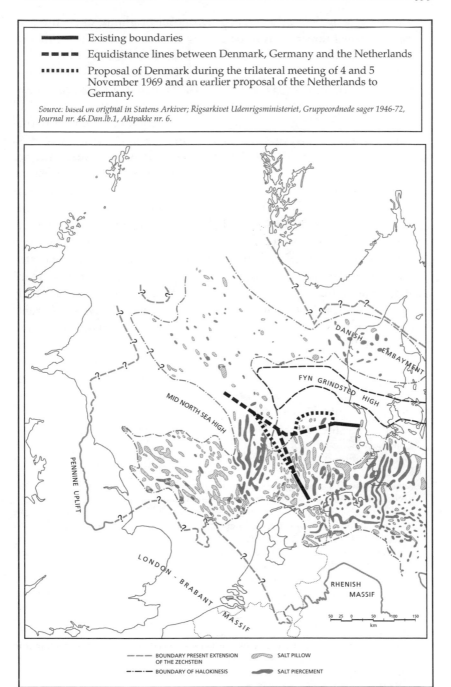

Figure 10.9 Danish proposal on a boundary with Germany based on geological considerations presented during the trilateral meeting of 4 and 5 November 1969

arguments that had been advanced in the negotiations, the proposal prepared by the ministry was to be preferred.[281]

The fourth trilateral meeting, which took place on 4 and 5 November 1969, brought the negotiations to a head. The German delegation unequivocally rejected the Dutch proposal. This obviously implied that the Danish proposal would befall a similar fate. On the other hand, Denmark and the Netherlands showed their dissatisfaction with the German approach.[282] At this point it was difficult to see how this dead-lock might be overcome. The Germans feared that Denmark and the Netherlands were looking for a political solution and would take advantage of the visit of the German President to The Hague on 24 November 1969 and Chancellor Brandt's visit to Copenhagen in the spring of 1970.[283] The Dutch delegation had already indicated that the matter would be brought up during the visit of Heinemann to The Hague.[284] For Denmark and the Netherlands there was one markedly positive aspect to this latest round of negotiations. The collision with Germany resulted in a better collaboration, as was recognized by both sides.[285] The fact that the Danish proposal concerned a similar area but was based on different methods than the Dutch proposal was considered to be favorable for the Dutch position.[286]

The German delegation's rejection of the Dutch proposal was accompanied by an explanation as to why it was not in conformity with the judgment.[287] The proposed method of adjusted equidistance had similar results as the equidistance method: it led to a cut-off effect in relation to Germany and the area concerned was too small. Obviously, the

---

[281] Ibid., p. 5.

[282] An account of this meeting is contained in Dutch Report November 1969 [NNA/31]; Report on the Danish-German-Dutch meeting in Bonn on 4 and 5 November 1969 concerning the delimitation of the continental shelf in the North Sea dated 19 November 1969 [DNA/144] (Danish Report November 1969 [DNA/144]).

[283] Note dated 13 November 1969 [B102/260035], p. 2.

[284] See note to the sections I A3; IA 5 dated 7 November 1969 [AA/74], p. 2; Dutch Report November 1969 [NNA/31], p. 24.

[285] See note 108/69 dated 7 November 1969 [NNA/31], p. 1; note; Shelf case dated 13 November 1969 [DNA/39], pp. 3–4. The bilateral meeting before the trilateral talks this time also had led to little controversy and mainly had been limited to an exchange of views (see also note 108/69 dated 7 November 1969 [NNA/31], p. 1; for a report on this bilateral meeting see Report on the Danish-Dutch preparatory meeting in Bonn on 5 November 1969 dated 19 November 1969 [DNA/144]).

[286] Note 108/69 dated 7 November 1969 [NNA/31], p. 1.

[287] Dutch Report November 1969 [NNA/31], p. 6; Danish Report November 1969 [DNA/144], p. 6.

Netherlands did not agree to these views. Its proposal was not concerned with square kilometers or areas but contained principles and methods for the delimitation and attempted to apply them in practice, as was required by the judgment. In any case, the Netherlands had not made a minimal offer. It would have been easy to find a justification for such an approach, but it had not done this to avoid a negative impact on the negotiations.[288] The Netherlands also presented a figure, which compared its proposal to the German claim before the Court and the solution which had been proposed by judge Ammoun in his separate opinion. Fack pointed out that the Dutch proposal was located between these two lines and that the proposal was more advantageous for Germany than judge Ammoun's solution.[289] Actually, the area of the Dutch proposal of 2,750 km^2 was only some 550 km^2 larger than the area between judge Ammoun's line and the equidistance line, while the German claim before the Court, which comprised 5,500 km^2 of the Dutch equidistance area, was twice as large as the Dutch proposal.[290] In reply, Groepper once more recalled the German position that its claim before the Court was no longer relevant at the present stage of the negotiations and judge Ammoun in his separate opinion had indicated that the Court had not accepted his view. Groepper found it incomprehensible that the Dutch delegation was not prepared to accept that it had made a minimal offer. The Dutch delegation might consider that it was the method that mattered, but no matter how good and equitable it seemed to be, its result also had to be equitable. How could one think that the German government could be satisfied with a mere 30,000 km^2 out of a total of 137,000 km^2?[291] This area would be almost 7,000 km^2 less than Germany had claimed before the Court. The German delegation did not yet comment on the Danish proposal during the meeting, but, as was to be expected, shortly after the meeting qualified it as unsatisfactory and not in conformity with the judgment.[292]

---

[288] Dutch Report November 1969 [NNA/31], pp. 8 and 11; Danish Report November 1969 [DNA/144], pp. 4 and 6.

[289] Dutch Report November 1969 [NNA/31], pp. 9–10. A figure comparing the three lines, entitled Nov 1969 Map no. II, is contained in [NNA/31].

[290] The figure of 2,200 km^2 for the area between judge Ammoun's line and the equidistance line is based on a comparison of this area to the area of the Dutch proposal as depicted on the figure Nov 1969 Map no. II [NNA/31].

[291] Dutch Report November 1969 [NNA/31], p. 12; Danish Report November 1969 [DNA/144], p. 7. Groepper reached the figure of 30,000 km^2 by setting the Dutch proposal at 3,000 km^2 and assuming a similar figure for Denmark and adding this to the German equidistance area (Dutch Report November 1969 [NNA/31], p. 12).

[292] Note to the sections I A3; IA 5 dated 7 November 1969 [AA/74], p. 1.

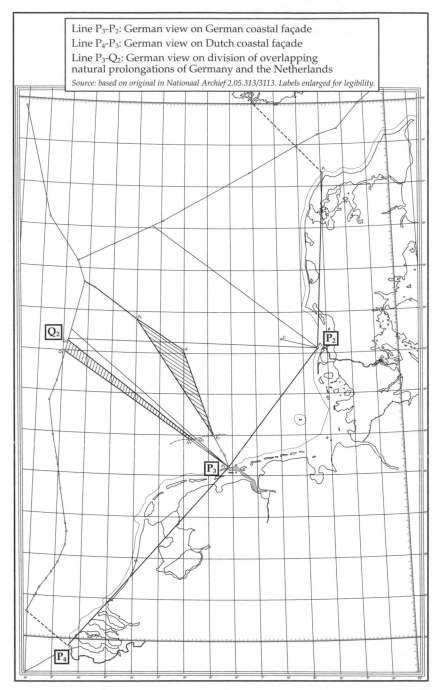

Line P₃-P₂: German view on German coastal façade
Line P₄-P₃: German view on Dutch coastal façade
Line P₃-Q₂: German view on division of overlapping
natural prolongations of Germany and the Netherlands

Source: based on original in Nationaal Archief 2.05.313/3113. Labels enlarged for legibility.

Figure 10.10   German view on natural prolongation as presented during the trilateral meeting of 4 and 5 November 1969

At the start of the second day of the meeting, the German delegation presented its view on the delimitation with the Netherlands.[293] This led to a cumbersome discussion. The German delegation at the outset observed that its proposal had to be seen as a delimitation on the basis of the principle of natural prolongation and should not be seen as a concession. It was understood that the proposals of Denmark and the Netherlands also had not been intended as a concession.[294] This characterization had two advantages. Germany could stick to its original position and implicitly was saying that now that all three parties had made known their understanding of the implications of the judgment it was time to start making concessions. As the Netherlands was in the most hurry to reach an agreement, Germany obviously did not need to go first.

Denmark and the Netherlands reacted as if stung to the German suggestion that the three parties had now presented their views on the implications of the judgment. The Dutch delegation submitted that it had already categorically rejected the German map during the June 1969 meeting. This map offered nothing new and the Dutch delegation refused to negotiate on this basis. Fack also submitted that he had made clear from the start that the Dutch proposal constituted a synthesis between the Dutch and German views and elements of the judgment. The Netherlands could not accept that its proposal was not viewed as a concession. Groepper in reply said that Germany had never retracted its map and rejected that the Dutch proposal constituted a concession. The area it purported to give to Germany was only part of the natural prolongation of Germany.[295] Because Germany had submitted that its map represented the German view on natural prolongation, Fack also introduced a new map to illustrate the Dutch view on natural prolongation.[296] According to this map only the north-facing coasts of Germany and the Netherlands were relevant for their bilateral delimitation. This map signaled that the Dutch delegation was not willing to make further concessions for the moment. This Dutch view was clearly contrary to the

---

[293] For a depiction of this proposal see Figure 10.10.

[294] Dutch Report November 1969 [NNA/31], p. 16; Danish Report November 1969 [DNA/144], pp. 8–9.

[295] Dutch Report November 1969 [NNA/31], pp. 17–18 and 21–22; Danish Report November 1969 [DNA/144], pp. 9–10.

[296] Dutch Report November 1969 [NNA/31], pp. 20–21. This map is included as Figure 10.11.

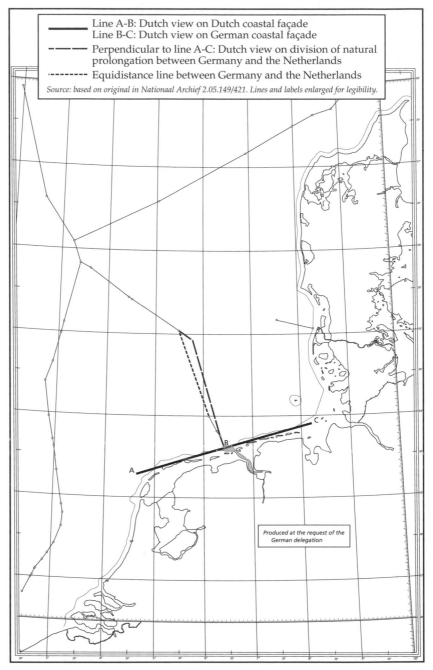

Figure 10.11   Dutch view on natural prolongation as presented during the trilateral meeting of 4 and 5 November 1969

judgment of the Court, as it implied that Germany's natural prolongation did not extend to the middle of the North Sea and implied a boundary which was even less advantageous to Germany than the equidistance line.[297]

Denmark insisted that Germany had undertaken to make a specific proposal. This was at first denied by Groepper, but after much toing and froing he admitted that the line on the German map represented a proposal to the Netherlands and that the proposal to Denmark would follow.[298] It can be questioned whether this was a fortunate admission, as it allowed the Netherlands and Denmark to maintain that Germany was taking an inflexible stand.[299] While discussing how to proceed, the parties agreed that they would exchange working papers to explain their proposals to each other and comment on the proposal of the other party as this might offer an avenue for progress.[300] At the very end of the meeting, the German delegation also recalled the map which had been handed by mistake to Fack during the June 1969 meeting.[301] That map illustrated that, had the Dutch delegation been prepared to accept the German approach, but not its actual implementation, the German delegation would have been prepared to make concessions.[302] As Figure 10.12 illustrates, if the methodology of equal division of overlapping prolongations of paragraph 101(C) would be applied on the basis of the second German map, the resulting boundary would be considerably closer to the German claim before the Court than to the line it had submitted during this last round of negotiations. At the same time, this concerned an area of more than twice the size of the part of the German sector claim on the Dutch side of the equidistance line, and still would have left a gap of over 8,500 km^2 with the Dutch opening bid of 2,750 km^2.

10.5.5 Bringing in the politicians

After the meeting in Bonn, Fack immediately drew up a memorandum for Luns, the Dutch Minister of Foreign Affairs. He painted a bleak picture of the negotiations. Although a solution had to be based in law,

---

[297] See also Danish Report November 1969 [DNA/144], p. 9.
[298] Dutch Report November 1969 [NNA/31], pp. 16–24.
[299] See also Danish Report November 1969 [DNA/144] p. 11.
[300] Dutch Report November 1969 [NNA/31], pp. 22–25.
[301] See also text at footnote 124 and following.
[302] Dutch Report November 1969 [NNA/31], p. 24. This map is included as Figure 10.6.

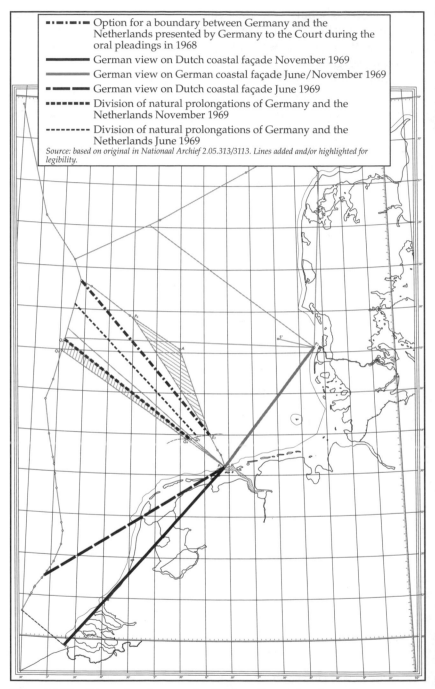

Figure 10.12   Comparison between the division of natural prolongations presented by Germany in November 1969, division of natural prolongations on map handed to the Dutch by mistake in June 1969 and the claim of Germany before the Court

politically speaking it was reasonable to expect that this solution would lie somewhere between the views that both parties had defended as being equitable before the Court. The latest negotiations had shown that the German delegation did not grasp this political context. It had presented a proposal to the Netherlands which was two to three times larger than the German claim before the Court. On the other hand, the Dutch delegation, which was not prepared to negotiate on this basis, had already made a proposal, which had taken into account the views of both parties and the guidelines of the Court. Fack suggested to Luns to talk to his German counterpart during the visit of the German President to The Hague later that month and tell him that the German delegation, contrary to the Dutch delegation, was pursuing a tactic of exaggerated claims, which was not only fruitless and irritating but also unnecessarily prolonged the negotiations, which might burden the bilateral relations. Luns might consider expressing the hope that the German delegation in the future would operate on the basis of the political expediency of arriving at an acceptable compromise.[303] No doubt this was a convincing plea to place the matter on the political agenda.[304] That this came at the cost of some tweaking of the facts was apparently taken for granted. It was certainly possible to arrive at different conclusions about the German claim, but the reference to a claim of three times the area Germany had claimed before the Court seems to purposely push the limits.[305] The whole notion that there was an expectation concerning the margin within which a compromise would be negotiated is tenuous at best. What the parties actually had agreed was that they differed about the legal rules applicable to the delimitation of the continental shelf. They had committed themselves to submit this dispute to the ICJ and to delimit their continental

---

[303] Note dated 7 November 1969 from the head of the Dutch delegation to the minister [NNA/31].

[304] The secretary-general of the ministry in a hand-written note in the margin of the memorandum strongly recommended the minister Fack's suggestions as "one of the best means to bring the recalcitrant elements in the German delegation to (political) reason" (note dated 7 November 1969 from the head of the Dutch delegation to the minister [NNA/31]). Translation by the author; the original text reads "een van de beste middelen om de recalcitrante elementen in de Duitse delegatie tot (politieke) rede te brengen."

[305] This figure was based on the proposal Germany had presented during the last meeting, but at that meeting Germany had also indicated that it would not insist on the consideration of this proposal.

shelf "by agreement in pursuance of the decision of the Court."[306] Fack's acclaim for the Dutch proposal was obviously misplaced as it was clearly contrary to the judgment. The proposal basically concerned an application of the method the Court had suggested in paragraph 98 of its judgment, but failed to make any adjustment to account for the convexity of the Dutch coast. The proposal moreover did not achieve an equal or comparable treatment of Germany and the Netherlands as was required by the judgment. The Dutch view on natural prolongation that was presented during the last meeting was a travesty of the Court's intention. The fact that the German delegation at the last meeting with Denmark and the Netherlands explicitly had indicated that there was a large scope for adjusting its position was conveniently disregarded by Fack.

Minister of Foreign Affairs Luns informed Scheel, his German colleague, about the "very unsatisfactory manner in which the German government was conducting itself concerning the continental shelf" during a private talk in The Hague on 25 November.[307] The ministers had a different perception of how the negotiations were progressing, which is hardly surprising in view of the one-sided account of Fack.[308] What counted most, however, was that the talk put the issue on the political agenda and was followed by further talks outside of the framework of the trilateral negotiations.

Prior to the trilateral meeting of December 1969, the parties exchanged working papers, as they had agreed in November.[309] The

---

[306] Special Agreements, article 1(2). As the Court's judgment indicates, it held that this reference implied that the parties would be negotiating "on the basis of [its] judgment" (judgment of 20 February 1969, [1969] ICJ Reports, p. 3, at p. 48, para. 87).

[307] Note 79/69 of Mr. J. M. A. H. Luns to the secretary-general dated 28 November 1969 [NNA/31]. Translation by the author; the original text reads "zeer onbevredigende wijze waarop de Duitse Regering zich opstelt inzake het continentaal plat."

[308] According to Luns, Scheel had said that he was certain that Germany had made big concessions to allow a speedy agreement with the Netherlands. Whereas Luns had felt it necessary to talk to Scheel because of the lack of progress at the last meeting, Scheel seemed to consider that the last meeting confirmed his views. Luns also indicated that Scheel told him that he must have gotten wrong information, because according to Scheel the German attitude was much more flexible than before the judgment (note 79/69 of Mr. J. M. A. H. Luns to the secretary-general dated 28 November 1969 [NNA/31]).

[309] This concerns Memorandum dated 3 December 1969 of the Danish delegation; Memorandum on the Dutch proposal of 23 September 1969 dated 1 December 1969; and Position of the German delegation in the negotiations concerning the delimitation of the continental shelf in the North Sea after the judgment of the International Court of Justice of 20 February 1969 as working basis for the negotiations to be continued in The Hague on 4 December 1969 (German paper December 1969 [NNA/31; DNA/144]). All three papers are contained in [NNA/31] and [DNA/144].

Dutch working paper recapitulated the statements on its proposal during the negotiations and further explained the technical construction of its proposed boundary line, but otherwise had little news to offer. The German paper gave a cursory explanation of the approach it had taken in respect of the Netherlands, limiting itself to stating that it applied paragraphs 101(C)(1) and (2) of the judgment.[310] It is likely that Germany did not want to commit itself too much to this approach. An analysis on this point in the legal department of the Foreign Office had argued that the judgment left considerable room for different interpretations and had concluded that if Germany wanted to attempt to reach a compromise it was not in its interest to argue that its maximal approach was the only possible way to deal with natural prolongation in accordance with the judgment.[311] The main objection of the German working paper against the Dutch proposal was that it cut Germany off from the middle of the North Sea and as such was not equitable and not in accordance with the judgment.[312] The Danish proposal gave geology a prominence that was not justified by the judgment. It was not possible to deduce the natural prolongation of the land territory by taking into account one geological structure which was related to only one particular geological era. The German paper in particular referred to paragraphs 101(C)(1) and (2) of the judgment read in conjunction with paragraphs 95 and 99.[313] Geological boundaries in any case did not provide a basis for determining natural prolongation, because they did not coincide with political boundaries. Finally, the Danish proposal implied that Germany's natural prolongation would also extend to an area to the north of the Dutch coast and up to the median line with the United Kingdom, as these areas had the same geological structure as the area Denmark was disposed to leave to Germany.[314] During the trilateral meeting of 4 and 5 December, the Danish delegation presented a paper in which it sought to refute both German contentions.[315] The paper pointed out that the geological map showed that the form of the salt structures in the geological structure on which Denmark relied were very different in shape and nature in the northern and southern part. This

---

[310] German paper December 1969 [NNA/31; DNA/144], pp. 1–2.
[311] Note dated 7 November 1969 [AA/74], pp. 1–4.
[312] German paper December 1969 [NNA/31; DNA/144], pp. 3–4.   [313] Ibid., p. 4.
[314] Ibid., p. 5.
[315] Geological answer to the German "Position Paper" (not dated; according to a handwritten note the paper was handed to the German delegation on 5 December 1969) [AA/74].

justified talking about different geological substructures. In the center of the North Sea these substructures more or less coincided with the equidistance line between Denmark and the Netherlands. The paper also remarked that the form and distribution of the salt structures was influenced by the underlying geological formations and that the salt structures influenced the location of the formation overlaying them. There thus was no basis for the German assertion that Denmark was relying on one specific geological structure in disregard of the overall geology of the shelf. This, at first sight, attractive argument overlooks that these older and newer geological structures might still have a completely different geographical extent than the structure on which Denmark was relying on.[316]

The Danish working paper contained by far the most detailed analysis of the positions of the parties and the judgment. The analysis attempted to make the most out of the fact that the judgment did not provide a clear basis for determining the natural prolongations of the parties. However, this was a problem that in particular concerned Germany. The Danish working paper considered that "it is relatively easy to see how geography can operate to indicate the natural prolongation of the Danish land territory – in a westerly direction."[317] For Germany, geography could not operate in the same positive way. The concavity of the German coast "prevented one from seeing a 'natural' prolongation of the German land territory in any specific geographical sense."[318] Although Denmark was prepared to accept that Germany's natural prolongation extended northwestward on a purely "geographic" basis – probably not that surprising in view of the indications to this effect in the judgment[319] – it was inherently imprecise and necessary to treat it with some reserve. Accordingly, a more realistic basis had to be found and it was necessary to consider other factors.[320] As the Court nowhere accepted geometry as the proper method,[321] the only other fact that remained was geology. According to the working paper, the Court had attached special

---

[316] See also text at footnote 438.
[317] Memorandum dated 3 December 1969 of the Danish delegation [NNA/31; DNA/144], p. 5.
[318] Ibid., pp. 4–5.
[319] See e.g. judgment of 20 February 1969, [1969] ICJ Reports, p. 3, at p. 17, para. 8.
[320] Memorandum dated 3 December 1969 of the Danish delegation [NNA/31; DNA/144], p. 5.
[321] Ibid.

importance to geology.[322] This is an altogether remarkable interpretation of the judgment. Because geography purportedly leads to somewhat imprecise results it is necessary to rely on geology, for which the judgment at no point suggests that it will lead to precise results. The Danish argument was also flawed for another reason. Denmark's natural prolongation based on geography completely overlapped with the area that, according to Denmark, constituted a natural prolongation of Germany on the basis of geology. The judgment did not provide a basis to attribute this whole area to Germany, as Denmark was proposing. As paragraph 101(C)(2) indicates, such areas of overlap had to be divided between the parties.

At the end of the working paper, Denmark set out the considerations which had given rise to its proposal. In this connection it once again recalled paragraph 93 of the judgment. As was set out previously, the interpretation of this provision provided by Denmark is not in line with the judgment of the Court.[323] The proposal reached an equitable balance by taking into account the following elements:

(a) it respected the existing partial boundary;
(b) it was influenced by geological data attributing an area north of the equidistance line to Germany. Another part of this structure further west was not attributed to Germany but to Denmark, because it had more similarities to the northern part of this structure located in the Danish shelf;
(c) it took into account the area in which the Danish concessionaire had drilled. This was equitable because of the priority of Danish activities coupled with the absence of a specific German claim prior to the proceedings before the Court and partly because of the considerations set out under (d) and (e) below
(d) it took into account the presence of the Fyn-Grindsted High which was lacking in oil and gas resources – this had a very direct bearing on the value of the Danish area and therefore could not be overlooked in satisfying the criteria of equity; and
(e) the comparative sizes and economic positions of Germany and Denmark had also been borne in mind. In particular, note had to be taken of the difference in the balance of payments positions of

---

[322] Ibid., p. 6, which in particular refers to paragraph 96 but also mentions paragraphs 43, 44, 85(c) and 101(C)(1). This argument is further developed in the subsequent pages of the memorandum.
[323] See Chapter 9, text at footnote 59 and following.

both States and the wealth of natural resources Germany already possessed, while Denmark possessed none.[324]

This balancing up of considerations had, as a result, that geographical considerations, which were central to the judgment, were virtually absent in the Danish approach.

During their bilateral meeting of 3 December 1969 Denmark and the Netherlands agreed on their approach in relation to Germany.[325] Fack reported that German Foreign Minister Scheel had indicated to Luns that Germany was willing to make significant concessions. This made it difficult to understand the German tactics. Possibly, Scheel had been informed by the Foreign Office about the outcome which was expected in the end. Both delegations agreed that the German working paper evaluating the proposals of the parties did not reflect this flexibility and did not offer any possibility to move the negotiations forward.[326] The time had come to get Germany to understand that Denmark and the Netherlands wanted a sensible solution. This should be resolved by a meeting between the three heads of delegation. Denmark supported the Dutch intention to make a demarche in Bonn if the present round of negotiations did not yield any results, but considered that before this step was taken Denmark and the Netherlands should clarify what area they were willing to give up.[327] Both delegations considered it possible to give up some 5,000 km². One difficulty lay in giving Germany access to the center of the North Sea. Sandager Jeppesen in this connection remarked that for Germany a boundary on the median line with the United Kingdom probably was a matter of political prestige. Denmark was not in a position to give up this area because of the location of the drill sites of the Danish concessionaire, but maybe the Dutch could take care of this part of the solution. However, Fack indicated that this might also be difficult for the Dutch, as the northern part of the Dutch equidistance area was possibly also not without interest.[328]

---

[324] Memorandum dated 3 December 1969 of the Danish delegation [NNA/31; DNA/144], pp. 21–22.

[325] An account of this meeting is contained in Report on the Danish-Dutch preparatory meeting in Bonn on 3 December 1969 dated 15 December 1969 [DNA/144].

[326] *Ibid.*, pp. 1–2.   [327] *Ibid.*, p. 2.

[328] *Ibid.*, pp. 2, 2a and 3. Page 2a of the report is attached to a letter of U. Engel to the members of the Continental shelf committee dated 29 December 1969 [DNA/144].

At the start of the trilateral meeting of 4 and 5 December 1969,[329] Fack provided a detailed refutation of the German arguments to reject the Dutch proposal.[330] Fack used his analysis to once more urge the German delegation to take into account constructive proposals, which were based on the views of both parties and a willingness to reach a political compromise, and above all to accord them the appropriate political value.[331]

In reply to Fack's arguments, Groepper focused on Fack's remark that the judgment did not talk about the size of the continental shelf of the parties and pointed out that paragraph 91 of the judgment did refer to this matter.[332] This focused the meeting on the fundamental questions of what the judgment said about the respective shares of the parties to the continental shelf and whether Germany was entitled to now claim a larger area of continental shelf than it had done before the Court. Fack at first submitted that paragraph 91 had to be read in conjunction with paragraph 98. Paragraph 98 required a reasonable degree of proportionality and indicated a specific method. The Netherlands had applied paragraph 98 and its proposal was in full agreement with the judgment.[333] The Dutch proposal actually implied that its share of the continental shelf would be almost twice as large as that of Germany.[334] Of course, the German position was also not in accordance with paragraph 91 of the judgment because it would give Germany a larger share of the continental shelf than the other two States.

[329] An account of this meeting is contained in Report on the fifth round of negotiations between Federal Republic of Germany on the one side and Denmark and the Netherlands on the other side on the delimitation of the continental shelf of the North Sea held in The Hague on 4 and 5 December 1969 [NNA/31] (Dutch Report December 1969 [NNA/31]); Report on the Danish-German-Dutch meeting in The Hague on 4 and 5 December 1969 concerning the delimitation of the continental shelf in the North Sea dated 15 December 1969 [DNA/144] (Danish Report December 1969 [DNA/144]).

[330] Dutch Report December 1969 [NNA/31], pp. 3–6; Danish Report December 1969 [DNA/144], pp. 3–4.

[331] Dutch Report December 1969 [NNA/31], p. 6; see also Danish Report December 1969 [DNA/144], p. 4.

[332] Dutch Report December 1969 [NNA/31], p. 6; Danish Report December 1969 [DNA/144], p. 4.

[333] Dutch Report December 1969 [NNA/31], p. 6.

[334] If the Netherlands and Denmark would both give Germany some 3,000 km² of their equidistance area, the Netherlands would have an area of more than 58,500 km² and Germany an area of less than 30,000 km².

Groepper submitted that paragraph 91 referred to a situation of "quasi-equality" and paragraph 98 did not detract from this. Groepper considered that this entitled the three parties to approximately the same share of the total area,[335] which would imply an area of about 45,500 km² for each of them. This remark led to another discussion concerning the relationship between the claim of Germany before the Court and its actual claim. Denmark and the Netherlands stressed the political impossibility of a larger claim. Sandager Jeppesen considered it impossible to negotiate further on this basis. For the Danish government it would be impossible to defend a deal in parliament which would give Germany a larger share than it had asked in Court. The German delegation indicated that it was willing to take into account political considerations, but it first had to be assessed whether Germany had given a correct interpretation of article 91 of the judgment.[336]

In view of the divide between Germany and the other two parties the meeting adjourned to allow the three heads of delegation to meet informally to discuss the further course of the negotiations. This clarified that the German proposal was not intended to delay the negotiations, but constituted a serious claim. The Danish and Dutch delegations refused to accept the proposal as a basis for negotiations. The German delegation would report this to Bonn.[337] Groepper also informed his counterparts that the German delegation had been prepared to make an offer to reduce its original claim by some 5,000km² to 7,000 km². After Fack informed Groepper that such a proposal would not serve any purpose because it still implied a larger claim than Germany had made before the Court, the German delegation refrained from formally submitting it during the current round of negotiations.[338] Fack also urged that Groepper would seek authorization in Bonn to negotiate within the margin set by the German claim during the proceedings before the

---

[335] Dutch Report December 1969 [NNA/31], pp. 7–8; Danish Report December 1969 [DNA/144], p. 4.

[336] Dutch Report December 1969 [NNA/31], pp. 7–9; Danish Report December 1969 [DNA/144], pp. 5–6.

[337] See note 120/69 dated 9 December 1969 [NNA/31], p. 2; see also Dutch Report December 1969 [NNA/31], p. 10.

[338] Note 120/69 dated 9 December 1969 [NNA/31], p. 2. For the consideration of this new proposal by the German authorities see note dated 20 November 1969 [AA/74]; note dated 28 November 1969 [AA/74]. The latter note refers to an area of 45,300 km² for Germany and respectively 48,985 km² and 42,263 km² for the Netherlands and Denmark (ibid., p. 1).

Court.[339] Groepper informally indicated that Germany would be prepared to refrain from claiming the promising Danish drill sites if it would be possible to find a solution that would satisfy Germany from the perspective of the area it would get.[340] Groepper's undertaking, in relation to Denmark, was further discussed in a confidential meeting with Sandager Jeppesen. According to Groepper, Germany attached more importance to getting a satisfactory share of the continental shelf than to getting access to the middle of the North Sea. When Sandager Jeppesen had asked Groepper what Germany was looking for in terms of area, he mentioned a figure of 40,000 km^2.[341]

During the second day of the meeting the discussions on paragraph 91 and the size of the German claim were continued. Professor Foighel of the Danish delegation gave a detailed review of paragraph 91 in the context of the subsequent paragraphs of the judgment, which had been discussed beforehand with the Dutch delegation.[342] The review did little more than setting out the content of these paragraphs and refrained from answering what paragraph 91, or for that matter paragraph 98, which had been invoked by Fack the previous day in connection with paragraph 91, actually might imply in practical terms.[343] Sandager Jeppesen did look at proportionality and submitted that on the basis of the judgment Germany certainly was not entitled to a third of the continental shelf, as the coastal length of Denmark, Germany and the Netherlands was respectively around 290, 200 and 350 kilometers. For Denmark this figure was based on a line starting at sea on the median line with Norway.[344] The figure for Germany can only be explained if the islands of Sylt and Borkum are discounted and, in the case of the Netherlands, only if its whole coast line is taken into account. Very different figures are also possible. If the actual western Danish coast, the German islands and

---

[339] Report on the meeting of the advisory committee for the case with Germany on the continental shelf (Continental shelf committee) held on Thursday 18 December 1969 (KSU/MR.16) dated 8 January 1970 [DNA/39] (Report Danish CSC December 1969 [DNA/39]), p. 3.

[340] Note to the head of department from S. Sandager Jeppesen dated 6 December 1969 [DNA/147].

[341] See Report Danish CSC December 1969 [DNA/39], p. 3; see also note; Shelf case dated 18 December 1969 [DNA/39], p. 3.

[342] On this latter point see note 120/69 dated 9 December 1969 [NNA/31], p. 2.

[343] Dutch Report December 1969 [NNA/31], pp. 13–15; Danish Report December 1969 [DNA/144], pp. 9–10.

[344] Dutch Report December 1969 [NNA/31], p. 15; Danish Report December 1969 [DNA/144], p. 10.

the Dutch north-facing coast are taken into account the respective lengths are 235, 250 and 130 kilometers.

As far as the fact that Germany was now claiming more than it had done before the Court was concerned, the Netherlands, apart from again referring to the fact that Jaenicke characterized the German claim before the Court as equitable, now also submitted that Germany was legally bound by its position before the Court. According to Fack, the judgment of the PCIJ in the *Free zones* case between France and Switzerland confirmed that a party remained bound by declarations it had made before the Court.[345] A review of the *Free zones* case indicates that it was concerned with a completely different situation. During the pleadings, the agent of the Swiss government had declared that:

> should the Court see fit to insert in its judgment provisions regarding the importation of French goods free of duty or at reduced rates across the line of the Federal customs, other than the provisions proposed in the Swiss plan, Switzerland, in her capacity as a Party to the present proceedings, here and now gives her consent, i.e. she will accept this decision of the Court as binding upon her. This declaration also is henceforward and unconditionally binding on Switzerland.[346]

As the judgment in the *Free zones* case observed, the French agent expressed certain doubts about the binding character of the Swiss declaration. The Court itself concluded that "having regard to the circumstances in which this declaration was made, the Court must however regard it as binding on Switzerland."[347] In its decision, the Court placed the declaration of the Swiss government before the Court on record.[348] In this light, it should be clear that reliance on the *Free zones* case is a non-starter. Germany did not make a statement similar to that of Switzerland and there thus was nothing the Court could place on record. The German delegation denied that it had made a specific claim and that the judgment did not talk about a German claim.[349] This led to another iteration between Fack and the German delegation about the significance of Jaenicke's statement before the Court in respect of the German claim.[350] When Fack was asked to say whether or not his position implied that he accepted that the judgment adopted the German vision

---

[345] Dutch Report December 1969 [NNA/31], pp. 13–14; Danish Report December 1969 [DNA/144] pp. 8–9.

[346] Case of the free zones of Upper Savoy and the District of Gex (France/Switzerland), judgment of 7 June 1932, PCIJ, Series A/B, p. 108.

[347] *Ibid.*, p. 170.   [348] *Ibid.*, p. 172.

[349] Dutch Report December 1969 [NNA/31], p. 16.   [350] *Ibid.*, pp. 16–17.

on the delimitation of the continental shelf and the parties should negotiate on that basis he refused to accept this logical consequence of his reliance on Jaenicke.[351]

10.5.6   Arguing the law, sort of

The Dutch Ministry of Foreign Affairs considered that the trilateral meeting of 4 and 5 December did not give reason to refrain from the demarche of the Dutch Embassy in Bonn with the German Foreign Office, which had already been proposed prior to the meeting.[352] In a meeting with Duckwitz, the Foreign State Secretary, on 14 January 1970, the Dutch ambassador pointed out that the claim which Germany had maintained during the last two meetings risked bringing the negotiations to a dead end. This would burden the bilateral relations. Political considerations should play a more important role and a compromise should lie between the claims that had been presented to the Court. It would not be acceptable for Dutch public opinion and parliament if the end result of the negotiations would give Germany more than its claim before the Court.[353] Treviranus, who accompanied Duckwitz and was a member of the German negotiating team, indicated that Germany had specified that it was willing to reach a compromise by making constructive proposals. It had to be understood that the German delegation had to take into account that the judgment had raised high expectations in Germany. Treviranus pointed out that it would be easier for the German delegation if the other side would refrain from continuously arguing that Germany could not claim more than it had done in front of the Court. The German delegation also understood the importance of a political settlement. Germany was particularly interested to avoid the negotiations resulting in a deadlock before the visit of Chancellor Brandt to Denmark in February 1970. Duckwitz concluded that it was too early to conduct the negotiations at a higher political level and proposed awaiting the outcome of the next round of negotiations. The Dutch ambassador hesitatingly agreed to this conclusion, and considered that the next

---

[351] *Ibid.*

[352] See letter of the MiFA to the Dutch ambassador in Bonn dated 22 December 1969 [NNA/31], p. 1. The instructions for the Dutch Embassy contained in letter are further supplemented by a letter of the MiFA to the Dutch ambassador in Bonn dated 6 January 1970 [NNA/32].

[353] Received message in cipher (ref. no. 1469) of the DEB to the MFA dated 15 January 1970 [NNA/32], p. 1; see also note dated 16 January 1970 [B102/260035], p. 1.

round of negotiations, which were scheduled for the next week, would clarify whether the German optimism was justified.[354]

On the German side there was indeed a more positive assessment of the last round of negotiations and it was concluded that especially the meeting between the heads of delegation had defused the tense atmosphere of the November meeting.[355] After the meeting of December 1969, Germany started to look for a new position, which had to be strong and defensible at the same time.[356] That position had to satisfy a number of requirements.[357] It had to be in accordance with the guidelines contained in the judgment, should not lead to a different treatment of Denmark and the Netherlands, and should give a result within the margin of what was possible on the basis of the judgment and acceptable to Germany. As far as the last point is concerned, it was maintained that the judgment allowed solutions which would give Germany a share of between 38,000 and 50,000 km^2, but it was admitted that this did not follow obligatorily from the judgment and that less advantageous interpretations could not be rejected.[358] A realistic negotiating margin for Germany was considered to lie between 36,000 and 43,000 km^2.[359] Just before the negotiations of January 1970, Groepper, the head of the German delegation, proposed to the Foreign Minister that the German delegation be allowed to negotiate between a margin of 36,000 and 42,000 km^2 and could eventually reach a settlement with Denmark on the drill sites of the Danish concessionaire, which would leave all the sites on the Danish

---

[354] Received message in cipher (ref. no. 1469) of the DEB to the MFA dated 15 January 1970 [NNA/32], pp. 1–2; note dated 16 January 1970 [B102/260035], pp. 1–2. The Danish authorities were kept informed of the Dutch demarche with the FO (see e.g. received message in cipher (ref. no. 1469) of the DEB to the MFA dated 15 January 1970 [NNA/32], p. 2; letter of the Danish ambassador in Bonn to the MFA dated 15 January 1970 [DNA/145]).

[355] Note dated 9 December 1969 [B102/260035], p. 1.

[356] See note dated 10 December 1969 [AA/74], p. 1.

[357] The following documents address these points: note dated 10 December 1969 [AA/74]; note dated 23 December 1969 [AA/74]; draft of a letter of the FO to G. Jaenicke dated 23 December 1969 [AA/74]; proposal for the negotiating round in Copenhagen on 20/21 January 1970 dated 12 January 1970 [B102/260035]; note dated 15 January 1970 [AA/97].

[358] Note dated 23 December 1969 [AA/74], pp. 1–2. The figure of 50,000 km^2 implied that Germany would be getting more than a third of the total continental shelf area of the three parties.

[359] Note dated 10 December 1969 [AA/74], p. 2; note dated 23 December 1969 [AA/74], p. 2.

shelf.[360] The German proposal only got its final form shortly before the trilateral meeting of January 1970. Initially, it was envisaged that the proposal would give Germany a share of some 42,000 km² and half of the drill sites of the Danish concessionaire. The latter point at that time was considered to be a tactical advantage in the, no doubt hard, negotiations in the period to come.[361] It had been difficult to find a solution which would be in accordance with the judgment and treated Denmark and the Netherlands similarly. A perfect solution in this respect was not considered to be possible.[362] To arrive at the German proposal various elements of the judgment were applied in a somewhat arbitrary manner,[363] but it could be said to be in accordance with the generally vague guidelines provided by the Court.[364] The proposal Germany eventually presented gave it a share of some 40,000 km² and only one of the promising sites was located on the German side of the proposed boundary. This adjustment was intended to address the mismatch between the continental shelf area that otherwise would fall to Denmark and the Netherlands.[365]

Groepper's undertaking in relation to Denmark's promising drill sites during the last round of the trilateral talks gave Denmark reason to consider a new proposal of its own. It was considered that it was not realistic to expect that a solution that would give Germany less than 35,000 km² would be possible. The last round of negotiations had indicated that Germany could, at a minimum, accept a shelf of around 40,000 km². At present it was not possible to estimate whether it would be possible to lower this figure, and if so to what extent. The Danish delegation considered that under the circumstances a constructive effort should be made to reach a solution. To that end, a new Danish proposal would offer Germany 6,000 km². This proposal enlarged the area that Denmark had previously offered and still did not affect the Danish drill sites.[366] This offer was about as far as Denmark would be willing to go, but there was still some room for a further concession to Germany

---

[360] Note dated 15 January 1970 [AA/97], p. 4.   [361] Ibid.
[362] Draft of a letter of the FO to G. Jaenicke dated 23 December 1969 [AA/74], p. 1.
[363] See the explanation of the German delegation (text after footnote 375).
[364] See note dated 23 December 1969 [AA/74], pp. 1–2.
[365] See further text after footnote 375; see also draft of a letter of the FO to G. Jaenicke dated 23 December 1969 [AA/74], p. 2.
[366] Note; Shelf case dated 18 December 1969 [DNA/39], pp. 3–4. For a discussion of the proposal see also Report Danish CSC December 1969 [DNA/39], pp. 3–5. The map depicting this proposal is included as Figure 10.14.

without affecting the drill sites.[367] If the Dutch would also be willing to offer 6,000 km^2 the German shelf would be 35,000 km^2.[368]

Fack was not particularly enthusiastic about the Danish ideas and apparently was not yet ready to make a similar move.[369] A difference between Denmark and the Netherlands was that Groepper, in the case of the former, had agreed to specific undertakings, which to a large extent would secure the Danish interests. This in any case was not possible in relation to the Netherlands, which did not have a similar interest in a specific area. Fack considered that it might be premature to make a new proposal before Germany had shown a willingness to negotiate within the margin of its claim before the Court. Second, the proposal Denmark wanted to make might require a similar step from the Netherlands. This would require the Netherlands to develop a new methodology to justify its proposal. Fack also doubted whether the Dutch authorities would be willing to double the Dutch offer to the Germans. The Dutch delegation itself rather preferred to make gradual concessions. Finally, there was a risk that the Germans would offer to split the difference between their claim and the Danish offer. In this way, while suggesting a willingness to compromise, Germany would get more than it had claimed in front of the Court.[370] The next round of bilateral talks between Denmark and the Netherlands indicated that the thinking of the Dutch was evolving rapidly on this point. The Dutch did not revisit their objections against the Danish approach and Fack indicated that the Netherlands might also be able to make a new proposal to Germany. That proposal would be based on geology and would not give Germany access to the central part of the North Sea.[371]

It is likely that the planned visit of Chancellor Brandt to Copenhagen in February 1970 was another reason for Denmark to make its bid to reach a final solution. The Danish authorities were well aware that Germany did not want the negotiations to break down shortly before the visit.[372] This point was also explained to the Dutch during their

---

[367] Note; Shelf case dated 18 December 1970 [DNA/39], p. 4; letter of S. Sandager Jeppesen to R. Fack dated 19 December 1969 [NNA/32].

[368] Note; Shelf case dated 18 December 1970 [DNA/39], p. 4.

[369] Letter of R. Fack to S. Sandager Jeppesen dated 12 January 1970 [NNA/32].

[370] Ibid.

[371] Report on the Danish-Dutch preparatory meeting in Copenhagen on 19 January 1970 dated 29 January 1970 [DNA/145], p. 3. The latter point is not explicitly mentioned in the report, but can be deduced from the way in which the boundary is described.

[372] See e.g. note dated 16 January 1970 [DNA/145].

bilateral preparation for the talks with Germany. Sandager Jeppesen indicated that his delegation wanted to press the German delegation as hard as possible during the upcoming round of negotiations while referring to the visit of Brandt.[373]

A first part of the trilateral negotiations of 20 and 21 January 1970 was concerned with a discussion on geology between a Danish and German geologist.[374] This discussion further elaborated the views of both parties on this point, without leading to any agreed conclusions.[375] Following an informal meeting between the heads of delegation, the meeting the next day immediately turned to the new German proposal.[376] The German delegation explained that it had first determined baselines for the three States from which to establish the natural prolongations. The major adjustment in comparison to the earlier German position concerned the line representing the Dutch coast, which from its original position had been moved much closer to a line along the north-facing coast of the Netherlands. As a second step, the areas of overlap had been divided equally between Denmark, Germany and the Netherlands, giving each of the former two 42,200 km^2 and the latter 53,100 km^2.[377] These areas had been corrected, taking into account coastal lengths. As a final step the line between Germany and Denmark had been shifted somewhat further south. This assured that all but one of the promising drill sites were located on the Danish continental shelf.[378] The resulting

---

[373] Report on the Danish-Dutch preparatory meeting in Copenhagen on 19 January 1970 dated 29 January 1970 [DNA/145], p. 2.

[374] An account of this meeting is contained in Report on the sixth round of negotiations between Federal Republic of Germany on the one side and Denmark and the Netherlands on the other side on the delimitation of the continental shelf of the North Sea held in Copenhagen on 20 and 21 January 1970 [NNA/32] (Dutch Report January 1970 [NNA/32]); Report on the Danish-German-Dutch meeting in Copenhagen on 20 and 21 January 1970 concerning the delimitation of the continental shelf in the North Sea dated 9 February 1970 [DNA/145] (Danish Report January 1970 [DNA/145]).

[375] Dutch Report January 1970 [NNA/32], pp. 1–2; see also Danish Report January 1970 [DNA/145], p. 2.

[376] The map depicting this proposal is included as Figure 10.13.

[377] These figures are mentioned at Dutch Report January 1970 [NNA/32], p. 3. The Danish Report January 1970 [DNA/145] at p. 2 refers to approximately 42,000 km^2 and 53,000 km^2.

[378] Dutch Report January 1970 [NNA/32], p. 3; Danish Report January 1970 [DNA/145], pp. 2–3. The German delegation actually submitted that all these sites would be on the Danish side of the boundary, but a comparison with the location of these sites on a Danish map which was presented at this meeting indicates that one site is to the south of this line (see also note dated 22 January 1970 [B102/260035], p. 1).

division would give around 40,000 km^2 to Germany, 44,000 km^2 to Denmark and 53,000 km^2 to the Netherlands.[379] Sandager Jeppesen in reply immediately introduced the Danish proposal and stressed that Denmark could only accept proposals which would leave all the promising drill sites on the Danish shelf.[380] Groepper in a tête-à-tête later assured Sandager Jeppesen that he would live up to his promise that all the sites would be left to Denmark.[381]

Sandager Jeppesen also reiterated that it remained politically impossible for the Danish Government to give more to Germany than it had claimed before the Court.[382] Sandager Jeppesen did not claim that Germany was legally bound by that claim, as the Netherlands did.[383] Denmark also considered that the German proposal used incorrect figures for coastal lengths and as a consequence did not meet the requirements of the proportionality test of the judgment. According to the Danish figures, Germany would be entitled to about 6,000 km^2 of the Danish continental shelf. This was accomplished by the new Danish proposal. If the German delegation would not accept the Danish proposal, Chancellor Brandt, during his visit to Denmark, would be informed about the attitude of the German delegation and the minimal progress in the negotiations.[384]

Next it was Fack's turn to deal with the German proposal: Germany was also barred from claiming more than it did before the Court for legal reasons; the German view on natural prolongation had nothing to do with the concept and he had already given the Dutch view in this respect; the Court had admitted that baselines could be useful in determining equidistance lines, but a baseline which had nothing to do with the natural course of the Dutch coast was completely unacceptable; the

---

[379] Figures included in map no. 3 presented to the meeting by the German delegation. The map is included in [NNA/32; DNA/145] and reproduced here as Figure 10.13.

[380] Dutch Report January 1970 [NNA/32], p. 4; Danish Report January 1970 [DNA/145], p. 3. The map depicting this proposal is included as Figure 10.14.

[381] Note; Shelf case of 4 February 1970 [DNA/39], p. 3.

[382] Dutch Report January 1970 [NNA/32], p. 4; Danish Report January 1970 [DNA/145], pp. 2–3.

[383] Sandager Jeppesen indicated that he in general could agree to the Dutch views, but that what counted most for him was the political impossibility of a German claim exceeding the German claim before the Court (Dutch Report January 1970 [NNA/32], p. 10; see also Danish Report January 1970 [DNA/145], p. 6).

[384] Dutch Report January 1970 [NNA/32], p. 4; see also note on the course of the continental shelf negotiations in Copenhagen on 20 and 21 January 1970 (not dated) [AA/78], p. 2.

P$_2$-S' and P$_3$-E$_3$: Existing boundaries
E$_3$-E$_4$ and S'-S'': German proposal for boundaries with the Netherlands
and Denmark
Source: based on original in Nationaal Archief 2.05.313/3102. Labels enlarged for legibility.

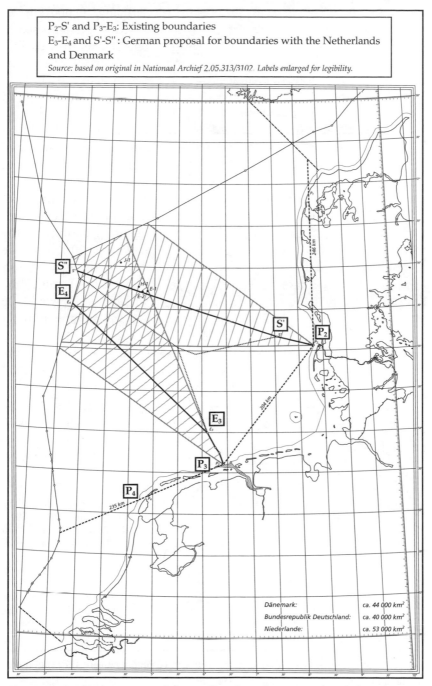

Figure 10.13 German proposal during the trilateral meeting of 20 and 21 January 1970

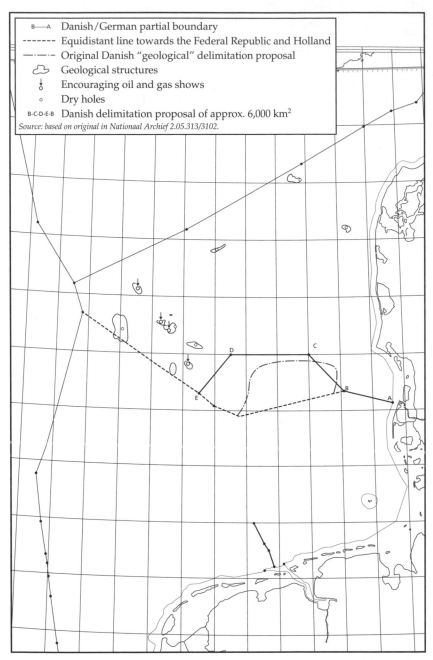

Figure 10.14 Danish proposal on a boundary with Germany presented during the trilateral meeting of 20 and 21 January 1970

Dutch coastline was much longer than the line Germany used in its proposal; and the German "Drang" to the middle of the North Sea had not been accepted by the Court.[385] Suggesting an analogy between Germany's claim to a median line with the United Kingdom to the concept of "Drang nach Osten," which had been used by Nazi Germany to justify its war of aggression in Eastern Europe, might be out of place, but was sure to bring home the point that a rejection of the Dutch demands would burden their bilateral relation. The legal arguments of Fack remained unconvincing.[386] A boundary up to or close to the median line with the United Kingdom would be in line with the judgment.[387] Second, the Dutch view on natural prolongation was contrary to the judgment.[388] Third, the German approach of using baselines in determining a bisector of overlapping areas, i.e. an equidistance line, was more in accord with paragraph 98 of the judgment than the Dutch approach. Whereas the latter approach treated the Dutch coast as a straight coast and gave full effect to it, the German baseline along the Dutch coast diminished the distorting effect of the – in the eyes of the Court – roughly convex Dutch coast on an equidistance line. Fourth, it would be possible to have different views on coastal lengths, but Fack in his statement on the length of the Dutch coast was relying on the western facing part of the Dutch coast, which in the Dutch view had no relevance for the delimitation with Germany. When Fack was later quizzed about this latter argument he indicated that he had no interest at all in baselines. The Dutch baseline in the German proposal was unacceptable, especially because it was not clear on what it was based.[389] One would be tempted to add "apart from the judgment."

After Fack had finished, Maas Geesteranus made a lengthy discourse on a number of Fack's points. Some discussion of Maas Geesteranus's presentation is useful in order to illustrate that this more detailed legal analysis did little to reinforce Fack's argument. To reject the German baseline along the Dutch coast, Maas Geesteranus referred to paragraph 91 of the judgment, which observed that the configuration of one of the States would deny one of them equal or comparable treatment if the equidistance method were to be used.[390] He failed to mention that

---

[385] Dutch Report January 1970 [NNA/32], p. 5; Danish Report January 1970 [DNA/145], pp. 3–4.

[386] The argument that Germany was bound by its claim during the proceedings before the Court is considered further in the text at footnote 391 and following.

[387] See further Chapter 9.4.   [388] See further text at footnote 297.

[389] Dutch Report January 1970 [NNA/32], p. 11.   [390] Ibid., p. 5.

the same paragraph subsequently referred to the combined effect of the concave coast of one State and the convex coast of another State or to consider the implications of paragraph 98 for the German and Dutch coasts. While arguing that Germany could not go beyond its claim before the Court, Maas Geesteranus referred to the fact that already in 1964 Germany had suggested a solution under which the Netherlands would have had to give up the relatively moderate figure of 4,800 km^2 of its equidistance area.[391] To better understand this offer, it would have been helpful if Maas Geesteranus would have provided some context. The German proposal was to give the Netherlands, Germany and Denmark respectively about 57,000, 44,000, and 40,000 km^2.[392] That is, a German share of over 7,000 km^2 more than it had claimed before the Court.

To substantiate the claim that Germany was bound by its position before the Court, Maas Geesteranus continued to rely on the *Free zones* case.[393] He admitted that the example was not completely analogous, but that there at the same time was a striking parallelism. In both cases "the Court did not reject the proposal as such, nor imposed it on the Parties."[394] As was set out above, any particular analogy between the two cases is missing.[395] Maas Geesteranus also submitted that if the Danish and Dutch governments "would accept today the proposals made on behalf of the Federal Republic before the Court in 1968, the Federal Government would be under the obligation to agree."[396] Even if the German proposal had still been on the table, this assertion would at least require significant qualifications.[397]

---

[391] *Ibid.*, p. 6.

[392] See Minutes German-Dutch negotiations of March 1964 [AA/38], p. 3. Somewhat different figures are contained in Report on the Dutch-German discussions concerning the delimitation of the continental shelf, held in Bonn on 3 and 4 March 1964 [NNA/24], p. 4.

[393] Dutch Report January 1970 [NNA/32], p. 8; Danish Report January 1970 [DNA/145], p. 5.

[394] Dutch Report January 1970 [NNA/32], p. 8.

[395] See text at footnote 345 and following.

[396] Dutch Report January 1970 [NNA/32], p. 6.

[397] For one thing, after its acceptance, the proposal would in any case have to be put in the form of an agreement. Negotiations might still break down over the exact content of an agreement and in case the agreement would be subject to approval by parliament and ratification it might never enter into force. In this context it is worth recalling an observation on proposals of the PCIJ in its judgment on the merits in the *Factory at Chorzów* case: "the Court cannot take into account declarations, admissions or proposals which the Parties may have made during direct negotiations between themselves, when such negotiations have not led to a complete agreement" (Factory at Chorzów (Merits), P.C.I.J., Series A, no. 17, p. 51). This finding was confirmed by the ICJ in the *Nuclear tests* cases [1974] ICJ Reports, p. 270, para. 54 and p. 476, para. 57.

However, during the negotiations following the judgment, Germany had made new proposals, which signified that the "proposal" it had made in 1968 was no longer on the table.

In the light of the Danish-Dutch barrage, the German reaction seems rather bland.[398] Jaenicke, who for the first time was part of the German delegation, and may have been brought in to refute the other side's reliance on his statement on the German claim before the Court,[399] even suggested that if the Dutch delegation would be prepared to make a proposal in agreement with that German sector claim, it would be difficult for the German delegation to reject it.[400]

At the end of the meeting, the heads of delegation again met *en comité restreint* on the request of Sandager Jeppesen in an attempt to get a positive German reply to the latest Danish proposal. The Germans only committed themselves to the promise that they would study the proposal.[401] They would in any case give a reply during the next trilateral meeting, but might already be able to reply during the visit of Chancellor Brandt to Denmark.[402] The Dutch in addition had pointed out that if Germany would accept the Danish proposal, the Netherlands would be prepared to make a similar proposal.[403] In a meeting between Sandager Jeppesen and Groepper, the former intimated that Denmark might offer a further 1,000 km^2, which would bring the total German area to 36,700 km^2, the figure of its claim before the Court.[404] This presupposed that the Netherlands would offer Germany almost the same area as Denmark.[405] However, the Dutch understood that their commitment to make a similar offer only implied an area of around 4,000 km^2.[406] An offer of 6,000 km^2 on the part of the Netherlands would, as a matter of

---

[398] See Dutch Report January 1970 [NNA/32], pp. 10–11; Danish Report January 1970 [DNA/145], p. 6.

[399] See also note no. 15/70 dated 22 January 1970 [NNA/32], p. 1.

[400] Dutch Report January 1970 [NNA/32], p. 11. Maas Geesteranus qualified Jaenicke's statement as "surprising" (note no. 15/70 dated 22 January 1970 [NNA/32], p. 1; see also note no. 15/70 dated 22 January 1970 [NNA/32], p. 4).

[401] Dutch Report January 1970 [NNA/32], p. 11.

[402] Note to the head of department from Sandager Jeppesen dated 22 January 1970 [DNA/147].

[403] Dutch Report January 1970 [NNA/32], p. 10.

[404] Note; Shelf case dated 22 January 1970 [DNA/145], p. 1.

[405] See also note dated 29 January 1970 [DNA/147], p. 1.

[406] This issue subsequently led to some controversy between Denmark and the Netherlands (see text at footnote 444 and following).

fact, have given Germany more than it had claimed from the Netherlands in 1968.[407]

10.5.7   The Dutch discovery of geology[408]

After the trilateral negotiations of January 1970, the Dutch delegation informed the secretary-general of the Ministry of Foreign Affairs that the demarche of the Dutch ambassador in Bonn had not yet led to any results. Although Germany had made a more modest proposal it was still claiming more than it did before the Court.[409] The memorandum also noted the faltering German reply to the Danish and Dutch legal arguments. The German delegation had failed to reply to Foighel's sharp analysis of the judgment during the December 1969 meeting, which had destroyed the legal basis of the previous German claim.[410] During the last meeting Maas Geesteranus himself had set out arguments in support of the thesis that Germany was legally bound by its claim in front of the ICJ. The only answer that had been provided by Jaenicke was that this "had been meant" as a minimum claim, but before the Court Jaenicke had not used the word "minimum."[411] Maas Geesteranus considered that the German position was mainly explained by political considerations. The German delegation had indicated that the judgment of the Court had led to expectations of a very large German shelf. This probably concerned the states and the Ministry for the Economy. The German delegation had not answered the question as to how these excessive expectations had come about.[412] This one-sided account of the progress of the negotiations confirmed the top of the ministry that Germany remained unreasonable. Foreign Minister Luns observed in the margin of the account of Maas Geesteranus: "highly improper German behavior!" and "utter nonsense!."[413] When a new Dutch proposal was

---

[407] This is explained by the fact that a larger part of the German sector claim was located in the Danish equidistance area (7,600 km^2 as compared to 5,500 km^2 for the Netherlands).

[408] For the last part of the negotiations, the relevant files at the German FO were not available (see further Chapter 1, footnote 6).

[409] Note no. 15/70 of Maas Geesteranus to the secretary-general via the directorate-general of political affairs dated 22 January 1970 [NNA/32], p. 1.

[410] Ibid., pp. 1–2. For a review of Foighel's analysis see text at footnote 342 and following.

[411] Ibid., p. 2.   [412] Ibid.

[413] Handwritten notes in the margin of ibid., p. 1. Translation by the author. The original text reads "Hoogst onbetamelijk duits [sic] optreden!" and "Grober Unfug!" The second remark of Luns in German also is the equivalence of the legal concept disturbance of the peace.

submitted to Luns for approval he found it very difficult to agree, but, because he was fed up, he was prepared to accept it reluctantly.[414]

The Dutch delegation, after the trilateral meeting, concluded that a Danish-German compromise was in the offing. This posed a problem for the Netherlands. Although the Danish proposal started to approximate the Danish share of the German sector claim of 1968, it did not give Germany access to the "much beloved" center of the North Sea. It was expected that Germany, as a consequence, would seek to accomplish this access via the Dutch side.[415] Although the German aspirations were treated with some ridicule,[416] the German claim in actual fact made sense and posed a risk to the perceived Dutch interests. The more western part of the Dutch equidistance area was considered to be the most promising from a resource perspective.[417] To make exploitation feasible, Germany could be expected to insist on a corridor of a certain width. A new Dutch proposal was specifically tailored to safeguard the Dutch interests in this respect.[418] The proposal was prepared on the basis of geological considerations, which allowed offering an area which did not extend to the western part of the Dutch equidistance area. The size of this area was some 4,100 km^2. This figure was based on the consideration that Denmark had offered 74 percent of the area Germany had claimed from it in 1968 and that the Netherlands should offer the same percentage. The Dutch delegation considered that this proposal could be said to implement the objectives of the judgment in a reasonable manner.[419] Although doubts about the reasonableness of this proposal certainly could be entertained – the judgment did not accord primacy to geology and indicated that Germany's continental shelf had to extend much further westward – it would be difficult for Germany to reject the relevance of the basis of the Dutch proposal since it had already accepted the Danish proposal based on geology.

The Danish delegation had a decidedly different view on the legal situation than the Dutch. Sandager Jeppesen considered that on all three

---

[414] Handwritten note in the margin of note 26 January 1970 [NNA/32], p. 1.

[415] Note no. 15/70 of Maas Geesteranus to the secretary-general via the directorate-general of political affairs dated 22 January 1970 [NNA/32], p. 3.

[416] See also the account of ambassador Fack, the head of the Dutch delegation (Fack, *Gedane zaken*, p. 80).

[417] See also note no. 15/70 of Maas Geesteranus to the secretary general via the directorate-general of political affairs dated 22 January 1970 [NNA/32], p. 3; note dated 26 January 1970 [NNA/32], p. 2.

[418] An explanation of the proposal is contained in *ibid.*, pp. 1–2.   [419] *Ibid.*, p. 2.

main issues at the current stage – the size of the area, the drilling sites and German access to the center of the North Sea – the Danish legal position was not particularly strong.[420] Denmark stood a better chance of realizing its objectives on the political level: agreement should be reached during the visit of Chancellor Brandt to Denmark.[421] The Danish concerns about the size of the German claim and the promising sites were brought up in a talk between Chancellor Brandt and the Danish Prime Minister Baunsgaard.[422] After Hartling, the Danish Foreign Minister had set out the Danish position, Brandt explained that the Federal government had to reach agreement with the coastal states. Once this process had been finished, a reasonable solution should be possible. Germany understood the Danish concerns about the promising sites. Sahm of the Federal Chancellery added that the German "claim" in 1968 had been intended to illustrate a possible solution. At the same time, the Danish proposal could be the basis of a solution if agreement could be reached on basic principles. This first of all concerned German access to the middle of the North Sea. A boundary with the United Kingdom was of major importance for Germany and a special arrangement, such as joint utilization, could be made for the southern drilling sites if they would be located in the German shelf. The area covered by the Danish offer was moreover artificial in form and completely uninteresting from a geological perspective. Sandager Jeppesen, who accompanied Baunsgaard and Hartling, explained that as a next step it would be very helpful if the Chancellor could issue a political directive that Germany would not seek an area larger than the 36,700 km^2 it had been asking before the Court. After Brandt had agreed to this suggestion on the condition that it would be agreeable to the coastal states, Sahm again raised the significance of German access to the center of the North Sea. Hartling replied that this did not follow from the judgment, to which Sandager Jeppesen added that this issue was bound up with the issue of the drilling sites. A possible solution was then suggested by Sahm. The boundary would not necessarily have to consist of straight lines, but Germany's access to the middle of the North Sea should be guaranteed.

---

[420] Report on the meeting of the advisory committee for the case with Germany on the continental shelf (Continental shelf committee) held on Wednesday 4 February 1970 (KSU/MR.17) dated 23 February 1970 [DNA/42], pp. 2–4.

[421] Ibid., p. 4.

[422] This talk is recorded in Discussions on the shelf case Saturday 14 February 1970 during the visit of the German Federal Chancellor [DNA/145].

Whether the German coastal states really were the cause of Germany's delay in accepting that it would not be able to go beyond its claim of 1968, as was among others suggested by Brandt's remarks, is not altogether certain. It may also have provided a convenient excuse to stall giving in on this point in order to extract further concessions on other points. The files that have been consulted in connection with this research do not suggest that the coastal states or, for that matter, the Ministry for the Economy, as was suspected by the Dutch Ministry of Foreign Affairs,[423] were still pressuring the Foreign Office to insist on a larger German share.

Before Brandt's visit to Copenhagen the German Foreign Office and other ministries had already begun to consider what a possible solution, within the margin of 36,700 km^2 set by the German claim before the Court, might look like. This consultation stressed the importance of securing German access to the center of the North Sea. The ministries considered that Germany should at least have a boundary of some fifteen nautical miles with the United Kingdom. In this connection it had to be kept in mind that continental shelf boundaries in the future might evolve into territorial boundaries.[424] In the light of the debates at the United Nations on the law of the sea, which at that time were taking place, it was certainly not unreasonable to expect that coastal States would acquire further rights beyond the territorial sea. A boundary with the United Kingdom would allow direct cooperation in respect of the laying of communication cables and the area was of interest for fisheries, navigation and military activities.[425] These interests also indicated that the boundary should be as simple as possible, but some deviation from straight lines was to be preferred to a Danish enclave around the southern drilling sites.[426]

The seventh trilateral meeting took place on 19 and 20 February 1970, shortly after Chancellor Brandt's visit to Copenhagen.[427] After a meeting

---

[423] See text at footnote 412.   [424] Note dated 16 February 1970 [B102/260035], p. 2.
[425] *Ibid.*   [426] *Ibid.*, pp. 2–3.
[427] An account of this meeting is contained in Report on the seventh round of negotiations between Federal Republic of Germany on the one side and Denmark and the Netherlands on the other side on the delimitation of the continental shelf of the North Sea held in Bad Godesberg on 19 and 20 February 1970 [NNA/32] (Dutch Report February 1970 [NNA/32]); Report on the Danish-German-Dutch meeting in Bonn on 19 and 20 February 1970 concerning the delimitation of the continental shelf in the North Sea dated 9 March 1970 [DNA/145] (Danish Report February 1970 [DNA/145]).

of the heads of delegation *en comité restreint*,[428] Fack introduced the new Dutch proposal to the plenary meeting.[429] Fack first reiterated a number of the Dutch objections to the German approach that had been presented at the last meeting and then explained that the Netherlands was prepared to make some adjustment to the previous Dutch proposal. However, the method that proposal had employed excluded giving Germany a larger share on the basis of geographical considerations.[430] As a matter of fact this would have been perfectly possible, but was not an option for the Dutch, because it would have given Germany access to an area that was considered to have greater resource potential. Fack explained that the Netherlands had assumed that Germany, which had accepted to negotiate on the basis of a Danish proposal based on geology, would have a similar willingness as far as the Netherlands was concerned.[431] The geological basis of the proposal was further explained by Dr. Thiadens of the Dutch Geological Service. Thiadens introduced a map which was based on data that had been obtained from oil companies and had been released.[432] A significant feature was the Central North Sea Graben. The fault lines of this feature could be found on maps of different geological eras. Paragraph 101(D)(2) enjoined the parties to take into account geological structures. The Central North Sea Graben was the first geological structure on the Dutch part of the North Sea coming from east. The eastern limit of this structure could serve as a continental shelf boundary. The boundary could first run along the center of the southeastern high and then follow the fault line in a northern direction.[433] After Fack and Thiadens had explained the proposal, Fack sang its praise. Not only did it give Germany some 75–80 percent of the area it had claimed in 1968, but the combined proposals of Denmark and the Netherlands guaranteed Germany a coherent continental shelf area. The Dutch proposal was based on geological natural prolongation and consequently in accordance with the judgment of the ICJ. This geological basis would also allow the Dutch government to defend the compromise

---

[428] Some points of the discussion *en comité restreint* are mentioned in the reports on the plenary or are reported upon in other documents.

[429] The map depicting this proposal is included as Figure 10.15.

[430] Dutch Report February 1970 [NNA/32], p. 3; Danish Report February 1970 [DNA/145], p. 3.

[431] Dutch Report February 1970 [NNA/32], p. 4.

[432] The map is included as Figure 10.16.

[433] Dutch Report February 1970 [NNA/32], pp. 4–5; Danish Report February 1970 [DNA/145], p. 3.

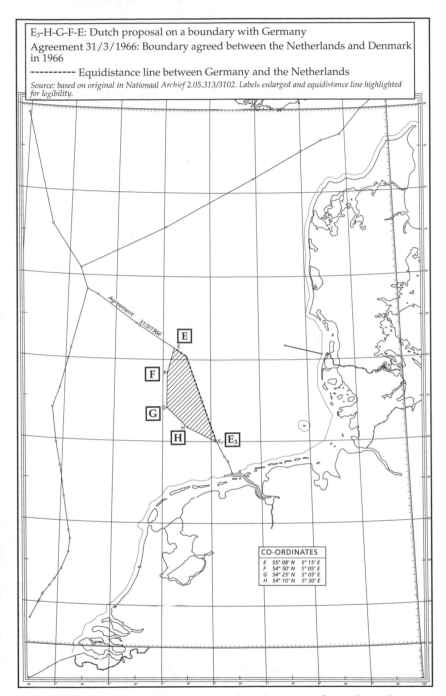

E₃-H-G-F-E: Dutch proposal on a boundary with Germany

Agreement 31/3/1966: Boundary agreed between the Netherlands and Denmark in 1966

---------- Equidistance line between Germany and the Netherlands

Source: based on original in Nationaal Archief 2.05.313/3102. Labels enlarged and equidistance line highlighted for legibility.

CO-ORDINATES

E	55° 08′ N	5° 15′ E
F	54° 50′ N	5° 05′ E
G	54° 25′ N	5° 05′ E
H	54° 10′ N	5° 30′ E

Figure 10.15 Dutch proposal during the trilateral meeting of 19 and 20 February 1970

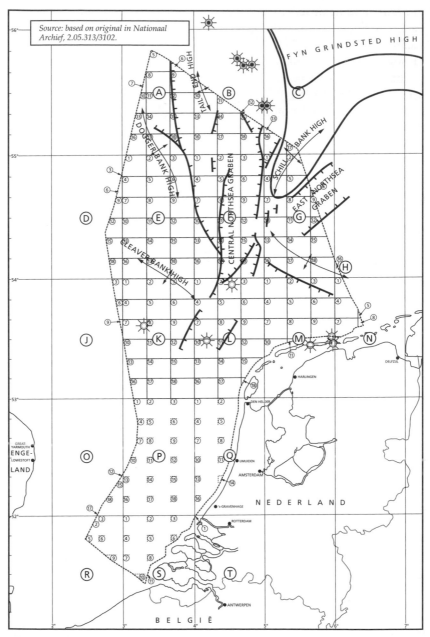

Figure 10.16 Map showing structural elements of the continental shelf; introduced by the Netherlands at the trilateral meeting of 19 and 20 February 1970

in parliament against criticism that it was too generous to Germany.[434] Fack concluded with a number of remarks, which signaled that this second offer would not be significantly altered. It had been made in response to the Danish proposal, which was very generous, and Germany's positive reply to it. The Netherlands had now made two proposals on the basis of two different methods. It did not know any other methods which might be employed. Finally, an additional condition was that a satisfactory solution should be found for the existing rights of the concessionaires of the area concerned.[435]

The German delegation recognized that it had given a favorable response to the Danish proposal, but remained of the opinion that geology could not be of decisive importance for the delimitation of the continental shelf. The explanations of the geological experts indicated that geology could lead to different results. Geology led further and further away from the basis of the judgment. Groepper then proposed to further discuss the Dutch proposal *en comité restreint*.[436] Before the plenary adjourned, Fack still submitted that on the basis of the Dutch proposal both parties were accorded structures which constituted a unit. Thiadens added that geology was a natural science and that in relation to natural prolongation geological lines were natural lines. It was not correct to claim that the geology allowed one to argue as he pleased. The geological data were a given.[437] During the meeting *en comité restreint* Boigk, the geologist on the German delegation, addressed the last point of Thiadens. Although there could be no doubt about the fact that the structures as such existed, they also unequivocally showed that structures with a north-south direction fused with structures going in a different direction. In particular, Boigk pointed out that the structures, on which the proposal was based, were linked inextricably with structures with a different direction. Boigk in particular recalled that the geological structures that could be said to represent the "basic plan of the North Sea," extended in a northwestern direction from the German coast.[438]

Even if the geological data on which the Dutch relied had been uncontroversial it would be difficult to conclude that they pointed to

---

[434] Dutch Report February 1970 [NNA/32], pp. 5–6; Danish Report February 1970 [DNA/145], p. 3.

[435] Dutch Report February 1970 [NNA/32], p. 6; Danish Report February 1970 [DNA/145], p. 3.

[436] Dutch Report February 1970 [NNA/32], p. 6.   [437] *Ibid.*, p. 7.

[438] Strictly confidential note dated 26 February 1970 [B102/260036], p. 3.

specific boundary lines in accordance with the judgment. For one thing, the fault lines of a graben do not indicate a boundary between two different geological structures in the horizontal plane, but a vertical break of horizontally continuous geological structures. In addition, the Dutch proposal did not accomplish what Fack and Thiadens said it did. In part, it did not propose to divide different geological units, but cut across a single geological unit. It is also difficult to accept that the data included in the map presented by Thiadens unequivocally point to a specific boundary. It could just as well be argued that some of the more southern fault lines indicated a boundary which would also be in line with natural prolongation in a geographical sense. In the final analysis, the Dutch proposal had the same defect as the Danish proposal. In contrast to the judgment, it relegated geographical factors to a secondary place.

The discussion of the Dutch proposal *en comité restreint* pointed out that the German delegation considered it a step forward but that it did not meet the German wishes as far as its location and size were concerned. The Netherlands had been prepared to enlarge its offer, but Germany considered that concession insufficient because it still did not give Germany access to the middle of the North Sea.[439] This led to a renewed discussion in the plenary as to whether or not the judgment justified this German demand. In the end, the Danish and Dutch delegations were prepared to accept it as a political fact, without agreeing on its legal merits. The Danish delegation indicated that it would be prepared to propose to the Danish government that apart from the area of 6,000 km^2, Germany should also get a 6 kilometer wide and 1,000 km^2 large corridor aligned with the Danish-Dutch equidistance line up to the median line with the United Kingdom.[440] Though agreeing to the size of the area involved, the German delegation considered that the Danish proposal led to too artificial a line and proposed one straight line from the terminus of the existing partial boundary to the median line with the United Kingdom with an indentation in the area of the promising drill

---

[439] Dutch Report February 1970 [NNA/32], pp. 7–8. This additional area of 900 km^2 was located to the south of the area the Netherlands had already offered (see the figure attached to note no. 31/70 dated 23 February 1970 [NNA/32]).

[440] Dutch Report February 1970 [NNA/32], pp. 8–11; Danish Report February 1970 [DNA/145], pp. 5–6. The figure of 1,000 km^2 is mentioned in memorandum no. 31/70 dated 23 February 1970 [NNA/32], p. 2.

sites. Denmark and Germany agreed to consider this matter during the next meeting.[441] A further meeting *en comité restreint* led to agreement on the size of the German area: of a total area of 35,700 km², 7,000 km² would be located in the Danish equidistance area and 5,000 km² in the Dutch equidistance area.[442]

10.5.8 Getting to the center of the North Sea

The Dutch delegation considered that the last round of negotiations had shown that all parties were now intent on finding an acceptable solution, notwithstanding the differences of opinion.[443] The new Dutch proposal had served a useful purpose and might continue to do so in the future. The proposal could be justified by reference to geology and thus was in accordance with the Court's ideas on natural prolongation. In addition, the proposal gave the Netherlands the last move in the tripartite game, as it had shown the Danes that the Netherlands was not prepared to save the Danish wells by giving Germany access to the center of the North Sea.[444] In the trilateral meeting of January 1970, Denmark itself had not been prepared to give Germany this access. The Netherlands had clashed with Denmark over this issue in their bilateral coordination prior to the next round of negotiations with Germany in February 1970. Sandager Jeppesen had considered it to be unrealistic that the Netherlands was not prepared to offer Germany access to the middle of the North Sea right away. According to Maas Geesteranus, Sandager Jeppesen had moreover been incensed about the limited area the Netherlands was willing to offer to Germany. He considered that the Dutch offer should not be proportional to the Danish offer, which would give Germany 4,100 km² or about 75 percent of the area it had claimed from the Netherlands before the Court, but should concern the same area as Denmark was now offering. Sandager Jeppesen felt that his Dutch friends had abandoned and betrayed him[445] – a piece of misplaced theatrics, according to Maas Geesteranus. He had talked to Sandager Jeppesen in January and, to be

---

[441] Report on the meeting of the advisory committee for the case with Germany on the continental shelf (Continental shelf committee) held on Wednesday 4 March 1970 (KSU/MR.18) dated 12 March 1970 [DNA/139] (Report Danish CSC March 1970 [DNA/139]), pp. 3–4.

[442] Note no. 31/70 dated 23 February 1970 [NNA/32], p. 2.

[443] Note no. 29/70 dated 21 February 1970 [NNA/32], p. 1.

[444] Note no. 31/70 dated 23 February 1970 [NNA/32], p. 3.   [445] *Ibid.*, pp. 1–2.

on the safe side, had made it clear that Fack's offer of a similar area referred to proportion and not absolute size.[446]

According to Maas Geesteranus, the Dutch refusal to give in to Denmark at the bilateral meeting had paid off. During the tripartite negotiations the Danish delegation had again made a common front with the Dutch and had shown itself prepared to give Germany access to the middle of the North Sea through the Danish equidistance area.[447] The Dutch delegation was now prepared to make a similar concession to the Germans, although it would have been preferable to only give Germany an area on the eastern part of the shelf. In exchange for this "foolish corridor" the Netherlands might still seek to retain 400 or 500 km^2 of the area of 5,000 km^2, which Germany in principle was going to get from the Netherlands.[448] The emphasis on the foolishness of the German demands in this respect is remarkable in the light of the Dutch assessment of the worth of this area and its attempts to keep Germany as far east as possible. This maybe also explains why it was concluded that all in all, the end result was a somewhat bitter pill. On the other hand, it was no worse than what had been expected one year ago after the judgment of the ICJ.[449]

Waldock, who had acted as counsel for Denmark and the Netherlands, but had not been involved in the negotiations, had a somewhat different perception of the outcome of the negotiations than Maas Geesteranus. Waldock considered that the fact that the promising locations were going to be on the Danish side of the boundary meant that Sandager Jeppesen and his colleagues "must have done extremely well in the negotiations."[450] In all likelihood, Waldock was not aware of the political dimension of the negotiations and had evaluated the outcome solely against the judgment of the ICJ.

The Danish offer to give Germany access to the middle of the North Sea along a narrow corridor led to some discussions between the Danish authorities and the Danish concessionaire. In a meeting on 2 March

---

[446] *Ibid.*, p. 2. There are no other reports on this meeting. A letter of Fack to Sandager Jeppesen rather suggests that the Netherlands considered that a similar offer referred to absolute size and not proportionality. The letter mentions that following the Danish lead would mean a doubling of the Dutch offer to the Germans (letter of R. Fack to S. Sandager Jeppesen dated 12 January 1970 [NNA/32], p. 1; see also note dated 29 January 1970 [DNA/147], p. 3). Doubling the offer would imply a new offer of between 5,500 and 6,000 km^2 and not an offer of 4,100 km^2.

[447] Note no. 31/70 dated 23 February 1970 [NNA/32], pp. 2–3.   [448] *Ibid.*, p. 3.

[449] *Ibid.* On this latter point see further text at footnote 72 and following.

[450] Letter of H. Waldock to S. Sandager Jeppesen dated 13 January 1970 [DNA/133].

1970, which considered various options in this respect, Kruse of A.P. Møller observed that it had come as somewhat of a shock that the negotiations had come to this point. Kruse thought that the concessionaire probably could accept the corridor solution, but in that case a solution for a number of less important structures would have to be found. This came as a surprise to the Ministry of Foreign Affairs, which thus far had not been informed of these structures, and it was considered unlikely that they could be saved.[451] At this point, Mærsk McKinney Møller, who headed the A.P. Møller Group, personally intervened with the ministry. Mærsk McKinney Møller regretted that the concessionaire had not been informed about the offer of a corridor and suggested that the ministry might consider a different solution, which would not affect the valuable area in the west.[452] In reply, Sandager Jeppesen gave a detailed explanation of the background to the negotiations and submitted that the corridor was an absolute German condition for an agreement. It moreover had to be recognized that the German interpretation that the judgment of the Court implied that it would get access to the middle of the North Sea could not be excluded. In view of the German position, the Danish delegation and the Continental shelf committee had concluded that Denmark should concede to the German wishes on this point.[453] The Danish delegation would seek to reach a solution for the Danish interests for those parts of the Danish equidistance area that would be located on the German side of a boundary. This in any case had been an objective from the start of the negotiations.[454] Mærsk McKinney Møller informed Sandager Jeppesen that the concessionaire could agree if two further conditions would also be fulfilled. The concessionaire would get a concession from Germany for the part of two specific structures ("Beth" and "John") that would be located on the German shelf and a satisfactory arrangement should be reached for any other structures which would be located in the corridor.[455] In a meeting with representatives of the concessionaire, the

---

[451] Note of 4 March 1970 [DNA/147], p. 3.

[452] Letter of A. Mærsk McKinney Møller to the MFA dated 3 March 1970 [DNA/147].

[453] Letter of S. Sandager Jeppesen to A. Mærsk McKinney Møller dated 6 March 1970 [DNA/147], pp. 1–2; on the latter point see also Report Danish CSC March 1970 [DNA/139], pp. 6–7.

[454] Letter of S. Sandager Jeppesen to A. Mærsk McKinney Møller dated 6 March 1970 [DNA/147], p. 3.

[455] Letter of A. Mærsk McKinney Møller to S. Sandager Jeppesen dated 9 March 1970 [DNA/147], pp. 1–2.

Ministry of Foreign Affairs pointed out that the Danish delegation would have to assess how to deal with this matter. Too inflexible a stance might jeopardize the negotiations. The concessionaire found it acceptable that a solution would give up one of the structures that had been mentioned by Mærsk McKinney Møller.[456]

The German authorities, after the meeting of February 1970, further considered Germany's options to get access to the middle of the North Sea. The Foreign Office held that it was absolutely necessary that Germany would get an appropriate access and all other directly involved ministries and the four coastal states concurred with this view.[457] In addition, the continental shelf of Germany should at least be between 35,000 km^2 and 36,000 km^2. A German shelf of 35,000 km^2 would be about one fourth of the total area under discussion and a smaller area would not be in accordance with the judgment of the Court.[458] Finally, the German continental shelf should have a reasonable configuration. The Foreign Minister proposed to the Cabinet that Germany should look for a solution that would consist of two straight lines running from the termini of the partial boundaries with Denmark and the Netherlands to two points on the equidistance line with the United Kingdom. In the case of Denmark, the line included an indentation to allow that the promising drilling sites would be attributed to Denmark.[459] This solution had the advantage that it would give Germany a relatively large share of the western part of the area in dispute and also had other advantages. For one thing, the judgment of the Court seemed to accord better with this solution than the proposals of Denmark and the Netherlands. Moreover, this proposal would not have any negative foreign policy consequences as it did not have anything unreasonable for the other two parties. Time was also playing to Germany's advantage. Denmark and the Netherlands were interested in continuing exploratory activities. Germany was not under a similar pressure.[460]

During their next bilateral meeting, Denmark and the Netherlands again collided over the size of the area the Netherlands was willing to offer to Germany. Fack argued that the Dutch offer of 5,000 km^2 had not taken into account that Germany wanted access to the middle of the

---

[456] Note dated 17 March 1970 [DNA/145], p. 2.
[457] Letter of the FM to the Head of the Federal Chancellery dated 16 March 1970 [B102/260036], p. 2.
[458] *Ibid.*   [459] *Ibid.*, p. 3. The map depicting this proposal is included as Figure 10.17.
[460] *Ibid.*, pp. 3–4.

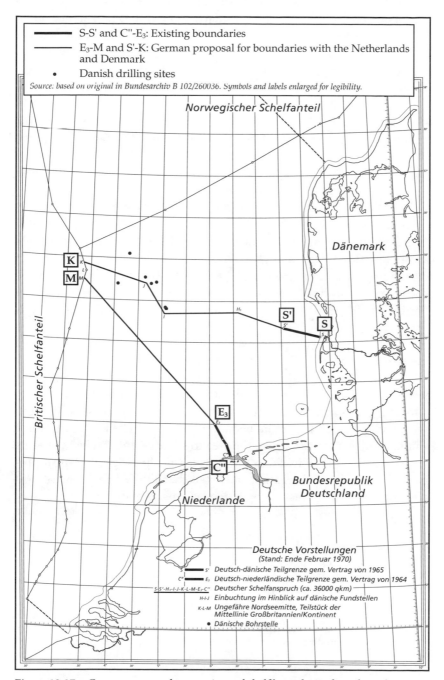

Figure 10.17   German proposal on continental shelf boundaries for trilateral meeting of 19 and 20 March 1970

North Sea. If that German wish had to be realized on the Dutch side, it would not be possible to go further than 4,500 km².[461] Sandager Jeppesen doubted whether this understanding had been clear and he moreover doubted that Germany was willing to accept less than 5,000 km² and access to the middle of the North Sea.[462] The rest of the meeting was concerned with other issues which would need to be considered once agreement would have been reached on the location of the continental shelf boundaries. This, among others, concerned a solution for the existing rights of concessionaires. Sandager Jeppesen suggested that the Netherlands might consider making its offer of 5,000 km² and access to the middle of the North Sea conditional on a satisfactory solution in respect of the existing rights of concessionaires.[463]

The March 1970 trilateral meeting in The Hague at first almost broke down as a result of the Dutch insistence that if the Netherlands would give Germany access to the middle of the North Sea, Germany should accept an offer of less than 5,000 km² from it.[464] Germany initially was demanding an area of 5,100 km.[465] After the German-Dutch discussion had come to a dead end, Denmark introduced its proposal consisting of a narrow corridor for Germany.[466] After Germany had indicated that this proposal was unacceptable for reasons of domestic politics, Denmark introduced a second proposal, which enlarged the corridor and reduced the nearshore area.[467] Denmark emphasized that this should result in a solution for the structures "Beth" and "John" through a concession or joint development. The negotiations were resumed on the second day after the parties had held internal deliberations. The discussions between the Dutch and Germans remained quite heated.[468] The Dutch were only willing to give in on the size of the area to some extent and were offering a corridor that was similar to the corridor of the original Danish proposal. The German delegation indicated that the Dutch proposal, just like the original Danish proposal, was unacceptable for reasons of

---

[461] Report on the Danish-Dutch preparatory meeting in The Hague of 18 March 1970 dated 9 April 1970 [DNA/145], p. 1.

[462] Ibid., p. 2.   [463] Ibid., p. 4.

[464] Report on the meeting of the advisory committee for the case with Germany on the continental shelf (Continental shelf committee) held on Wednesday 1 April 1970 (KSU/MR.19) dated 9 April 1970 [DNA/139] (Report Danish CSC April 1970 [DNA/139]), p. 1.

[465] Note no. 42/70 dated 25 March 1970 [NNA/32], p. 1.

[466] For a figure illustrating this proposal see Report Danish CSC April 1970 [DNA/139], Annex 1.

[467] For a figure illustrating this proposal see ibid., Annex 2.   [468] Ibid., p. 2.

domestic politics. The second Danish proposal also remained very difficult to accept because it gave Germany too large a share of the nearshore
area and too deep an indentation around the promising Danish drilling
sites.[469] The Dutch and German delegations then started to consider
whether the matter should be referred to the foreign ministers.[470]
Denmark considered that this was not particularly expedient and made
a further proposal to meet the German objections to its second proposal.
According to Sandager Jeppesen, a solution was reached after Fack "all of
a sudden and rather unexpectedly" declared to be prepared to offer
5,000 km^2 and consider a different configuration.[471] This compromise
in principle was translated into precise boundary lines by cartographers
of the delegations, which were agreed upon in the plenary.[472] For
Denmark its agreement on the boundary was conditional on agreement
on a concession for the Danish concessionaire for a part of the area
concerned. The Netherlands attached a similar condition: the rights of
concessionaires of existing concessions had to be protected.[473] While the
delimitation had been considered by the heads of delegation *en comité
restreint*, the delegations in the plenary had worked out a list of topics
which still had to be resolved before the negotiations could be finalized.
This, among others, concerned the question of how to deal with existing
concessions, whether the parties should (also) conclude a trilateral
agreement or two bilateral agreements, whether the provisions of the
agreements on the partial boundaries should also be incorporated in the
new agreement(s); and what kind of interim solution might be
required.[474] Fack and Sandager Jeppesen proposed to institute a working

---

[469] *Ibid.*; see also memorandum no. 42/70 dated 25 March 1970 [NNA/32], p. 1.

[470] Report Danish CSC April 1970 [DNA/139], p. 2; see also memorandum no. 42/70 dated 25 March 1970 [NNA/32], p. 1.

[471] Report Danish CSC April 1970 [DNA/139], p. 2; translation by the author. The original text reads "pludselig og ret overraskende"; see also memorandum no. 42/70 dated 25 March 1970 [NNA/32], p. 1.

[472] Report on the Danish-German-Dutch meeting in The Hague on 19 and 20 March 1970 concerning the delimitation of the continental shelf in the North Sea dated 9 April 1970 [DNA/145] (Danish Report March 1970 [DNA/145]), p. 3. For a depiction of the boundaries and equidistance lines see Figure 10.18.

[473] *Ibid.*

[474] For the full list of topics see Report on the eighth round of negotiations between Federal Republic of Germany on the one side and Denmark and the Netherlands on the other side on the delimitation of the continental shelf of the North Sea held in The Hague on 19 and 20 March 1970 [NNA/32] (Dutch Report March 1970 [NNA/32]), pp. 2–3; Danish Report March 1970 [DNA/145], Annex I.

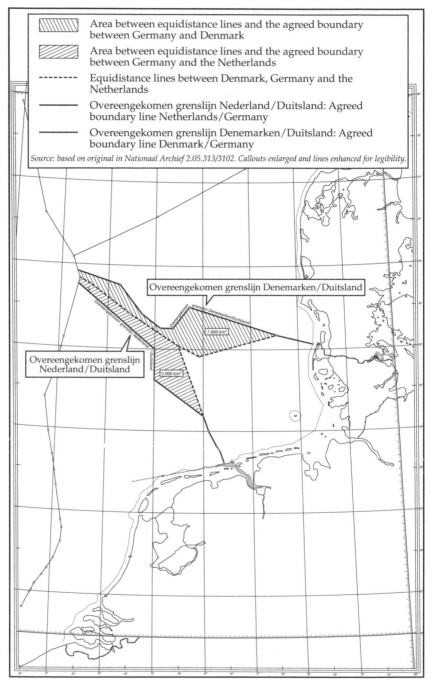

Figure 10.18   Outcome of the negotiations between Denmark, Germany and the Netherlands after the judgment of the ICJ

group, which could meet regularly to discuss these points, but Groepper was not yet prepared to accept this.[475]

10.5.9   A bearable outcome

The Danish and Dutch governments readily accepted the outcome of the trilateral talks of 19 and 20 March 1969 as a basis for further negotiations.[476] The only issue for Denmark was the size of the German concession the Danish concessionaire would get. About 50 percent of the Dutch equidistance area which would be part of the German continental shelf was part of existing concessions. The Danish concessionaire assumed that if Germany would accept that these existing concessions would be respected, Denmark could attain a similar outcome.[477] A discussion in the Danish Continental shelf committee pointed out that this would be a difficult proposition, both from a legal and political perspective.[478] Krog-Meyer, the head of the legal and political department of the Ministry of Foreign Affairs, submitted that in his view Germany considered that it had already made major concessions and might take this further Danish demand as a welcome pretext to extract itself from the existing compromise.[479] In the end, Sandager Jeppesen concluded that whether this issue would be taken up with Germany would depend on the course of the further negotiations.[480]

For the German authorities the outcome of the last round of trilateral talks was hard to swallow. The Foreign Office concluded that this was "a compromise that was still bearable."[481] A better result for Germany would only have been possible if Germany would have been willing to face a political confrontation with Denmark and the Netherlands.[482] The Foreign Office considered that the configuration of the German

---

[475] Dutch Report March 1970 [NNA/32]; Danish Report March 1970 [DNA/145], p. 4.

[476] See note dated 2 April 1970 [DNA/147]; letter of the MiFA to the Prime Minister dated 2 April 1970 [NNA/32]; and Report on the ninth round of negotiations between Federal Republic of Germany on the one side and Denmark and the Netherlands on the other side on the delimitation of the continental shelf of the North Sea held in Lyngby on 16 and 17 April 1970 [NNA/32], p. 1.

[477] Letter of A. Mærsk McKinney Møller to S. Sandager Jeppesen dated 25 March 1970 [DNA/147].

[478] See Report Danish CSC April 1970 [DNA/139], pp. 3–5.   [479] *Ibid.*, p. 3.

[480] *Ibid.*, p. 5.

[481] Annex to the proposal to the Cabinet of the FO dated 17 April 1970 [B102/260036], p. 3. Translation by the author. The original text reads "ein noch tragbarer Kompromiß."

[482] *Ibid.*

continental shelf and the demands in respect of the protection of the existing rights were the most problematical aspect of the compromise. In March, the German delegation had done everything to get a better configuration of the boundary lines, but had met with the intransigence of the other side. Further concessions on other points to get a better outcome were not an option.[483]

The German North Sea consortium considered that the additional German continental shelf area was completely uninteresting.[484] Not surprisingly, it was concluded that the form of the area could only be understood if one knew the geology of the shelf. This made it clear to what extent the oil interests of the Danish and Dutch delegations had been successful.[485] The German coastal states agreed to the draft agreements with reservation. The area concerned was too small, in view of the existing geological knowledge of little interest, and German firms only would get access to the least promising areas because the rights of existing concessionaires would be respected. Because the Federal government had already committed itself to the agreements, the states considered that it was not politically possible for them to reject them.[486]

A final squabble over the location of the continental shelf boundary took place during the trilateral talks of 16 and 17 April 1970. The German government had not yet given its agreement to the outcome of the previous negotiating round and was still looking for a modification of the course of the boundary with Denmark. On its part, Denmark brought up the size of the concession area the Danish concessionaire would get from Germany. The German delegation indicated that it could not make further concessions on this point for internal reasons. The discussion on this point prevented Germany from looking for further modifications of the course of its boundary with Denmark.[487] The German delegation also agreed to the Dutch demand that the rights of existing

---

[483] *Ibid.*, pp. 2–3.   [484] Draft note dated 14 April 1970 [B102/260036], p. 1.
[485] Letter of Preussag to the ME dated 21 April 1970 [B102/260036], p. 2.
[486] Letter of the Minister for the Economy and Public Works of Lower Saxony to the Prime Minister of Lower Saxony dated 15 October 1970 [B102/260038], pp. 3–4; see also Annex 11 to that letter, pp. 17–20.
[487] Report on the Danish-German-Dutch meeting in Bonn on 16 and 17 April 1970 concerning the delimitation of the continental shelf in the North Sea dated 23 April 1970 [DNA/145], p. 7; memorandum 50/70 dated 20 April 1970 [NNA/32], p. 2.

concessionaires would be respected, although it had serious objections because this arrangement had not been part of the political deal Brandt had reached in Copenhagen in February 1970.[488]

After the trilateral meeting of April 1970 all issues in respect of the location of the continental shelf boundaries of Germany with Denmark and the Netherlands had been resolved. The remaining issues were handled without major difficulties and the two bilateral delimitation agreements were initialed October 1970. Their parliamentary approval was secured without any difficulty. The explanatory memoranda to the bills for approval in the Dutch and German case differed significantly. In the case of the Netherlands, the conformity with the judgment was stressed, while pointing to the significance of natural prolongation in the geological sense. The memorandum only had difficulty in giving a legal explanation for the last part of the boundary. The memorandum indicated that the knowledge about the geology of the seabed in this area was limited.[489] The memorandum did not inform parliament that this most western area was considered to be of most interest from a resource perspective. Apparently, it was also a bridge too far to recognize that natural prolongation in a geographical sense might explain this part of the boundary. The German Explanatory memorandum in its overall conclusion stressed the political considerations, which Germany had to take into account in respect of both of its neighbors. In that light, the outcome could be considered to be an acceptable compromise solution.[490]

The ratification of the two agreements on 7 December 1972 finally settled the continental shelf boundaries between the three States.

10.6   Sorting out the roles of politics and law

When they submitted their disputes to the ICJ, Denmark, Germany and the Netherlands committed themselves to delimit the continental shelf in

---

[488] Letter of the Minister for the Economy and Public Works of Lower Saxony to the Prime Minister of Lower Saxony dated 15 October 1970 [B102/260038], Annex 11, p. 15.

[489] See Explanatory memorandum for the bill for approval of the Treaty between the Kingdom of the Netherlands and the Federal Republic of Germany on the delimitation of the continental shelf under the North Sea concluded in Copenhagen on 28 January 1971 (K. II (BH) 11314 (1970–1971), no. 3).

[490] See memorandum dated 3 March 1972 in Drucksache VI/3224; Deutscher Bundestag – 6. Wahlperiode, pp. 25–28, at p. 27; see also Von Schenk, "Die vertragliche Abgrenzung," pp. 389–390.

the North Sea in pursuance of the decision by the Court.[491] Did the parties live up to that commitment in the subsequent negotiations? Serious doubts have been voiced in this respect. As has been observed by one commentator:

> The agreements concluded by the Federal Republic of Germany with its neighbours following the Court's 1969 judgment, even though they claim to be on the basis ... the judgment, rest in fact on considerations of economics and expediency unconnected with the Court's legal guidelines, in particular, those relating to natural prolongation.[492]

Another analysis refers to the boundaries of the two agreements as respectively "a purely pragmatic solution" and a "pragmatic line."[493] Von Schenk concluded shortly after the negotiations that the outcome was at least as much influenced by political and economic considerations as the judgment of the Court.[494]

The detailed account of the negotiations following the judgment in this chapter confirms these assessments and moreover makes it possible to establish how this result came about.[495] It also indicates that the fact that the judgment is not reflected in the agreements did not imply that it had no impact at all. The main reason that the judgment actually played such a limited role is that Denmark and the Netherlands consistently sought to avoid its negative implications and Germany was unwilling to risk damaging the bilateral relations with its two neighbors. Germany did not throw in its political weight to back up its legal arguments and most of the time did not seriously seek to expose the weaknesses in the arguments of its adversaries. Although this only came out in the open a couple of times, Germany's restraint can largely be explained by a fear

---

[491] Special Agreements, article 1(2). As the Court's judgment indicates, it held that this reference implied that the parties would be negotiating "on the basis of [its] judgment" (judgment dated 20 February 1969, [1969] ICJ Reports, p. 3, at p. 48, para. 87).

[492] Weil, *Reflections*, pp. 112–113. In a footnote to this quotation Weil refers to an assessment by Wenger who reaches an almost identical conclusion (A. Wenger, *Pétrole et gaz naturel en Mer du Nord; Droit et Économie* (Paris: Editions Technip, 1971), pp. 133 and 135); see also Reynaud, *Les différends*, pp. 176–179.

[493] D. H. Anderson, "Denmark-Federal Republic; Report Number 9–8" in J. I. Charney and L. M. Alexander (eds), *International Maritime Boundaries* Vol. II (Dordrecht: Martinus Nijhoff Publishers, 1993), pp. 1801–1805, at p. 1804 and D. H. Anderson, "Federal Republic-The Netherlands; Report Number 9–11" in Charney and Alexander (eds), *International Maritime Boundaries*, pp. 1835–1839, at p. 1838.

[494] Von Schenk, "Die vertragliche Abgrenzung," p. 390.

[495] The account in *ibid.*, pp. 378–382 and 388–391 in general coincides with the present analysis.

that too assertive a stance might raise the specter of Germany's Nazi past. Germany's initial refusal to submit the disputes to the ICJ and its apparent relief about the course of the proceedings also illustrate that Germany was far from confident about its standing in the international community. In a subsequent account of the negotiations ambassador Fack was quite frank about how he had approached the negotiations. When he learned that Germany, on the basis of the judgment, might not be prepared to negotiate a compromise solution between the claims of the parties before the Court:

> I immediately informed my Danish colleague and we agreed that any aggressive German passes would be kicked back ferociously by the two of us and to corner our German colleague, who had been appointed in the meantime, until some sign of reasonableness would be visible on the German side ... Eventually, the Germans, who initially had made preposterous demands, were forced to back down somewhat. They obviously were embarrassed by the verbal sarcastic castigations of the Dane and me.[496]

Obviously, this later account, to stick to Fack's sports imagery, is a vintage example of gallery play that sits uncomfortably with what actually went on.

Germany's opening bid for the negotiations could be said to have immediately put the difficulties of Denmark and the Netherlands in focus. Germany's view of natural prolongation certainly was open to criticism, but it would have been difficult for Denmark and the Netherlands to engage Germany on this point. For the Netherlands, any attempt to determine its natural prolongation by a straight line to redress the convexity of its coasts was bound to significantly diminish its continental shelf. When the Netherlands, at a later stage of the negotiations, finally submitted its view on natural prolongation, it basically confirmed this point. As was set out in Chapter 10.5.4, the Dutch view was in clear contradiction with the judgment. This also provided an

---

[496] Fack, *Gedane zaken*, p. 79. Translation by the author. The original text reads "Ik lichtte mijn Deense collega meteen in en we kwamen overeen getweeën alle Duitse agressieve ballen keihard terug te trappen en onze inmiddels benoemde Duitse collega in de tang te nemen totdat een eerste schijnsel van redelijkheid zichtbaar zou worden ... De Duitsers, die aanvankelijk zo bespottelijk hadden overvraagd, bonden gaandeweg wat in. Ze waren zichtbaar in verlegenheid gebracht door de verbale sarcastische striemen van de Deen en mij."

example of Germany's failure to seriously engage its opponents. Next to nothing was done to demonstrate the tenuousness of the Dutch view on natural prolongation. Denmark's problems were different from those of the Netherlands. The German approach to natural prolongation probably would not have given Denmark a bad deal in terms of area, but almost certainly would place the promising drill sites of the Danish concessionaire on the German side of a continental shelf boundary.

Denmark and the Netherlands had different views on the handling of the trilateral negotiations. Whereas Denmark sought to shift the focus of the talks from natural prolongation to other aspects of the judgment, the Netherlands wanted to get directly down to the discussion of specific delimitation proposals. This led to friction between the two sides because Denmark felt that the Dutch approach did not take the Danish interests sufficiently into account. The Netherlands was indeed prepared to risk a break with Denmark if it could get a good deal from Germany. It probably did not come to a break between the two parties because Germany did not immediately categorically reject the Dutch proposal. This allowed Denmark to come up with a proposal of its own, which eventually secured the promising drill sites of its concessionaire. The Danish and Dutch proposals had little to do with the judgment. As much was admitted by the Dutch head of delegation, who repeatedly stressed that a solution first of all had to be based on political considerations. The Danish proposal was completely based on geological considerations, giving them a prominence that went beyond any reasonable interpretation of the judgment. The different bases of the Danish and Dutch proposals and Germany's unwillingness to first force them to reach agreement on a common interpretation of the judgment had a clear advantage for Denmark and the Netherlands. While the judgment was based on the premise that the two delimitations involved a single situation, Denmark and the Netherlands after they had submitted their first proposals were basically engaged in separate bilateral negotiations with Germany. This limited the risk of conflict, but each State remained intent on advancing its own interests, where necessary by putting pressure on the other.

The first Dutch proposal had one serious drawback. Any further adjustment on the basis of its methodology would have to recognize that the Dutch coast was convex and would give Germany a larger part of the western part of the disputed area, which was considered to be of most interest from a resource perspective. The fact that Germany had accepted a Danish proposal based on geological considerations allowed the

Netherlands to also play that card. The German acceptance of this proposal is another illustration of its unwillingness to hold the other two States to the judgment of the Court.

The final compromise was only achieved after an agreement on a number of underlying assumptions had been reached at the highest political level. Although it might at first sight appear that this political intervention was primarily needed to convince Germany to tone down its demands, on closer consideration it also was conducive to the acceptance by Denmark and the Netherlands that Germany would get access to the middle of the North Sea. The last stage of the negotiations showed that neither Denmark nor the Netherlands were really willing to accord Germany access to the middle of the North Sea on their own side of their mutual equidistance line, although the agreement on principles that had been reached at the highest political level entailed that they both would have to make certain concessions in this respect.

Although Germany did not put its political weight behind its attempt to negotiate a settlement on the basis of the judgment, and Denmark and the Netherlands, to put it mildly, had no interest in following the judgment too strictly, this does not mean that the judgment was without impact. The insistence of Denmark and the Netherlands that Germany after the judgment could not claim a larger share of the continental shelf than it had done during the judgment is a telling illustration. If the judgment really would have curtailed a possible German claim, it would not have been necessary to stress the political side of this argument to such a large extent. In other words, Denmark and the Netherlands implicitly acknowledged that the law did not assist them. The judgment also curbed the possibilities of Denmark and the Netherlands to simply stick to their own views. For instance, although the judgment was far from clear, there was little support for the argument that Germany should not get access to the middle of the North Sea. Without saying that the legal argument was decisive in this respect, it did back up and work in the same direction as Germany's political priorities. This probably points to one of the most significant conclusions of this case study. The impact of international law always has to be assessed in a matrix involving other considerations and interests. As the case of Germany shows, sound legal arguments without sufficient political resolve backing them up may remain largely ineffective. At the same time, even in this situation, the other side will be aware of the relevant legal framework and will have to consider how far it wants to go in testing the political resolve of the other party.

The record of the internal deliberations in Denmark supports the view that Denmark and the Netherlands were working towards a political solution that would allow escaping from the full impact of the judgment of the Court. This involved using legal arguments that internally were recognized to be of dubious value. For example, Denmark in the negotiations insisted on a solution based on geology, but in internal discussions it was recognized that geography formed the basis of the judgment. The Danish delegation extensively argued the significance of Sylt but in an internal document referred to it as Denmark's best red herring.

The German internal deliberations also express the belief that Denmark and the Netherlands were looking for a political settlement. On the other hand, the reports of the Dutch delegation present the image of its unfaltering commitment to the judgment and exaggerated demands on the part of Germany.[497] As will probably be more often the case, the reasons for this attempt to rewrite the history of the negotiations are not documented. Was there a sincere conviction that this constituted a fair portrayal of what transpired? In view of the experience of those involved, such a thought would seem to be somewhat naïve. Two other explanations might be possible. One advantage of this approach was that Minister of Foreign Affairs Luns could be expected to give his full political support to convince the Germans to be reasonable. In addition, internal documentation would show that the Netherlands had been convinced that it had acted in a reasonable way, as had been required by the judgment of the Court. Whether this would have been of much assistance to the Netherlands if it would have had to justify itself before the Court is another question. For instance, it would have been obvious to any neutral observer that the Dutch delegation's view on natural prolongation presented during the trilateral meeting of November 1969 was in clear opposition to the judgment, and was only intended to signal to Germany that the Netherlands was not willing to discuss the implications of the judgment on this point.

---

[497] This same image is conveyed by Fack in his subsequent account of the negotiations. Fack among others recounts that at the end of the negotiations his German colleague confided that at the start of the negotiations he had had little hope to obtain a reasonable result. Fack reports to have replied that this indicated that he did not really know the Dutch. There had always been a willingness to arrive at a reasonable compromise and the Netherlands had accepted the judgment of the Court beforehand (*ibid.*, pp. 81–82).

11

The outcomes of the case study in a broader perspective

11.1 Introduction

This final chapter further discusses what the case study on the continental shelf delimitation between Denmark, Germany and the Netherlands tells us about the impact of international law on State behavior. To this end, the next section will first of all provide a brief description of a number of key episodes of the case study. How did the parties assess the content of the law in those instances and how did they take the law into account in formulating their positions and in reacting to the arguments of their adversaries? This analysis is not intended to repeat the more detailed conclusions on the case study that are included at the end of each individual chapter. The next two sections will put these findings in a broader perspective. The relationship between international law and State behavior has led to the development of a considerable number of theoretical perspectives. Chapter 11.3 describes a number of these perspectives and Chapter 11.4 assesses the fit between these perspectives and the findings of the case study. Before embarking on that analysis I would like to briefly revisit the research question I formulated at the start of this project.

My main interest was to establish the role of international law in the negotiations between Denmark, Germany and the Netherlands on the delimitation of their continental shelf in the North Sea. During my research I came across a study from the early 1980s that observed that it had often been said Germany up to the 1970s did not dare to translate its economic power into political power, and submitted that this was not completely correct as far as the bilateral relations with the Netherlands were concerned. The delimitation of the continental shelf was considered to be one case in point.[1] The outcome of the negotiations was

---

[1] H. J. G. Beunders and H. H. Selier, *Argwaan en Profijt; Nederland en West-Duitsland* (Amsterdam: Historisch Seminarium van de Universiteit van Amsterdam, 1983), p. 46.

explained by a number of factors. The German demand was probably considered to be reasonable, there were no vested interests and the Netherlands moreover had to resign itself to the decision of the ICJ. The faith in supranational institutions was still big enough to sacrifice, if necessary, certain national interests. This allowed that the dispute could be settled by experts, without involving the heads of government.[2] Reading this assessment in the light of the negotiating record, it is clear that it needs to be adjusted in many respects. However, that is not my main point. The authors at that time did not have access to the negotiating record and only could base themselves on publicly available sources. Rather, a comparison of the present analysis to this previous assessment effectively illustrates that only a detailed study will allow determination of the relation between international law and State behavior in specific instances.[3] What may seem obvious explanations in the light of outcomes are only partial explanations, or totally beside the point, on closer scrutiny of the negotiating record. Exactly what the present case study allows us to determine in this respect will be further considered in the remainder of this chapter.

11.2   Salient points of the case study

In general terms, the regime of the continental shelf and the rules for its delimitation to a large extent shaped the positions of Denmark, Germany and the Netherlands and their negotiations. When the continental shelf regime started to develop in the late 1940s and early 1950s, Denmark and the Netherlands had little difficulty in accepting it. The idea of rejecting the new regime, which was attracting broad support in State practice and was part of the ILC's work on the law of the sea, was never seriously entertained and both States readily accepted that the continental shelf regime had become part of customary international law. Things were different for Germany. For a considerable time, Germany opposed the

---

[2] *Ibid.*, p. 48.

[3] A similar point is made by Bethlehem, a former principal legal adviser of the United Kingdom Foreign and Commonwealth Office, who in commenting on the role of international law in political decision-making, observes: "All of this practice and these appreciations are simply not visible outside of government, save in the relatively unusual instances in which it becomes visible because it is leaked or because it is disclosed in court proceedings or because it is disclosed in some other manner" (D. Bethlehem, "The Secret Life of International Law," 2012 (1)1 *Cambridge Journal of International and Comparative Law*, pp. 23–36, at p. 30).

newly developing regime and had serious doubts that it had become part of customary international law. Although it is difficult to pinpoint when the continental shelf regime exactly became part of general international law, the German position in the 1950s in any case was based on too optimistic an assessment of the possibilities to halt its development. The German attempt at the 1958 Conference to replace the continental shelf regime by an internationalized regime was introduced at a very late stage and there had been no prior contacts with other States beforehand to gain support for it. The fact that a large number of States had an entrenched interest in the continental shelf regime made it unrealistic to expect that delegations at the 1958 Conference would even be willing to discuss the German initiative. Germany's approach in the 1950s illustrates how a State's interests may be affected by failing to properly assess the development of international law. As a consequence, Germany missed out on the opportunity of trying to have a meaningful impact on the development of the continental shelf regime, including the rules for its delimitation. In large part, this failure is explained by the fact that the German Foreign Office, which was responsible for dealing with legal aspects of this issue, refused to take the lead in addressing this matter.

Germany voted against the adoption of the Convention on the continental shelf at the 1958 Conference, but subsequently signed the Convention and considered the possibility of becoming a party to it. That process was never finalized because of the interpretation of the rule on the delimitation of the continental shelf contained in article 6 of the Convention by Denmark and the Netherlands. Germany was finally forced to take a position on the status of the continental shelf regime in 1964 in the face of the possibility that an American oil company would start activities in the putative German continental shelf without German permission. Despite the prior rejection that the regime had become part of general international law, Germany now had no problem with relying on general international law to adopt a proclamation on its continental shelf. This was a matter of expediency and not a result of a significant change in the law.

The equidistance principle, which was contained in article 6 of the Convention on the continental shelf, provided Denmark, Germany and the Netherlands a focal point for assessing the delimitation of their continental shelf. In the early 1950s Denmark and the Netherlands were among the States who urged the ILC to adopt a substantive rule on the delimitation of the continental shelf. Denmark in this connection specifically referred to the equidistance method. At the time, the main issue for Denmark was not that much its delimitation with Germany, in which case the equidistance method produced a satisfactory outcome for Denmark, but its delimitation with Norway. Denmark had to consider whether or not the Norwegian Trough was relevant to determining the extent of the continental shelf of Denmark and Norway or provided an argument for a boundary different from the equidistance line. In both cases, Denmark would have had a significantly larger continental shelf than on the basis of the equidistance method. In dealing with this issue, which resurfaced in the first half of the 1960s, legal arguments and other considerations both played a role in the Danish acceptance of the equi-distance line as a boundary. Denmark did attempt to extract a commit-ment from Norway that this boundary might be renegotiated in the future, but subsequently dropped this demand to avoid weakening its position in relation to Germany by signaling that Denmark was not consistently committed to the equidistance line.

In its negotiations with Germany, in the period before the judgment of the ICJ, Denmark, just like the Netherlands, relied on an interpretation of article 6 of the Convention on the continental shelf that accorded special circumstances a limited role as compared to equidistance. Internally, there may have been less confidence in this interpretation. For instance, Sørensen, the legal advisor of the Ministry of Foreign Affairs, considered that there was a 50 percent chance that a judgment of the Court would go against Denmark and the Netherlands and that the outcome of the cases was completely unpredictable. There is no evidence that a detailed assessment of the possible interpretations of article 6 of the Convention on the continental shelf was made. Denmark also took into account one other aspect of article 6 in determining its course of action. It was considered necessary to conduct negotiations with Germany to comply with the requirement that boundaries had to be established by agreement. The Netherlands also subscribed to this view.

In the 1950s the Netherlands, in considering the regime proposed by the ILC, also found the equidistance principle acceptable for the delim-itation of its continental shelf. Interestingly, at the same time it was concluded the equidistance principle accorded Belgium and Germany a limited part of the continental shelf. It was acknowledged that the reference to special circumstances in such cases was intended to achieve an equitable solution, but there was no further assessment of the exact implications of the special circumstances clause. When the delimitation with Germany had to be further considered in 1964, a memorandum of the Minister of Foreign Affairs to the Council of Ministers pointed to significant uncertainties involved in the interpretation of the special circumstances clause. The memorandum was nonetheless confident that the configuration of the German coast did not constitute a special circumstance, but admitted that the outcome of a decision by the ICJ or an arbitration was uncertain. The memorandum proposed that the Netherlands could first seek a partial agreement with Germany and then seek to achieve a negotiated settlement at a multilateral conference or a decision of the ICJ for the remainder of the boundary. It is not known what changed the appreciation of the delimitation involving Germany and the Netherlands between the middle of the 1950s and 1964. The legal and geographical context had not changed. Possibly, the change is explained by the fact that the matter was considered by different persons, and that in 1964 concrete interests were involved. At that time, it had become clear that the area between the equidistance line and a possible claim of Germany was likely to contain oil and gas.

The Council of Ministers agreed on a different course of action than the memorandum of the Minister of Foreign Affairs proposed. The Netherlands should first seek to delimit its continental shelf boundaries with its other neighbors in the North Sea. This decision relegated legal considerations to a secondary place. The decision made sense from a political perspective: it would show that Germany stood alone in its rejection of equidistance. From a legal perspective the conclusion of these agreements would do next to nothing in reinforcing the Dutch position. These bilateral agreements would not be opposable to Germany as it was not a party to them. Germany moreover could argue that the fact that equidistance was applied in these cases did not imply that it was a reasonable solution between itself and the Netherlands. Although the Council of Ministers would subsequently agree to further negotiations with Germany, the policy of concluding agreements with the other Dutch neighbors in the North Sea was also pursued. Germany did take care to ensure that the agreements between the United Kingdom and Denmark and the Netherlands could not be viewed as prejudicing its position.

The initial Dutch and Danish interpretation of article 6 of the Convention on the continental shelf, to the effect that it entitled them to unilaterally define their boundaries and that it was on Germany to prove the presence of special circumstances, probably was not conducive to presenting a convincing case later on. As was argued in Chapter 3.5 this position is not easily reconciled with article 6 itself and Denmark and the Netherlands had difficulties in defending this position before the

Court. This is illustrated by their reliance on the Court's judgment in the *Anglo-Norwegian fisheries* case during the pleadings. The analogy between that case, which was concerned with the establishment of baselines by the coastal State, and the delimitation of the continental shelf, which requires agreement between the States concerned, is hard to see. It is unlikely that such arguments convinced the Court that Denmark and the Netherlands had a convincing case. This episode illustrates that the initial choice for a not too credible legal argument may have continued repercussions for a party.

For Germany, the way equidistance worked out for it in the North Sea was an important factor to oppose the developing continental shelf regime in the 1950s. Notwithstanding this importance of the delimitation rule, there was little attention for the formulation of a different rule of delimitation. For the opponents of the continental shelf regime this would have made little sense as it might have weakened their position that the regime as such had to be opposed. Proponents of the continental shelf regime suggested some options to address this matter without coming up with alternative rules of delimitation. At the 1958 Conference Germany supported a proposal to limit the rule on delimitation to a requirement that this had to be done by agreement, but again did not suggest alternatives for a substantive rule of delimitation. The Foreign Office, which had the primary responsibility for formulating the German position, in general took little initiative and rather waited for other ministries to provide input. This stance contributed significantly to Germany's passivity before and after the 1958 Conference.

After the Conference, Germany at first relied on the rule contained in article 6 of the Convention on the continental shelf to define a position and to reject that equidistance provided an appropriate solution for its delimitation with Denmark and the Netherlands. Alternative methods for the delimitation of the continental shelf were also developed and presented, first to the Netherlands and subsequently also to Denmark. Until the proceedings before the ICJ, Germany failed to translate these proposals into a specific claim on the location of its continental shelf boundaries. That failure would, just like Germany's minimalist claim before the Court would later on, to a large extent, frustrate Germany's attempts to arrive at a settlement that would have taken full advantage of the Court's favorable judgment. What explains this failure?

The applicable law might seem to offer a good explanation as to why Germany failed to come up with a specific claim. Article 6 of the Convention on the continental shelf did not contain any specific indications on how to operationalize the special circumstances clause and customary international law, on which Germany relied subsequently, was certainly not more specific. This no doubt is part of the explanation. If the law apart from equidistance had provided for other delimitation methods, Germany could have formulated a claim on the basis of those alternatives. At the same time, the absence of such alternatives did not imply that a State was barred from determining a specific claim line. As much is illustrated by the German claim before the Court. The judgment of the Court does not contain any indication that such a claim was considered to be inappropriate. In the end, there can be little doubt that it was not the law, but the unwillingness of the Foreign Office to present a specific claim that for a long time resulted in its absence. The justification the Foreign Office at one time gave for refraining from presenting a specific position, namely that a claim had to be fully sound from a legal perspective, is not convincing. First of all, the indeterminate nature of the law implied that it was impossible to determine one specific line in the case of Germany once it had been concluded that the equidistance method was not applicable. Second, the Foreign Office did accept that Denmark and the Netherlands had determined their claim unilaterally, even though Germany did not accept that the equidistance principle was applicable. Denmark and the Netherlands did not have a position that was above criticism from a legal perspective. Just as a German claim could be found to be contrary to the law, the same fate could befall the Danish and Dutch claims. All in all, it has to be concluded that the Foreign Office was not willing to commit itself to a specific line because this would have put the German position in much sharper focus and might have led to stronger negative reactions in Denmark and the Netherlands, burdening the bilateral relations.[4]

---

[4] At first sight, another bilateral matter between Germany and the Netherlands seems to suggest that a fear of negative reactions in Dutch (or Danish) public opinion and damage to their bilateral relations is not the main or only explanation for the German stance. This concerned the so-called Generalbereinigung (general settlement). The settlement comprised a number of issues that were pending between the Netherlands and Germany in the aftermath of the Second World War and the German occupation of the Netherlands, resulting in a number of agreements that were treated as a package. The Dutch government considered that it had made significant concessions to Germany, and parliament and public opinion were critical about the outcome (see e.g. Beunders and Selier, *Argwaan*, pp. 19–22; W. H. Weenink, *Johan Willem Beyen, 1897–1976: Bankier van de Wereld, bouwer van Europa* (Amsterdam/Rotterdam: Prometheus/NRC Handelsblad, 2005), pp. 407–417; F. Wielenga, *Van Vijand tot bondgenoot; Nederland en Duitsland na 1945* (Amsterdam: Boom, 1999), pp. 232–254). Two considerations indicate that care

After Germany had reached partial delimitation agreements with Denmark and the Netherlands, the focus of the discussions shifted to the choice of the appropriate framework to settle the remaining boundaries: arbitration or the ICJ. Germany initially had a preference for arbitration and Denmark and the Netherlands for the ICJ. Germany's preference to a large extent was explained by political considerations. Denmark and the Netherlands favored the ICJ because they considered that the Court was more likely to find that the equidistance principle was applicable between them and Germany. In the absence of agreement on the appropriate forum, from a legal perspective the existing treaty relations of Germany with Denmark and the Netherlands would be the determining factor for establishing the forum to which the dispute could be submitted. The analysis by all three parties pointed out that different outcomes might be possible on the basis of these rather complex treaty relations. A number of factors contributed to the agreement to submit the two disputes to the ICJ. First of all, the Netherlands postured to take the initiative for further talks to prevent a German initiative on arbitration. Second, the Netherlands and Denmark not only rejected arbitration but positively expressed themselves about their willingness to cooperate with Germany in going to the ICJ to prevent that the submission of the disputes would lead to political fallout. Another reason for Germany to accept the ICJ was that it was not certain that the Court would reject that it had jurisdiction if it would be seized on that matter.

In connection with the submission of the disputes to the Court, the parties also had to agree on an interim arrangement for the disputed area. As was set out in Chapter 6.6, although the arrangement was satisfactory to Germany from a purely legal perspective, Germany, unlike Denmark, failed to take the political implications of this regime sufficiently into account. As a result, Denmark in the negotiations following the judgment could politically take advantage of the location of a number of promising drill sites of its concessionaire in the disputed area, even though this was irrelevant from a legal perspective.

---

should be taken in drawing conclusions from the apparent differences in the German approach to these two cases. Wielenga considers that a comparison of the outcome of the negotiations on the Generalbereinigung to the Dutch desiderata indicate that the Netherlands reached a satisfactory result (*ibid.*, p. 247). Even more importantly, the publications mentioned above are mostly based on Dutch sources, and thus provide little direct insight in the German perception as regards the course of the negotiations. The present case study confirms that there may be a chasm between how a party sees itself and how its negotiating partner looks at that same behavior.

Legal arguments obviously play a central role in the arguments of parties in proceedings before the ICJ. It would, however, be a mistake to assume that the whole process is only concerned with lining up the relevant legal argument. For one thing, the parties to the *North Sea continental shelf* cases in various ways sought to undermine the credibility of the other party. For example, Denmark intimated that Germany had only become interested in the area beyond the equidistance line after it had become aware that this area was promising from a resource perspective. This argument carried little weight for determining the applicable law, but did suggest that Germany had changed its mind because of economic interests and previously had been satisfied with its own equidistance area. Another example is provided by Germany's reference to the delimitation between Suriname and Guyana. In this case, the Court was told that the Kingdom of the Netherlands was being inconsistent. What the Netherlands was denying to Germany, it was itself claiming in relation to Guyana. Whether the two situations were really comparable – in the case of Guyana and Suriname there was a protruding headland and this was not the case between Germany and the Netherlands – and whether one situation legally speaking formed a relevant precedent for the other, was something Germany did not address. The reaction of the Dutch agent, who notwithstanding the record, denied that the Kingdom of the Netherlands had ever claimed something different from equidistance in relation to Guyana, shows the significance he attached to projecting an image of consistency.

After the judgment of the Court the parties conducted further negotiations to delimit the remainder of their continental shelf boundaries. Legal arguments figured prominently in the discussions, but in the end, the judgment only had a limited impact on the outcome of the negotiations. This is mostly explained by the fact that Denmark and the Netherlands exploited the political sensitivities of Germany. Germany remained fearful that too assertive a stance might damage its bilateral relations with its partners. As a result, arguments with limited to no legal significance carried the day. Germany's shelf area was kept slightly under the claim it had presented to the Court and the promising drill sites of the Danish concessionaire were located in the Danish continental shelf. At the same time, the judgment was not completely ineffective. It excluded that Denmark and the Netherlands could keep Germany significantly under its claim before the Court or keep Germany from getting some access to the middle of the North Sea. It would have been too obvious that a solution along those lines would not be in accordance with the judgment and it would have been impossible for the German government to credibly justify it. If Denmark and the Netherlands would have been too inflexible, Germany might also have opted to return to the ICJ, where it was likely to do better and Denmark and the Netherlands risked criticism from the Court once more for their approach to the negotiations.

The preceding analysis indicates that the law, at least on the surface, played a significant role in shaping the arguments of the parties. All claims of the parties were underpinned by legal arguments, just as these claims were rejected by reference to the law. As the preceding analysis also indicates, this reliance on the legal argument does not, however, mean that the law always had any real influence on the behavior of the parties in the sense that it constrained them in selecting preferred outcomes or forcing them to accept outcomes that were not to their liking. Is it possible to say something more about the kind of constraints the law may have imposed in this respect? For one thing, throughout the negotiations, the parties accepted that the delimitation of the areas off their coasts had to be carried out in accordance with the applicable legal regime. They accepted that their rights over this area were defined by the regime of the continental shelf and this regime implied an obligation to determine boundaries between neighboring States. An open rejection of this basic premise, safe by reference to other rules of international law, would seem to be excluded, as it would reject the whole fabric on which interstate relations are built. Beyond this general framework, the impact of the legal regime at first sight might seem to have been very limited. During the first phase of the negotiations, Denmark and the Netherlands insisted on the obligatory nature of the equidistance principle and did not want to engage in a debate on the applicable law, although their own assessment pointed out that there was a large measure of uncertainty about the content of the law. They also sought to build a web of bilateral treaty relations with their other North Sea neighbors to reinforce their position in relation to Germany. These treaties could be used to argue that Germany was isolated in its rejection of equidistance. Such arguments might carry political and psychological weight but these treaties did not have legal consequences for Germany. After the judgment of the Court, Denmark and the Netherlands successfully disengaged themselves from a meaningful discussion of the implications of the judgment and exerted political pressure on Germany to accept a compromise solution. The legal arguments they advanced most of the time were mere window dressing, but Germany basically refused to expose the weakness of its adversaries on this point. This hardly suggests that the law had a significant impact on the behavior of the parties beyond requiring them to pay lip service to it. On closer consideration, a less dismal picture emerges.

During the first phase of the negotiations, Denmark and the Netherlands did acknowledge that it would be inevitable to settle their dispute with Germany and they acknowledged that a decision on legal grounds might lead to a boundary other than the equidistance line. Their unwillingness to reach a compromise solution with Germany without going to the Court may be explained by a number of reasons. These indicate that the content of the law did play a role in determining their choice. First of all, the law hardly allowed determining with any certainty whether the German coastal configuration constituted a special circumstance and, if it did, what kind of implications this would have for Germany's continental shelf boundaries. This would make it difficult for Denmark and the Netherlands to explain why a result different from equidistance was an outcome in accordance with the law. This was all the more the case because the Danish and Dutch governments had publicly committed themselves to the equidistance principle as the applicable standard to delimit the North Sea. This had led to a perception in public opinion that Germany was claiming part of an area that rightfully belonged to the Netherlands and Denmark because of the resource potential of the area.[5] A compromise with Germany might give the impression that they had given in to their powerful neighbor. Second, the fact that the delimitation primarily concerned a legal question also provided a forceful argument to go to the Court. A compromise might have led to the criticism that Denmark and the Netherlands had forsaken the possibility to vindicate their convincing legal position by a judgment of the Court. Finally, the Netherlands had reached the conclusion that legal settlement of the dispute offered a better chance of success than a

---

[5] For instance, the Dutch newspaper *Haagsche Courant* on 24 March 1964 reported on the first talks between the Netherlands and Germany on its front page using the following headline: "Now that gas and oil riches has been shown: Germans want a part of our seabed. Should the Netherlands give in to save friendship?" Translation by the author. The original text reads: "Nu gas- en olierijkdom blijkt: Duitsers willen stuk van onze zeebodem. Moet Nederland toegeven om vriendschap te redden?" See also e.g. "Bonn blijft stuk van Nederlands plat eisen," *De Waarheid*, 17 July 1964, p. 1; "Twijfel over verdrag Duitsland; Hoe werd lijn getrokken over continentaal plat?," *Leeuwarder Courant*, 3 December 1964, p. 3; "Vat Den Haag continentaal plat-verdeling te licht op? Concessie aan Bonn: rijk stuk van Noordzeebodem wordt nog niet geëxploreerd," *Het Vrije Volk*, 19 March 1965, p. 11.

negotiated compromise. The Netherlands in any case was not prepared to reach a bilateral settlement and feared that if the other North Sea States would also be involved they would rather put pressure on the Netherlands to give in to Germany than the other way around. As was set out above, as far as Germany was concerned, the content of the applicable law provides part of the explanation for its approach to the first stage of the negotiations.

After the judgment, political arguments relegated legal considerations to a secondary place. However, although this in first instance might seem to be contradictory, the fact that Denmark and the Netherlands had to rely to such an extent on political arguments rather attests to the potential relevance of the law for shaping State behavior in this particular case. The judgment of the Court had provided a new legal framework for the negotiations. The fact that Denmark and the Netherlands did not want to engage Germany in a serious debate on the judgment indicates that they were well aware of its potential implications. Germany for its part was well aware that this was the case, but did not want to force the issue. The reason that the law did not play a more significant role in the outcome of the negotiations was that Germany was not willing to spend too much political capital to back up its legal case and not that the judgment provided the parties with too little guidance. The position of Denmark and the Netherlands prior to the conclusion of the Special Agreements to submit the disputes to the ICJ indicates a similar relationship between political and legal arguments. They wanted that the questions to be submitted to the Court would allow it to clarify the content of the law in order to prevent Germany being able to use ambiguities in the judgment to get a good outcome. Obviously, the fear of such a German approach implied that it would also be using other than legal arguments.

The review of the internal deliberations on the approach to the negotiations of each party sheds further light on the significance of the law. Whereas the negotiations in general remained limited to an exchange of legal justifications without any real debate, the internal deliberations many times focused on the strengths and weaknesses of a party's own legal positions and those of its adversary's and how this fitted in with that party's continental shelf policy. Obviously, if international law would be considered to be largely irrelevant to the formulation of that policy, there would have been little need to take it into account to such an extent. So, what explains the inflexibility and, at times, apparent disregard for the law during the negotiations?

There are a number of reasons why the nuanced views on the content of the law expressed in internal deliberations did not find their way into the negotiating process. Many times the law is indeterminate and leaves room for different interpretations and it is difficult to predict how an authoritative decision maker would rule. In such cases, it is to be expected that States will opt to argue the interpretation of the law that is best aligned with their interest and will maximize their outcome. The rules on the delimitation of the continental shelf are an excellent example of such indeterminate rules.

The nature of the delimitation process, moreover, implies that there is little incentive to adjust a position on the content of the law during negotiations. Unlike treaties in other fields, a boundary treaty does not require the formulation of detailed rules of law to regulate future behavior, but it suffices to define a specific boundary.[6] However, the elaboration on the agreements to submit the two disputes to the ICJ indicates that in other cases there also may not be a need to work out agreement on all controversial legal points. The primary aim of negotiations is to reach workable arrangements to facilitate international cooperation, not to debate fine points of law. Moreover, giving in on a specific point before a compromise has been secured is not helpful. If the compromise does not materialize, this may have unnecessarily weakened a State's position.

Second, the fact that a State maintains its position on the interpretation of a legal rule does not necessarily imply that this position informs its decisions on whether or not to accept a specific compromise solution. Rather, the possibility that a specific rule leaves room for another interpretation, or actually is difficult to bring in line with a State's preferred interpretation can be one of the factors contributing to shaping a compromise. In this respect reference was already made to the acceptance by Denmark and the Netherlands that the continental shelf of Germany would extend to the median line between the continent and the United Kingdom and the role of the judgment of the Court in that connection. This example also illustrates that it is only possible to determine how individual factors contribute to the outcome of negotiations in general terms. It is possible to establish that a number of different factors are pulling in the same direction, and others possibly in a different direction, but it is not possible to attach a specific weight to each

---

[6] See also Oude Elferink, *Maritime boundary delimitation*, p. 372.

of these factors.[7] It is clear that the judgment of the ICJ did assist Germany in getting to the middle of the North Sea. If the judgment would have indicated that the law excluded this solution, Germany would only have been able to reach this goal by making even more significant concessions on other points than it actually did.

The case study also shows that States will, at times, consciously present an interpretation of the law that is implausible. An example is provided by the Dutch view on natural prolongation offered during the negotiations following the judgment of the Court. The Dutch view implied that Germany would get a lesser share of the continental shelf than it would get in accordance with equidistance. That outcome cannot be reconciled with the judgment of the Court as far as the implications of natural prolongation are concerned. The fact that States offer implausible interpretations of the law does not necessarily mean that the law is ineffective and can be flouted without consequences. What is however required is that the other State concerned is willing to expose the deficiencies of such a legal argument. This will not happen if other considerations indicate that it is not worthwhile to pursue this course of action. In this particular case, the Dutch presentation of its view on natural prolongation would have been an appropriate moment for Germany to suggest breaking off the negotiations and going back to the ICJ. In this case, the Netherlands would have had to consider whether it would like to stick to this untenable position. An eventual review by the Court might have been rather embarrassing for the Netherlands. Germany's lukewarm reply to the Dutch argument suggests that other interests were heavily stacked against returning to the Court.[8] The fact that States at times offer implausible interpretations of the law does raise the question as to why States bother to make such arguments to start with. This issue will be further considered in the next sections.

Another significant point that results from the case study is the importance that in many cases is attached to consistency in the sense that it is preferred to make relatively straightforward legal arguments that allow subsuming different situations under one general rule rather than making a more complex argument that allows distinguishing between different situations. For example, during the proceedings before the Court, the Netherlands did not want to be identified with an argument that islands constituted a typical example of

---

[7] For a detailed account of a model to explain the pull of different factors on decision making see B. D. Lepard, *Customary International Law; A New Theory with Practical Applications* (Cambridge University Press, 2010), pp. 47–76.

[8] See also von Schenk, "Die vertragliche Abgrenzung," p. 382.

special circumstances. This was explained by the fact that the Netherlands was confronted by a claim of Venezuela that the island of Aruba constituted a special circumstance and that the Netherlands did not want to weaken its position in this case. The Dutch stance allowed it to continue to categorically reject that Aruba was a special circumstance, instead of having to distinguish between different types of islands. The Dutch treatment of the delimitation between Suriname and Guyana, which was discussed in Chapter 7.3.2, provides another example of the importance attached to consistency and avoiding complicating your legal argument. Avoiding unnecessary sophistication of the law prevents that at a later point a State will be confronted with further subcategories of cases under a specific rule or refinements of the rule. It also avoids that a State will lose its credibility because it seems to be inconsistent in its application of the rules to serve its own interests.

Apart from the three States, the companies holding concessions for exploratory activities on the continental shelf had an interest in its delimitation. The case study shows that they had a considerable involvement during most of the negotiations. This was less the case in the Netherlands than in Germany and particularly Denmark. This difference can be mostly explained by the licensing policy of the three States. The Netherlands had divided its continental shelf in blocks and had opened specific blocks for bidding by companies. Denmark and Germany had given a concession for their entire shelf to one party, respectively the A.P. Møller Group and the North Sea consortium. In determining the blocks to be opened for bidding, the Netherlands did take into account its interests in its delimitation dispute with Germany. Some blocks extending up to the equidistance lines with Germany and Denmark were included and in working out a compromise solution with Germany after the judgment, the Netherlands sought to safeguard the interest of the companies holding a license for blocks that would be located beyond the Dutch continental shelf. The involvement of companies in the formulation of the Dutch delimitation policy was limited. Representatives of Shell and its affiliates did discuss some aspects of the dispute with Germany with the Ministry of Foreign Affairs and also passed on some information on Germany's intentions but the ministry did not seek to involve them in policy formulation. Government officials also considered that the oil companies had not always shared all information that might have been of interest to the government in determining its position on boundary matters. The Dutch interest in a quick settlement was explained by its interest in attracting companies to work on its continental shelf and prevent that they would turn elsewhere.

The German North Sea consortium was more directly involved in formulating Germany's approach to the delimitation with the Netherlands. The interest of the consortium in exploratory activities in the area north of the land boundary provided an important impetus to conclude a partial boundary with the Netherlands and the German authorities were in close contact with the consortium concerning whether drilling at specific locations was opportune in view of the ongoing negotiations with the Netherlands. Germany tried to pressure the Netherlands by telling it that it might not be able to keep the consortium from operating beyond the equidistance line – the consortium had a license for the entire German shelf that at this time was not defined by specific boundaries – but there actually are no indications that the consortium was contemplating such a step and the German government advised against it. After the partial boundary agreements with the Netherlands and Denmark had been concluded, the North Sea consortium initially was interested in an early settlement of the remainder of Germany's boundary. Later on it attached more importance to ensuring access to a larger area of continental shelf, preferably including areas further to the northwest. This change of position was taken into account by the German government in determining its position in relation to Denmark and the Netherlands. After the judgment of the Court, the consortium's views were taken into account in considering Germany's options in relation to the boundaries to be delimited. The interests of Germany and the consortium coincided in this respect.

The Danish A.P. Møller Group, had obtained a concession for the entire continental shelf of Denmark in October 1963. The A.P. Møller Group collaborated with Shell and Gulf in the DUC to carry out activities in its concession area. Earlier in 1963, Director Hoppe of A.P. Møller had already submitted to the Danish government that it should seek to effect a delimitation with Norway on the basis of the Norwegian Trough. After the Ministry of Foreign Affairs had explained why it did not support this approach and the dispute with Germany became its main concern, the DUC continued to provide the government with input. The DUC retained Elihu Lauterpacht to provide it with legal advice. His opinions were shared with the government on a regular basis. The Danish and Dutch agents agreed to a suggestion of Waldock that he liaise with Lauterpacht to see if he could be of assistance during the preparation of the pleadings in the cases against Denmark before the ICJ and Lauterpacht was part of the Danish team during the oral proceedings before the Court. The DUC was represented in the Continental shelf committee that had been set up to advise the Ministry of Foreign Affairs on the case with Germany through the A.P. Møller Group.

The DUC no doubt had its biggest impact on the dispute through its exploratory work on the Danish continental shelf after Denmark and Germany had agreed to submit their dispute to the ICJ. The DUC drilled a number of wells in the potentially disputed area and a couple of these indicated the presence of potentially exploitable quantities of oil and gas. The first drilling in the potentially disputed area was initially postponed after Germany reacted to a Danish notification, but was recommenced after the DUC had obtained advice on the legal and policy implications. The record suggests that these activities were not actively encouraged by the Danish government, which indicated that the DUC was acting on its own risk. At the same time, the Danish government did seek to use these activities to its advantage. The Danish agent in a meeting with the Registrar of the Court in September 1967 referred the activities of the DUC in an attempt to convince the Court of the interest of the parties in an early decision. Denmark used the successful drillings in its pleadings before the ICJ and one of its main objectives in the negotiations after the Court's judgment was to ensure that these sites would be located in the Danish continental shelf.

The interests of the Danish government and the DUC to a large extent ran parallel. Denmark was interested in maximizing the extent of its continental shelf and securing potential resources and the DUC in max-imizing the extent of its license area covering the entire Danish shelf and securing its promising sites. However, the commonality of interests was not complete. For instance, as one note prepared in the Ministry of Foreign Affairs concluded, the Danish authorities lacked experience and knowledge in respect of oil exploration and exploitation and its financing and consequently were not in a position to judge whether joint exploitation was more advantageous to Denmark than a compromise on the boundary. The DUC obviously had this knowledge but had an interest of its own in this matter.

11.3 Theoretical perspectives on the relation between international law and State behavior

This study thus far has focused on describing the negotiations on the delimitation of the continental shelf between Denmark, Germany and the Netherlands and the impact of international law on this process. The preceding section indicates that this impact can be captured in a limited number of statements. The present section seeks to determine how these statements fit in with different theoretical perspectives on the relation between international law and State behavior. There exists a wide array of theoretical perspectives that offer widely diverging views on this relation. These diverging views can be placed along two axes. One concerns the impact of international law on State behavior. On the one hand, certain perspectives submit that international law induces States to behave in a specific way. States accept that international law limits their scope of action. Theoretical perspectives at the other end of this axis submit that international law is merely used to justify State behavior. Under these perspectives, international law has no real impact on behavior. A specific course of action is chosen on the basis of a balancing up of the interests of a State, among which a perceived need to comply with international law does not figure. On the other axis, it is possible to distinguish perspectives according to the autonomy of international law in relation to behavior. At one end of this axis, international law is a set of rules that is exogenous and prior to behavior. Rules of international law will limit the scope for action of a State and the kind of arguments a State can make in connection with the selected action. At the other end of the spectrum, international law does not have a separate existence but is endogenous to behavior. If a specific course of action has been selected, legal argument will be adapted to that course of action. The flexibility and indeterminacy of the law make it possible to justify virtually any course of action with reference to the law.

It will be obvious that there is a relationship between the poles of both axes. A perspective that international law induces compliance implies the existence of rules that are prior to State behavior and cannot be adapted indiscriminately by States. On the other hand, the perspective that international law primarily is used to justify behavior rather implies that States will adapt legal argument to their needs. All perspectives raise the question what induces States to take into account the phenomenon of international law and to engage in legal arguments in their relations with other States. Having set out these preliminary considerations, let us look in some more detail at a number of specific perspectives on the relation between international law and State behavior. It is not intended to provide an exhaustive overview of the existing literature as this would go beyond the limited purpose of putting the case study in a broader perspective. Instead I will provide a brief discussion of a number of specific authors that offer different perspectives on the relationship between the law and State behavior. These perspectives will in turn be compared to the outcomes of the case study.

A cogent argument that international law has little sway over the behavior of States has recently been made by Goldsmith and Posner.[9] Goldsmith and Posner offer four possible models for behavioral regularities in inter-State relationships: coincidence of interest, coordination, cooperation and coercion.[10] While acknowledging that these models do not cover all types of international interaction, they posit that taken together these models:

> offer a different explanation for the state behaviors associated with international law than the explanation usually offered in international law scholarship. The usual view is that international law is a check on state interests, causing a state to behave in a way contrary to its interests. In our view, the causal relationship between international law and state interests runs in the opposite direction. International law emerges from states' pursuit of self-interested policies on the international stage. International law is, in this sense, *endogenous* to state interests.[11]

Goldsmith and Posner explain that this does not mean that international law is irrelevant. Especially treaties can help in "clarifying what counts as cooperation in or coordination in interstate interactions."[12] However, under their theory:

> international law does not pull states towards compliance contrary to their interests and the possibilities for what international law can achieve are limited by the configurations of state interests and the distribution of state power.[13]

If it is assumed that international law exercises normative force and thus constrains the pursuit of State interest, cooperation or coordination under a specific set of rules may continue if the conditions that made it possible will change.[14] Goldsmith and Posner consider that "cooperation and coordination will last only as long as the conditions that made them possible in the first place."[15] Under this view existing rules will not exercise any independent pull towards the continuation of cooperation and coordination.

Goldsmith and Posner also seek to answer the question why international discourse has the content it has. Why do States make "moralistic

---

[9] J. L. Goldsmith and E. A. Posner, *The Limits of International Law* (Oxford University Press, 2005).
[10] *Ibid.*, pp. 10–12.   [11] *Ibid.*, p. 13 (emphasis in the original).   [12] *Ibid.*   [13] *Ibid.*
[14] See also *ibid.*, p. 165.   [15] *Ibid.*

and legalistic claims rather than simply say that they are cooperative or something similar"?[16] On the basis of a number of game theoretical models they explain why the interests of States are served by engaging in talk.[17] The language of legal obligation is used in this connection because this is the language of cooperation:[18] "[w]hen states cooperate in their self-interest, they naturally use the moralistic language of obligation rather than the strategic language of interest."[19] Adopting this view implies that there is no need to explain the use of legal language by the existence of a conviction of States that they are bound by the norms they refer to.

International legal scholars who consider that international law is exogenous to State interests and constrains State behavior have also offered explanations as to why this is the case. Part of these explanations is concerned with the nature of international law. International law as a system backs up and reinforces the efficacy of individual rules. Another part of these explanations focuses on the functions international law has in facilitating international politics. Both these strands are present in the work of Louis Henkin, who has made the oft-quoted observation that "[i]t is probably the case that almost all nations observe almost all principles of international law and almost all of their obligations all of the time."[20] While it is easy to demonstrate that States in general observe the law, it is not easy to explain why States generally observe the law.[21] Part of this difficulty is explained by a lack of a sufficient theoretical framework: "we have barely begun to develop a 'psychology' and 'sociology' of State behavior or even the behavior of human beings in their capacity as government officials."[22] In addition, there is a problem of proving causality. Acting consistently with the law does not necessarily imply that this concordance is the result of deference to the law.[23] In developing a framework to explain why States observe the law, Henkin distinguishes between "internal motivations" and "external inducements" to comply.[24] External inducements are considered to be the more impor-

---

[16] *Ibid.*, p. 175.   [17] *Ibid.*, pp. 175–181.   [18] *Ibid.*, p. 183.   [19] *Ibid.*, p. 184.

[20] L. Henkin, *How Nations Behave; Law and Foreign Policy*, 2nd edn (New York: Columbia University Press, 1979), p. 47 (emphasis suppressed).

[21] L. Henkin, *International Law: Politics and Values* (Dordrecht: Martinus Nijhoff Publishers, 1995), p. 48.

[22] *Ibid.*   [23] Henkin, *How Nations Behave* p. 48.   [24] Henkin, *International Law*, p. 49.

tant of the two.[25] Internal motivations mostly have to do with a "sense of commitment to order and stability in the international system."[26] Self-interest plays an important role in this respect at two levels. The maintenance of basic norms of the international system is in the interest of every State as "States recognize that stability, law and order, reliability (and a warranted reputation for reliability) are in their national interest."[27] At the level of individual agreements or norms in whose creation a State participated, there ordinarily will be an interest which explains this and therefore there also generally will be an interest to maintain them.[28] This stress on self-interest seems to leave little room for the law as an independent variable constraining States. This in particular concerns the adherence to individual agreements and norms. Henkin's submission on that point seems to echo the view of Goldsmith and Posner that rules defining cooperation or collaboration will not outlast the conditions that made them possible in the first place. However, Henkin takes a more nuanced view on the relationship between compliance in individual cases and national interest. This follows from the importance he attaches to such factors as the maintenance of the international system and the need States have for a reputation of reliability. It would be a mistake for policy-makers to only take into account immediate interests and consider observance of the law as a cost as this approach may become "an inducement to favor the immediate tangible interest as the dominant national interest and to depreciate the national interest in observing legal obligations."[29] Although interests thus might seem to play an important if not dominant role in determining whether a State should opt to comply with its legal obligations or not, Henkin also submits that States have a preference for compliance. The rules themselves exert a pull to compliance.[30]

External inducements to comply with the law are primarily concerned with "horizontal enforcement" by a State that considers itself to be a

---

[25] *Ibid.*, p. 50. Henkin submits that internal motivations are weaker and less numerous in the international system than in states, although "they are not insignificant" (*ibid.*, p. 49).

[26] *Ibid.*, p. 49. For a detailed assessment of the internal motivations that contribute to compliance with the law see Henkin, *How Nations Behave*, pp. 49–68.

[27] Henkin, *International Law*, p. 49.   [28] *Ibid.*

[29] *Ibid.*, p. 62, quoting from Henkin, *How Nations Behave*, p. 332.

[30] Henkin, *How Nations Behave*, pp. 49–50; Henkin, *International Law*, p. 49. On this latter point see also e.g. A. Chayes and A. Handler Chayes, "On Compliance," 1993 (47) *International Organization*, pp. 175–205, at p. 178; O. Schachter, *International Law in Theory and Practice* (Dordrecht: Martinus Nijhoff Publishers, 1991), p. 7.

victim of a violation of a rule of international law.[31] There is a political and legal aspect to this horizontal enforcement. States that repudiate legal obligations can expect that other States will be willing to invest resources in reacting if they have suffered negative consequences.[32] The international legal system itself contains a number of mechanisms to respond to perceived violations of its norms, such as countermeasures and recourse to third party dispute settlement.[33]

Guzman offers a theoretical model to explain the impact of law on the behavior of States that does not rely on the assumption that States have a preference for complying with international law.[34] Instead Guzman is intent on explaining when compliance will come about and when not.[35] One fundamental starting point to understanding international law is that it is just one of the factors that affect the behavior of States. Other factors are also relevant and some of them may contribute to compliance while others may contribute to non-compliance.[36] Guzman's model is based on a number of rational choice assumptions: "States are assumed to be rational, self-interested, and able to identify and pursue their interests."[37] These assumptions imply that States will only cooperate if there is a pay-off. As a result a model that explains cooperation under these assumptions can also be expected to work under assumptions that lead to cooperation more easily. A second reason for adopting these assumptions is that they are employed by many social scientists and international law scholars.[38]

Guzman deduces three main reasons for compliance by looking at cooperation in a difficult setting. His example concerns the 1972 Anti-Ballistic Missiles Treaty between the United States and the Soviet Union,[39] which provides a classic example of prisoner's dilemma. Cooperation results in the highest payoff for each party, but each party has an incentive to defect regardless of what the other party does, because if it does not defect and the other party does, it achieves the worst possible result. However, as Guzman observes that is not what happened and the degree of cooperation that was going on is impossible to reconcile with the predicted outcome of a one-time prisoner's dilemma.[40] The critical difference was the interest of both States in

---

[31] Henkin, *International Law*, p. 50.   [32] *Ibid.*, p. 50–51.   [33] *Ibid.*, pp. 51–61.
[34] A. T. Guzman, *How International Law Works; A Rational Choice Theory* (Oxford University Press, 2008), pp. 16 and 20.
[35] *Ibid.*, pp. 15–16.   [36] *Ibid.*, p. 15.   [37] *Ibid.*, p. 17.   [38] *Ibid.*
[39] Treaty between the United States of America and the Union of Soviet Socialist Republics on the Limitation of Anti-Ballistic Missile Systems of 26 May 1972 (944 UNTS, p. 14).
[40] Guzman, *International Law*, pp. 31–32.

future cooperation. Such a future interest leads to three reasons for compliance: reciprocity, reputation and retaliation.[41] Reciprocity, which Guzman considers to be perhaps the most important reason, entails that a violation by one State is likely to lead to a violation by the other State after which the treaty would collapse and the parties would again be without regulatory framework for their interactions. Reciprocity will be effective if the one-time gain of violation will be less beneficial that continued cooperation. Reputation is built by maintaining one's commitments and may achieve that other States are more likely to accept future commitments of that State. This will make the conclusion of future agreements less costly. Retaliatory action may further increase the cost of breach and under a rational actor model retaliation will be used to end ongoing violations or signaling that future violations will also be punished.[42] Especially how reputation may work is set out in considerable detail.[43] The availability of information plays an important role in this respect. One informational uncertainty is concerned with the content of the applicable legal rules. Such uncertainty could also be said to have existed for most of the relevant rules in respect of the delimitation of the continental shelf in the North Sea. If the law is uncertain, it may be difficult for a State to determine if it will suffer a reputational sanction for its conduct.[44] Even if a State acts in a way that it considers itself to be legal, another State may judge that the behavior of the acting State is illegal and this will affect the reputation of the acting State. A State in any case in general will have an interest in arguing that it is observing the law. Another State may either conclude that the acting State itself considered being in breach of the law and by arguing the legality of its actions was only trying to muddy the waters. However, it also cannot be excluded that it was expressing a genuine belief that it was observing the law. If another State reaches the latter conclusion the reputational loss of the acting State will be less. The more uncertain the law is, the more difficult it will be for a State to reach an unequivocal conclusion in this respect.[45] In other words, uncertainty, be it concerning the law or other issues, "reduces incentives to comply by reducing the cost of a violation."[46] The compliance pull of reputation may also be limited for other reasons. This concerns different forms of

---

[41] *Ibid.*, p. 32.   [42] *Ibid.*, pp. 32, 42 and 211.
[43] The three factors explaining compliance are further elaborated in *ibid.*, pp. 33–48. The working of reputation is further explained at *ibid.*, pp. 71–117.
[44] *Ibid.*, p. 93.   [45] See *ibid.*, pp. 94–95.   [46] *Ibid.*, p. 97.

compartmentalization of reputation. For instance, a change in reputation in one issue area may not affect a reputation in another issue area.
Non-compliance with environmental obligations does not necessarily
imply that a State will not comply with its obligations in the field of
security.[47]

Shirley Scott's model to explain the relationship between international
law and international politics, and the significance of the law in this
respect, takes as its starting point that ideas have power: "[t]he power of
an idea can be said to reside in people's acceptance of that idea as the
basis for action."[48] Her reason for choosing this starting point is that
existing approaches looking at the relationship between international
law and politics fail to account adequately why the perception that
international law is a coherent body of binding rules is so persistent.[49]
To overcome these inadequacies, it is necessary to "deal explicitly with
the relationship of international law to the broader structures of
power."[50] In explaining this relationship she draws on studies that seek
to explain the relationship between power structures and ideologies. One
assumption in this respect is that a political structure always has one
particular ideology that is integral to it.[51] The relationship between the
two is further elaborated as follows:

> The power structure is more than that ideology, but without the ideology,
> the structure would collapse. Members of the power structure commu-
> nicate on the basis of the implicit acceptance of the ideology. The
> ideology has existence only in its repeated expression and in allusions
> to it. An ideology can be brought into service by those in positions of
> domination, but ideology is a tool at the disposal of all members of a
> political structure.[52]

So, how does this general proposition translate to the specific relationship between international law and politics? If international law is
viewed in terms of an ideology underpinning the international State
system, support for that ideology is an essential aspect of a rational
foreign policy and two of its aspects must in particular be upheld:

---

[47] See further *ibid.*, pp. 100–111.
[48] S. V. Scott, "International Law as Ideology: Theorizing the Relationship between
International Law and International Politics," 1994 (5) *European Journal of
International Law*, pp. 313–325, at p. 317. As Scott points out, the proposition that
ideas have power cannot be definitively proven within a positivist epistemology. It is
however possible to point to instances in which changes in ideas preceded major
historical developments (*ibid.*).
[49] *Ibid.*, p. 316.   [50] *Ibid.*, p. 317.   [51] *Ibid.*, p. 318.   [52] *Ibid.*

The first is the legal-non-legal dichotomy which is the essence of the ideology. Acceptance of this is reflected in the categorization of political positions as either "legal" or "illegal". A second aspect of the ideology is the notion that international law is a set of binding rules. This is reinforced whenever an international actor emphasizes the legality of its own position.[53]

The need to uphold the ideology of international law in respect of these aspects also offers States the opportunity to use the ideology to their own advantage in pursuing their political interests. A State can improve its position on a particular issue by demonstrating that there is a discrepancy between specific legal rules and the basic notions of international law. To address such a discrepancy, either the specific rules will need to be adjusted or there has to be a substantial modification of the ideology. The importance of the ideology to powerful States at times may permit that less powerful States can accomplish an adjustment of specific rules or regimes.[54] Similarly, a State will seek to explain its own behavior as being in accordance with international law and to highlight the discrepancies between the behavior of its adversary and the law.[55] This type of argument can be employed to persuade an opponent that it needs to adjust its behavior to its own disadvantage to play its part in upholding the ideology.[56] Compliance with the law is most likely if there is little doubt about what would constitute a breach of the law because "to refrain from acting according to widely recognized law is to deny acceptance of the law's binding quality."[57]

11.4 The case study and the theoretical perspectives

All of the theoretical perspectives that were discussed in the preceding section to a larger or lesser extent fit with the various episodes of the case study. That should not come as a surprise. All these perspectives seek to explain the role of international law in international politics. At the same time, it should be clear that it is difficult to be too specific about the explanatory power of the different perspectives. International law is only one of the variables that are relevant to explaining State behavior. In general, it will be possible to assign a negative or positive value to these various variables in the sense that they contribute towards or work

---

[53] Ibid., p. 323 (footnote omitted).   [54] Ibid., p. 324.
[55] R. Withana, Power, Politics, Law: International Law and State Behaviour during International Crises (Leiden: Martinus Nijhoff Publishers, 2008), pp. 77–78.
[56] Ibid., p. 88.   [57] Scott, "International Law," p. 324.

against a specific outcome. Assigning specific values or even estimates in most instances does not seem realistic. The different variables are too diverse in nature and even comparing the relative weight of different legal arguments that feed into decision making and negotiations may be problematic.

The case study itself points to another difficulty in determining the role of different variables in internal decision making and negotiations. Although there exist thousands of pages of primary sources that allow piecing together a detailed account of almost all episodes of the negoti- ating history, the reasons for specific choices, including key decisions, at times are hardly documented or even completely undocumented, making it difficult to determine the considerations that feed into the decision-making process with certainty. A part of the explanation of this phenomenon may be that these considerations have been internalized by decision-makers and they may not feel a need to make the obvious explicit. At other times the choice to refrain from recording the decision- making process is likely to have been intentional.[58] For instance, it is more than likely that the Dutch side intentionally refrained from doc- umenting the reasons for the discrepancy between the Dutch reports that the Dutch delegation faithfully adhered to the judgment of the Court during the second stage of the negotiations with Germany and the actual Dutch positions that at times clearly contradicted the judgment. The lack of information on the considerations for dealing with the law in a particular way is a serious drawback in studying the interaction between international law and State behavior. Still, even in cases in which infor- mation is incomplete certain inferences are possible.

Before turning to the various models that seek to explain the relation- ship between international law and decision processes and negotiations, it is useful to have a brief look at the positivist view on international law. That view does not inquire into the considerations why States comply with or disregard the law, but offers a tool to assess which of the two actually is the case.[59] In that sense, this type of legal analysis provides the

---

[58] Historian Christ Klep even submits that government agencies are apt to refrain from documenting key decisions (J. Eijsvoogel, "De waarheid komt niet naar boven; Historicus Klep over nutteloos onderzoek naar Irak-besluit," *NRC Handelsblad Zaterdag & Cetera*, 28 February 2009 and 1 March 2009, p. 15).

[59] It should be recognized that this approach has been the object of criticism to the effect that it does not allow to determine the content of the law with certainty (see e.g. M. Koskenniemi, *From Apology to Utopia: The Structure on International Legal Argument* (Helsinki: Finnish Lawyers' Publishing Company, 1989)). Although it is not

basis for any inquiry into the relationship between international law and decision processes and negotiations. Without a sound understanding of the law, it is impossible to say anything of consequence about this relationship. The presentation of the case study has taken this perspective as its frame of reference in order to outline the legal arguments that the parties considered and actually made and how these relate to the content of the positive law on the continental shelf and international law generally. That analysis led to a mixed picture as far as conformity to the law is concerned. Certain arguments are no doubt in full accordance with the positive law and in other instances the opposite is the case. In other instances, it is not possible to draw a firm conclusion because the law is indeterminate. This analysis of the law as it so to speak "is" offers the necessary framework for assessing the explanatory value of the different perspectives on the relationship between international law and decision processes and negotiations.

All the theoretical perspectives provide a useful tool for probing the relationship between international law and State behavior. For instance, Scott's international law as ideology perspective posits that a State can advance its interests by advancing arguments that show its strong commitment to international law and the lesser commitment in this respect of its opponent. This assumption shows an interesting fit with the pleadings of the parties in the *North Sea continental shelf* cases. Germany's arguments in respect of delimitation were founded in the concepts of equality of treatment and fairness, which are structural components of the international legal order.[60] Denmark and the Netherlands in general relied on more technical and formalistic arguments that were grounded in a unilateralist approach to the delimitation. The international law as ideology perspective goes a long way in explaining why the Court basically adopted the German view. Of course, if the law would have contained a more clearly elaborated technical rule, the Court might have reached a different outcome. This does however not

suggested that legal positivism always leads to incontrovertible conclusions, in specific cases it is possible to assess how statements of States on the law relate to legal sources such as treaties and judgments. This view is also reflected in the work of lawyers providing advice to government agencies.

[60] Another example of such reasoning is offered by the second round argument of Riphagen before the Court. Riphagen attempted to ground the bilateral nature of delimitations in the decentralized nature of power and authority in the present state of international law (see further Chapter 8, footnote 343). This was intended to show the Dutch and Danish commitment to the international legal order and Germany's disregard for it.

detract from the relevance of this example. In that case, the German arguments would have demonstrated lesser commitment to international law and could have been viewed as an attempt to get out under a clear and specific obligation.

The perspective of Goldsmith and Posner has as its basic tenet that international law does not have any real impact on the behavior of States. States do not comply with international law if this is contrary to their interests. States only resort to international law because in this way they can signal their willingness to cooperate. The case study could be said to offer significant support for this perspective. So let us try to describe it in the terms of the interests and seek to determine to what extent the law could be said to be just window dressing.

The support of Denmark and the Netherlands for the continental shelf regime in the 1950s can be explained by the fact that the equidistance method gave them a considerable part of the North Sea. The broad acceptance of the regime also could be said to have contributed to this acceptance. Opposition to the regime would have stood little chance of success. Germany's rejection of the regime can be explained by the fact that it did not give Germany a satisfactory part of the continental shelf. The large amount of support for the regime and the broader interest of Germany of appearing to be a cooperative member of the international community meant that its opposition was rather weak. Germany eventually accepted the regime of the continental shelf in 1964. If Germany would not have made a claim to a continental shelf in 1964, it would even have been worse off as it risked that it would not even benefit from oil and gas exploitation in the area in the immediate vicinity of its coast. Its virtual volte-face as regards the customary law status of the continental shelf regime in this connection also could be said to illustrate that views on the law are easily adapted in the light of changing circumstances and interests.

The negotiations on the delimitation of the continental shelf between the three States can also without difficulty be framed in the terms of interests. They all were interested in starting the exploration of the oil and gas resources of the North Sea. In the absence of agreement on the delimitation of the continental shelf they would not have been able to secure access to specific areas and exploration would not have been possible. International law offered a convenient language to maintain a dialogue and agreement on partial boundaries was possible because both sides considered that this served their interests. The law did not contribute to reaching an agreement on the final part of the boundaries because interests diverged. Instead, Denmark and the Netherlands pursued a policy of bilateral boundary delimitations with other North Sea States to present Germany with a *fait accompli*, although from a legal perspective these boundaries did not have any effect for Germany.

The main reason to submit the disputes to the ICJ was the continued interest of the parties to settle the matter once and for all and allow exploration activities further offshore. A judgment of the Court would offer the parties a justification for a final settlement. During the proceedings before the Court, the parties obviously availed themselves of legal argument. The divergence in the arguments and proposed outcomes of the parties provides further illustration of the malleability of the law to bring it in line with the specific interests of each party.

The negotiations after the judgment could be said to provide the best illustration that the law provides a convenient language for negotiations but does not exercise any compliance pull on the parties. The attempts of Germany to engage Denmark and the Netherlands in a dialogue on the implications of the judgment were futile and Denmark and the Netherlands used elements of the judgment in a haphazard way to arrive at an outcome that secured all their main objectives. For its part, Germany did not press harder for a different outcome on the basis of the law because of its overriding interest in good bilateral relations with its two direct neighbors. The outline of a compromise was arrived at during a meeting at the highest political level by mutually accommodating the interests of the parties in relation to specific aspects of the overall settlement.

Based on the above assessment, it would seem that there is little need to look beyond the interests of the parties to explain the course and outcome of their negotiations on the delimitation of the continental shelf. For various reasons, the three States were prepared to negotiate within the framework set by the continental shelf regime, but the components of that regime had little impact on shaping the outcome of the negotiations. However, if we also take into account the role of international law in the internal deliberations of the parties a somewhat different picture emerges.[61] Obviously, interests also played a role in shaping

---

[61] Goldsmith and Posner in at least one instance also indicate that the law is an independent variable. They submit that States will also engage in statements to clarify their own actions or to protest the actions of other States to negotiate over what actions count as proper. Argument is used to establish the meaning of words they use and thus control the consequences of their announcements (Goldsmith and Posner, *Limits of*

these internal debates. At the same time, they show that international law was taken into account as an independent variable that did play a role in influencing positions. In general, if the law would be considered to be completely irrelevant, it would not make any sense for the parties to devote so much attention to this matter.[62] A couple of examples may further illustrate the significance of legal considerations. When the regime of the Convention on the continental shelf was negotiated, among others Denmark and the Netherlands argued that the rule on delimitation should not be limited to a reference to agreement between the interested States, as this would put weaker States in a difficult position. This explanation clearly indicates that these States considered that a more specific rule could put a break on demands of more powerful States. The content of the rule did matter. Later on, the content of the rule on delimitation contained in the Convention on the continental shelf, coupled with the interpretation of this rule by Denmark and the Netherlands, led Germany to refrain from ratifying the Convention. If the rules contained in the Convention would not have any compliance pull at all, there would not have been any reason for Germany to refrain from taking this step. To the contrary, by ratifying the Convention Germany would have confirmed its commitment to participate as a full member in the international community. In the negotiations with Denmark and the Netherlands it could easily have referred to the special circumstances clause to reject the obligatory force of the equidistance rule.

The Danish internal deliberations after the judgment indicate that it was considered that the law did not offer Denmark prospects of a favorable outcome. For this reason, Denmark aimed for a political compromise. At first sight, this might reaffirm the irrelevance of international law. The outcome was a political deal that safeguarded all of the Danish interests. On closer consideration, this episode confirms that international law did have an impact on the negotiations. Both Denmark and Germany were aware of the potential implications of the law. To avoid these implications, Denmark (like the Netherlands) was prepared to bring in political arguments. If Germany would have withstood this political pressure and would have insisted on dealing with the law seriously, the outcome would have been different. These examples indicate that interests play a significant role, but the law is not irrelevant. Rather, it is one of the

International Law, p. 184). Such an approach only makes sense if it is accepted that States will place reliance on prior agreement on the meaning of words.

[62] See also Guzman, International Law, p. 13.

considerations pulling in a certain direction and in this way it can contribute to or work against specific outcomes together with other relevant variables. An illustration of this weighing up of different factors is provided by Denmark's handling of the issue of the Norwegian Trough. Denmark had to consider whether or not the Norwegian Trough was relevant to determining the extent of the continental shelf of Denmark and Norway or provided an argument for a boundary different from the equidistance line. In comparison to the equidistance method, Denmark would have had a significantly larger continental shelf if it would have based itself on a boundary on the basis of the Norwegian Trough. In dealing with this issue, which first was taken up in the early 1950s and resurfaced in the first half of the 1960s, legal arguments and other considerations both impacted on Denmark's decision to accept the equidistance line as a boundary.

During the first phase of the negotiations between Denmark, Germany and the Netherlands the law also can be said to have had a considerable impact. On the one hand, the specific mention of the equidistance principle in the Convention on the continental shelf made it possible for Denmark and the Netherlands to rely on this principle during their negotiations with Germany, without seriously entertaining the need for making concessions. Germany could even be depicted as seeking to divest Denmark and the Netherlands of a part of continental shelf that rightfully belonged to them. If the rule contained in the Convention would have stressed the need for an equitable outcome without referring explicitly to the equidistance principle, Denmark and the Netherlands could not have taken this approach. On the other hand, the content of the law contributed to Germany's at times hesitant defense of a claim to a larger part of the continental shelf. Germany was struggling to come up with a coherent legal defense of a satisfactory claim that would serve its interests in the development of the continental shelf in the North Sea.

The perspective of Henkin offers a more nuanced view on the relationship between international law and interests and submits that international law exerts a compliance pull on States. Does this perspective allow taking on board the arguments on the impacts of international law set out in the preceding paragraphs? To a certain extent it does. International law has an impact on State behavior. Whether the notion of compliance captures the essence of the nature of this impact is another question. As the case study demonstrates, an assessment by a State that the most likely interpretation of the law points towards a certain outcome does not necessarily imply it will accept that outcome. In such a case a State may still seek at least in first instance to offer a different view on the law that is in line with its own preferred outcome. In other words, there is not a binary opposition between compliance and non-compliance, but there rather is a sliding scale between these two extremes. States will seek to exploit the law to their advantage and the interaction with its negotiating partners will determine their success in this respect.[63]

The case study also offers considerable support for the idea that States comply with the law because of their broader interest in the maintenance of the international legal order or cooperation. Germany's hesitance to push its interests in the negotiations is in large part explained by the perception that this would negatively impact on its standing in the international community. The participation of Germany in the proceedings before the ICJ was also considered to be successful by Germany because it had shown that Germany, like any other State, had been able to play its part and there had not been any repercussions of its recent dark past. On the other hand, the approach of Denmark and the Netherlands rather suggests that broader considerations played a very limited role in defining their negotiating positions. In that respect, the idea that States have different reputations with respect to different issues offers (part of) an explanation.[64] Denmark and the Netherlands apparently considered that their inflexible stand in the negotiations would not damage their reputation as regards cooperation with Germany on other issues. As a matter of fact, they rather used the interest of Germany in good bilateral relations to arrive at a satisfactory outcome on this particular issue without having to pay too much attention to the applicable law.

Like Henkin, Guzman looks at international law in terms of compliance and the same considerations as set above are relevant in this respect. His model to explain compliance on the basis of reciprocity, reputation and retaliation does offer interesting insights into the case study. In discussing his perspective, it was submitted that the rules on delimitation provide an example of informational uncertainty, in which case States have less reason to fear reputational loss if they adopt a potentially controversial interpretation of the law. Reputation and reciprocity also explain the Dutch and Danish approach to their other delimitations.

---

[63] This is of course not to deny that the law is often routinely complied with. Although these cases of routine compliance constitute a large majority, they are of little interest in looking at the question why States carefully look at the law and factor it into their decision-making processes when their interests will be affected if they accept a specific view on the law.

[64] See e.g. Guzman, *International Law*, pp. 100 and following.

They both wanted to avoid the impression that they were dealing with Germany on terms different to their other neighbors. Such difference in treatment would affect their reputation for compliance as it would signal to Germany that they were not willing to live up to the rule they were seeking to impose on Germany. At the same time this also concerns reciprocity in a broader sense. The ideas of reputation and reciprocity also explain the emphasis of Denmark and the Netherlands during the pleadings on alleged inconsistencies in Germany's position and the suggestion that Germany only became interested in the area in dispute after it had learned about its resource potential. They among others implied that Germany had profited from equidistance in the nearshore area but did not want to adopt the same method in the offshore and that the German arguments offered to the Court should not be taken seriously as they were based on interests and not on the law.

Scott's perspective of international law as ideology assumes that as all States have an interest in upholding international law as the necessary ideological support of the international State system, they are intent on explaining their behavior in terms of consistency with this ideological support and that of their opponents in terms of inconsistency with it. In this way, international law can be an effective tool to advance the interests of States. Compliance with specific rules according to Scott is most likely if there is relatively little doubt about what would constitute the "legally correct course of action," because to do otherwise would indicate that a State is denying that the law has binding quality. The case study indicates that international law is part of the argumentative structure that States use in seeking to advance their interests. As was already observed previously, the case study suggests that there does not exist a binary opposition between compliance with the law and non-observance, but rather that there is a sliding scale between these two extremes and that States will seek to exploit the law to their advantage and that the interaction with its negotiating partners will determine their success in this respect.

As is likely to be the case for any explanatory model, the case study also contains episodes that do not fit Scott's perspective in the sense that a State makes little effort to uphold the ideology of international law. In the negotiations following the judgment of the Court, in particular the Netherlands in comparison to Germany did not argue strongly in terms of adherence to international law. Notwithstanding the fact that the Special Agreement required the parties to negotiate on the basis of the judgment of the Court, the Netherlands emphasized the political nature of the negotiations and offered legal arguments of dubious quality. This shows that there obviously is a limit to credibly arguing the law to one's own advantage and different means may in that case be necessary. On the other hand, this episode could also be said to confirm the central tenets of the international law as ideology perspective. The success of the strategy of Denmark and the Netherlands can be largely explained by the fact that Germany chose not to present a more convincing legal case to expose their opponents' disregard for the law, even though that would certainly have been an option.

On the basis of one case study dealing with one specific issue, even if it is composed of a considerable number of separate episodes, it is not feasible to be too specific about the explanatory value of the different perspectives. They are all built on the observation of certain regularities in State behavior and their relationship to international law. It thus is virtually excluded that a perspective would have no explanatory value whatsoever. Looking at the case study from different angles enhances our overall understanding of it. Points that may be neglected under one perspective may be highlighted under another one.

The case study demonstrates that it is essential to not only look at public pronouncements on the law, but to also analyze the internal deliberations by the government agencies involved. Due to the limited access to the relevant documents and/or a poorly documented record this is one of the main challenges facing this field of research. The case study clearly does allow establishing that international law is an independent variable impacting on State behavior. The exact nature of that impact will depend on the broader setting in which it operates.

BIBLIOGRAPHY

Archival materials

In connection with this study archives in Denmark, Germany, the Netherlands and the United Kingdom have been consulted. References to the files of these archive holders have been provided in abbreviated form in the footnotes to the text. The present section of the bibliography provides the full titles of the files that are mentioned in the study.

1. Denmark

Danish State Archives – Danish National Archives, Ministry of Foreign Affairs

DNA/1   Udenrigsministeriet, Gruppeordnede sager 1946–72, Journal nr. 46.D.25

DNA/2   Udenrigsministeriet, Gruppeordnede sager 1946–72, Journal nr. 46.Dan.1.b.1, Aktpakke nr. 1

DNA/3   Udenrigsministeriet, Gruppeordnede sager 1946–72, Journal nr. 46.Dan.1.b.1, Aktpakke nr. 2

DNA/4A   Udenrigsministeriet, Gruppeordnede sager 1946–72, Journal nr. 46.Dan.1.b.1, Aktpakke nr. 3

DNA/4B   Udenrigsministeriet, Gruppeordnede sager 1946–72, Journal nr. 46.Dan.1.b.1, Aktpakke nr. 4

DNA/4C   Udenrigsministeriet, Gruppeordnede sager 1946–72, Journal nr. 46.Dan.1.b.1, Aktpakke nr. 5

DNA/6   Udenrigsministeriet, Gruppeordnede sager 1946–72, Journal nr. 46.D.24

DNA/8   Udenrigsministeriet, Gruppeordnede sager 1946–72, Journal nr. 119.N.2/3a/1c, Aktpakke nr. 1

DNA/9   Udenrigsministeriet, Gruppeordnede sager 1946–72, Journal nr. 119.N.2/3a/1a, Aktpakke nr. 9

DNA/12   Udenrigsministeriet, Gruppeordnede sager 1946–72, Journal nr. 119.N.2/3a/1b, Aktpakke nr. 1

DNA/18	Udenrigsministeriet, Gruppeordnede sager 1946–72, Journal nr. 46.DAN.50.C.2
DNA/20	Udenrigsministeriet, Gruppeordnede sager 1946–72, Journal nr. 46.DAN.50.D.2.B
DNA/21	Udenrigsministeriet, Gruppeordnede sager 1946–72, Journal nr. 46.DAN.50.E
DNA/22	Udenrigsministeriet, Gruppeordnede sager 1946–72, Journal nr. 46.DAN.50.F.1
DNA/24	Udenrigsministeriet, Gruppeordnede sager 1946–72, Journal nr. 46.DAN.50.F.3
DNA/25	Udenrigsministeriet, Gruppeordnede sager 1946–72, Journal nr. 46.D.13.A, Aktpakke nr. 1
DNA/29	Udenrigsministeriet, Gruppeordnede sager 1946–72, Journal nr. 46.D.13.A, Aktpakke nr. 4
DNA/30	Udenrigsministeriet, Gruppeordnede sager 1946–72, Journal nr. 46.D.13.C, Aktpakke nr. 1
DNA/34	Udenrigsministeriet, Gruppeordnede sager 1946–72, Journal nr. 46.D.32, Aktpakke nr. 2
DNA/36	Udenrigsministeriet, Gruppeordnede sager 1946–72, Journal nr. 46.D.26
DNA/39	Udenrigsministeriet, Gruppeordnede sager 1946–72, Journal nr. 46.Dan.1b.1, Aktpakke nr. 6
DNA/40	Udenrigsministeriet, Gruppeordnede sager 1946–72, Journal nr. 46.DAN.53.D.1, Aktpakke nr. 5
DNA/42	Udenrigsministeriet, Gruppeordnede sager 1946–72, Journal nr. 46.D.27, Aktpakke nr. 1
DNA/44	Udenrigsministeriet, Gruppeordnede sager 1946–72, Journal nr. 46.DAN.53.B
DNA/49	Udenrigsministeriet, Gruppeordnede sager 1946–72, Journal nr. 46.DAN.51.H.1, Aktpakke nr. 3
DNA/50	Udenrigsministeriet, Gruppeordnede sager 1946–72, Journal nr. 46.DAN.51.H.1, Aktpakke nr. 2
DNA/53	Udenrigsministeriet, Gruppeordnede sager 1946–72, Journal nr. 46.DAN.51.H.1/Sup
DNA/56	Udenrigsministeriet, Gruppeordnede sager 1946–72, Journal nr. 46.DAN.51.A
DNA/58	Udenrigsministeriet, Gruppeordnede sager 1946–72, Journal nr. 46.DAN.51.B
DNA/59	Udenrigsministeriet, Gruppeordnede sager 1946–72, Journal nr. 46.DAN.51.H.1, Aktpakke nr. 1
DNA/60	Udenrigsministeriet, Gruppeordnede sager 1946–72, Journal nr. 46.DAN.51.H.2, Aktpakke nr. 1

DNA/61	Udenrigsministeriet, Gruppeordnede sager 1946–72, Journal nr. 46.DAN.51.H.2, Aktpakke nr. 2
DNA/67	Udenrigsministeriet, Gruppeordnede sager 1946–72, Journal nr. 46.DAN.51.H.3, Aktpakke nr. 3
DNA/68	Udenrigsministeriet, Gruppeordnede sager 1946–72, Journal nr. 46.DAN.51.H.3, Aktpakke nr. 4
DNA/79	Udenrigsministeriet, Gruppeordnede sager 1946–72, Journal nr. 46.DAN.51.I
DNA/81	Udenrigsministeriet, Gruppeordnede sager 1946–72, Journal nr. 46.DAN.51.J, Aktpakke nr. 1
DNA/82	Udenrigsministeriet, Gruppeordnede sager 1946–72, Journal nr. 46.DAN.51.J, Aktpakke nr. 2
DNA/85	Udenrigsministeriet, Gruppeordnede sager 1946–72, Journal nr. 119.N.2/3.A, Aktpakke nr. 1
DNA/89	Udenrigsministeriet, Gruppeordnede sager 1946–72, Journal nr. 119.N.2/3.A, Aktpakke nr. 6
DNA/119	Udenrigsministeriet, Gruppeordnede sager 1946–72, Journal nr. 46.DAN.50.M.2
DNA/123	Udenrigsministeriet, Gruppeordnede sager 1946–72, Journal nr. 46.DAN.50.B.2, Aktpakke nr. 1 and 2
DNA/132	Udenrigsministeriet, Gruppeordnede sager 1946–72, Journal nr. 46.DAN.50.D.1
DNA/133	Udenrigsministeriet, Gruppeordnede sager 1946–72, Journal nr. 46.DAN.52.C.2
DNA/134	Udenrigsministeriet, Gruppeordnede sager 1946–72, Journal nr. 46.DAN.52.C.2, Bilagspakke
DNA/137	Udenrigsministeriet, Gruppeordnede sager 1946–72, Journal nr. 46.DAN.52.B.1
DNA/138	Udenrigsministeriet, Gruppeordnede sager 1946–72, Journal nr. 46.DAN.52.A.2, Aktpakke nr. 1
DNA/139	Udenrigsministeriet, Gruppeordnede sager 1946–72, Journal nr. 46.DAN.52.A.2, Aktpakke nr. 3
DNA/142	Udenrigsministeriet, Gruppeordnede sager 1946–72, Journal nr. 46.DAN.52.A.2, Aktpakke nr. 2
DNA/143	Udenrigsministeriet, Gruppeordnede sager 1946–72, Journal nr. 46.DAN.53.D.1, Aktpakke nr. 1
DNA/144	Udenrigsministeriet, Gruppeordnede sager 1946–72, Journal nr. 46.DAN.53.D.1, Aktpakke nr. 2
DNA/145	Udenrigsministeriet, Gruppeordnede sager 1946–72, Journal nr. 46.DAN.53.D.1, Aktpakke nr. 3
DNA/147	Udenrigsministeriet, Gruppeordnede sager 1946–72, Journal nr. 46.DAN.53.A, Aktpakke nrs. 1–3

2. Germany

2.1 Political Archive of the Foreign Office

AA/2   PA AA, Aktenbestand B 80, Band 247
AA/3   PA AA, Aktenbestand B 80, Band 394
AA/7   PA AA, Aktenbestand B 80, Band 637
AA/12   PA AA, Aktenbestand B 80, Band 622
AA/32   PA AA, Aktenbestand B 80, Band 395
AA/33   PA AA, Aktenbestand B 80, Band 601
AA/34   PA AA, Aktenbestand B 80, Band 604
AA/35   PA AA, Aktenbestand B 80, Band 603
AA/37   PA AA, Aktenbestand B 80, Band 828
AA/38   PA AA, Aktenbestand B 80, Band 605
AA/39   PA AA, Aktenbestand B 80, Band 606
AA/40   PA AA, Aktenbestand B 80, Band 779
AA/41   PA AA, Aktenbestand B 80, Band 778
AA/42   PA AA, Aktenbestand B 80, Band 780
AA/43   PA AA, Aktenbestand B 80, Band 684
AA/44   PA AA, Aktenbestand B 80, Band 762
AA/45   PA AA, Aktenbestand B 80, Band 702
AA/46   PA AA, Aktenbestand B 80, Band 738
AA/47   PA AA, Aktenbestand B 80, Band 739
AA/48   PA AA, Aktenbestand B 80, Band 965
AA/51   PA AA, Aktenbestand B 80, Band 723
AA/57   PA AA, Aktenbestand B 80, Band 713
AA/59   PA AA, Aktenbestand B 80, Band 776
AA/61   PA AA, Aktenbestand B 80, Band 722
AA/62   PA AA, Aktenbestand B 80, Band 724
AA/63   PA AA, Aktenbestand B 80, Band 763
AA/64   PA AA, Aktenbestand B 80, Band 764
AA/73   PA AA, Aktenbestand B 80, Band 946
AA/74   PA AA, Aktenbestand B 80, Band 966
AA/84   PA AA, Zwischenarchiv, Band 230193
AA/88   PA AA, Aktenbestand B 80, Band 394
AA/91   PA AA, Aktenbestand B 80, Band 636
AA/94   PA AA, Aktenbestand B 80, Band 241
AA/96   PA AA, Aktenbestand B 80, Band 634
AA/97   PA AA, Aktenbestand B1, Band 534

2.2   Bundesarchiv Koblenz

Ministry for the Economy

B102/119551   BArch B102/119551
B102/119553   BArch B102/119553
B102/119554   BArch B102/119554
B102/119556   BArch B102/119556
B102/119557   BArch B102/119557
B102/260030   BArch B102/260030
B102/260031   BArch B102/260031
B102/260032   BArch B102/260032
B102/260033   BArch B102/260033
B102/260034   BArch B102/260034
B102/260035   BArch B102/260035
B102/260036   BArch B102/260036
B102/260037   BArch B102/260037
B102/260038   BArch B102/260038
B102/260124   BArch B102/260124
B102/260125   BArch B102/260125
B102/260126   BArch B102/260126
B102/260127   BArch B102/260127
B102/260131   BArch B102/260131
B102/260133   BArch B102/260133
B102/260135   BArch B102/260135
B102/260136   BArch B102/260136
B102/260138   BArch B102/260138
B102/260173   BArch B102/260173
B102/318712   BArch B102/260712
B102/318713   BArch B102/260713
B102/318714   BArch B102/260714

Ministry for the Interior

B106/53685   BArch B106/53685

Ministry of Justice

B141/22020   BArch B141/22020
B141/22023   BArch B141/22023
B141/22024   BArch B141/22024
B141/22027   BArch B141/22027
B141/22028   BArch B141/22028
B141/22029   BArch B141/22029

3. The Netherlands

3.1 National Archive

NNA/1   Archiefinventaris 2.02.05.02, inventarisnummer 411 (Council of Ministers)

NNA/2   Archiefinventaris 2.02.05.02, inventarisnummer 757 (Council of Ministers)

NNA/3   Archiefinventaris 2.02.05.02, inventarisnummer 784 (Council of Ministers)

NNA/4   Archiefinventaris 2.02.05.02, inventarisnummer 785 (Council of Ministers)

NNA/5   Archiefinventaris 2.02.05.02, inventarisnummer 815 (Council of Ministers)

NNA/6   Archiefinventaris 2.05.117, inventarisnummer 2638 (Ministry of Foreign Affairs)

NNA/7   Archiefinventaris 2.05.117, inventarisnummer 2639 (Ministry of Foreign Affairs)

NNA/8   Archiefinventaris 2.05.118, inventarisnummer 2945 (Ministry of Foreign Affairs)

NNA/8A   Archiefinventaris 2.05.118, inventarisnummer 3397 (Ministry of Foreign Affairs)

NNA/9   Archiefinventaris 2.05.118, inventarisnummer 3399 (Ministry of Foreign Affairs)

NNA/10   Archiefinventaris 2.05.118, inventarisnummer 3401 (Ministry of Foreign Affairs)

NNA/11   Archiefinventaris 2.05.118, inventarisnummer 3402 (Ministry of Foreign Affairs)

NNA/12   Archiefinventaris 2.05.118, inventarisnummer 3404 (Ministry of Foreign Affairs)

NNA/13   Archiefinventaris 2.05.118, inventarisnummer 3430 (Ministry of Foreign Affairs)

NNA/14   Archiefinventaris 2.05.118, inventarisnummer 3434 (Ministry of Foreign Affairs)

NNA/15   Archiefinventaris 2.05.118, inventarisnummer 3442 (Ministry of Foreign Affairs)

NNA/16   Archiefinventaris 2.05.118, inventarisnummer 3444 (Ministry of Foreign Affairs)

NNA/17   Archiefinventaris 2.05.118, inventarisnummer 27131 (Ministry of Foreign Affairs)

NNA/18   Archiefinventaris 2.05.118, inventarisnummer 27137 (Ministry of Foreign Affairs)

NNA/19   Archiefinventaris 2.05.118, inventarisnummer 27138 (Ministry
            of Foreign Affairs)
NNA/20   Archiefinventaris 2.05.171, inventarisnummer 257 (Ministry of
            Foreign Affairs)
NNA/21   Archiefinventaris 2.05.149, inventarisnummer 3025 (Ministry of
            Foreign Affairs)
NNA/21A   Archiefinventaris 2.05.149, inventarisnummer 3026 (Ministry of
            Foreign Affairs)
NNA/22   Archiefinventaris 2.05.149, inventarisnummers 419–421
            (Ministry of Foreign Affairs)
NNA/23   Archiefinventaris 2.05.313, inventarisnummer 3058 (Ministry of
            Foreign Affairs)
NNA/24   Archiefinventaris 2.05.313, inventarisnummer 3059 (Ministry of
            Foreign Affairs)
NNA/25   Archiefinventaris 2.05.313, inventarisnummer 3060 (Ministry of
            Foreign Affairs)
NNA/26   Archiefinventaris 2.05.313, inventarisnummer 3097 (Ministry of
            Foreign Affairs)
NNA/27   Archiefinventaris 2.05.313, inventarisnummer 3098 (Ministry of
            Foreign Affairs)
NNA/28   Archiefinventaris 2.05.313, inventarisnummer 3190 (Ministry of
            Foreign Affairs)
NNA/29   Archiefinventaris 2.05.313, inventarisnummer 3099 (Ministry of
            Foreign Affairs)
NNA/30   Archiefinventaris 2.05.313, inventarisnummer 3100 (Ministry of
            Foreign Affairs)
NNA/31   Archiefinventaris 2.05.313, inventarisnummer 3101 (Ministry of
            Foreign Affairs)
NNA/32   Archiefinventaris 2.05.313, inventarisnummer 3102 (Ministry of
            Foreign Affairs)
NNA/33   Archiefinventaris 2.05.313, inventarisnummer 3106 (Ministry of
            Foreign Affairs)
NNA/34   Archiefinventaris 2.05.313, inventarisnummer 3107 (Ministry of
            Foreign Affairs)
NNA/35   Archiefinventaris 2.05.313, inventarisnummer 3149 (Ministry of
            Foreign Affairs)
NNA/36   Archiefinventaris 2.05.313, inventarisnummer 3150 (Ministry of
            Foreign Affairs)
NNA/37   Archiefinventaris 2.05.313, inventarisnummer 3196 (Ministry of
            Foreign Affairs)

NNA/38	Archiefinventaris 2.05.313, inventarisnummer 3219 (Ministry of Foreign Affairs)
NNA/39	Archiefinventaris 2.05.313, inventarisnummer 3388 (Ministry of Foreign Affairs)
NNA/40	Archiefinventaris 2.05.313, inventarisnummer 25055 (Ministry of Foreign Affairs)
NNA/41	Archiefinventaris 2.06.087, inventarisnummer 803 (Ministry of Economic Affairs)
NNA/42	Archiefinventaris 2.06.087, inventarisnummer 810 (Ministry of Economic Affairs)
NNA/43	Archiefinventaris 2.10.26, inventarisnummer 1086 (Governor of Suriname: Cabinet)
NNA/44	Archiefinventaris 2.10.26, inventarisnummer 1087 (Governor of Suriname: Cabinet)
NNA/45	Archiefinventaris 2.10.26, inventarisnummer 2872 (Governor of Suriname: Cabinet)
NNA/46	Archiefinventaris 2.10.54, inventarisnummer 10468 (Ministry for the Colonies: Archive of files)
NNA/47	Archiefinventaris 2.10.54, inventarisnummer 10470 (Ministry for the Colonies: Archive of files)
NNA/48	Archiefinventaris 2.21.351, inventarisnummer 1137 (Archive of Dr. J. M. A. H. Luns)

3.2 Ministry of Foreign Affairs

NMFA/1	ddi/dve / code 3 / ord. 311, inventarisnummer dve / 1945–1984 / 00046
NMFA/2	ddi/dve / code 3 / ord. 311, inventarisnummer dve / 1945–1984 / 00057
NMFA/3	ddi/dve / code 3 / ord. 311, inventarisnummer dve / 1945–1984 / 00058
NMFA/4	ddi/dve / code 3 / ord. 311, inventarisnummer dve / 1945–1984 / 00067

3.3 Ministry of Economic Affairs

| NMEA/1 | 1.823.3 07.76 (Negotiations with Denmark concerning the continental shelf) |
| NMEA/2 | 1.823.3 07.76 (Negotiations with Germany concerning the continental shelf) |

4. United Kingdom

The National Archives

FO 371/176337
FO 371/176338
FO 371/181320
FO 371/181321
FO 371/181322
FCO 14/366

Books and articles[1]

Alexander, L. M., "Canada-Denmark (Greenland); Report Number 1–1" in J. I. Charney and L. M. Alexander (eds), *International Maritime Boundaries*, Vol. I (Dordrecht: Martinus Nijhoff Publishers, 1993), pp. 371–378

Anderson, D. H., "Denmark-Federal Republic; Report Number 9–8" in J. I. Charney and L. M. Alexander (eds), *International Maritime Boundaries*, Vol. II (Dordrecht: Martinus Nijhoff Publishers, 1993), pp. 1801–1805

"Federal Republic-The Netherlands; Report Number 9–11" in J. I. Charney and L. M. Alexander (eds), *International Maritime Boundaries*, Vol. II (Dordrecht: Martinus Nijhoff Publishers, 1993), pp. 1835–1839

Antunes, N. M., *Towards the Conceptualisation of Maritime Delimitation* (Leiden: Martinus Nijhoff Publishers, 2003)

Bethlehem, D., "The Secret Life of International Law," 2012 (1)1 *Cambridge Journal of International and Comparative Law*, pp. 23–36

Beunders, H. J. G. and Selier, H. H., *Argwaan en Profijt; Nederland en West-Duitsland* (Amsterdam: Historisch Seminarium van de Universiteit van Amsterdam, 1983)

Bøgh, N., *Hækkerup* (Aschehoug Dansk Forlag A/S, 2003)

Böhmert, V., "Meeresfreiheit und Schelfproklamationen," (1956) 6 *Jahrbuch für internationales Recht*, pp. 7–99

"Bonn blijft stuk van Nederlands plat eisen," *De Waarheid*, 17 July 1964, p. 1

Brown, E. D., *Sea-bed Energy and Minerals: The International Legal Regime* (Dordrecht: Martinus Nijhoff Publishers, 1992)

Chayes, A. and Handler Chayes, A., "On Compliance," 1993 (47) *International Organization*, pp. 175–205

Czempiel, E., *Macht und Kompromiß; Die Beziehungen der Bundesrepublik Deutschland zu den Vereinten Nationen 1956–1970* (Düsseldorf: Bertelmans Universitätsverlag, 1971)

---

[1] Documents contained in the archives mentioned in the preceding section of the bibliography are not listed separately in this section.

492 BIBLIOGRAPHY

Dröge, H., Münch, F. and Von Puttkamer, E., *The Federal Republic of Germany and the United Nations* (New York: Carnegie Endowment for International Peace, 1967)

Durante F. and Rodino, W. (eds), *Western Europe and the Development of the Law of the Sea* (Dobbs Ferry, NY: Oceana Publications 1979–...)

Eijsvoogel, J., "De waarheid komt niet naar boven; Historicus Klep over nutteloos onderzoek naar Irak-besluit," *NRC Handelsblad Zaterdag & Cetera*, 28 February 2009 and 1 March 2009, p. 15

Evans, M. D., "Maritime Boundary Delimitation: Where Do We Go from Here?" in D. Freestone, R. Barnes and D. Ong (eds), *The Law of the Sea; Progress and Prospects* (Oxford University Press, 2006), pp. 137–160

Fack, R., *Gedane zaken; Diplomatieke verkenningen* (Amsterdam: Sijthoff, 1984)

Feith, J. R., *De betekenis van de "Continental-Shelf" theorie voor de exploitatie van onderzeese gebieden* (Mededelingen van de Nederlandse Vereniging voor Internationaal Recht, no. 26 (1948))

Foighel, I., "The North Sea Continental Shelf Case; Judgment by the International Court of Justice of 20 February 1969," 1969 (39) *Nordisk tidsskrift for international ret*, pp. 229–240

Franckx, E., "Belgium-Netherlands; Report Number no. 9–21" in J. I. Charney and L. M. Alexander (eds), *International Maritime Boundaries*, Vol. IV (The Hague: Martinus Nijhoff Publishers, 2002), pp. 2921–2934

Friedman, W., "The North Sea Continental Shelf Cases – A critique," 1970 (64) *AJIL*, pp. 229–240

Frowein, J., "Verfassungsrechtliche Probleme um den deutschen Festlandsockel," 1965 (25/1) *ZaöRV*, pp. 1–25

Gill, T. D. (ed.), *Rosenne's The World Court; What it is and How it Works* (The Hague: Martinus Nijhoff Publishers, 2003)

Goldie, L. F. E., "Australia's Continental Shelf: Legislation and Proclamations," 1954 (3) *ICLQ*, pp. 535–575

Goldsmith, J. L. and Posner, E. A., *The Limits of International Law* (Oxford University Press, 2005)

Grisel, E., "The Lateral Boundaries of the Continental Shelf and the Judgment of the International Court of Justice in the North Sea Continental Shelf Cases," 1970 (64) *AJIL*, pp. 562–593

Guzman, A. T., *How International Law Works; A Rational Choice Theory* (Oxford University Press, 2008)

Hahn-Pedersen, M., *A.P. Møller and the Danish Oil* (Copenhagen: Schultz Forlag, 1999)

Henkin, L., *How Nations Behave; Law and Foreign Policy*, 2nd edn (New York: Columbia University Press, 1979)

International Law: Politics and Values (Dordrecht: Martinus Nijhoff Publishers, 1995)

Heybroek, P., Haanstra, U. and Erdman, D. A., "Observations on the Geology of the North Sea Area" in *Origin of Oil, Geology and Geophysics – Seventh World Petroleum Congress, Mexico*, 1967; *Proceedings*, Vol. 2, pp. 905–916

Hutchinson, D. N., "The Seaward Limit to Continental Shelf Jurisdiction in Customary International Law," 1985 (56) *British Yearbook of International Law*, pp. 111–184

"Is There Oil or Gas in the North Sea?," *The Financial Times*, 2 April 1963, p. 2

Jaenicke, G., "Der Streit über den Nordsee-Festlandsockel; Grundsätzliches Urteil des Internationalen Gerichtshofs," *Neue Zürcher Zeitung*, 27 March 1969, p. 3

Jansen, P. E. L., *Willem Riphagen 1919–1994* (Den Haag: T.M.C. Asser Instituut, 1998)

Jennings, R. Y. and Watts, A. (eds), *Oppenheim's International Law*, 9th edn (Harlow: Longman, 1992)

de Jong, J. and Koper, A., "Staat, bodemschatten en energiepolitiek: een analyse van de strijd om de Mijnwet continentaal plat," 1978 (2(1)) *Tijdschrift voor politieke ekonomie*, pp. 7–51

Judd, A. and Hovland, M., *Seabed Fluid Flow* (Cambridge University Press, 2007)

Koelmans, A., "Van pomp tot put in honderd jaar; Bijdrage tot de geschiedenis van de voorziening van Nederland met aardolieprodukten," Doctoral dissertation (University of Amsterdam, 1970)

Kölble, J., "Bundesstaat und Festlandsockel," 1964 (17) *Die öffentliche Verwaltung*, pp. 217–225

Koskenniemi, M., *From Apology to Utopia: The Structure on International Legal Argument* (Helsinki: Finnish Lawyers' Publishing Company, 1989)

Laursen, F., *Small Powers at Sea; Scandinavia and the New International Marine Order* (Dordrecht: Martinus Nijhoff Publishers, 1993)

Lepard, B. D., *Customary International Law; A New Theory with Practical Applications* (Cambridge University Press, 2010)

Menzel, E., "Der deutsche Festlandsockel in der Nordsee und seine rechtliche Ordnung," 1965 (90/1) *Archiv des öffentlichen Rechts*, pp. 1–61

"Der Festlandsockel der Bundesrepublik Deutschland und das Urteil des Internationalen Gerichtshofs vom 20. Februar 1969," 1969 (14) *Jahrbuch für internationales Recht*, pp. 14–100

Gutachten zur Frage des kontinentalen Schelfs in der Nordsee; dem Deutschen Nordsee-Konsortium (Kiel: March 1964)

Meyer-Lindenberg, H., "Das Genfer Übereinkommen über den Festlandsockel vom 29. April 1958," 1959 (20) *ZaöRV*, pp. 5–35

Ministerie van Buitenlandse Zaken, 56 *Verslag over de Conferentie van de Verenigde Naties over het Zeerecht* (The Hague: Staatsdrukkerij – en uitgeverijbedrijf, 1958)

Mouton, M. W., *The Continental Shelf* (The Hague: Nijhoff, 1952)

"The Continental Shelf," 1954 (85) *Recueil des Cours*, pp. 347–463

Münch, F., "Das Urteil des Internationalen Gerichtshofes vom 20. Februar 1969 über den deutschen Anteil am Festlandsockel in der Nordsee," 1969 (29) ZaöRV, pp. 455–475

"Die Internationale Seerechtkonferenz in Genf 1958," 1959–1960 (8) Archiv des Völkerrechts, pp. 180–208

von Münch, I., "Die Ausnutzung des Festlandsockels vor der deutschen Nordseeküste," 1963–1964 (11) Archiv des Völkerrechts, pp. 391–416

"Nederlandse en Duitse belangen in botsing; Prof. François bepleit overleg van drie landen," Utrechtsch Nieuwsblad, 28 April 1964

"Nu gas- en olierijkdom blijkt: Duitsers willen stuk van onze zeebodem. Moet Nederland toegeven om vriendschap te redden?," Haagsche Courant, 24 March 1964, p. 1

Nweihed, K. G., "The Netherlands (Antilles)-Venezuela; Report Number 2–12" in J. I. Charney and L. M. Alexander (eds), International Maritime Boundaries, Vol. I (Dordrecht: Martinus Nijhoff Publishers, 1993), pp. 615–629

Oda, S., International Control of Sea Resources (Leyden: A.W. Sythoff, 1963)

Oude Elferink, A. G., "Article 76 of the LOSC on the Definition of the Continental Shelf: Questions Concerning its Interpretation from a Legal Perspective," 2006 (21) International Journal for Marine and Coastal Law, pp. 269–285

The Law of Maritime Boundary Delimitation: A Case Study of the Russian Federation (Dordrecht: Martinus Nijhoff Publishers, 1994)

"Third States in Maritime Delimitation Cases: Too Big a Role, Too Small a Role or Both?" in A. Chircop, T. L. McDorman, S. J. Rolston (eds), The Future of Ocean Regime-building; Essays in Tribute to Douglas M. Johnston (Leiden: Martinus Nijhoff Publishers: 2009), pp. 611–641

Oxman, B. H., "The Preparation of Article 1 of the Convention on the Continental Shelf," 1972 (3) Journal of Maritime Law and Commerce, pp. 245–305, 445–472 and 683–723

Reynaud, A., Les Différends du Plateau Continental de la Mer du Nord devant la Cour Internationale de Justice (Paris: Librairie générale de droit et de jurisprudence, 1975)

Rosenne, S., The International Court of Justice (Leyden: A.W. Sythoff, 1961)

Rothwell, D. R. and Stephens, T., The International Law of the Sea (Oxford: Hart Publishing, 2010)

von Schenk, D., "Die Festlandsockel-Proklamation der Bundesregierung vom 20. Januar 1964" in W. J. Schütz (ed.), Aus der Schule der Diplomatie; Beiträge zu Außenpolitik, Recht, Kultur, Menschenführung (Düsseldorf: Econ-Verlag, 1965), pp. 485–498

"Die vertragliche Abgrenzung des Festlandsockels unter der Nordsee zwischen der Bundesrepublik Deutschland, Dänemark und den Niederlanden nach dem Urteil des Internationalen Gerichtshofes vom 20. Februar 1969," 1971 (15) *Jahrbuch für internationales Recht*, pp. 370–391

Scott, S. V., "International Law as Ideology: Theorizing the Relationship between International Law and International Politics," 1994 (5) *European Journal of International Law*, pp. 313–325

"The Inclusion of Sedentary Fisheries Within the Continental Shelf Regime," 1992 (41) *ICLQ*, pp. 788–807

Tanja, G. J., "A New Treaty Regime for the Ems-Dollard Region," 1987 (2) *International Journal of Estuarine and Coastal Law*, pp. 123–142

"Twijfel over verdrag Duitsland; Hoe werd lijn getrokken over continentaal plat?," *Leeuwarder Courant*, 3 December 1964, p. 3

"Vat Den Haag continentaal plat-verdeling te licht op? Concessie aan Bonn: rijk stuk van Noordzeebodem wordt nog niet geëxploreerd," *Het Vrije Volk*, 19 March 1965, p. 11

Waldock, H. M., *The International Court and the Law of the Sea* (The Hague: T.M.C. Asser Institute, 1979)

Weenink, W. H., *Johan Willem Beyen, 1897–1976: Bankier van de Wereld, bouwer van Europa* (Amsterdam/Rotterdam: Prometheus/NRC Handelsblad, 2005)

Weil, P., *The Law of Maritime Delimitation: Reflections* (Cambridge: Grotius Publications, 1989)

Wenger, A., *Pétrole et gaz naturel en Mer du Nord; Droit et Économie* (Paris: Editions Technip, 1971)

Werners, S. E., "Complicaties bij een grensgeschil," 1968 Issue 9 *Nederlands Juristenblad*, pp. 224–225

Whittemore Boggs, S., "Delimitation of Seaward Areas under National Jurisdiction," 1951 (45) *AJIL*, pp. 240–266

Wielenga, F., *Van Vijand tot bondgenoot; Nederland en Duitsland na 1945* (Amsterdam: Boom, 1999)

Willecke, R., "Der Festlandsockel – seine völker- und verfassungsrechtliche Problematik," 1966 (81/13) *Deutsches Verwaltungsblatt*, pp. 461–468

Withana, R., *Power, Politics, Law: International Law and State Behaviour during International Crises* (Leiden: Martinus Nijhoff Publishers, 2008)

Zimmerman, A., "International Courts and Tribunals, Intervention in Proceedings," *Max Planck Encyclopedia of Public International Law* (online) www.mpepil.com

INDEX